# Linear Algebra and Geometry

**AMS / MAA** | TEXTBOOKS

VOL **46**

# Linear Algebra and Geometry

**Al Cuoco**
**Kevin Waterman**
**Bowen Kerins**
**Elena Kaczorowski**
**Michelle Manes**

MAA PRESS An Imprint of the AMERICAN MATHEMATICAL SOCIETY

Providence, Rhode Island

2010 *Mathematics Subject Classification.* Primary 08-01, 15-01, 15A03, 15A04, 15A06, 15A09, 15A15, 15A18, 60J10, 97-01.



For additional information and updates on this book, visit
**www.ams.org/bookpages/text-46**

**Library of Congress Cataloging-in-Publication Data**

Names: Cuoco, Albert, author.
Title: Linear algebra and geometry / Al Cuoco [and four others].
Description: Providence, Rhode Island : MAA Press, an imprint of the American Mathematical Society, [2019] | Series: AMS/MAA textbooks ; volume 46 | Includes index.
Identifiers: LCCN 2018037261 | ISBN 9781470443504 (alk. paper)
Subjects: LCSH: Algebras, Linear–Textbooks. | Geometry, Algebraic–Textbooks. | AMS: General algebraic systems – Instructional exposition (textbooks, tutorial papers, etc.). msc | Linear and multilinear algebra; matrix theory – Instructional exposition (textbooks, tutorial papers, etc.). msc | Linear and multilinear algebra; matrix theory – Basic linear algebra – Vector spaces, linear dependence, rank. msc | Linear and multilinear algebra; matrix theory – Basic linear algebra – Linear transformations, semilinear transformations. msc | Linear and multilinear algebra; matrix theory – Basic linear algebra – Linear equations. msc | Linear and multilinear algebra; matrix theory – Basic linear algebra – Matrix inversion, generalized inverses. msc | Linear and multilinear algebra; matrix theory – Basic linear algebra – Determinants, permanents, other special matrix functions. msc | Linear and multilinear algebra; matrix theory – Basic linear algebra – Eigenvalues, singular values, and eigenvectors. msc | Probability theory and stochastic processes – Markov processes – Markov chains (discrete-time Markov processes on discrete state spaces). msc | Mathematics education – Instructional exposition (textbooks, tutorial papers, etc.). msc
Classification: LCC QA184.2 .L5295 2019 | DDC 512/.5–dc23
LC record available at https://lccn.loc.gov/2018037261



♾ The paper used in this book is acid-free and falls within the guidelines established to ensure permanence and durability.
Visit the AMS home page at `https://www.ams.org/`

10 9 8 7 6 5 4 3 2 1 24 23 22 21 20 19

National Science Foundation

This material was produced at Education Development Center based on work supported by the National Science Foundation under Grant No. DRL-0733015. Any opinions, findings, and conclusions or recommendations expressed in this material are those of the author(s) and do not necessarily reflect the views of the National Science Foundation.

Education Development Center

**Education Development Center, Inc.**
**Waltham, Massachusetts**

*Linear Algebra and Geometry* was developed at Education Development Center, Inc. (EDC), with the support of the National Science Foundation.

***Linear Algebra and Geometry* Development Team**
***Authors:*** **Al Cuoco**, Kevin Waterman, Bowen Kerins, Elena Kaczorowski, and Michelle Manes
***Others who contributed include:*** Doreen Kilday, Ken Levasseur, Stephen Maurer, Wayne Harvey, Joseph Obrycki, Kerry Ouellet, and Stephanie Ragucci
***Core Mathematical Consultants:*** Thomas Banchoff, Roger Howe, and Glenn Stevens

# Contents

# Acknowledgments

A Note from Al Cuoco

It's no exaggeration to say that *Linear Algebra and Geometry* has been under development for over three decades.

In the early 1970s, with all of two years of teaching under my belt, I participated in an NSF program for high school teachers at Bowdoin. Jim Ward assembled an astounding faculty for this four-summer delight—Ken Ireland, Jon Lubin, Dick Chittham, and A. W. Tucker, among others. Jim taught several of the courses himself, including a course in concrete linear algebra. It was immediately clear to me that his approach and this material would be accessible to high school students.

We instituted a linear algebra course in my school—Woburn High school in Massachusetts—in the mid 1970s. As enrollment grew, I was joined by my colleague Elfreda Kallock, one of the most expert teachers I've ever known, and together we organized the course into daily problem sets. These were polished for another two decades, revised almost weekly to reflect what had happened in class. Elfreda and I had a great deal of fun as we created the problem sets, sequenced the problems, made sure that the numbers in the problems uncovered the principles we wanted to expose, learned TEX, and wrote corny jokes that the kids learned to love. I've posted three samples of the sets at

`https://go.edc.org/woburn-high-samples`

After coming to EDC, and with support from NSF, my colleagues and I began work on creating a course from these notes. Working closely with advisors and teachers, we refined the materials, added exposition, and added solutions and teaching notes. And we ran summer workshops for teachers. Originally designed for high school students who were looking for an elective, it became evident that teachers found the materials valuable resources for themselves. Many told us that they wished that they had learned linear algebra with this approach (an approach described in the introduction—essentially based in the extraction of general principles from numerical experiments).

So, we revised again, this time aiming at a dual audience—preservice (and inservice) teachers and fourth year high school students. Throughout all these revisions and changes, we kept to the original philosophy of developing linear algebra with a dual approach based in reasoning about calculations and generalizing geometric ideas via their algebraic characterizations.

There are so many people to thank for this effort—all of the folks listed in the title page have been inspirations and have left their indelible stamps on the program. Stan Seltzer and his team at MAA reviewed the manuscript, working every problem, finding errors, and suggesting fixes. The AMS team: Kerri Malatesta, Steve Kennedy, Chris Thivierge, and Sergei Gelfand helped in so many ways, from design to catching more mistakes. I read the entire MS more than once. Any mistakes that remain are the responsibility of the other authors.

Thanks to Paul Goldenberg for the design of the cover graphic (and for contributing to the ideas in Chapter 3), June Mark and Deb Spencer for help with so many things, large and small, and Stephanie Ragucci for help with the teaching notes, for piloting the course at Andover High, and for being such a wonderful kid when she was a student in my original Woburn High course in the 1980s.

# Introduction

Welcome to *Linear Algebra and Geometry*. You probably have an idea about the "Geometry" in the title, but what about "Linear Algebra"?

It's not so easy to explain what linear algebra is about until you've done some of it. Here's an attempt:

You may have studied some of these topics in previous courses:

- Solving systems of two linear equations in two unknowns and systems of three linear equations in three unknowns
- Using matrices to solve systems of equations
- Using matrices and matrix algebra for other purposes
- Using coordinates or vectors to help with geometry
- Solving systems of equations with determinants
- Working with reflections, rotations, and translations

Linear algebra ties all these ideas together and makes connections among them.

And it does much more than this. Much of high school algebra is about developing tools for solving equations and analyzing functions. The equations and functions usually involve one, or maybe two, variables. Linear algebra develops tools for solving equations and analyzing functions that involve *many* variables, all at once. For example, you'll learn how to find the equation of a plane in space and how to get formulas for rotations about a point in space (these involve three variables). You'll learn how to analyze systems of linear equations in many variables and work with matrices of any size—many applications involve matrices with thousands or even millions of entries.

Here are some quotes from two people who use linear algebra in their professions:

> *Linear algebra is a powerful tool in finance. Innovations are often developed within the world's most sophisticated financial firms by those fluent in the language of vectors and matrices.*

> *Linear algebra is not only a valuable tool in its own right, but it also facilitates the applications of tools in multivariate calculus and multivariate statistics to many problems in finance that involve risks and decisions in multiple dimensions. Studying linear algebra first, before those subjects, puts one in the strongest position to grasp and exploit their insights.*
>
> — Robert Stambaugh,
> Professor of Finance
> Wharton School

> *Students who will take such a course have probably had the equivalent of two years of algebra and a year of geometry, at least if they come from a fairly standard program. They will have seen some analytic geometry, but not enough to give them much confidence in the relationship between algebraic and geometric thinking in the plane, and even less in three-space. Linear algebra can bring those subjects together in ways that reinforce both. That is a goal for all students, whether or not they have taken calculus, and it can form a viable alternative to calculus in high school. I would love to have students in a first-year course in calculus who already had thought deeply about the relationships between algebra and geometry.*
>
> — Thomas Banchoff,
> Professor of Mathematics
> Brown University

⟵ Dr. Banchoff is one of the core consultants to this book.

When you finish the core program (Chapters 1–5), you'll have the language, the tools, and the habits of mind necessary to understand many questions in advanced mathematics and its applications to science, engineering, computer science, and finance.

It takes some time, effort, and practice to develop these skills.

- The **language** of linear algebra speaks about two kinds of mathematical objects—*vectors* and *matrices*—as well as special functions—*linear mappings*—defined on these objects. One of the core skills in the language of linear algebra is to learn how to use geometric and algebraic images interchangeably. For example, you'll refer to the set of solutions to the equation $x - 2y + 3x + w = 0$ as a "hyperplane in four dimensions."

⟵ If you don't know what a vector or matrix is, don't worry—you soon will.

- The **tools** of linear algebra involve developing a new kind of algebraic skill—you'll be calculating with vectors and matrices, solving equations, and learning about algorithms that carry out certain processes. You may have met matrix multiplication or matrix row reduction in other courses. These are examples of the kinds of tools you'll learn about in this course.

⟵ If you don't know about these operations, don't worry—you soon will.

- The **habits of mind** in linear algebra are the most important things for you to develop. These involve being able to imagine a calculation—with matrices, say—without having to carry it out, making use of a general kind of distributive law (that works with vectors and matrices), and extending an operation from a small set to a big set by preserving the rules for calculating. An example of the kind of mathematical thinking that's important in linear algebra is the ability to analyze the following question *without finding two points on the graph*:

If $(a, b)$ and $(c, d)$ are points on the graph of $3x + 5y = 7$, is $(a + c, b + d)$ on the graph?

Mathematical habits are just that—habits. And they take time to develop. The best way to develop these habits is to work carefully through all the problems.

This book contains many problems, more than in most courses. That's for a reason. All the main results and methods in this book come from generalizing numerical examples. So, a problem set that looks like an extensive list of calculations is there because, if you carefully work through the calculations and ask yourself what's common among them, a general result (*and* its proof) will be sitting right in front of you.

The authors of this book took care never to include extraneous problems. Usually, the problems build on each other like the stories of a tower, so that you can climb to the top a little at a time and then see something of the whole landscape. Many of these problem sets have evolved over several decades of use in high school classrooms, gradually polished every year and influenced by input from a couple of generations of students.

This is all to say that linear algebra is an important, useful, and beautiful area of mathematics, and it's a subject at which you can become very good by working the problems—and analyzing your work—in the chapters ahead.

Before you start, the authors of this program have some advice for you:

← The authors include teachers, mathematicians, education professionals, and students; most of them fit into more than one of these categories.

*The best way to understand mathematics is to work really hard on the problems.*

If you work through these problems carefully, you'll never wonder why a new fact is true; you'll know because you discovered the fact for yourself. Theorems in linear algebra spring from calculations, and the problem sets ask you to do lots of calculations that highlight these theorems.

The sections themselves provide examples and ideas about the ways people think about the mathematics in the chapters. They are designed to give you a reference, but they probably won't be as complete as the classroom discussions you'll have with your classmates and your teacher. In other words, you still have to pay attention in class.

But you'll have to pay attention a lot less if you do these problems carefully. That's because many of the problems are previews of coming attractions, so doing them and looking for new ideas will mean fewer surprises when new ideas are presented in class.

This approach to learning has been evolving for more than 40 years—many students have learned, *really* learned, linear algebra by working through these problems. You are cordially invited to join them.

CHAPTER

# 1 Points and Vectors

One of the real breakthroughs in mathematics happened when people realized that algebra could be joined with geometry. By setting up a coordinate system and assigning coordinates to points, mathematicians were able to describe geometric phenomena with algebraic equations and formulas.

This process allowed mathematicians and physicists to develop, over long periods of time, intuitions about geometric objects in dimensions greater than three. Through what this book refers to as *the extension program*, geometric ideas that are tangible in two (and three) dimensions can be extended to higher dimensions via algebra. Doing so will help you develop geometric intuitions for higher dimensions you cannot physically experience.

**By the end of this chapter, you will be able to answer questions like these:**

**1.** How can you describe adding and scaling vectors in geometric terms?

**2.** How can you use vectors to describe lines in space?

**3.** Let $A = (3, 2)$ and $B = (-1, 4)$.
   **a.** How do you calculate and graph the following: $A + B$, $2A$, $-3B$, $2A - 3B$?
   **b.** What is the value of $\|2A - 3B\|$?

**You will build good habits and skills for ways to**

- generalize from numerical examples
- use algebra to extend geometric ideas
- connect the rules of arithmetic to an algebra of points
- use different forms for different purposes

**Vocabulary and Notation**

- coordinates
- direction
- equivalent vectors
- extension program
- initial point (tail)
- length $\|X\|$
- linear combination
- magnitude
- $n$-dimensional Euclidean space
- opposite direction
- ordered $n$-tuple
- point
- same direction
- scalar multiple
- spanned
- terminal point (head)
- unit vector
- vector
- vector equation
- zero vector

## 1.1 Getting Started

In this chapter, you'll develop an "algebra of points." Before things get formal, here's a preview of coming attractions.

- To *add* two points in the coordinate plane, add the corresponding coordinates: $(3,2)+(5,7)=(8,9)$ and, more generally, $(x,y)+(z,w)=(x+z,y+w)$.
- To *scale* a point in the coordinate plane by a number, multiply both coordinates of that point by that number: $5(3,2)=(15,10)$ and, more generally, $c(x,y)=(cx,cy)$.

## Exercises

**1.** Suppose $A=(1,2)$. On one set of axes, plot these points:

**a.** $2A$ **b.** $3A$ **c.** $5A$
**d.** $(-1)A$ **e.** $(-3)A$ **f.** $(-6.5)A$

**2.** Here's a picture of a point $A$, with an arrow drawn from the origin to $A$.

⟵ The arrow in the figure is called a *vector*.

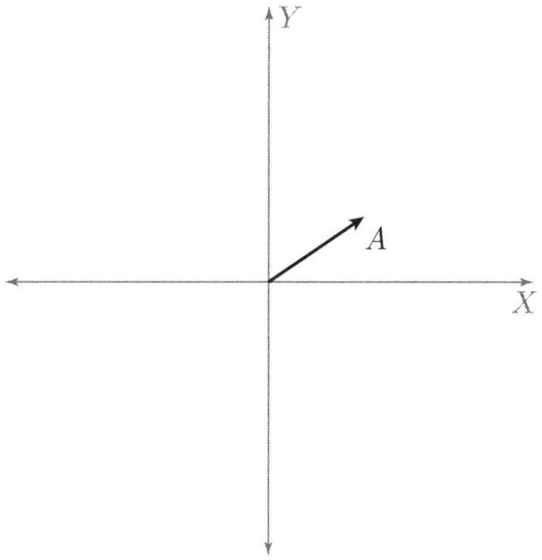


Draw these vectors, all on the same axes:

**a.** $2A$ **b.** $3A$ **c.** $5A$
**d.** $(-1)A$ **e.** $(-3)A$ **f.** $(-6.5)A$

**3.** Here's a picture of a point $A$, with an arrow drawn from the origin to $A$.

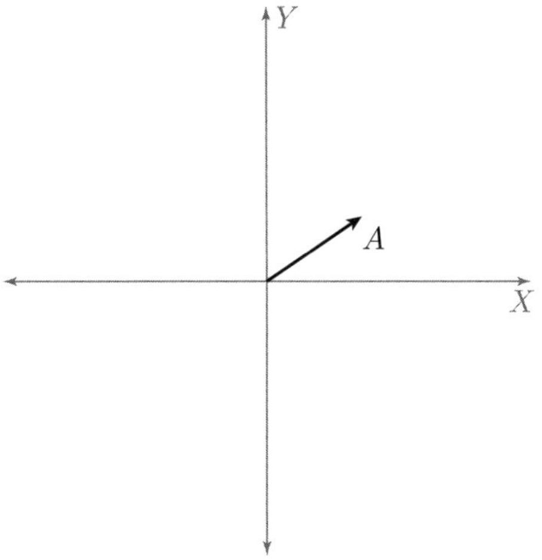


**a.** Describe in words the set of all multiples $tA$, where $t$ ranges over all real numbers.

**b.** If $A = (r, s)$, find a coordinate equation for the set of multiples $tA$, where $t$ ranges over all real numbers.

$\longleftarrow$
A **coordinate equation** in $\mathbb{R}^2$ is an equation of the form $ax + by = c$.

**4.** Suppose $O = (0, 0)$, $A = (5, 3)$, and $B = (3, -1)$. Show that $O$, $A$, $B$, and $A + B$ lie on the vertices of a parallelogram. It may be helpful to draw a picture.

**5.** Suppose $A = (a_1, a_2)$ and $B = (b_1, b_2)$. Show that $O$, $A$, $B$, and $A+B$ lie on the vertices of a parallelogram. Again, it may be helpful to draw a picture.

$\longleftarrow$
$O = (0, 0)$.

**6.** Suppose $A = (5, 3)$ and $B = (3, -1)$. Find and plot these points, all on the same axes:

**a.** $A + B$ **b.** $A + 3B$ **c.** $A + 5B$
**d.** $A + (-1B)$ **e.** $A + (-3B)$ **f.** $A + (-6.5B)$

**Habits of Mind**
Draw a picture!!

**7.** Suppose $A = (5, 3)$ and $B = (3, -1)$. Find a coordinate equation for the set of points $X$ that is generated by $A + tB$, where $t$ ranges over all real numbers.

**8.** Here's a picture of two points, $A$ and $B$, with arrows drawn to each from the origin.

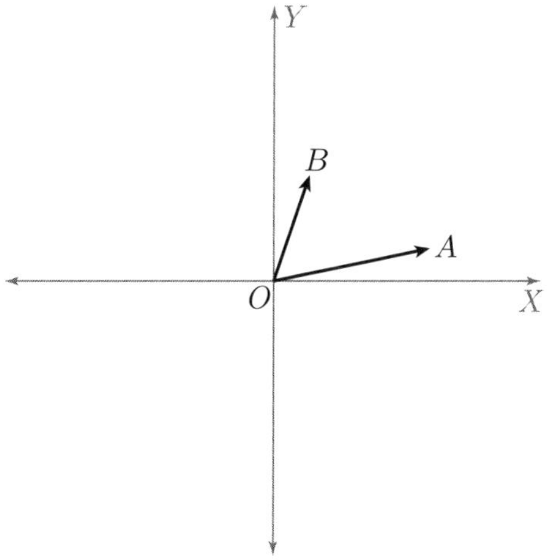


Draw these vectors, all on the same axes:

**a.** $A+B$ **b.** $A+3B$ **c.** $A+5B$
**d.** $A+(-1B)$ **e.** $A+(-3B)$ **f.** $A+(-6.5B)$

**9.** Here's a picture of two points, $A$ and $B$, with arrows drawn to each from the origin.

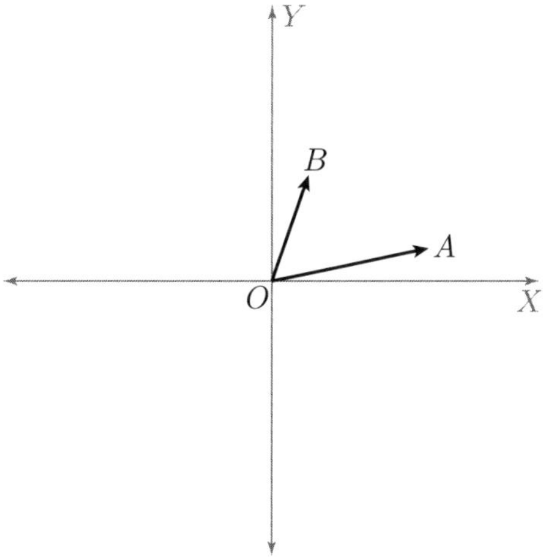


Draw a picture of the set of all points $X$ that is generated by $A+tB$, where $t$ ranges over all real numbers.

$\longleftarrow$ You can think of the equation $X = A + tB$ as a *point-generator*: different numbers $t$ generate different points $X$.

**10.** Suppose $A = (5, 4)$ and $B = (-1, 3)$. Find numbers $c_1$ and $c_2$ if

**a.** $c_1A + c_2B = (4, 7)$ **b.** $c_1A + c_2B = (9, 11)$
**c.** $c_1A + c_2B = (11, 5)$ **d.** $c_1A + c_2B = (-13, 1)$
**e.** $c_1A + c_2B = (1, \frac{-5}{2})$ **f.** $c_1A + c_2B = (-\frac{11}{5}, -1)$

**Habits of Mind**

Draw a picture!!!

**11.** Here's a picture of two points, $A$ and $B$, with arrows drawn to each from the origin, as well as some other points.

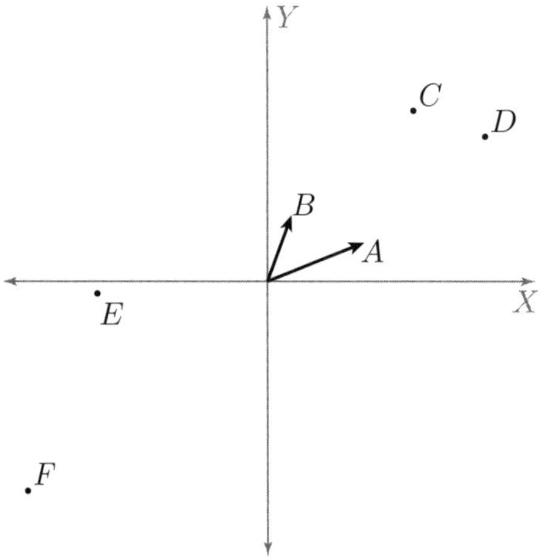


Estimate the values for $c_1$ and $c_2$ if

**a.** $c_1A + c_2B = C$ **b.** $c_1A + c_2B = D$
**c.** $c_1A + c_2B = E$ **d.** $c_1A + c_2B = F$

**12.** Find the length of each vector.

**a.** $A = (5, 12)$ **b.** $B = (3, 4)$ **c.** $C = (-2, 10)$
**d.** $P = (4, 1, 8)$ **e.** $P = (4, 1, 9)$ **f.** $D = (a, b)$
**g.** $Q = (a, b, c)$

←
Draw a picture.

←
In each part the vector goes from the origin to the labeled point.

**13.** Find the lengths of the sides of $\triangle AOB$ if

**a.** $A = (5, 12), B = (-27, 36)$
**b.** $A = (21, 20), B = (21, 220)$
**c.** $A = (48, 64), B = (15, 8)$
**d.** $A = (4, 4), B = (4, -4)$
**e.** $A = (4, 0), B = (2, 2\sqrt{3})$
**f.** $A = (3\sqrt{2}, 3\sqrt{2}), B = (-4\sqrt{2}, 4\sqrt{2})$
**g.** $A = (-14, 29, 22), B = (-126, 45, -18)$

**14.** Here's a picture of a *three*-dimensional coordinate system.

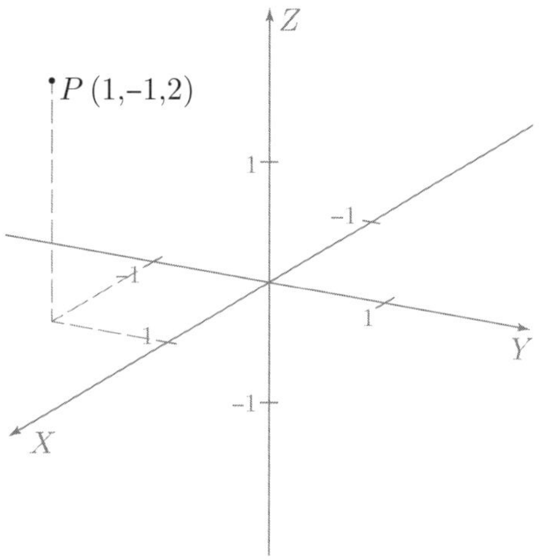


Find the equation of

**a.** the $x$–$y$ plane
**b.** the $x$–$z$ plane
**c.** the $y$–$z$ plane
**d.** the plane parallel to the $x$–$y$ plane that contains the point $(1, -1, 2)$

**Remember**

Equations are point-testers: a point is on the graph of your equation if and only if the coordinates of the point make the equation true.

## 1.2 Points

In most of your high school work so far, the equations and formulas have expressed facts about the *coordinates* of points—the variables have been placeholders for numbers. In this lesson, you will begin to develop an *algebra of points*, in which you can write equations and formulas whose variables are points in two and three dimensions.

**In this lesson, you will learn how to**

- locate points in space and describe objects with equations
- use the algebra of points to calculate, solve equations, and transform expressions, all in $\mathbb{R}^n$
- understand the geometric interpretations of adding and scaling

You probably studied the method for building number lines (or "coordinatized lines") in previous courses. Given a line, you can pick two points $O$ and $A$ and assign the number 0 to $O$ and 1 to $A$.

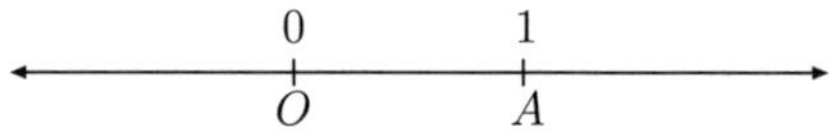


This sets the "unit" of the number line, and you can now set up a one-to-one correspondence between the set of real numbers, denoted by $\mathbb{R}$, and the set of points on the number line.

⟵ *One-to-one correspondence* means that for every point, there is exactly one real number (its coordinate) and for every real number, there is exactly one point (its graph).

- Suppose $P$ is a point on the number line that is located $x$ units to the right of $O$. Then $x$ is called the **coordinate** of $P$, and $P$ is called the **graph** of $x$.
- Suppose $Q$ is a point on the number line that is located $x$ units to the *left* of $O$. Its distance from $O$ is still $x$, but it's not the same point as $P$. In this case, $-x$ is the coordinate of $Q$, and $Q$ is the graph of $-x$.

The figure below shows several points and their coordinates.

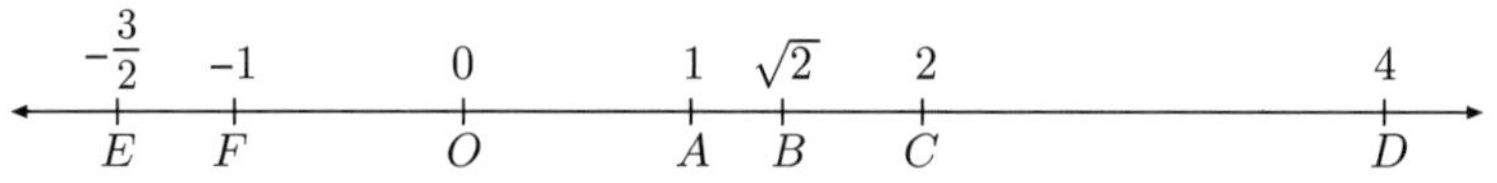


This idea of relating the set of all points on a line with the real numbers goes back to antiquity, but it was not until the 17th century that mathematicians (notably Descartes and Fermat) had a clear notion of how to coordinatize a plane: draw two perpendicular coordinatized lines (usually the scale is the same on each) that intersect at their common origin. These lines are called the **$x$-axis** and **$y$-axis**. You can now uniquely identify every point on the plane using an ordered pair of numbers. If the point $P$ corresponds to the ordered pair $(x, y)$, $x$ and $y$ are the **coordinates** of $P$.

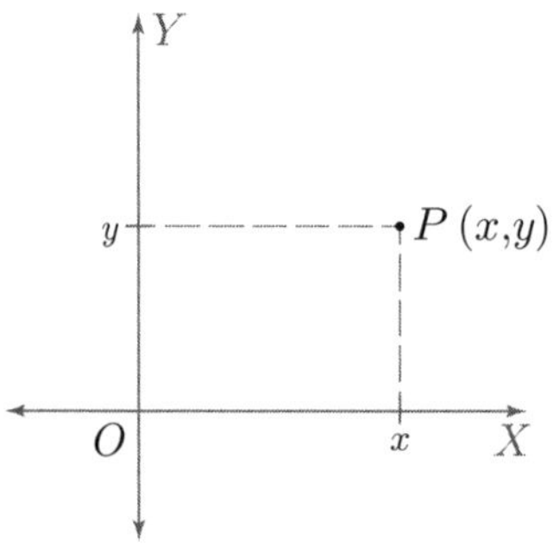


⟵ In fact, the idea of coordinatizing the plane does not require that the two axes be perpendicular, only that each point lies on a unique pair of lines parallel to a given pair of axes.

The set of all ordered pairs of real numbers is denoted by $\mathbb{R}^2$. Because of the correspondence between $\mathbb{R}^2$ and points on a plane, you can think of $\mathbb{R}^2$ as the set of points on a coordinatized plane, so that statements like "the point $(5,0)$ is the same distance from the point $(0,0)$ as the point $(3,4)$" make sense.

Identifying$\mathbb{R}^2$ with a plane provides a way to use algebra to describe geometric objects. Indeed, this is the central theme of analytic geometry.

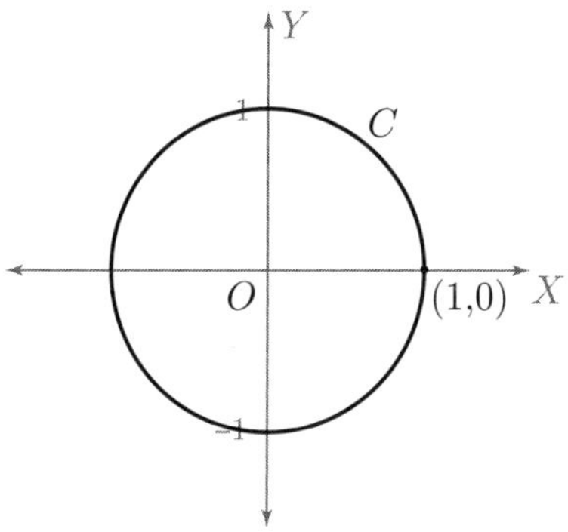


Consider the circle $C$ on the left. You can describe $C$ geometrically by saying that $C$ consists of all points in the plane that are 1 unit from $O$. However, you can also describe $C$ algebraically in terms of the coordinates of the points that lie on $C$: $C$ is the set of points $(x, y)$ so that $x^2 + y^2 = 1$.

The connection between the geometric description ("$C$ consists of all points . . . ") and the equation ("$x^2 + y^2 = 1$") is that *the equation is a* **point-tester** *for the geometric definition.* This means that you can test a point to see if it's on the circle by seeing if its coordinates satisfy the equation. For example,

- $(1,0)$ is on $C$ because $1^2 + 0^2 = 1$
- $\left(\frac{1}{2}, \frac{1}{3}\right)$ is not on $C$ because
$$\left(\frac{1}{2}\right)^2 + \left(\frac{1}{3}\right)^2 \neq 1$$
- $\left(\frac{\sqrt{3}}{2}, -\frac{1}{2}\right)$ is on $C$ because
$$\left(\frac{\sqrt{3}}{2}\right)^2 + \left(-\frac{1}{2}\right)^2 = 1$$

Many people make statements like "$C$ is the circle $x^2 + y^2 = 1$"; this statement is shorthand for "$C$ is the circle whose equation is $x^2 + y^2 = 1$."

Think about how you get the equation of the circle in the first place: you take the geometric description—"all the points that are 1 unit from the origin"—and use the distance formula to translate that into algebra.

$$\begin{aligned} P = (x, y) \text{ is on } C &\Leftrightarrow \text{the distance from } P \text{ to } O \text{ is } 1 \\ &\Leftrightarrow \sqrt{x^2 + y^2} = 1 \quad \text{(the distance formula)} \\ &\Leftrightarrow x^2 + y^2 = 1 \end{aligned}$$

The symbol "$\Leftrightarrow$" means "the two statements are equivalent." You can read it quickly by saying "if and only if."

### For You to Do

1. Find five points on the circle of radius 5 centered at $(3,4)$. Find the equation of this circle.
2. **Take It Further.** Find five points on the sphere of radius 5 centered at $(3,4,2)$. Find the equation of this sphere.

### Developing Habits of Mind

**Explore multiple representations.** All of plane geometry *could* be carried out using the algebra of $\mathbb{R}^2$ without any reference to diagrams or points on a plane. For example, you could define a line to be the set of pairs $(x, y)$ that satisfy an equation of the form $ax + by = c$ for some real numbers $a$, $b$, and $c$. The fact that two distinct lines intersect in, at most, one point would then be a fact about the solution set of two equations in two unknowns. This would be silly when studying two- or three-dimensional geometry—the pictures help so much with understanding—but you will see shortly that characterizing geometric properties algebraically makes it possible to generalize many of the facts in elementary geometry to situations for which there is no physical model.

The method for coordinatizing three-dimensional space is similar. Choose three mutually perpendicular coordinatized lines (all with the same scale) intersecting at their origin. Then set up a one-to-one correspondence between points in space and ordered triples of numbers $(x, y, z)$. In the following figure, point $P$ has coordinates $x = 1$, $y = -1$, and $z = 2$.

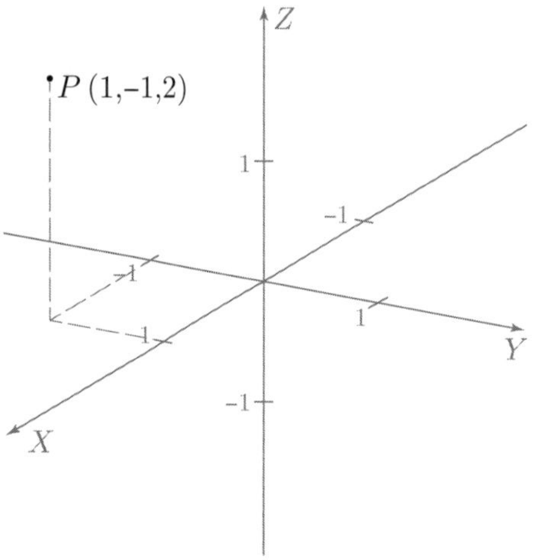


The set of ordered triples of real numbers is denoted by $\mathbb{R}^3$, and the elements of $\mathbb{R}^3$ are spoken of as points. In the next figure, the line through $O = (0, 0, 0)$ and $A = (1, 1, 0)$ makes an angle of 45° with the $x$- and $y$-axes and an angle of 90° with the $z$-axis.

$\longleftarrow$ If you're not convinced, stay tuned . . . you'll revisit the notion of angles in $\mathbb{R}^3$ in Chapter 2.

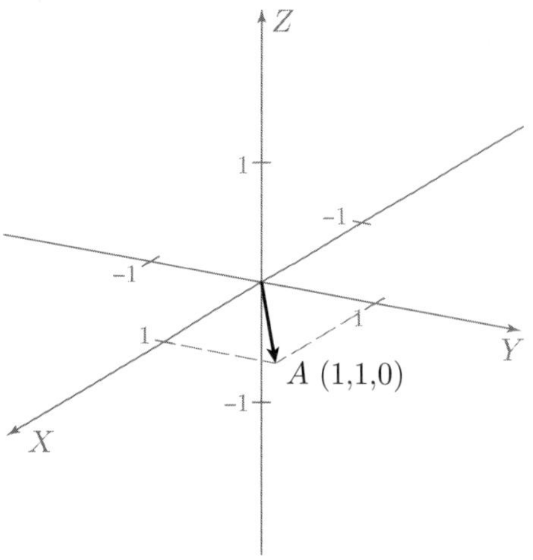

**Minds in Action** Episode 1

*Tony and Sasha are two students studying Linear Algebra. They are thinking about how to use the point-tester idea to describe objects in space.*

TONY: What would the equation of the $x$–$y$ plane in $\mathbb{R}^3$ be?

SASHA: Don't you remember using point-tester way back in Algebra 1 when we were finding equations of lines? First, think about some points on the $x$–$y$ plane.

TONY: Well, $(0, 0, 0)$ is on that plane. So is $(1, 0, 0)$ and $(2, 3, 0)$. There are a lot, Sasha. How long do you want me to go on for?

SASHA: Until you see the pattern, of course! But this one's easy. In fact, all the points on the $x$–$y$ plane have one thing in common: the $z$-coordinate is 0.

TONY: Yes! So that's easy. The equation would be $z = 0$. But isn't that the equation of a line?

SASHA: I guess it's not in $\mathbb{R}^3$. It has to describe a plane.

TONY: So what does the equation of a line look like in $\mathbb{R}^3$?

SASHA: Here, let's try an easy line, like the $x$-axis ... Well, all the points would look like (something, $0, 0$). So the $y$-coordinate is always 0 and the $z$-coordinate is always 0. How do I say that in one equation?

TONY: I don't know ... I guess the best we can do for now is to say the line is given by *two* equations: $y = 0$ and $z = 0$.

SASHA: Wait a second ... what about $y^2 + z^2 = 0$?

TONY: Sasha, where do you get these ideas? It works, but that's not a linear equation, is it?

**For You to Do**

**3.** Find the equation of

**a.** the $x$–$z$ plane

**b.** the plane parallel to the $x$–$z$ plane that contains the point $(3, 1, 4)$

In the middle of the 19th century, mathematicians began to realize that it was often convenient to speak of quadruples of numbers $(x, y, z, w)$as points of "four-dimensional" space. It seemed very natural to speak of $(1, 3, 2, 0)$ as being a point on the graph of $x+2y-z+w = 5$ rather than saying, "One solution to $x + 2y - z + w = 5$ is $x = 1, y = 3, z = 2$, and $w = 0$." Defining $\mathbb{R}^4$ as the set of all quadruples of real numbers, you can call its elements "points" in $\mathbb{R}^4$. Although there is no physical model for $\mathbb{R}^4$ (as there was for $\mathbb{R}^2$ and $\mathbb{R}^3$), you can borrow the geometric language used for $\mathbb{R}^2$ and speak of $O = (0, 0, 0, 0)$ as the origin in $\mathbb{R}^4$, $A = (1, 0, 0, 0)$ as a point on the $x$-axis in $\mathbb{R}^4$, and so on. This is, for now, just an analogy: $\mathbb{R}^4$ "is" the set of all ordered quadruples of numbers, and geometric statements about $\mathbb{R}^4$ are simple analogies with $\mathbb{R}^2$ and $\mathbb{R}^3$.

← In 1884, E. A. Abbott wrote a book that captured what it might be like to visualize four dimensions. The book has been adapted in an animated film: see `flatlandthemovie.com`

Of course, there is no need to stop here. You can define $\mathbb{R}^5$ as the set of all ordered quintuples of numbers, and so on . . . .

#### Habits of Mind

**Think like a mathematician.** Like many mathematicians, after awhile you may develop a sense for picturing things in higher dimensions. This happens when the algebraic descriptions become identified with the geometric descriptions, deep in your mind.

#### Definition

If $n$ is a positive integer, an **ordered $n$-tuple** is a sequence of $n$ real numbers $(x_1, x_2, \ldots, x_n)$. The set of all ordered $n$-tuples is called **$n$-dimensional Euclidean space** and is denoted by $\mathbb{R}^n$.

An ordered $n$-tuple will be referred to as a **point** in $\mathbb{R}^n$.

#### Facts and Notation

- Capital letters (such as $A$, $B$, or $P$) are often used for points.
- If $A = (a_1, a_2, \ldots, a_n)$, the numbers $a_1, a_2, \ldots, a_n$ are called the **coordinates** of $A$.
- Two points $A = (a_1, a_2, \ldots, a_n)$ and $B = (b_1, b_2, \ldots, b_n)$ in $\mathbb{R}^n$ are **equal** if their corresponding coordinates are equal; that is, $A = B$ means $a_1 = b_1, a_2 = b_2, \ldots,$ and $a_n = b_n$.
- By analogy with $\mathbb{R}^2$, the origin of $\mathbb{R}^n$ is the point $O = (0, 0, 0, \ldots, 0)$.

$\longleftarrow$ Note that instead of writing $a_1 = b_1, a_2 = b_2, \ldots,$ and $a_n = b_n$, you can use shorthand notation "$a_i = b_i$ for each $i = 1, 2, \ldots, n$."

To extend a definition from geometry to $\mathbb{R}^n$, you must first characterize the geometric notions in terms of coordinates of points. You can accomplish this goal most easily by defining several operations on $\mathbb{R}^n$.

The first of these operations is addition. If you were asked to decide what $(2, 3) + (6, 1)$ should be, you might naturally say "$(8, 4)$, of course." It turns out that this definition is very useful: you add points by adding their corresponding coordinates.

#### Definition

If $A = (a_1, a_2, \ldots, a_n)$ and $B = (b_1, b_2, \ldots, b_n)$ are points in $\mathbb{R}^n$, the **sum** of $A$ and $B$, written $A + B$, is

$$A + B = (a_1 + b_1, a_2 + b_2, \ldots, a_n + b_n).$$

#### Developing Habits of Mind

**Make strategic choices.** This is a *definition*—$(2, 3) + (6, 1)$ equals $(8, 4)$, not because of any intrinsic reason. It isn't forced on you by the laws of physics or the basic rules of algebra, for example. Mathematicians have defined the sum in this way because it has many useful properties. One of the most useful is that there is a nice geometric interpretation for this method for adding in $\mathbb{R}^2$ and $\mathbb{R}^3$.

### Example 1

Consider the points $A = (3, 1)$, $B = (1, 4)$, and $A + B = (4, 5)$. If you plot these three points, you may not see anything interesting, but if you throw the origin into the figure, it looks as if $O$, $A$, $B$, and $A + B$ lie on the vertices of a parallelogram.

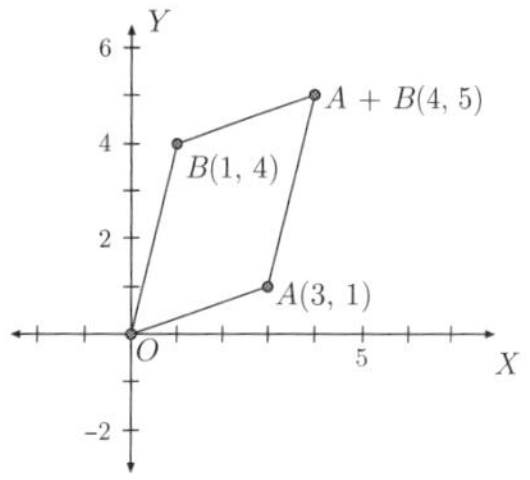


Example 1 suggests the following theorem.

**Theorem 1.1 (The Parallelogram Rule)**

*If $A$ and $B$ are* any *points in $\mathbb{R}^2$, then $O$, $A$, $A + B$, and $B$ lie on the vertices of a parallelogram.*

You can refer to this parallelogram as "the parallelogram determined by $A$ and $B$."

### For You to Do

**4.** **a.** Show that $(0, 0)$, $(3, 1)$, $(1, 4)$, and $(4, 5)$ from Example 1 form the vertices of a parallelogram.
**b.** While you're at it, explain why Theorem 1.1 must be true.

⟵
First generate some examples, then try to generalize them.

In linear algebra, it is customary to refer to real numbers (or the elements of any number system) as **scalars**. The second operation to define on $\mathbb{R}^n$ is the multiplication of a point by a scalar, and it is called *multiplication of a point by a scalar*.

⟵
In this book, the terms *scalar* and *number* will be used interchangeably. Scalar has a geometric interpretation—see below.

**Definition**

Let $A = (a_1, a_2, \ldots, a_n)$ be a point in $\mathbb{R}^n$ and suppose $c$ is a scalar. The **scalar multiple** $cA$ is

$$cA = (ca_1, ca_2, \ldots, ca_n).$$

In other words, to multiply a point by a number, you simply multiply each of the point's coordinates by that number. So, $3(1, 4, 2, 0) = (3, 12, 6, 0)$.

**Habits of Mind**

Try proving that one of these statements is true. For example, show that $2A$ is collinear with $O$ and $A$ and is twice as far from $O$ as $A$ is.

Why are numbers called "scalars"? Scalar multiplication can be visualized in $\mathbb{R}^2$ or $\mathbb{R}^3$ as follows: if $A = (2, 1)$, then $2A = (4, 2)$, $\frac{1}{2}A = (1, \frac{1}{2})$, $-1A = (-2, -1)$, and $-2A = (-4, -2)$.

**Habits of Mind**

In previous courses, you saw that if you view $\mathbb{R}$ as a number line, multiplication by 3 stretches points by a factor of 3.

From this figure, you can see that if $c$ is any real number, $cA$ is collinear with $O$ and $A$; $cA$ is obtained from $A$ by stretching or shrinking—so *scaling*—the distance from $O$ to $A$ by a factor of $|c|$. If $c > 0$, $cA$ is in the same "direction" as $A$; multiplying by a negative reverses direction.

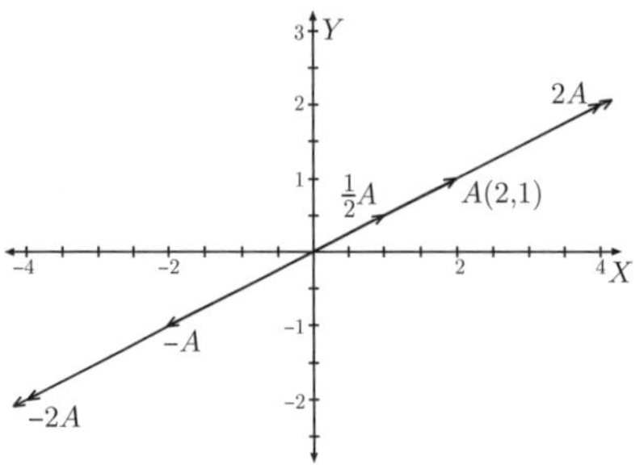


### For You to Do

**5.** Let $c$ be a real number, and let $A$ be a point in $\mathbb{R}^2$.

**a.** Show that $cA$ is collinear with $O$ and $A$.

**b.** Show that if $c \geq 0$, $cA$ is obtained from $A$ by scaling the distance from $O$ to $A$ by a factor of $|c|$. If $c < 0$, $cA$ is obtained from $A$ by scaling the distance from $O$ to $A$ by a factor of $|c|$ and reversing direction.

So, now you have two operations on points: addition and scalar multiplication. How do the operations behave?

**Theorem 1.2 (The Basic Rules of Arithmetic with Points)**
*Let*

$$\begin{aligned} A &= (a_1, a_2, \ldots, a_n) \\ B &= (b_1, b_2, \ldots, b_n) \text{ and} \\ C &= (c_1, c_2, \ldots, c_n) \end{aligned}$$

*be points in* $\mathbb{R}^n$*, and let* $d$ *and* $e$ *be scalars. Then*

(1) $A + B = B + A$

(2) $A + (B + C) = (A + B) + C$

(3) $A + O = A$

(4) $A + (-1)A = O$

(5) $(d + e)A = dA + eA$

(6) $d(A + B) = dA + dB$

(7) $d(eA) = (de)A$

(8) $1A = A$

⟵ Because of property ((4)), $(-1)A$ is called the negative of $A$ and is often written $-A$.

***Proof.*** The proofs of these facts all use the same strategy: reduce the property in question to a statement about real numbers. To illustrate, here are proofs for ((1)) and ((7)). The proofs of the other facts are left as exercises.

$$\begin{aligned}
((1))\quad A + B &= (a_1, a_2, \dots, a_n) + (b_1, b_2, \dots, b_n) \\
&= (a_1 + b_1, a_2 + b_2, \dots, a_n + b_n) && \text{(definition of addition in } \mathbb{R}^n\text{)} \\
&= (b_1 + a_1, b_2 + a_2, \dots, b_n + a_n) && \text{(commutativity of addition in } \mathbb{R}\text{)} \\
&= (b_1, b_2, \dots, b_n) + (a_1, a_2, \dots, a_n) && \text{(definition of addition in } \mathbb{R}^n\text{)} \\
&= B + A
\end{aligned}$$

$$\begin{aligned}
((7))\quad d(eA) &= d\Big(e(a_1, a_2, \dots, a_n)\Big) \\
&= d(ea_1, ea_2, \dots, ea_n) && \text{(definition of scalar multiplication)} \\
&= \Big(d(ea_1), d(ea_2), \dots, d(ea_n)\Big) && \text{(definition of scalar multiplication)} \\
&= \Big((de)a_1, (de)a_2, \dots, (de)a_n\Big) && \text{(associativity of multiplication in } \mathbb{R}\text{)} \\
&= (de)(a_1, a_2, \dots, a_n) && \text{(definition of scalar multiplication)} \\
&= (de)A
\end{aligned}$$

■

**Subtraction** for points is defined by the equation

$$A - B = A + (-B)$$

← . . . and "$-B$" means $(-1)B$.

### Developing Habits of Mind

**Use coordinates to prove statements about points.** The strategy of reducing a statement about points to one about coordinates will be used throughout this book.

But how do you come up with valid statements about points in the first place? One way is to see if analogous statements are true in one dimension—with numbers. So, $2 + 3 = 3 + 2$ might give you a clue that $A + B = B + A$ for points. Once you have a clue, *try it with actual points*. Does $(7, 1) + (9, 8) = (9, 8) + (7, 1)$? Yes. And why? You might reason by writing things out and not simplifying until the end:

$$(7, 1) + (9, 8) = (7 + 9, 1 + 8) \quad \text{and}$$
$$(9, 8) + (7, 1) = (9 + 7, 8 + 1)$$

Since $7 + 9 = 9 + 7$ and $1 + 8 = 8 + 1$, $(7, 1) + (9, 8) = (9, 8) + (7, 1)$. And this gives you an idea for how a proof in general will go.

← This habit of "writing things out and not simplifying until the end" is an important algebraic strategy, often called *delayed evaluation*.

### Example 2

**Problem.** Find $A$ if $A$ is in $\mathbb{R}^3$ and $2A + (-3, 4, 2) = (5, 2, 2)$.

**Solution.** Here are two different ways to find $A$.

**Habits of Mind**

Fill in the reasons.

1. Suppose $A = (a_1, a_2, a_3)$ and calculate as follows.

$$\begin{aligned}
2(a_1, a_2, a_3) + (-3, 4, 2) &= (5, 2, 2) \\
(2a_1, 2a_2, 2a_3) + (-3, 4, 2) &= (5, 2, 2) \\
(2a_1 - 3, 2a_2 + 4, 2a_3 + 2) &= (5, 2, 2) \\
2a_1 - 3 = 5,\ 2a_2 + 4 = 2,\ 2a_3 + 3 &= 2 \\
a_1 = 4,\ a_2 = -1,\ a_3 = 0; \qquad A &= (4, -1, 0)
\end{aligned}$$

2. Instead of calculating with coordinates, you can also use Theorem 1.2.

$$\begin{aligned} 2A + (-3, 4, 2) &= (5, 2, 2) \\ 2A &= (8, -2, 0) && \text{(subtract } (-3,4,2) \text{ from both sides)} \\ A &= (4, -1, 0) && \text{(multiply both sides by } \tfrac{1}{2}\text{)} \end{aligned}$$

---

**Minds in Action** Episode 2

*Tony and Sasha are working on the following problem:*

Find points $A$ and $B$ in $\mathbb{R}^2$, where $A + B = (3, 11)$ and $2A - B = (3, 1)$.

SASHA: In Example 2, we solved the equation with points just like any other equation. So, here we have two equations and two unknowns . . .

TONY: So we can use elimination. And look, it's easy—if we add both the equations together, the $B$'s cancel out and we get $3A = (6, 12)$.

SASHA: So we divide both sides by 3 to get $A = (2, 4)$. We can plug that into the first equation . . .

TONY: . . . and subtract $(2, 4)$ from both sides to get $B = (1, 7)$.

SASHA: Smooth, Tony. I wonder how much harder it would be to use coordinates. What if we say $A = (a_1, a_2)$ and $B = (b_1, b_2)$. We can then work it through like the first part of Example 2.

TONY: Have fun with that, Sasha.

**Habits of Mind**

Make sure that Sasha and Tony's calculations are legal. Theorem 1.2 gives the basic rules.

---

**Developing Habits of Mind**

**Find connections.** After using Theorem 1.2 for a while to calculate with points and scalars, you might begin to feel like you did in Algebra 1 when you first practiced solving equations like $3x + 1 = 7$: you can forget the meaning of the letters and just proceed formally, applying the basic rules.

---

**Example 3**

**Problem.** Find scalars $c_1$ and $c_2$ so that

$$c_1(1, 4, -1) + c_2(3, -1, 2) = (-1, 9, -4)$$

**Solution.** Simplify the left-hand side to get

$$(c_1 + 3c_2, 4c_1 - c_2, -c_1 + 2c_2) = (-1, 9 - 4, ), \text{ or}$$

$$\begin{aligned} c_1 + 3c_2 &= -1 \\ 4c_1 - c_2 &= 9 \\ -c_1 + 2c_2 &= -4 \end{aligned}$$

So, you are looking for a solution to this system of equations.

Solve the first two equations simultaneously to find the solution $c_1 = 2$, $c_2 = -1$. This solution works in the third equation also, so 2 and $-1$ are the desired scalars. Because $(-1, 9, -4)$ can be written as $2(1, 4, -1) + -1(3, -1, 2)$, $(-1, 9, -4)$ is a **linear combination** of $(1, 4, -1)$ and $(3, -1, 2)$.

⟵ In Chapter 3, you will study other methods for solving systems of linear equations.

## Exercises

**1.** Let $A = (3, 1), B = (2, -4)$, and $C = (1, 0)$. Calculate and plot the following:

**a.** $A + 3B$
**b.** $2A - C$
**c.** $A + B - 2C$
**d.** $-A + \frac{1}{2}B + 3C$
**e.** $\frac{1}{2}(A + B) + \frac{1}{2}(A - B)$

**2.** For each choice of $U$ and $V$, find $U + V$ and $3U - 2V$.

**a.** $U = (4, -1, 2), V = (1, 3, -2)$
**b.** $U = (3, 0, 1, -2), V = (1, -1, 0, 1)$
**c.** $U = (3, 7, 0), V = (0, 0, 2)$
**d.** $U = (1, \frac{1}{2}, 3), V = 2U$

**3.** Let $A = (3, 1)$ and $B = (-2, 4)$. Calculate each result and plot your answers.

**a.** $A + B$
**b.** $A + 2B$
**c.** $A + 3B$
**d.** $A - B$
**e.** $A + \frac{1}{2}B$
**f.** $A + 7B$
**g.** $A - \frac{1}{3}B$
**h.** $A + \frac{5}{2}B$
**i.** $A - 4B$

**4.** Let $A = (5, -2)$ and $B = (2, 5)$. Calculate each result and plot your answers.

**a.** $A + B$
**b.** $A + 2B$
**c.** $2A + 3B$
**d.** $2A - 3B$
**e.** $\frac{1}{2}A + \frac{1}{2}B$
**f.** $\frac{1}{3}A + \frac{2}{3}B$
**g.** $\frac{1}{10}A + \frac{9}{10}B$
**h.** $-3A - 4B$
**i.** $A - 4B$

**5.** Let $A = (5, -2)$ and $B = (2, 5)$. Calculate each result and plot your answers.

**a.** $A + (B - A)$
**b.** $A + 2(B - A)$
**c.** $A + 3(B - A)$
**d.** $A - 3(B - A)$
**e.** $A + \frac{1}{2}(B - A)$
**f.** $A + \frac{2}{3}(B - A)$
**g.** $A + \frac{9}{10}(B - A)$
**h.** $A - 4(B - A)$
**i.** $A + 4(B - A)$

**6.** Let $A = (a_1, a_2)$ and $B = (b_1, b_2)$. Find an expression for the area of the parallelogram whose vertices are $O$, $A$, $A + B$, and $B$.

**7.** In $\mathbb{R}^3$, find the equation of each of the following:

**a.** the $y$–$z$ plane
**b.** the plane through $(-3, 5, -1)$ parallel to the $y$–$z$ plane
**c.** the plane through $(-3, 5, -1)$ parallel to the $x$–$y$ plane
**d.** the sphere with center $(0, 0, 0)$ and radius 1
**e.** the sphere with center $(2, 3, 6)$ and radius 1

**8.** Find the point $A = (a_1, a_2, a_3, a_4, a_5)$ in $\mathbb{R}^5$ if $a_j = j^2$ for each $j = 1, 2, \ldots, 5$.

**9.** For each of the following equations, solve for $A$.

**a.** $3A - (4, 7) = (-1, -4)$
**b.** $2A + 3(2, -1, 3, 6) = 4A + (2, -1, 3, 2)$
**c.** $2A - (4, 6, 2) = O$
**d.** $5A - (-1, 7, 1) = 3A + 4(8, -1, 2)$

**10.** Find $A$ and $B$ if $A + B = (4, 8)$ and $A - B = (-2, -6)$.

**11.** For each of the following equations, find $c_1$ and $c_2$.

**a.** $c_1(2, 3, 9) + c_2(1, 2, 5) = (1, 0, 3)$
**b.** $c_1(2, 3, 9) + c_2(1, 2, 5) = (0, 1, 1)$

**12.** Show that there are no scalars $c_1$ and $c_2$ so that

$$c_1(4, 1, 2) + c_2(-8, -2, -4) = (3, 1, 2)$$

$\longleftarrow$ What does the set of all points of the form $c_1(4,1,2)+c_2(-8,-2,-4)$ look like in $\mathbb{R}^3$?

**13.** Find nonzero scalars $c_1, c_2$, and $c_3$ so that

$$c_1(1, 5, 1) + c_2(2, 0, 3) + c_3(3, 5, 4) = (0, 0, 0)$$

**14.** Show that if $c_1(3, 2) + c_2(4, 1) = (0, 0)$, then $c_1 = c_2 = 0$.

**15.** Prove (2), (3), and (4) in Theorem 1.2.

**16.** Prove (5), (6), and (8) in Theorem 1.2.

## 1.3 Vectors

The real number system evolved in an attempt to measure physical quantities like length, area, and volume. Certain physical phenomena, however, cannot be characterized by a single real number. For example, there are two equally important pieces of information that specify the velocity of an object: the speed (or magnitude of the velocity) and the direction. You can represent velocity using a single object: a *vector*.

← Unless, of course, the speed is 0.

**In this lesson, you will learn how to**

- test vectors for equivalence using the algebra of points
- prove simple geometric theorems with vector methods
- develop a level of comfort moving back and forth between points and vectors
- think of linear combinations geometrically

A vector is a directed line segment that is usually represented by drawing an arrow. The arrow has a length (or magnitude), and one end has an arrowhead that denotes the direction the arrow is pointing. In this figure, the two endpoints of the line segment are labeled $A$ and $B$. If you know the two endpoints, you can completely describe the vector. This vector starts at $A$ and ends at $B$, so it is denoted by $\overrightarrow{AB}$. The point $A$ is called the **tail** (or **initial point**) of $\overrightarrow{AB}$, and the point $B$ is called the **head** (or **terminal point**) of $\overrightarrow{AB}$.

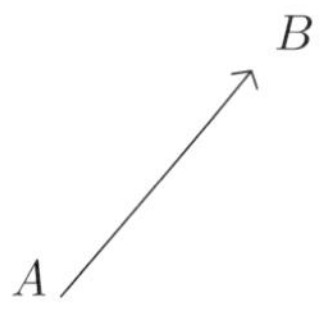


← Note that $\overrightarrow{AB}$ is not the same vector as $\overrightarrow{BA}$.

In fact, you can completely describe any vector by specifying just its tail and its head, so you do not have to rely on a drawing. The following definition works for any dimension.

#### Definition

If $A$ and $B$ are points in $\mathbb{R}^n$, the **vector** with tail $A$ and head $B$ is the ordered pair of points $[A, B]$. You can denote the vector with tail $A$ and head $B$ by $\overrightarrow{AB}$.

#### Facts and Notation

There's no real agreement about the definition of "vector." Many books insist that a vector must have its tail at the origin, calling vectors that don't start at the origin "located vectors" or "free vectors." While there are good reasons for making such fine distinctions, they are not necessary at the start. *This* book will soon concentrate on vectors that start at the origin too, but for now, think of a vector as an arrow or an ordered pair of points.

← There's plenty of time for formalities later.

## Developing Habits of Mind

**Use algebra to extend geometric ideas.** There are many ways to think about vectors. Physicists talk about quantities that have a "magnitude" and "direction" (like velocity, as opposed to speed). Football coaches draw arrows. Some people talk about "directed" line segments. Mathematics, as usual, makes all this fuzzy talk precise: a vector is nothing other than an*ordered pair of points*.

$\longleftarrow$ The gain in precision is accompanied by a loss of all these romantic images carried by the arrows and colorful language.

But the geometry is essential: a central theme in this book is to start with a geometric idea in $\mathbb{R}^2$ or $\mathbb{R}^3$, find a way to characterize it with algebra, and then use that algebra as the *definition* of the idea in higher dimensions. The details of how this theme is carried out will become clear over time. The next discussion gives an example.

**Habits of Mind**

"The geometry" referred to here is the regular Euclidean plane geometry you studied in earlier courses. Later, you may study just how many of these ideas can be extended if you start with, say, geometry on a sphere.

In $\mathbb{R}^2$ or $\mathbb{R}^3$, two vectors are called **equivalent** if they have the same magnitude (length) and the same direction. For example, in the figure to the right,

$$A = (1, -1)$$
$$B = (4, 1)$$
$$C = (2, 3) \text{ and}$$
$$D = (5, 5)$$

$\overrightarrow{AB}$ is equivalent to $\overrightarrow{CD}$.

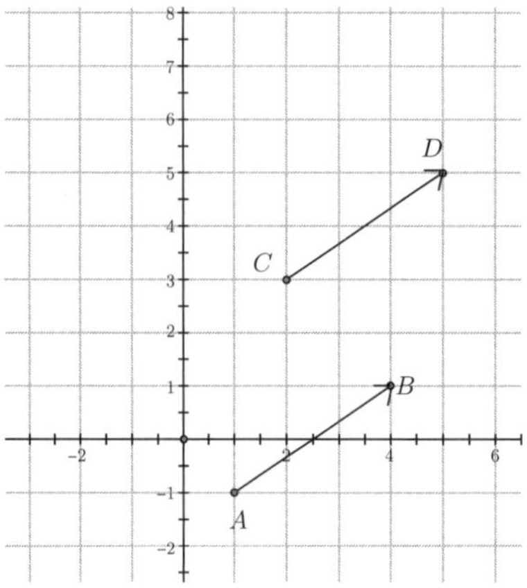


## For You to Do

**1.** Show that vectors $\overrightarrow{AB}$ and $\overrightarrow{CD}$ have the same length.

## What's Wrong Here?

**2.** Derman calculates the slope from $A$ to $B$ as $\frac{2}{3}$. But he also remembers that the slope from $B$ to $A$ is also $\frac{2}{3}$, so he thinks that $\overrightarrow{AB}$ is equivalent to $\overrightarrow{BA}$. Can that be right?

So, there is this geometric idea of equivalent vectors. To define equivalence of vectors in $\mathbb{R}^n$ in a way that agrees with this notion of equivalence in $\mathbb{R}^2$, you need to characterize equivalent vectors in $\mathbb{R}^2$ without using words like "magnitude" or "direction." Suppose $A = (a_1, a_2), B = (b_1, b_2), C = (c_1, c_2)$, and $B$ is to the right and above $A$ in $\mathbb{R}^2$ as in the following figure.

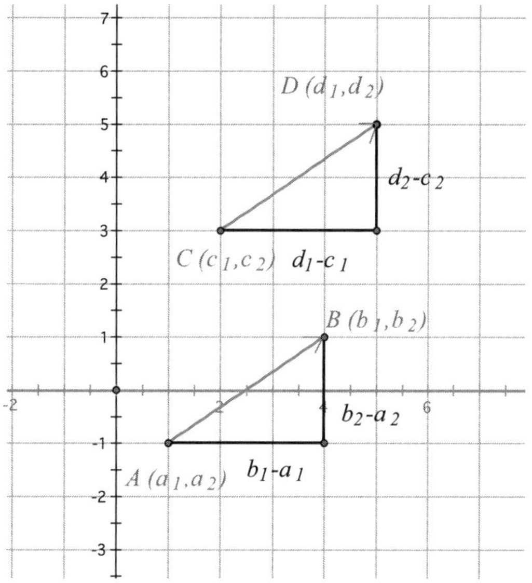


To find a point $D = (d_1, d_2)$ so that $\overrightarrow{AB}$ is equivalent to $\overrightarrow{CD}$, starting from $C$, move to the right a distance equal to $b_1 - a_1$, and then move up a distance equal to $b_2 - a_2$. In other words, $d_1 = c_1 + (b_1 - a_1)$ and $d_2 = c_2 + (b_2 - a_2)$. Therefore, $d_1 - c_1 = b_1 - a_1$ and $d_2 - c_2 = b_2 - a_2$.

This can be written as $(d_1 - c_1, d_2 - c_2) = (b_1 - a_1, b_2 - a_2)$, or, using the algebra of points,

$$D - C = B - A$$

You can call this the "head minus tail test" in $\mathbb{R}^2$.

**Habits of Mind**

What are the slopes of $\overrightarrow{AB}$ and $\overrightarrow{CD}$?

### For You to Do

**3.** In the figure above, show that if $D - C = B - A$, the distance from $A$ to $B$ is the same as the distance from $C$ to $D$ and that the slope from $A$ to $B$ is the same as the slope from $C$ to $D$.

⟵ In the *CME Project* series, the slope from $A$ to $B$ is written as $m(A, B)$.

**Theorem 1.3 (Head Minus Tail Test)**
*In $\mathbb{R}^2$, the vectors $\overrightarrow{AB}$ and $\overrightarrow{CD}$ are equivalent if and only if*

$$D - C = B - A$$

The discussion leading up to Theorem 1.3 makes its result seem plausible, but there are other details to check.

1. The preceding argument for finding point $D$ depends on a particular orientation of the two vectors—$B$ is to the right and above $A$. A careful proof would have to account for other cases.

2. That argument shows that if $\overrightarrow{AB}$ is equivalent to $\overrightarrow{CD}$, then $B - A = D - C$. A careful proof would also show the converse: if $B - A = D - C$, then $\overrightarrow{AB}$ is equivalent to $\overrightarrow{CD}$.

Both of these details can be handled with some careful analytic geometry.

It also can be shown (using analytic geometry in three dimensions) that the Head Minus Tail (HmT) Test works in $\mathbb{R}^3$. Since this characterization of equivalence makes no use of geometric language, it makes sense in $\mathbb{R}^n$.

⟵ Much more attention will be given to the geometry of $\mathbb{R}^3$ in the next chapter.

**Habits of Mind**

**Use algebra to extend geometric ideas.** This definition of "equivalent" uses the algebra you developed in Theorem 1.3 and extends that algebra to any dimension.

### Definition

If $A, B, C$, and $D$ are points in $\mathbb{R}^n$, the vectors $\overrightarrow{AB}$ and $\overrightarrow{CD}$ are said to be **equivalent** if $B - A = D - C$.

### Example 1

**Problem.** Is $\overrightarrow{(\frac{1}{2},3)\,(3,5)}$ equivalent to $\overrightarrow{(\frac{17}{8},0)\,(\frac{35}{8},2)}$?

**Solution.** You could check slopes and (directed) distances, but both of those are checked in the HmT Test. For the first vector, HmT yields $\left(\frac{5}{2},2\right)$; for the second, you get $\left(\frac{9}{4},2\right)$. So the vectors are not equivalent.

**Habits of Mind**

Draw a picture.

### Example 2

**Problem.** In $\mathbb{R}^4$, if $X = (-1,2,3,1), Y = (1,-2,5,4)$, and $Z = (3,1,1,0)$, find $W$ so that $\overrightarrow{XY}$ is equivalent to $\overrightarrow{ZW}$.

**Solution.** By *definition*, this means $W - Z = Y - X$ or

$$W = Z + Y - X = (5,-3,3,3)$$

### Developing Habits of Mind

**Use algebra to extend geometric ideas.** The process that led to the definition of equivalent vectors in $\mathbb{R}^n$ is important.

- First, equivalent vectors in $\mathbb{R}^2$ are defined using geometric ideas.
- Next, equivalent vectors in $\mathbb{R}^2$ are characterized by an equation involving only the operation of subtraction of points.
- Finally, this *equation* is used as the *definition* of equivalent vectors in $\mathbb{R}^n$.

⟵ This process will be called the *extension program* from now on.

This theme will be used throughout the book, and it will allow you to generalize many familiar geometric notions from the plane (and in the next chapter, from three-dimensional space) to $\mathbb{R}^n$.

### For You to Do

4. Show that in $\mathbb{R}^2$ every vector is equivalent to a vector whose tail is at $O$.

The same result—every vector is equivalent to a vector whose tail is at $O$—is true in $\mathbb{R}^n$, and the proof may seem surprising.

**Theorem 1.4**

*Every vector in $\mathbb{R}^n$ is equivalent to a vector whose tail is at $O$. In fact, $\overrightarrow{AB}$ is equivalent to $\overrightarrow{O(B-A)}$.*

***Proof.*** $B - A = (B - A) - O$. ■

The following figure illustrates Theorem 1.4 for several vectors.

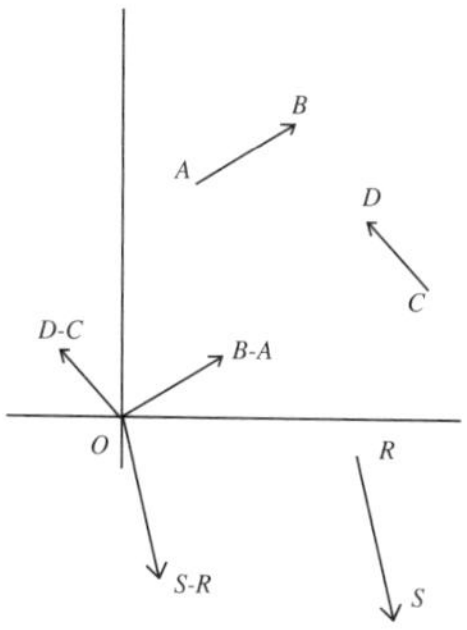


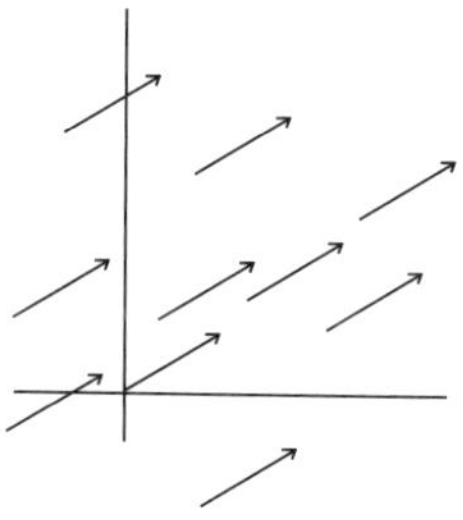

**Facts and Notation**

The vectors in $\mathbb{R}^n$ break up into "classes": two vectors belong to the same "class" if and only if they are equivalent. Every nonzero point in $\mathbb{R}^n$ determines one of these classes. That is, the point $A$ determines the class of vectors equivalent to $\overrightarrow{OA}$. Theorem 1.4 shows that every class of vectors is obtained in this way. Furthermore, you can show (see Exercise 5) that if $\overrightarrow{OA}$ is equivalent to $\overrightarrow{OB}$, then $A = B$.

Because of this, the following convention will be in force for the rest of this book: from now on, an ordered $n$-tuple $A \neq O$ will stand for *either* a point in $\mathbb{R}^n$ or the vector $\overrightarrow{OA}$. You can also consider $O$ as a vector (the **zero vector**). The context will always make it clear whether an element in $\mathbb{R}^n$ is considered a point or a vector.

**Example 3**

**Problem.** Show that the points $A = (-2, 4)$, $B = (2, 5)$, $C = (4, 3)$, and $D = (0, 2)$ lie on the vertices of a parallelogram.

**Solution Method 1.** Translate the quadrilateral to the origin; that is, slide the parallelogram so that one of the vertices (say, $A$) lands at the origin, and translate the other three points similarly.

← A **translation** is a transformation that slides a figure without changing its size, its shape, or its orientation. If $A$ is a point, subtracting $A$ from each vertex of a polygon translates that polygon (why?).

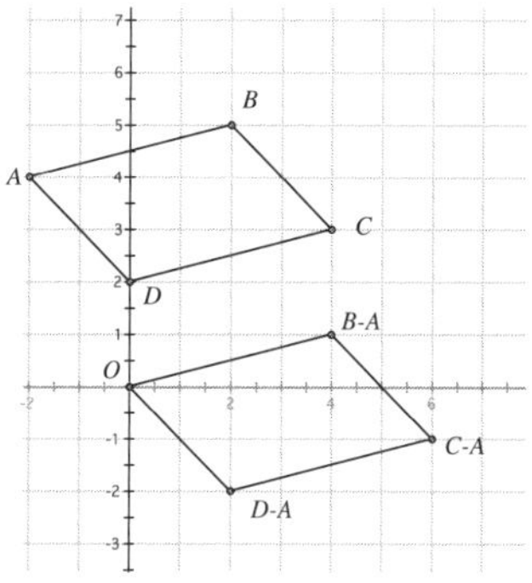

Since $A - A = O$, you can translate the other three points by subtracting $A$. More precisely, $\overrightarrow{AB}$ is equivalent to $B - A = (4, 1)$, $\overrightarrow{AC}$ is equivalent to $C - A = (6, -1)$, and $\overrightarrow{AD}$ is equivalent to $D - A = (2, -2)$. Since $(4, 1) + (2, -2) = (6, -1)$, $C - A = (B - A) + (D - A)$, and you can say that, by the Parallelogram Rule (Theorem 1.1 from Lesson 1.2), $O, B - A, C - A$, and $D - A$ lie on the vertices of a parallelogram. Since the translation affects only the position of the figure, $A, B, C$, and $D$ must also lie on the vertices of a parallelogram.

**Solution Method 2.** $\overrightarrow{AB}$ is equivalent to $\overrightarrow{DC}$, since $B - A = C - D = (4, 1)$. Since they are equivalent, you know they have the same length (magnitude), so the line segments are congruent. They also have the same direction, so the line segments have the same slope and are thus parallel.

$\longleftarrow$ A convex quadrilateral is a parallelogram if *one* pair of opposite sides is both parallel and congruent, so there's no need to show that $\overrightarrow{BC}$ is equivalent to $\overrightarrow{AD}$.

### For You to Do

**5.** Derman tried to show that he had a parallelogram if $A = (2, 1)$, $B = (4, 2)$, $C = (6, 3)$, and $D = (10, 5)$, and he ended up scratching his head. Help him figure out why these points do not form a parallelogram.

### Developing Habits of Mind

**Use vectors to describe geometric ideas.** If points in $\mathbb{R}^2$ and $\mathbb{R}^3$ are viewed as vectors, the geometric description of addition and scalar multiplication is much easier.

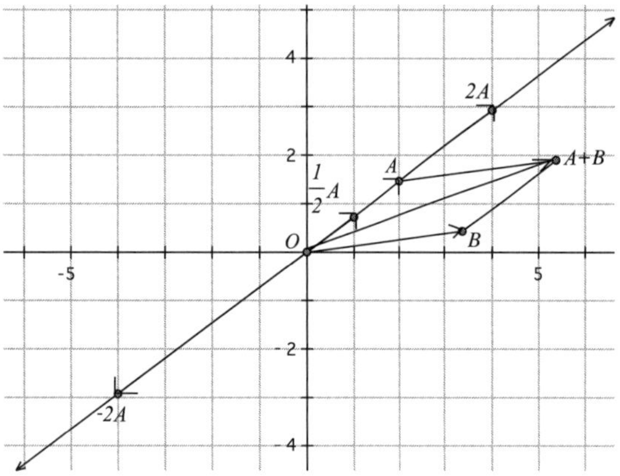


If $A$ and $B$ are vectors in $\mathbb{R}^2$, $A + B$ is the diagonal of the parallelogram whose sides are $A$ and $B$. Multiplying $A$ by $c$ yields a vector whose length is the length of $A$ multiplied by $|c|$; $cA$ has the same direction as $A$ if $c > 0$; $cA$ has the opposite direction of $A$ if $c < 0$.

In Exercise 1 from Lesson 1.1, you plotted several points that were scalar multiples of a point $A = (1, 2)$. What you might have noticed was that all of the resulting points ended up on the same line:

$\longleftarrow$ In Exercise 6, you'll show that if $P$ and $Q$ are nonzero vectors in $\mathbb{R}^2$ or $\mathbb{R}^3$, and that if $O$, $P$, and $Q$ are collinear, then $Q = cP$ for some real number $c$. In fact, the set of points collinear with the points $O$ and $P$ is the collection of multiples $cP$ of $P$.

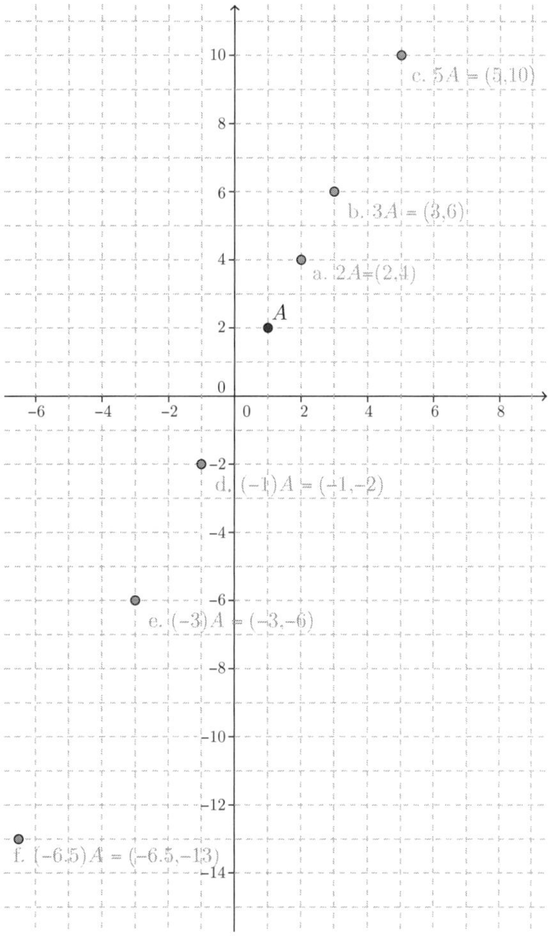


The fact that scalar multiples of points are collinear hints at an algebraic way to characterize parallel vectors in $\mathbb{R}^2$. In the following figure, $\overrightarrow{AB}$ and $\overrightarrow{CD}$ have the same direction, $\overrightarrow{AB}$ and $\overrightarrow{EF}$ have opposite directions, and all three vectors appear to be parallel.

**Remember**

A point $A$ corresponds to the vector $\overrightarrow{OA}$.

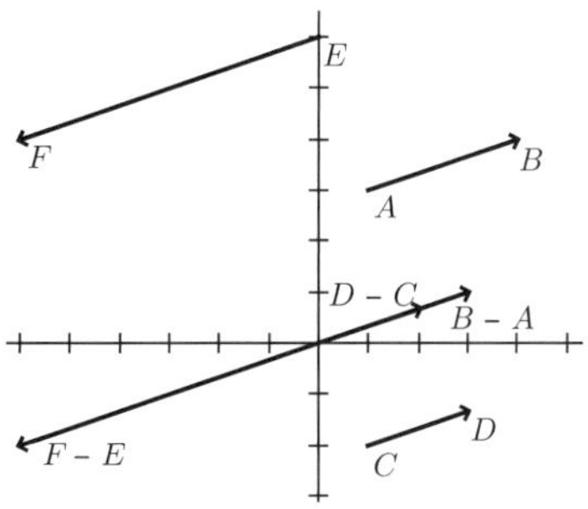


If you construct equivalent vectors starting at $O$, vectors $B - A$, $D - C$, and $F - E$ are all scalar multiples of each other. More precisely, $D - C = k(B - A)$ for some $k > 0$, and $F - E = k(B - A)$ for some $k < 0$. This statement was developed in $\mathbb{R}^2$, but it makes sense in $\mathbb{R}^n$ for any $n$.

### Definition

Two vectors $\overrightarrow{AB}$ and $\overrightarrow{CD}$ in $\mathbb{R}^n$ are said to be **parallel** if there is a nonzero real number $k$ so that

$$B - A = k(D - C)$$

- If $k > 0$, $\overrightarrow{AB}$ and $\overrightarrow{CD}$ have the **same direction**.
- If $k < 0$, $\overrightarrow{AB}$ and $\overrightarrow{CD}$ have **opposite directions**.

The zero vector $O$ is parallel to every vector (with no conclusion about same or opposite direction).

**Habits of Mind**

**Use algebra to extend geometric ideas.** The definition takes an algebraic characterization—$B - A = k(D - C)$—of a geometric property—$\overrightarrow{AB}$ is parallel to $\overrightarrow{CD}$—and makes it the *definition* of the geometric property in $\mathbb{R}^n$.

### For You to Do

**6.** Let $A = (3, -1, 2, 4), B = (1, 2, 0, 1), C = (3, 2, -3, 5)$, and $D = (7, -4, 1, 11)$. Show that $\overrightarrow{AB}$ is parallel to $\overrightarrow{CD}$.

### Minds in Action Episode 3

*Tony and Sasha are working on the following problem.*

**Problem.** Suppose that $P = (2, 4), A = (3, 1)$, and $S$ is the set of points $Q$ so that $\overrightarrow{PQ}$ is parallel to $A$. Draw a picture of $S$ and find an equation for it.

TONY: From the definition, if $\overrightarrow{PQ}$ is parallel to $A$, then there is a real number $k$ so that $Q - P = kA$. Okay, now what?

SASHA: Well, let's rewrite that as

$$Q = P + kA$$

I can draw what $kA$ looks like for all real values of $k$: it's a line through the origin and the point $A$.

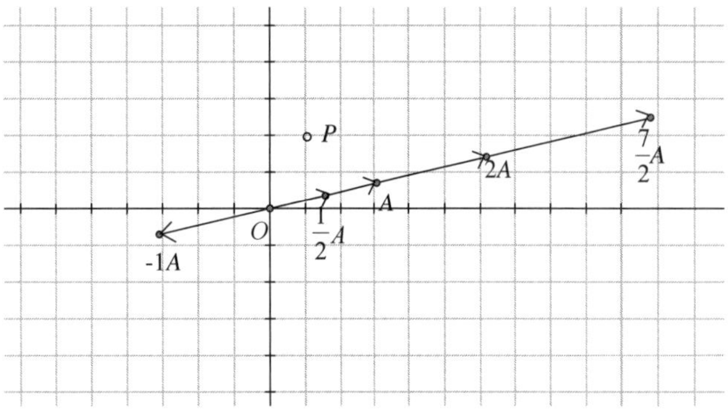


TONY: Great, but what about the $P$?

SASHA: Well, we have to add $P$ to each of the multiples.

TONY: That's a lot of parallelograms! All right, let's try it.

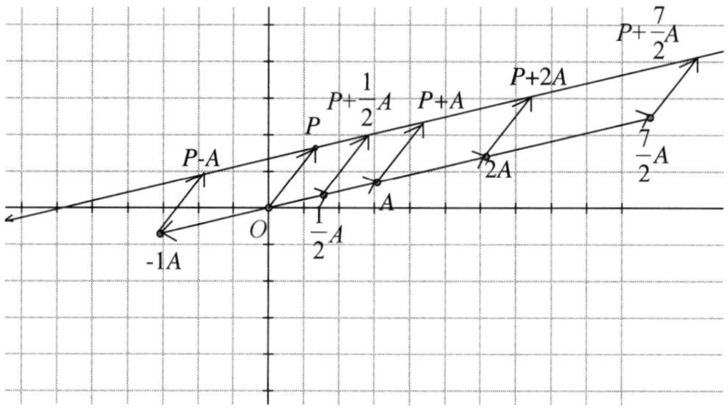


Ah, I see it. That works . . . it's a line that goes through $P$ and is parallel to $A$. So is $Q = P + kA$ the equation of the line?

SASHA: Well, it's not *the* equation for the line, but I guess it's *an* equation for the line. But it's different than ones we've used before, because it has vectors in it and not coordinates. And, there's this $k$ in there, standing for a real number.

TONY: There's probably a special name for it, then.

---

As Tony and Sasha found out, the graph of $Q = P + kA$ is a straight line through $P$ in the direction of $A$. An equation of the form

$$X = P + kA$$

is called a **vector equation of a line**. It works as a point-tester, too, in the sense that a point $X$ is on this line if and only if there is a number $k$ so that $X = P + kA$—although you'll have to do some algebra to see if such a $k$ exists in specific cases. But it also works as a *point-generator* (see the Developing Habits of Mind below).

⟵ Context clues: Tony and Sasha are thinking of $P$ as a point and $A$ as a vector. Why?

⟵ The equation $X = P + kA$ is also called a **parametric equation** for the line, when you want to emphasize the role of the "parameter" $k$.

To find a more familiar coordinate equation for $S$, replace the vector letters with coordinates. If $X = (x, y)$, you can say

$$(x, y) = (2, 4) + k(3, 1)$$

and thus

$$x = 2 + 3k$$
$$y = 4 + k$$

⟵ In the next chapter, you'll become very familiar with this vector equation, so you'll have less need to move to the linear coordinate equation.

Multiply the second equation by 3 and subtract from the first to obtain

$$x - 3y = -10$$

This kind of equation probably looks more like equations of lines than you are used to, so it is pretty clear that $S$ is in fact a line.

### Developing Habits of Mind

**Use the vector equation as a point-generator.** You can use the equation $X = P + kA$ to generate points: consider a "slider" of length $k$ that you can control with your mouse. As you change the length of the slider, $A$ gets scaled by $k$ and added to $P$. The varying $P + kA$ traces out $\ell$.

← You can create a drawing like this with geometry software.

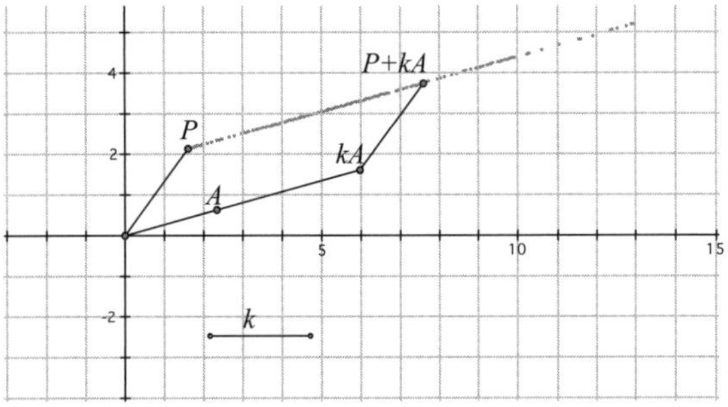


Every value of $k$ generates a point on the line, so the equation $X = P + kA$ is a kind of "function machine" that takes in numbers $k$ and produces points on the line through $P$ in the direction of $A$.

### For You to Do

**7.** Tony and Sasha worked on multiples of a single vector in $\mathbb{R}^2$. What if, in the equation $X = P + kA$, $X$, $P$, and $A$ are in $\mathbb{R}^3$? Would the equation still describe a line? Why or why not?

### Minds in Action Episode 4

*Tony and Sasha are working on the following problem.*

**Problem.** In $\mathbb{R}^3$, let $A = (2, 3, 9)$ and $B = (1, 2, 5)$. What do all the linear combinations of $A$ and $B$ describe?

TONY: I'm on it, Sasha. "Linear combination" means we multiply $A$ by something, and $B$ by something, and then add them together. In other words, we want something that looks like $k_1A + k_2B$. Should I start plugging away?

SASHA: Well, let's think about it for a second. We can find three points easily enough: $A$, $B$, and $O$.

TONY: Oh. $O$?

SASHA: Yes, $O$, because $O = 0A + 0B$. As I was saying, we have three points: $A$, $B$, and $O$. Those three points aren't on the same line, so they determine a plane. Let's call that plane $E$.

← How does Sasha know that $O$, $A$, and $B$ are not collinear? What would happen if they were?

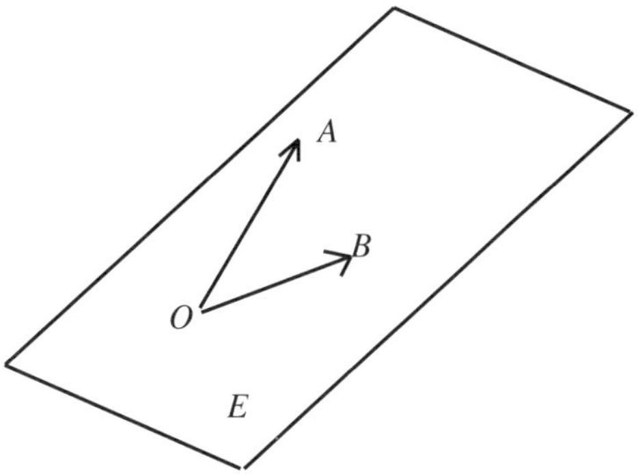


TONY: But I can pick anything for $k_1$ and anything for $k_2$, so won't that give us infinitely many points? What does that determine?

SASHA: Hold on. Say both $k_1$ and $k_2$ are 1. Then $A+B$ is another point. But we already know that completes a parallelogram, right? And if that's the case, $A+B$ should be on the same plane as the other three points, otherwise, it wouldn't make a parallelogram, but some weird twisted 3D shape.

TONY: I gotcha. That makes sense.

SASHA: And look at this. $2B$ would have to be on the same plane, too, right? I mean, it's on the line $\overleftrightarrow{OB}$, and if $O$ and $B$ are both on the plane $E$, all of the line $\overleftrightarrow{OB}$ has to be on $E$.

TONY: Same deal with multiples of $A$, like $\frac{1}{2}A$ has to be on $E$, too.

SASHA: And by the Parallelogram Rule again, their sum, $\frac{1}{2}A+2B$, has to be, too.

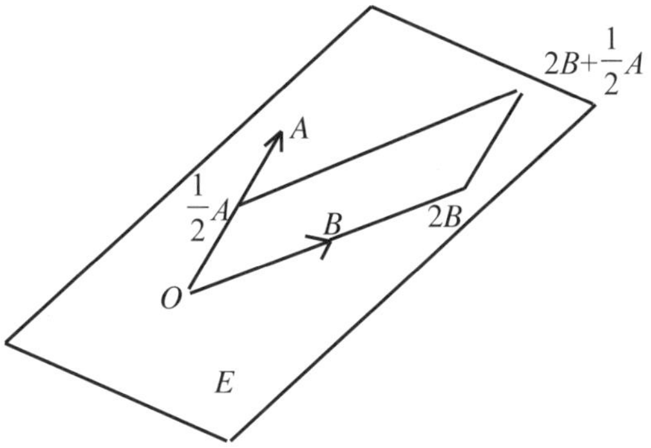


TONY: Then all these other points will have to stay on the plane $E$, right? Because any point $k_1A$ will be on the line $\overleftrightarrow{OA}$, so it's on $E$. And any point $k_2B$ will be on the line $\overleftrightarrow{OB}$, so it's on $E$, too. And the sum of any of those two points is part of a parallelogram where we know three of the points are on one plane, so the fourth has to be too.

SASHA: Brilliant, Tony! So an equation for $E$ could be $X = k_1A + k_2B$, or, better yet, $X = k_1(2,3,9) + k_2(1,2,5)$. Wait . . . uh oh . . .

← $E$ is called the plane **spanned** by $A$ and $B$.

TONY: What now?

SASHA: Well, we showed that any linear combination of $A$ and $B$ must be on $E$. But we *didn't* show it the other way . . . must every point on $E$ be a linear combination of $A$ and $B$?

TONY: It's not the same thing?

SASHA: No, we have to make sure that taking all the linear combinations doesn't leave holes in the plane.

*They both sit quietly for a while, thinking.*

I think I got it. Say $Q$ is a point on $E$. I can draw a line through $Q$ that's parallel to $A$—that line will be on $E$. I can also draw a line through $Q$ parallel to $B$, also on $E$.

TONY: Good job, Sasha. The line through $Q$ parallel to $A$ will intersect $\overleftrightarrow{OB}$ somewhere. And the line through $Q$ parallel to $B$ will intersect $\overleftrightarrow{OA}$ somewhere, too. And that makes a parallelogram.

SASHA: That's what I was thinking. Say $S$ is on $\overleftrightarrow{OA}$ and $\overrightarrow{SQ}$ is parallel to $B$, and say $R$ is on $\overleftrightarrow{OB}$ and $\overrightarrow{RQ}$ is parallel to $A$. Here, look at my picture.

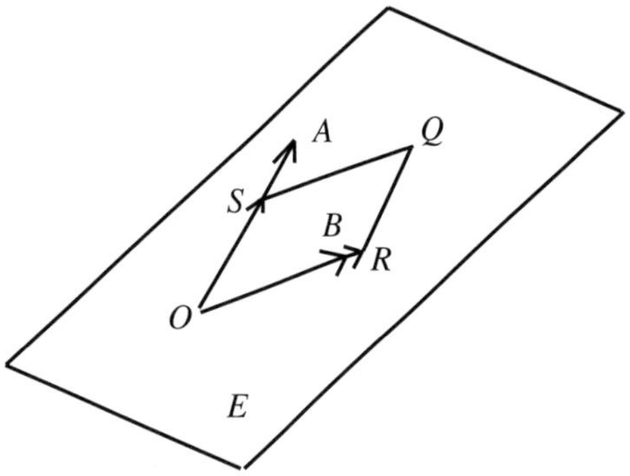


TONY: Yep. So $Q = S + R$, by the Parallelogram Rule. But since $S$ is on $\overleftrightarrow{OA}$, it has to equal $k_1A$ for some $k_1$, and $R$ being on $\overleftrightarrow{OB}$ means it has to equal $k_2B$ for some $k_2$. So, $Q = k_1A + k_2B$. And we're done.

SASHA: Great work, Tony.

TONY: Awesome . . . I think we've got a vector equation of a plane: $X = k_1A + k_2B$.

SASHA: Maybe, but something's missing, I think.

### For Discussion

**8.** **a.** Suppose $X$, $A$, and $B$ are vectors in $\mathbb{R}^2$, and $A$ and $B$ are *not* parallel. What does $X = k_1A + k_2B$ describe?

**b.** Suppose $X$, $A$, and $B$ are vectors in $\mathbb{R}^4$, and $A$ and $B$ are *not* parallel. What does $X = k_1A + k_2B$ describe?

**9.** Sasha thinks something is missing from Tony's "vector equation of a plane." What do you think she means?

So, the point-tester for the plane $E$ is "a point $Q$ is on the plane $E$ if and only if $Q$ is a linear combination of $A$ and $B$." This point-tester leads to the vector equation

$$X = c_1 A + c_2 B$$

You can use this equation to test any point $X$ in $\mathbb{R}^3$ to see if it is on the plane $E$.

$\longleftarrow$ $X = c_1 A + c_2 B$ is also a point-generator. Why?

### Example 4

**Problem.** Is $U = (3, 5, 14)$ on Sasha's plane $E$ (from Episode 4)?

**Solution.** You might spot that $U = A + B$, so $U$ is a linear combination of $A$ and $B$, and hence it's on $E$. If you didn't spot that, you could set up the equations

$$U = c_1 A + c_2 B \quad \text{or}$$
$$(3, 5, 14) = c_1(2, 3, 9) + c_2(1, 2, 5)$$

Now look at it coordinate by coordinate:

$$\begin{aligned} 3 &= 2c_1 + c_2 \\ 5 &= 3c_1 + 2c_2 \\ 14 &= 9c_1 + 5c_2 \end{aligned}$$

Solve the first two equations for $c_1$ and $c_2$; $(1, 1)$ works. And $(1, 1)$ works in the last equation, too, so $c_1 = 1$ and $c_2 = 1$ is a solution. Hence, $U = 1A + 1B$, so $U$ is on $E$.

**Habits of Mind**

It's a good idea to train yourself to check if a point is a linear combination of some other points by playing around with the numbers in your head.

### For You to Do

**10.** Check to see if the following points are on $E$:

**a.** $(3, 5, 13)$ **b.** $(5, 8, 23)$ **c.** $(10, 16, 46)$

### Example 5

**Problem.** Tony and Sasha showed that the plane $E$ spanned by $A = (2, 3, 9)$ and $B = (1, 2, 5)$ has a vector equation

$$X = k_1 A + k_2 B$$

Find a coordinate equation for $E$.

**Solution.** A coordinate equation is just a point-tester for $E$ whose test is carried out by calculating with coordinates and not vectors. Start out with the vector equation

$$X = k_1 A + k_2 B$$

and substitute $A = (2, 3, 9), B = (1, 2, 5)$, and $X = (x, y, z)$ to get

$$(x, y, z) = (2k_1 + k_2, 3k_1 + 2k_2, 9k_1 + 5k_2)$$

So,

$$x = 2k_1 + k_2$$
$$y = 3k_1 + 2k_2$$
$$z = 9k_1 + 5k_2$$

You want a relation between $x$, $y$, and $z$ without using any of the $k$'s. One way to start it is to eliminate $k_2$ from two pairs of equations:

1. Multiply the first equation by 2 and subtract the second from the result to get
$$2x - y = k_1$$
2. Multiply the second equation by 5 and the third equation by 2 and then subtract to get
$$5y - 2z = -3k_1$$

Now substitute the left-hand side of the equation into the right-hand side of the last equation to get
$$5y - 2z = -3(2x - y)$$
This equation simplifies to
$$6x + 2y - 2z = 0$$
or
$$3x + y - z = 0 \tag{1}$$
This is a coordinate equation for $E$.

---

The derivation of the coordinate equation in this example should feel like the calculation shown in Example 4. Compare the system from Example 4 with the system from Example 5. To *prove* that the last equation is a coordinate equation for $E$, you'd have to show that a point $(x, y, z)$ is on $E$ if and only if $(x, y, z)$ satisfies it. There are some details to be filled in (what are they?).

In the next chapter, you'll develop a much more efficient way to find coordinate equations for planes that will build on the technique developed in this example.

## Exercises

**1.** For each set of vectors, determine whether the pairs of vectors $\overrightarrow{AB}$ and $\overrightarrow{CD}$ are equivalent, parallel in the same direction, or parallel in the opposite direction.

**a.** $A = (3, 1)$, $B = (4, 2)$, $C = (-1, 4)$, $D = (0, 5)$
**b.** $A = (3, 1)$, $B = (4, 2)$, $C = (0, 5)$, $D = (-1, 4)$
**c.** $A = (3, 1, 5)$, $B = (-4, 1, 3)$, $C = (0, 1, 0)$, $D = (14, 1, 4)$
**d.** $A = (-4, 1, 3)$, $B = (3, 1, 5)$, $C = (0, 1, 0)$, $D = (14, 1, 4)$
**e.** $A = (1, 3)$, $B = (4, 1)$, $C = (-2, 3)$, $D = (13, -7)$
**f.** $A = (3, 4)$, $B = (5, 6)$, $C = B - A$, $D = O$

$\longleftarrow$ Look for shortcuts. For instance, notice how solutions to earlier problems can help with later ones.

**g.** $A = O$, $B = (4, 7)$, $C = (5, 2)$, $D = B + C$
**h.** $A = (-1, 2, 1, 5)$, $B = (0, 1, 3, 0)$, $C = (-2, 3, 2, 1)$, $D = (-1, 2, 4, -4)$
**i.** $A = (-1, 2, 1, 5)$, $B = (-2, 3, 2, 1)$, $C = (0, 1, 3, 0)$, $D = (-1, 2, 4, -4)$

**2.** For each set of vectors from Exercise 1, parts **a**–**g**, sketch the points $A$, $B$, $C$, $D$, $B - A$, and $D - C$. Use a separate coordinate system for each set.

**3.** Find a point $P$ if $\overrightarrow{PQ}$ is equivalent to $\overrightarrow{AB}$, where

$$A = (2, -1, 4), B = (3, 2, 1), \text{ and } Q = (1, -1, 6)$$

**4.** In $\mathbb{R}^4$, suppose $A = (3, 1, -1, 4)$, $B = (1, 3, 2, 0)$, and $C = (1, 1, -1, 3)$. If $D = (-3, a, b, c)$, find $a, b,$ and $c$ so that $\overrightarrow{AB}$ is parallel to $\overrightarrow{CD}$.

**5.** In $\mathbb{R}^n$, show that if $\overrightarrow{AB}$ is equivalent to $\overrightarrow{AC}$, then $B = C$.

**6.** In $\mathbb{R}^2$, show that if $A$, $B$, and $O$ are collinear, then $B = cA$ for some number $c$.

**7.** Suppose $A = (1, 2, 3)$ and $B = (4, 5, 6)$. Is each point a linear combination of $A$ and $B$? If so, give the coefficients. If not, explain why.

⟵ Try to do this problem in your head. If you get stuck, write down equations.

| | | |
|---|---|---|
| **a.** $(5, 7, 9)$ | **b.** $(3, 3, 3)$ | **c.** $(-5, -7, -9)$ |
| **d.** $(10, 14, 18)$ | **e.** $(8, 10, 12)$ | **f.** $(7, 8, 9)$ |
| **g.** $(7, 8, 10)$ | **h.** $(1, 2, 3)$ | **i.** $(1, 2, 4)$ |

**8.** Some people, especially physicists, talk about adding vectors "head to toe" in the following way: to find $\overrightarrow{AB} + \overrightarrow{CD}$, move $\overrightarrow{CD}$ to an equivalent vector starting at $B$, say $\overrightarrow{BQ}$. Then

$$\overrightarrow{AB} + \overrightarrow{CD} = \overrightarrow{AQ}$$

**a.** Draw a picture of how this works.
**b.** Show that $\overrightarrow{AQ}$ is equivalent to $(B - A) + (D - C)$.

**9.** In $\mathbb{R}^n$, if $\overrightarrow{AB}$ is equivalent to $\overrightarrow{CD}$, show that $\overrightarrow{AC}$ is equivalent to $\overrightarrow{BD}$. Illustrate geometrically in $\mathbb{R}^2$ or $\mathbb{R}^3$.

**10.** In $\mathbb{R}^2$, let $\ell$ be the line whose equation is $5x + 4y + 20 = 0$. If $P = (-4, 0)$ and $A = (-4, 5)$, show that $\ell$ is the set of all points $Q$ so that $\overrightarrow{PQ}$ is parallel to $A$.

**11.** Let $P = (3, 0)$ and $A = (1, 5)$. If $\ell$ is the set of all points $Q$ so that $\overrightarrow{PQ}$ is parallel to $A$, find a vector equation and a coordinate equation for $\ell$.

**12.** In $\mathbb{R}^3$, let $A = (1, 0, 2)$ and $B = (0, 1, 3)$.

**a.** Find a vector equation and a coordinate equation for the plane spanned by $A$ and $B$.

**b. Take It Further.** Find a vector equation and a coordinate equation for the plane parallel to the one you found in part **a** that passes through the point $C = (1, 1, 1)$.

**13.** Show that the following definition for the midpoint of a vector in $\mathbb{R}^n$ agrees with the usual midpoint formula in $\mathbb{R}^2$: the **midpoint** of $\overrightarrow{AB}$ is the point $\frac{1}{2}(A + B)$.

**14.** In $\mathbb{R}^4$, let $A = (-3, 1, 2, 4)$, $B = (5, 3, 6, -2)$, and $C = (1, 1, -2, 0)$. If $M$ is the midpoint (see Exercise 13) of $\overrightarrow{AB}$ and $N$ is the midpoint of $\overrightarrow{BC}$, show that $\overrightarrow{MN}$ is parallel to $\overrightarrow{AC}$. This exhibits a generalization of what fact in plane geometry?

**15.** If $A$, $B$, and $C$ are points in $\mathbb{R}^n$, $M$ is the midpoint (see Exercise 13) of $\overrightarrow{AB}$, and $N$ is the midpoint of $\overrightarrow{BC}$,

**a.** show that $M - A = B - M$

**b.** show that $M - N = \frac{1}{2}(A - C)$

**c.** prove that $\overrightarrow{MN}$ is parallel to $\overrightarrow{AC}$

# 1.4 Length

In the previous lesson, you saw that the two key attributes of a vector are its *magnitude* and *direction*, but you didn't spend much time on either one. This lesson focuses on magnitude (length).

⟵
You'll explore direction in the next chapter.

**In this lesson, you will learn how to**

- calculate the length of a vector and apply the algebraic properties described in the theorems
- give geometric interpretations of algebraic results that involve length
- understand how the extension program is used to define length in higher dimensions
- identify a unit vector

---

If $A = (a_1, a_2)$ is a vector in $\mathbb{R}^2$, the length of $A$ can be calculated as the distance between the origin $O$ and $A$ using the distance formula (which is derived from the Pythagorean Theorem): $\sqrt{a_1^2 + a_2^2}$. So the length of $(3, 4)$ is 5 and the length of $(5, 1)$ is $\sqrt{26}$.

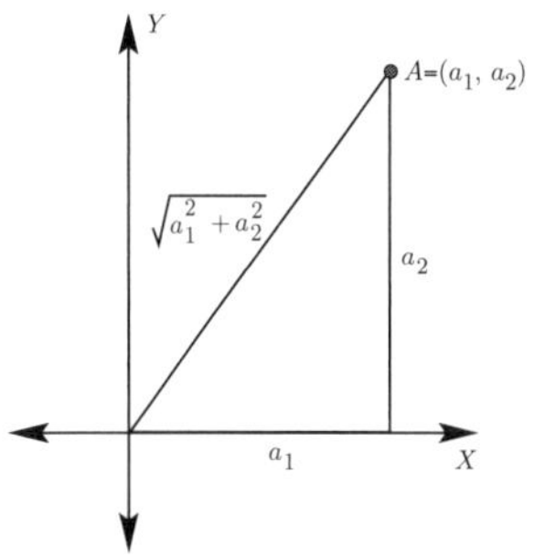


You can find a similar formula in $\mathbb{R}^3$.

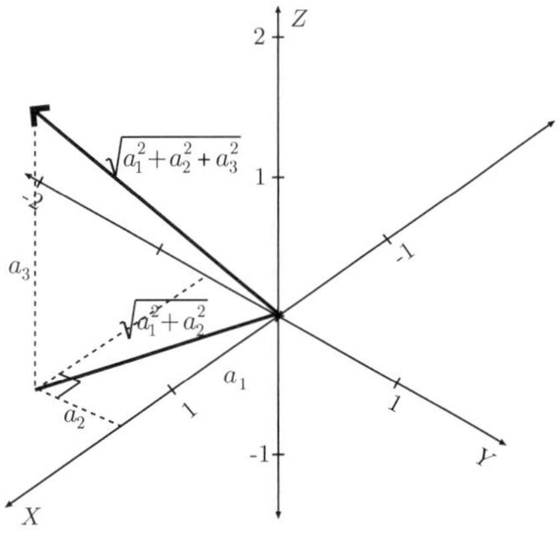

If $A = (a_1, a_2, a_3)$, repeated applications of the Pythagorean Theorem shows that

$$\text{the length of } A = \sqrt{\left(\sqrt{a_1^2 + a_2^2}\right)^2 + a_3^2}$$
$$= \sqrt{a_1^2 + a_2^2 + a_3^2}$$

You can define length in $\mathbb{R}^n$ by continuing the process that used the Pythagorean Theorem to go from two to three dimensions.

### Definition

Let $A = (a_1, a_2, \ldots, a_n)$ be a vector in $\mathbb{R}^n$. The **length** of $A$, written $\|A\|$, is given by the formula

$$\|A\| = \sqrt{a_1^2 + a_2^2 + \cdots + a_n^2}$$

Notice that while $A$ is a vector, $\|A\|$ is a **number**.

**Remember**

If $a$ is a positive real number, $\sqrt{a}$ is defined as the *positive* root of the equation $x^2 = a$.

### For You to Do

**1.** **a.** Find the length of $(5, 3, 1)$. **b.** Find $\|(9, 3, 3, 1)\|$.

There are some fundamental properties of length that are inspired by geometry and proved by algebra.

### Theorem 1.5

*Let $A$ and $B$ be vectors in $\mathbb{R}^n$ and let $c$ be a real number. Then*

(1) $\|A\| \geq 0$, *and* $\|A\| = 0$ *if and only if* $A = O$

(2) $\|cA\| = |c| \, \|A\|$

(3) $\|A + B\| \leq \|A\| + \|B\|$

***Proof.*** Here are the proofs of parts ((1)) and ((2)); the proof of ((3)) will be given in the next chapter.

← Or, you can try to prove it yourself, right now.

(1) Since $\|A\| = \sqrt{a_1^2 + a_2^2 + \cdots + a_n^2}$, $\|A\|$ is the square root of the sum of squares, hence it is nonnegative. And $\|A\| = 0$ if and only if $\sqrt{a_1^2 + a_2^2 + \cdots + a_n^2} = 0$. But for the sum of a set of nonnegative numbers to equal 0, each number must equal 0, and thus all the coordinates $a_1, a_2, \ldots, a_n$ must be 0. In other words, $A = O$.

(2) If $A = (a_1, a_2, \ldots, a_n)$, then $cA = (ca_1, ca_2, \ldots, ca_n)$. So

$$\begin{aligned}\|cA\| &= \sqrt{(ca_1)^2 + (ca_2)^2 + \cdots + (ca_n)^2} \\ &= \sqrt{c^2a_1^2 + c^2a_2^2 + \cdots + c^2a_n^2} \\ &= \sqrt{c^2(a_1^2 + a_2^2 + \cdots + a_n^2)} \\ &= \sqrt{c^2} \cdot \sqrt{a_1^2 + a_2^2 + \cdots + a_n^2} \\ &= |c| \, \|A\|\end{aligned}$$

■

Part ((3)) of Theorem 1.5 is known as the **Triangle Inequality** because, in $\mathbb{R}^2$, it says that if you form a triangle with vertices $O$, $A$, and $A + B$, the length of $A + B$ is less than or equal to the sum of the lengths of the other two sides:

← You'll see a proof of the Triangle Inequality in $\mathbb{R}^n$ in Chapter 2.

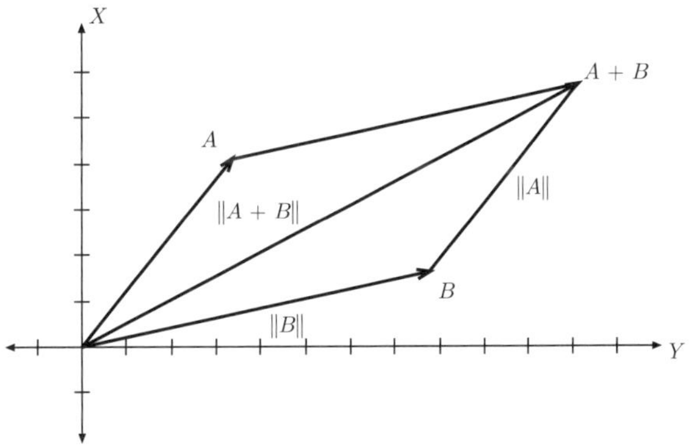


### Unit Vectors

For every nonzero vector $A$ in $\mathbb{R}^2$ or $\mathbb{R}^3$, there is a vector with the same direction as $A$ with length 1. If $A = (3, 4)$, this vector is $\left(\frac{3}{5}, \frac{4}{5}\right)$.

This same phenomenon occurs in $\mathbb{R}^n$, and, as usual, the proof is algebraic. Suppose $A$ is a nonzero vector in $\mathbb{R}^n$, and suppose $\|A\| = k$ (so that $k > 0$). The vector $\frac{1}{k}A$ is in the same direction as $A$, and

$$\begin{aligned}\left\|\tfrac{1}{k}A\right\| &= \left|\tfrac{1}{k}\right| \|A\| \\ &= \tfrac{1}{k} \|A\| \\ &= \tfrac{1}{k}k = 1\end{aligned}$$

← Where is Theorem 1.5 used here?

That is, by dividing each component of $A$ by $\|A\|$, you get a vector in the same direction as $A$ but with length 1.

#### Theorem 1.6

*Let $A$ be a nonzero vector in $\mathbb{R}^n$. There is a vector in the same direction as $A$ with length 1; in fact, this vector is $\frac{1}{\|A\|}A$.*

The vector $\frac{1}{\|A\|}A$ is called the **unit vector** in the direction of $A$.

← Can there be more than one unit vector in any given direction?

### Example 1

**Problem.** In $\mathbb{R}^4$, find the unit vector in the direction of $(5, 10, 6, 8)$.

**Solution.**

$$\|(5, 10, 6, 8)\| = 15$$

so the unit vector in the direction of $(5, 10, 6, 8)$ is $\frac{1}{15}(5, 10, 6, 8)$ or

$$\left(\frac{1}{3}, \frac{2}{3}, \frac{2}{5}, \frac{8}{15}\right)$$

### For You to Do

2. **a.** If $A \neq O$ is a vector in $\mathbb{R}^n$ and $c > 0$ is a number, show that the unit vector in the direction of $cA$ is the same as the unit vector in the direction of $A$.
   **b.** What if $c < 0$?
   **c.** What if $A = O$?

### Distance

In $\mathbb{R}^2$, the distance between $A$ and $B$ is the length of the vector $\overrightarrow{AB}$, which is the same as $\|B - A\|$. You can use this characterization of distance as a *definition* in $\mathbb{R}^n$.

### Definition

The **distance between two points** $A$ and $B$ in $\mathbb{R}^n$, written $d(A, B)$, is defined by the equation

$$d(A, B) = \|B - A\|$$

← The extension program again.

### Example 2

**Problem.** In $\mathbb{R}^4$, show that the triangle with the following vertices is isosceles.

$$A = (476, -306, -932, 1117)$$
$$B = (-1060, -690, 220, -995)$$
$$C = (140, -210, 580, 205)$$

**Solution.** Compute the lengths of the three sides.

← A calculator will help with the arithmetic.

$$d(A, B) = \|B - A\| = \|(-1536, -384, 1152, -2112)\| = 2880$$
$$d(B, C) = \|C - B\| = \|(1200, 480, 360, 1200)\| = 1800$$
$$d(A, C) = \|C - A\| = \|(-336, 96, 1512, -912)\| = 1800$$

Since $d(B, C) = d(A, C)$, the triangle is isosceles.

The next theorem gives some important properties of the distance function. These properties generalize from the geometry of $\mathbb{R}^2$ and $\mathbb{R}^3$. To prove them in $\mathbb{R}^n$, use Theorem 1.5.

**Theorem 1.7**

*Let $A$, $B$, and $C$ be points in $\mathbb{R}^n$.*

(1) $d(A,B) \geq 0$, *and* $d(A,B) = 0$ *if and only if* $A = B$.

(2) $d(A,B) = d(B,A)$.

(3) $d(A,C) \leq d(A,B) + d(B,C)$.

⟵ You should illustrate each of these properties in $\mathbb{R}^2$ or $\mathbb{R}^3$ with a sketch.

## Exercises

**1.** Find $\|A\|$ for each of the following:

**a.** $A = (3,6)$ **b.** $A = (4,3,0)$ **c.** $A = (-1,3,4,1)$
**d.** $A = (1,0,1,0)$ **e.** $A = (4,-1,3,5)$ **f.** $A = (0,0,0,1)$

**2.** Let $A = (3,-1,4)$ and $B = (4,2,-1)$. Find

**a.** $\|A+B\|$ **b.** $\|A-B\|$
**c.** $\|2A+2B\|$ **d.** $\|A-2B\|^2$

**3.** Find the unit vector in the direction of $A$ if

**a.** $A = (1,1)$ **b.** $A = (6,8)$ **c.** $A = (440,-539,330,598)$
**d.** $A = (1,3,0,-1)$ **e.** $A = (1,0,0)$ **f.** $A = (3,4,5)$

**4.** **Write About It.** Prove that, in $\mathbb{R}^2$ and $\mathbb{R}^3$, the distance between two points $A$ and $B$ is the length of $B - A$. Give a formula for the distance between $A = (a_1, a_2, a_3)$ and $B = (b_1, b_2, b_3)$.

**5.** In each exercise, find $d(A,B)$.

**a.** $A = (3,1), B = (4,2)$
**b.** $A = (1,0,1), B = (0,1,0)$
**c.** $A = (1,3,2), B = (4,1,3)$
**d.** $A = (1,3,-1,4), B = (2,1,3,8)$

**6.** Find the length of the sides of $\triangle ABC$ if

$$A = (-120,-1680,-115,465)$$
$$B = (680,1120,485,865)$$
$$C = (240,1659,155,267)$$

⟵ Once again, a calculator is useful for these problems.

**7.** Which triangles are isosceles?

**a.** $\triangle PQR$, where

$$P = (-1791,11089,-279)$$
$$Q = (5954,16991,7835)$$
$$R = (1234,-1209,235)$$

**b.** $\triangle MNP$, where

$$M = (-120, -1680, -115, 465)$$
$$N = (680, 1120, 485, 865)$$
$$P = (240, 1659, 155, 267)$$

**c.** $\triangle ABC$, where

$$A = (1, 1, 3, -2)$$
$$B = (-2, 4, 6, 4)$$
$$C = (5, 3, 3, 2)$$

**8.** Prove Theorem 1.7.

**9.** If $A$ and $B$ are points in $\mathbb{R}^n$ and $M$ is the midpoint of $\overrightarrow{AB}$ (see Exercise 13 from Lesson 1.3), show that

$$d(A, M) = d(M, B)$$

**10.** Let $A = (1, 2)$, $B = (13, 4)$, and $C = (7, 10)$. Suppose $M$ is the midpoint of $\overrightarrow{AB}$ and $N$ is the midpoint of $\overrightarrow{BC}$. Show that $d(M, N) = \frac{1}{2}d(A, C)$.

**11.** Let $A = (1, -2, 2)$ and $B = (7, -5, 4)$. If $M$ is the midpoint of $\overrightarrow{OA}$ and $N$ is the midpoint of $\overrightarrow{OB}$, show that $d(M, N) = \frac{1}{2}d(A, B)$.

**12.** If $A$, $B$, and $C$ are distinct points in $\mathbb{R}^n$, $M$ is the midpoint of $\overrightarrow{AB}$, and $N$ is the midpoint of $\overrightarrow{BC}$, show that

$$d(M, N) = \tfrac{1}{2}d(A, C)$$

What fact from plane geometry does this generalize?

**13.** The centroid of a triangle is the point where its three medians meet.

←
A median of a triangle is a line segment from a vertex to the midpoint of the opposite side.

**a.** Find the centroid of the triangle whose vertices are $O$, $A = (3, 1)$, and $B = (6, 5)$.
**b.** Show that this centroid is $\frac{A+B}{3}$.
**c.** Find the centroid of the triangle whose vertices are $Q = (4, 1)$, $Q + A = (7, 2)$, and $Q + B = (10, 6)$.
**d.** **Take It Further.** Show that in $\mathbb{R}^2$ the centroid of a triangle whose vertices are $M$, $N$, and $R$ is $\frac{M+N+R}{3}$.

**14.** **a.** Pick three cities, say Boston, New York, and Cleveland. Approximately where is the centroid of the triangle that has your three cities as vertices?
**b.** What is a reasonable definition of the "population center" for three cities?
**c.** Find the population center for your three cities using your definition.

**15.** Show that the points $(5, 7, 4)$, $(7, 3, 2)$, and $(3, 7, -2)$ all lie on a sphere with center $(1, 3, 2)$.

**16.** If $A$ and $B$ are nonzero vectors in $\mathbb{R}^n$ and $a = \|A\|$ and $b = \|B\|$, show that $\|bA\| = \|aB\|$.

**17.** Suppose $A$ and $B$ are nonzero vectors in $\mathbb{R}^2$. Let $U = \|B\|\, A$ and $V = \|A\|\, B$. Show that the parallelogram determined by $U$ and $V$ is a rhombus.

$\longleftarrow$ Draw a picture.

## Chapter 1 Mathematical Reflections

These problems will help you summarize what you have learned in this chapter:

**1.** For each of the following equations, solve for $A$.

**a.** $2A + (1, 6) = (-3, 4)$
**b.** $-A - (5, 3, -4) = O$
**c.** $4A + (1, -3, -2) = 2A - (0, -5, 2)$

**2.** For each set of vectors, determine whether the pairs of vectors $\overrightarrow{AB}$ and $\overrightarrow{CD}$ are equivalent, parallel in the same direction, or parallel in the opposite direction.

**a.** $A = (2, 5)$, $B = (6, 3)$, $C = (-5, 6)$, $D = (-1, 4)$
**b.** $A = (2, 5)$, $B = (6, 3)$, $C = (-5, 6)$, $D = (3, 2)$
**c.** $A = (-4, 2, 0)$, $B = (7, -1, 9)$, $C = (0, 0, 2)$, $D = (-11, 3, -7)$
**d.** $A = (2, 9, 5, -7)$, $B = (4, -3, 8, 0)$, $C = (-1, -9, 0, 5)$, $D = (1, -21, 3, 12)$

**3.** The vector equation of line $\ell$ is $X = (3, 5) + k(4, 2)$.

**a.** Find $X$ if $k = 2$.
**b.** Find another point on line $\ell$.
**c.** Find a coordinate equation for line $\ell$. Verify that the points from parts **a** and **b** are on line $\ell$ using the coordinate equation.

**4.** Let $A = (4, 2)$ and $B = (6, -8)$. Find

**a.** $\|A\|$ **b.** $\|B\|$ **c.** $\|A + B\|$ **d.** $d(A, B)$

**5.** Let$A = (4, 2)$ and $B = (6, -8)$. $O$, $A$, $B$, and $A + B$ form the vertices of a parallelogram. What additional information about the parallelogram do $\|A + B\|$ and $d(A, B)$ give you?

**6.** How can you describe adding and scaling vectors in geometric terms?

**7.** How can you use vectors to describe lines in space?

**8.** Let $A = (3, 2)$ and $B = (-1, 4)$.

**a.** Calculate and graph the following: $A + B$, $2A$, $-3B$, $2A - 3B$.
**b.** Calculate $\|2A - 3B\|$.

**Vocabulary**

In this chapter, you saw these terms and symbols for the first time. Make sure you understand what each one means, and how it is used.

- coordinates
- direction
- equivalent vectors
- extension program
- initial point (tail)
- length $\|X\|$
- linear combination
- magnitude
- $n$-dimensional Euclidean space
- opposite direction
- ordered $n$-tuple
- point
- same direction
- scalar multiple
- spanned
- terminal point (head)
- unit vector
- vector
- vector equation
- zero vector

←
$A$ and $B$ are the same as in Exercise 4.

# Chapter 1 Review

In Lesson 1.2, you learned to

- locate points in space and describe objects with equations
- use the algebra of points to calculate, solve equations, and transform expressions, all in $\mathbb{R}^n$
- understand the geometric interpretations of adding and scaling

The following problems will help you check your understanding.

**1.** Given $A = (2, 3)$, $B = (4, -3)$, and $C = (-5, -4)$. Calculate and plot the following:

**a.** $A + B$ **b.** $A + 2B$ **c.** $A + 3B$
**d.** $2 \cdot (A + B)$ **e.** $A + B + C$ **f.** $A + B - C$

**2.** In $\mathbb{R}^3$, find the equation of each of the following:

**a.** the $x$–$y$ plane **b.** the $x$–$z$ plane
**c.** the plane through $(-2, 3, 4)$ parallel to the $x$–$y$ plane
**d.** the plane through $(-2, 3, 4)$ parallel to the $x$–$z$ plane

**3.** For each of the following equations, solve for $A$.

**a.** $4A - (-4, 9) = (2, -5)$
**b.** $A + (-1, -7, 8) = 3A - (-11, 1, -8)$
**c.** $(1, 15, 2, -5) - 5A = 2(3, 0, -4, 10)$

**4.** For each of the following equations, find $c_1$ and $c_2$.

**a.** $c_1(2, -5, 3) + c_2(4, 1, 8) = (0, -11, -2)$
**b.** $c_1(2, -5, 3) + c_2(4, 1, 8) = (4, 12, 10)$

In Lesson 1.3, you learned to

- test vectors for equivalence using the algebra of points
- prove simple geometric theorems with vector methods
- think of linear combinations geometrically

The following problems will help you check your understanding.

**5.** For each set of points $A$, $B$, and $Q$, find a point $P$ if $\overrightarrow{PQ}$ is equivalent to $\overrightarrow{AB}$.

**a.** $A = (-2, 8), B = (3, -1)$, and $Q = (2, 5)$
**b.** $A = (3, 5, 7), B = (-1, -2, 4)$, and $Q = (-6, 8, 3)$

**6.** In $\mathbb{R}^3$, suppose $A = (2, -2, 1)$, $B = (3, 4, -2)$, and $C = (5, -1, 6)$. If $D = (3, a, b)$, find $a$ and $b$ so that $\overrightarrow{AB}$ is parallel to $\overrightarrow{CD}$. Are the vectors parallel in the same direction or in the opposite direction? How do you know?

**7.** Let $P = (2, -1)$ and $A = (3, 4)$. If $\ell$ is the set of all points $Q$ so that $\overrightarrow{PQ}$ is parallel to $A$, find a vector equation and a coordinate equation for $\ell$.

**8.** In $\mathbb{R}^3$, let $A = (3, -2, 1)$ and $B = (2, 4, 0)$.

**a.** Is $(14, 12, 2)$ a linear combination of $A$ and $B$? If so, give the coefficients. If not, explain why not.
**b.** Is $(14, 12, 1)$ a linear combination of $A$ and $B$? If so, give the coefficients. If not, explain why not.
**c.** Find a vector equation and a coordinate equation for the plane spanned by $A$ and $B$.

In Lesson 1.4, you learned to

- calculate length and distance and apply the algebraic properties described in the theorems
- understand how the extension program is used to define length in higher dimensions
- identify a unit vector

The following problems will help you check your understanding:

**9.** Let $A = (-4, -3)$ and $B = (2, 2)$. Find

**a.** $\|A\|$ **b.** $\|B\|$ **c.** $\|A + B\|$
**d.** $\|2A\|$ **e.** $\|A - B\|$ **f.** $\|B - A\|$

**10.** Find the unit vector in the direction of $A$ if

**a.** $A = (5, -12)$ **b.** $A = (-3, -3)$
**c.** $A = (2, 1, -3)$ **d.** $A = (1, 1, 1, 1)$

**11.** Find $d(A, B)$ if

**a.** $A = (-2, 4), B = (0, 5)$
**b.** $A = (1, 1, 1), B = (2, 2, 2)$
**c.** $A = (0, 3, 1, 2), B = (-2, 1, -3, 0)$

**12.** Find the perimeter of $\triangle ABC$ if $A = (2, 3)$, $B = (5, 9)$, and $C = (8, 0)$.

# Chapter 1 Test

## Multiple Choice

**1.** Let $A = (3, -6)$ and $B = (-4, 5)$. Which is equivalent to $A + 2B$?

**A.** $(-5, 4)$ **B.** $(-2, -2)$ **C.** $(-1, -1)$ **D.** $(2, -7)$

**2.** Which is the equation of the plane through $(1, -2, 6)$ parallel to the $x$–$y$ plane?

**A.** $x = 1$
**B.** $y = -2$
**C.** $z = 6$
**D.** $x + y + z = 5$

**3.** Let $A = (4, 0)$, $B = (-2, -1)$, and $P = (-5, -3)$. If $\overrightarrow{AB}$ is equivalent to $\overrightarrow{PQ}$, which are the coordinates of $Q$?

**A.** $(-11, -4)$ **B.** $(-1, 2)$ **C.** $(1, -2)$ **D.** $(11, 4)$

**4.** In $\mathbb{R}^3$, suppose $A = (1, -4, 2)$, $B = (-5, 7, 3)$, $P = (2, -2, 3)$, and $Q = (-10, 20, 5)$. Which of these statements is true?

**A.** $\overrightarrow{AB}$ and $\overrightarrow{PQ}$ are equivalent.
**B.** $\overrightarrow{AB}$ and $\overrightarrow{PQ}$ are parallel in the same direction.
**C.** $\overrightarrow{AB}$ and $\overrightarrow{PQ}$ are parallel in the opposite direction.
**D.** None of the above.

**5.** Let $P = (4, -7)$ and $A = (5, 2)$. If $\ell$ is the set of all points $Q$ so that $\overrightarrow{PQ}$ is parallel to $A$, which is a coordinate equation for $\ell$?

**A.** $2x - 5y - 43 = 0$
**B.** $2x + 5y + 27 = 0$
**C.** $4x - 7y - 6 = 0$
**D.** $4x + 7y + 33 = 0$

**6.** Let $A = (2, -4, 7)$ and $B = (-2, 1, 5)$. What is $\|A - B\|$?

**A.** $\sqrt{11}$ **B.** $\sqrt{13}$ **C.** $3\sqrt{5}$ **D.** $3\sqrt{17}$

## Open Response

**7.** Solve each of the following equations for $A$.

**a.** $2A + (6, -1) = (8, 4)$
**b.** $3A - (4, 6, -2) = -A + 2(0, 5, 4)$

**8.** For each equation, find scalars $c_1$ and $c_2$. If it is not possible, explain why.

**a.** $c_1(1, -2, -1) + c_2(2, -3, 4) = (1, -3, -7)$
**b.** $c_1(1, -3, 4) + c_2(2, -6, 8) = (5, -1, 2)$

**9.** Let $A = (2,5)$, $B = (4,9)$, $C = (10,11)$, and $D = (8,7)$.

**a.** Translate the quadrilateral $ABCD$ so that $A$ is at the origin.

**b.** Use the Parallelogram Rule to show that $A$, $B$, $C$, and $D$ lie on the vertices of a parallelogram.

**c.** Sketch both parallelograms.

**10.** In $R^3$, let $A = (-2,0,3)$ and $B = (1,4,2)$. Find a vector equation and a coordinate equation for the plane spanned by $A$ and $B$.

**11.** In $\mathbb{R}^4$, find the unit vector in the direction of $A$ if $A = (-5,1,-7,5)$.

**12.** In $R^3$, let $A = (-4,1,5)$, $B = (2,3,4)$, and $C = (2,-1,6)$. Show that $\triangle ABC$ is isosceles.

CHAPTER

# 2 Vector Geometry

The algebra of points and vectors that you learned about in Chapter 1 gives you the basic tools with which you can implement the extension program.

- Take a familiar geometric idea in two and three dimensions.
- Find a way to describe it algebraically with vectors.
- Use the algebra as the definition of the idea in higher dimensions.

In this chapter, you'll use vectors to describe and extend ideas like perpendicularity and angle. You'll also learn to describe lines (in $\mathbb{R}^2$ and $\mathbb{R}^3$) and planes (in $\mathbb{R}^3$) with *vector* equations. These vector equations are often much more useful than the coordinate equations you learned about in other courses, and they allow you to extend the ideas of lines and planes to higher dimensions. Along the way, you'll encounter some simple ways to calculate area and volume using vector methods.

**By the end of this chapter, you will be able to answer questions like these:**

1. How can you determine whether two vectors (of any dimension) are orthogonal?
2. How can you find a vector orthogonal to two given vectors in $\mathbb{R}^3$?
3. Let $A = (2, -1, 3)$, $B = (1, 1, 2)$, and $C = (2, 0, 5)$. What is an equation of the hyperplane $E$ containing $A$, $B$, and $C$?

**You will build good habits and skills for ways to**

- use algebra to extend geometric ideas
- use vectors to prove facts about numbers
- generalize from numerical examples
- use different forms for different purposes

### Vocabulary and Notation

- angle (between two vectors)
- component
- cross product
- determinant
- direction vector of a line
- dot product
- hyperplane
- initial point
- lemma
- linear equation
- normal
- orthogonal
- projection
- right-hand rule
- standard basis vectors
- vector equation of a line

## 2.1 Getting Started

### Exercises

1. Suppose $A = (3, 1)$. Find an equation for each of the following lines:

   **a.** the line through the origin perpendicular to $A$
   **b.** the line through $P = (-3, 2)$ perpendicular to $A$
   **c.** the line through $P = (3, -2)$ perpendicular to $A$
   **d.** the line through $P = (6, -4)$ perpendicular to $A$
   **e.** the line through $P = (0, 6)$ perpendicular to $A$
   **f.** the line through $P = (0, 6)$ that's parallel to $A$

← Think of $A$ as a vector here. Write your equations in the form $ax + by = c$.

← There's a point to these problems. Look for patterns.

2. Find a nonzero vector $Q$ perpendicular to

   **a.** $A = (5, 1)$
   **b.** $A = (3, 2)$
   **c.** $A = (-2, 10)$
   **d.** $A = (6, 4)$
   **e.** $A = (a, b)$
   **f.** both $A = (5, 0, 0)$ and $B = (0, 0, -3)$

**Remember**

Unless you are told otherwise, a vector starts at the origin.

Suppose $A$ and $B$ are vectors. If you drop a perpendicular from the head of $A$ to the line along $B$, it will hit that line at a point $P$ that's called the *projection* of $A$ on $B$.

← If $B = O$, the convention is that the projection of $A$ on $B$ is $O$.

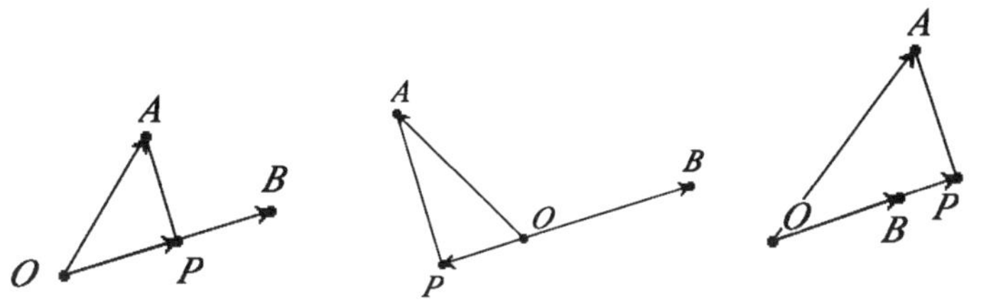


3. Find the projection of $A$ on $B$ if

   **a.** $A = (2, 9)$, $B = (10, 0)$
   **b.** $A = (2, 9)$, $B = (-10, 0)$
   **c.** $A = (2, 9)$, $B = (0, 6)$
   **d.** $A = (2, 9)$, $B = (6, 4)$
   **e.** $A = (2, 9)$, $B = (12, 8)$
   **f.** $A = (2, 9)$, $B = (-6, -4)$
   **g.** $A = (2, 9)$, $B = (4, 8)$
   **h.** $A = (-8, 4)$, $B = (4, 8)$

4. **Write About It.** Given vectors $A$ and $B$, describe a method for finding the projection of $A$ on $B$.

5. Find the angle between each pair of vectors.

   **a.** $A = (5, 5), B = (10, 0)$
   **b.** $A = (5, 5), B = (-10, 0)$
   **c.** $A = (1, \sqrt{3}), B = (0, 6)$
   **d.** $A = (2, 9), B = (6, 4)$
   **e.** $A = (2, 9), B = (12, 8)$
   **f.** $A = (2, 9), B = (-6, -4)$
   **g.** $A = (2, 9), B = (4, 8)$
   **h.** $A = (-8, 4), B = (4, 8)$

**6.** Find the shortest distance from point $P$ to line $\ell$.

**a.** $P = (5, 6)$, $\ell$ is the $x$-axis in $\mathbb{R}^2$
**b.** $P = (5, 6, 7)$, $\ell$ is the $y$-axis in $\mathbb{R}^3$
**c.** $P = (5, 0)$, $\ell$ is the graph of $x = y$ in $\mathbb{R}^2$
**d.** $P = (5, 0)$, $\ell$ is the graph of $x = -y$ in $\mathbb{R}^2$
**e.** $P = (8, -24)$, $\ell$ is the graph of $7x + 4y = 25$ in $\mathbb{R}^2$
**f.** $P = (13, -49)$, $\ell$ is the graph of $7x + 4y = 25$ in $\mathbb{R}^2$
**g.** $P = (-387, 651)$, $\ell$ is the graph of $7x + 4y = 25$ in $\mathbb{R}^2$
**h.** $P = (3, 1)$, $\ell$ is the graph of $7x + 4y = 25$ in $\mathbb{R}^2$
**i.** $P = (0, 0)$, $\ell$ is the graph of $7x + 4y = 25$ in $\mathbb{R}^2$

**7. Write About It.** Given a point $P$ and the equation for a line $\ell$, describe a method for finding the distance from $P$ to $\ell$.

**8.** Suppose you had a sketch like this:

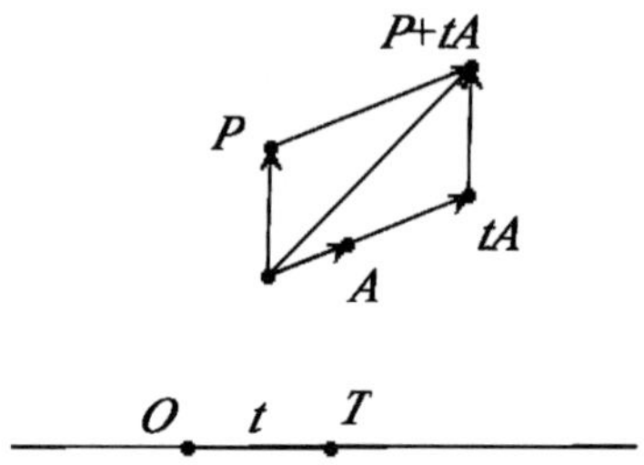


Here, $P$ and $A$ are fixed vectors. (You may instead think of $P$ as a point and $A$ as a vector.) $O$ is a fixed point—the origin of a coordinate system. $T$ has coordinates $(t, 0)$. The number $t$ is used as a scale factor to construct $tA$ and then $P + tA$.

Find a coordinate equation for the path of $P + tA$ as $t$ ranges over $\mathbb{R}$ if

**a.** $P = (3, 5)$ and $A = (6, 1)$
**b.** $P = (5, -7)$ and $A = (6, 1)$
**c.** $P = (6, 10)$ and $A = (6, 1)$
**d.** $P = (0, 0)$ and $A = (6, 1)$
**e.** $P = (5, -7)$ and $A = (12, 2)$
**f.** $P = (5, -7)$ and $A = (-12, -2)$
**g.** $P = (p_1, p_2)$ and $A = (a_1, a_2)$

## 2.2 Dot Product

In Chapter 1, you explored addition of vectors and multiplication of a vector by a scalar. Dot product is another operation on vectors whose calculation may look familiar to you.

**In this lesson, you will learn how to**

- find the dot product of two vectors of any dimension
- determine whether two vectors are orthogonal
- use the basic properties of dot product to prove statements and solve problems

Much of the study of geometry involves lengths and angles in two dimensions. You can extend these ideas to higher dimensions by looking at them algebraically. To do so, characterize these ideas in terms of vectors. Luckily, much of the groundwork for this process has been established in analytic geometry and in trigonometry.

⟵ For instance, you spent a good deal of time proving congruence by comparing equal lengths and angle measures, and you proved similarity by comparing proportional lengths and congruent angles.

### Developing Habits of Mind

**Use the Pythagorean Theorem.** How can you tell if two vectors are perpendicular? In $\mathbb{R}^2$, you can use slope: two lines in $\mathbb{R}^2$ are perpendicular if their slopes are negative reciprocals.In $\mathbb{R}^2$, a line is determined by its slope and a point on it.

⟵ Except that this doesn't work for horizontal and vertical lines.

Unfortunately, the idea of slope isn't quite so simple in $\mathbb{R}^3$. Sure, you can come up with ways to describe a line in $\mathbb{R}^3$ by its "steepness,"but there isn't a single number that would uniquely characterize the line.

⟵ Take two pencils. Hold one vertically and the other angled out with the erasers touching. Rotate the two without changing the angle between the two pencils. The slope appears to change as you rotate it.

Fortunately, you already know another way to test for perpendicularity: use the converse of the Pythagorean Theorem. If the side-lengths of a triangle are $a$, $b$, and $c$, and if $a^2 + b^2 = c^2$, then the angle opposite the side of length $c$ is a right angle.

The Pythagorean Theorem assumes that the triangle lies in a two-dimensional plane. Here, you have two vectors that share the same tail point. So whether those vectors are in $\mathbb{R}^2$ or $\mathbb{R}^3$, there must be a plane that contains them both. So the theorem works in three dimensions as well.

⟵ You may have heard this stated as "three points determine a plane." You'll see later that another variation is "two vectors determine a plane."

In $\mathbb{R}^3$, two vectors $A = (a_1, a_2, a_3)$ and $B(b_1, b_2, b_3)$ will be perpendicular if and only if

$$\|A\|^2 + \|B\|^2 = \|A - B\|^2$$

⟵ Why $\|A - B\|$? This is $d(B, A)$.

Using the definitions of length and distance (see Lesson 1.4), this can be stated as

$$a_1^2 + a_2^2 + a_3^2 \; + \; b_1^2 + b_2^2 + b_3^2 = (a_1 - b_1)^2 + (a_2 - b_2)^2 + (a_3 - b_3)^2$$

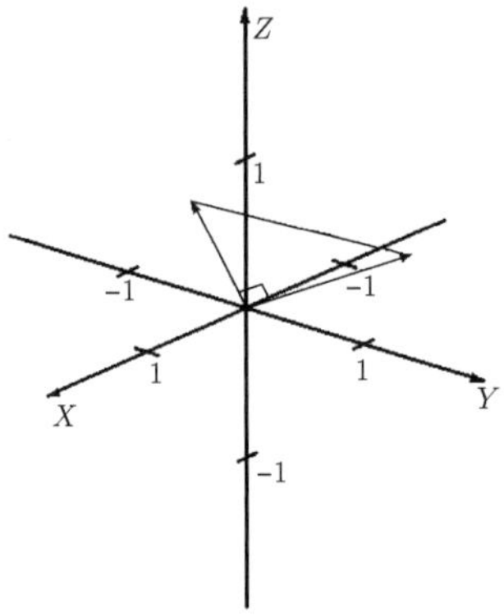


When you expand the right-hand side, you will see that $a_1^2 + a_2^2 + a_3^2 + b_1^2 + b_2^2 + b_3^2$ ends up on both sides of the equation. Subtract that from both sides, and you get

$$-2a_1b_1 - 2a_2b_2 - 2a_3b_3 = 0$$

Divide both sides by $-2$, and the equation simplifies to

$$a_1b_1 + a_2b_2 + a_3b_3 = 0$$

All the steps in these calculations are reversible. So, $A \perp B$ if and only if the sum of the products of the corresponding coordinates is 0.

$\longleftarrow$
Make sure you check that all the steps are reversible. Start from "the sum of the products of the corresponding coordinates is 0" and work back to the statement about equal lengths.

### For You to Do

**1.** Show that two vectors in $\mathbb{R}^2$, say $A = (a_1, a_2)$ and $B(b_1, b_2)$, are perpendicular if and only if

$$a_1b_1 + a_2b_2 = 0$$

So, now you have an algebraic description of what it takes for two vectors in $\mathbb{R}^2$ or $\mathbb{R}^3$ to be perpendicular: the sum of the products of the corresponding coordinates has to equal 0. Because this sum is such a useful computation—not only in $\mathbb{R}^2$ and $\mathbb{R}^3$, but in any dimension—it has a name: dot product.

### Definition

Let $A = (a_1, a_2, \ldots, a_n)$ and $B = (b_1, b_2, \ldots, b_n)$ be points in $\mathbb{R}^n$. The **dot product** of $A$ and $B$, written $A \cdot B$, is defined by the formula

$$A \cdot B = a_1b_1 + a_2b_2 + \cdots + a_nb_n$$

Note carefully that while $A$ and $B$ are *vectors*, $A \cdot B$ is a *number*.

### For You to Do

**2.** Let $A = (3, -1, 2, 4)$ and $B = (1, 5, -1, 6)$. Find $A \cdot B$.

So, now you can say that vectors $A$ and $B$ in $\mathbb{R}^2$ or $\mathbb{R}^3$ are perpendicular if and only if $A \cdot B = 0$. One more refinement:

### Facts and Notation

While it's traditional to use the word "perpendicular" when talking about *lines* that meet at a right angle, it is more common to use the word **orthogonal** when talking about two *vectors*.

⟵ "Perpendicular" is from Latin and "orthogonal" is from Greek.

So, two vectors $A$ and $B$ in $\mathbb{R}^2$ or $\mathbb{R}^3$ are orthogonal if and only if their dot product, $A \cdot B$, equals 0. Because the dot product is defined for vectors in any dimension, you can use the extension program to define orthogonal vectors in *any* dimension.

**Remember**

The extension program: Take a familiar geometric idea in two and three dimensions, find a way to describe it with vectors, and then use the algebra as the definition of the idea in higher dimensions.

**Habits of Mind**

The origin $O$ is orthogonal to every vector. Why?

### Definition

Two vectors $A$ and $B$ in $\mathbb{R}^n$ are said to be **orthogonal** if and only if their dot product is 0. In symbols,

$$A \perp B \Leftrightarrow A \cdot B = 0$$

### For Discussion

**3.** Why is the above definition a definition rather than a theorem?

### Example 1

In $\mathbb{R}^4$, the vectors $A = (-1, 3, 2, 1)$, $B = (1, 1, -1, 0)$, and $C = (6, -2, 4, 4)$ are **mutually orthogonal**; that is, $A \cdot B = 0$, $A \cdot C = 0$, and $B \cdot C = 0$.

### For You to Do

In $\mathbb{R}^3$, the vectors $(1, 0, 0)$, $(0, 1, 0)$, and $(0, 0, 1)$ are mutually orthogonal.

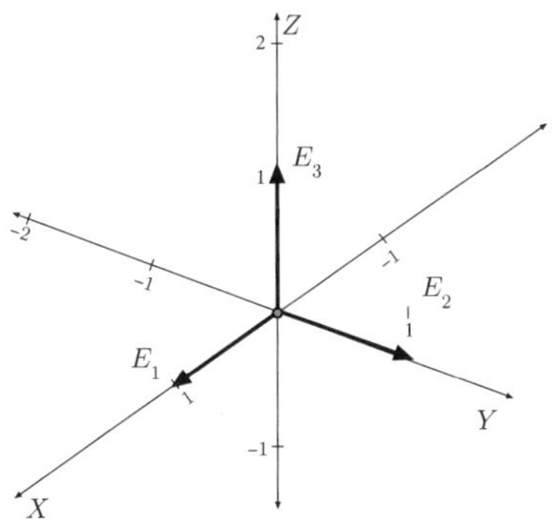


**4.** Show that the dot product of any two of these vectors is 0.

**5.** In fact, in $\mathbb{R}^n$, let $E_i$ be the vector whose $i^{\text{th}}$ coordinate is 1 and whose remaining coordinates are all 0. So, $E_1 = (1,0,\dots,0), E_2 = (0,1,0,\dots,0),\dots,E_n = (0,0,\dots,0,1)$.

**a.** Show that $E_1, E_2, \dots, E_n$ form a set of $n$ mutually orthogonal vectors.

**b.** If $A = (a_1, a_2, \dots, a_n)$ is an arbitrary vector in $\mathbb{R}^n$, show that $A \cdot E_i = a_i$, the $i^{\text{th}}$ coordinate of $A$.

**c.** Use parts **a** and **b** to prove that $A$ is orthogonal to $E_i$ if and only if its $i^{\text{th}}$ coordinate is zero.

⟵ It follows that the only vector orthogonal to all of the $E_i$ is $O$.

---

**Minds in Action** Episode 5

*Tony and Sasha are trying to describe all the vectors that are orthogonal to both $A = (1,1,-1)$ and $B = (-1,-2,4)$.*

TONY: The vectors $A$ and $B$ determine a plane, and the vectors we are looking for are those vectors $X$ that are perpendicular to this plane. That's got to be a line.

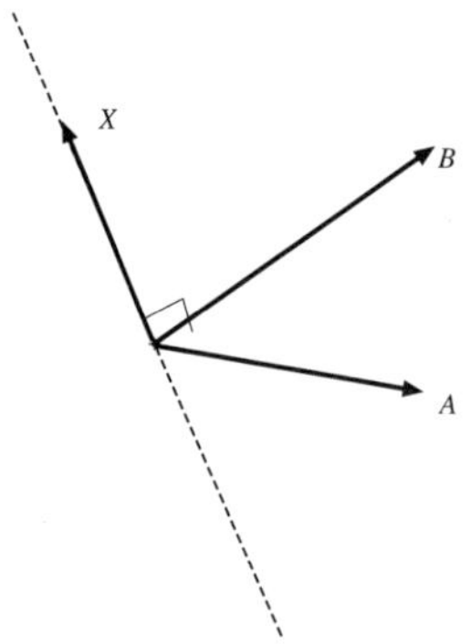


*Tony draws on the board.*

SASHA: So, if we find one vector $X$ that's orthogonal to $A$ and $B$, we can just take all multiples of it. Let's see what the algebra tells us.

⟵ Notice how Sasha says "orthogonal" instead of "perpendicular."

TONY: If $X = (x,y,z)$ is orthogonal to both $A$ and $B$, then $A \cdot X = 0$ and $B \cdot X = 0$. Writing this out, we have the system of two equations in three unknowns.

$$\begin{aligned} x + y - z &= 0 \\ -x - 2y + 4z &= 0 \end{aligned}$$

Now what?

SASHA: Let's just see what the algebra tells us.

*Sasha starts writing on the board.*

Solve the first equation for $x$: $x = -y + z$; substitute this for $x$ in the second equation and simplify. We get $y = 3z$. Since $x = -y + z$ and $y = 3z$, we have $x = -2z$. That is, any vector $X = (x,y,z)$ where $x = -2z$ and $y = 3z$ will be orthogonal to both $A$ and $B$. For example, letting $z = 1$, we have $x = (-2,3,1)$ as a solution. The general solution is $(-2z, 3z, z)$, where $z$ can be anything it likes.

TONY: Looks messy.

SASHA: Hey! No, it's very simple: $(-2z,3z,z) = z(-2,3,1)$. So the set of all vectors $X$ orthogonal to both $A$ and $B$ is the set of all multiples of $(-2,3,1)$. Got it?

TONY: That's a line through the origin. Ohh . . . that makes sense—look at my picture.

### For You to Do

**6.** Describe all the vectors that are orthogonal to both $C = (2, -1, 1)$ and $D = (-1, 3, 0)$.

### Example 2

**Problem.** Characterize the set of all vectors $X$ in $\mathbb{R}^3$ that are orthogonal to $A = (1, 3, 0)$, $B = (1, 4, 1)$, and $C = (3, 10, 2)$.

**Solution.** Let $X = (x, y, z)$ be a solution to the problem. Then $A \cdot X = 0$, $B \cdot X = 0$, and $C \cdot X = 0$. Writing this out, you get the system:

$$\begin{aligned} x + 3y &= 0 \\ x + 4y + z &= 0 \\ 3x + 10y + 2z &= 0 \end{aligned}$$

Solving this system, you obtain $x = y = z = 0$, so $X = O$.

⟵ If you draw a picture, it seems that the only vector in $\mathbb{R}^3$ that is orthogonal to three given vectors is $O$. But the algebra lets you know for sure.

### Example 3

**Problem.** Characterize the set of vectors $X$ that are orthogonal to $A = (1, 1, -1)$, $B = (-1, -2, 4)$, and $C = (1, 0, 2)$.

**Solution.** Again, you may expect the only solution to be $O$. Use algebra to make sure. Let $X = (x, y, z)$, and the system of equations becomes

$$\begin{aligned} x + y - z &= 0 \\ -x - 2y + 4z &= 0 \\ x + 2z &= 0 \end{aligned}$$

The third equation is twice the first equation plus the second equation. So the last equation is unnecessary and this system is equivalent to (that is, has the same solutions as) the following system:

$$\begin{aligned} x + y - z &= 0 \\ -x - 2y + 4z &= 0 \end{aligned}$$

Sasha solved this system of equations in Episode 5: it is satisfied by any multiple of $(-2, 3, 1)$, so any vector of the form $k(-2, 3, 1)$ is orthogonal to $A$ and $B$, which is a line. But, since the two systems have the same solutions, that line is also orthogonal to $C$.

⟵ Show that any vector of the form $k(-2, 3, 1)$ is orthogonal to $C$.

### For Discussion

**7.** In Example 3, you saw that the line formed by multiples of $(-2, 3, 1)$ is orthogonal to three vectors. How is this possible geometrically?

⟵ Draw a picture.

### Example 4

**Problem.** Suppose $A = (3, 1)$ and $B = (5, 2)$. Find a vector $X$ in $\mathbb{R}^2$ such that $A \cdot X = 4$ and $B \cdot X = 2$.

**Solution.** Let $X = (x, y)$. Then $A \cdot X = 3x + y$ and $B \cdot X = 5x + 2y$. So, the *vector* equations $A \cdot X = 4$ and $B \cdot X = 2$ can be written as a system of two equations in two unknowns:

$$\begin{aligned} 3x + y &= 4 \\ 5x + 2y &= 2 \end{aligned}$$

Solve this system to get $X = (6, -14)$.

---

The dot product is a new kind of operation: it takes two *vectors* and produces a *number*. Still, it has some familiar-looking algebraic properties that allow you to calculate with it.

**Theorem 2.1 (The Basic Rules of Dot Product)**
*Let $A = (a_1, a_2, \ldots, a_n), B = (b_1, b_2, \ldots, b_n)$, and $C = (c_1, c_2, \ldots, c_n)$ be vectors in $\mathbb{R}^n$, and let $k$ be a real number. Then*

(1) $A \cdot B = B \cdot A$

(2) $A \cdot (B + C) = A \cdot B + A \cdot C$

(3) $A \cdot kB = kA \cdot B = k(A \cdot B)$

(4) $A \cdot A \geq 0$, *and* $A \cdot A = 0$ *if and only if* $A = O$

***Proof.*** Here are the proofs of (1), (3), and (4). The proof of (2) is left as an exercise.

$$\begin{aligned} (1) \qquad A \cdot B &= (a_1, a_2, \ldots, a_n) \cdot (b_1, b_2, \ldots, b_n) \\ &= a_1 b_1 + a_2 b_2 + \cdots + a_n b_n \\ &= b_1 a_1 + b_2 a_2 + \cdots + b_n a_n \\ &= (b_1, b_2, \ldots, b_n) \cdot (a_1, a_2, \ldots, a_n) \\ &= B \cdot A \end{aligned}$$

$$\begin{aligned} (3) \qquad A \cdot kB &= (a_1, a_2, \ldots, a_n) \cdot k(b_1, b_2, \ldots, b_n) \\ &= (a_1, a_2, \ldots, a_n) \cdot (kb_1, kb_2, \ldots, kb_n) \\ &= a_1(kb_1) + a_2(kb_2) + \cdots + a_n(kb_n) \\ &= k(a_1 b_1) + k(a_2 b_2) + \cdots + k(a_n b_n) \\ &= k(a_1 b_1 + a_2 b_2 + \cdots + a_n b_n) \\ &= k(A \cdot B) \end{aligned}$$

The proof that $kA \cdot B = k(A \cdot B)$ is exactly the same.

$$\begin{aligned} (4) \qquad A \cdot A &= (a_1, a_2, \ldots, a_n) \cdot (a_1, a_2, \ldots, a_n) \\ &= a_1^2 + a_2^2 + \cdots + a_n^2 \end{aligned}$$

Now, the sum of squares of real numbers is nonnegative, and such a sum is 0 if and only if each $a_i = 0$. ■

In the proof of part (4), you see the equation

$$A \cdot A = a_1^2 + a_2^2 + \cdots + a_n^2$$

**Habits of Mind**

Note that in the equation $kA \cdot B = k(A \cdot B)$, the insertion of parentheses changes the object that is being multiplied by $k$. On the left side, you are multiplying $k$ by $A$, a vector in $\mathbb{R}^n$; on the right side, you are multiplying $k$ by $A \cdot B$, a real number.

The right-hand side of that equation should look familiar—you saw it in the definition of the length of a vector,

$$\|A\| = \sqrt{a_1^2 + a_2^2 + \cdots + a_n^2}$$

So you can substitute $A \cdot A$ for $a_1^2 + a_2^2 + \cdots + a_n^2$ to get a more efficient way to write the length of a vector.

**Theorem 2.2**

*If $A$ is a vector in $\mathbb{R}^n$, $\|A\| = \sqrt{A \cdot A}$.*

**Example 5**

**Problem.** Show that if $A$ and $B$ are vectors in $\mathbb{R}^n$,

$$(A + B) \cdot (A + B) = A \cdot A + 2(A \cdot B) + B \cdot B$$

**Solution.** The proof of this fact is exactly the same as the proof from elementary algebra that established the identity $(a + b)^2 = a^2 + 2ab + b^2$.

$$\begin{aligned}
(A + B) \cdot (A + B) &= (A + B) \cdot A + (A + B) \cdot B \quad &&\text{(Theorem 2.1 (2))}\\
&= A \cdot (A + B) + B \cdot (A + B) \quad &&\text{(Theorem 2.1 (1))}\\
&= A \cdot A + A \cdot B + B \cdot A + B \cdot B \quad &&\text{(Theorem 2.1 (2))}\\
&= A \cdot A + A \cdot B + A \cdot B + B \cdot B \quad &&\text{(Theorem 2.1 (1))}\\
&= A \cdot A + 2(A \cdot B) + B \cdot B
\end{aligned}$$

**Example 6**

**Problem.** Let $A = (1, 4, 0, 1)$. For what values of $c$ is $cA \cdot cA = 72$?

**Solution.** Using part (3) of Theorem 2.1, you have

$$cA \cdot cA = c(A \cdot cA) = c^2(A \cdot A)$$

Since $A \cdot A = 18$, this becomes $18c^2 = 72$, so $c = \pm 2$.

**Example 7**

**Problem.** Let $A$ and $B$ be vectors in $\mathbb{R}^n$, with $B \neq O$. Show that

$$\left(A - \left(\tfrac{A \cdot B}{B \cdot B}\right) B\right) \cdot B = 0$$

← This example will be important in the next lesson.

**Solution.** An equation like this can be confusing, since it mixes operations between vectors and numbers. It may help to first read through the equation to check that the operations are working on the right kind of input.

- $A$ and $B$ are vectors.
- $A \cdot B$ and $B \cdot B$ (each the dot product of two vectors) are both numbers.
- Thus, $\frac{A \cdot B}{B \cdot B}$ (the quotient of two numbers) is also a number.
- $\left(\frac{A \cdot B}{B \cdot B}\right) B$ (a scalar multiple of a vector) is a vector.
- That means $A - \left(\frac{A \cdot B}{B \cdot B}\right) B$ (the difference of two vectors) is a vector.
- Finally, $\left(A - \left(\frac{A \cdot B}{B \cdot B}\right) B\right) \cdot B$ (the dot product of two vectors) is a number.

$\longleftarrow$ Note that since $B \neq O$, $B \cdot B > 0$, so division by $B \cdot B$ is okay.

To see that this number is 0, use Theorem 2.1.

$$\begin{aligned} \left(A - \left(\tfrac{A \cdot B}{B \cdot B}\right) B\right) \cdot B &= A \cdot B - \left(\left(\tfrac{A \cdot B}{B \cdot B}\right) B\right) \cdot B \\ &= A \cdot B - \left(\tfrac{A \cdot B}{B \cdot B}\right)(B \cdot B) \\ &= A \cdot B - A \cdot B = 0 \end{aligned}$$

The next two examples show how the basic rules for dot product can be applied to geometry.

### Example 8

**Problem.** Suppose $A$ and $B$ are nonzero orthogonal vectors in $\mathbb{R}^n$ and $c_1A + c_2B = O$. Show that $c_1 = c_2 = 0$.

**Solution.** Take the equation $c_1A + c_2B = O$ and dot both sides with $A$:

$$\begin{aligned} A \cdot (c_1A + c_2B) &= A \cdot O \\ A \cdot (c_1A) + A \cdot (c_2B) &= 0 \\ c_1(A \cdot A) + c_2(A \cdot B) &= 0 \end{aligned}$$

Since $A$ is orthogonal to $B$, $A \cdot B = 0$, so this last equation becomes $c_1(A \cdot A) = 0$. Since $A \neq O$, $A \cdot A > 0$. Since $c_1(A \cdot A) = 0$, it follows that $c_1 = 0$. To prove $c_2 = 0$, take the equation $c_1A + c_2B = O$ and dot both sides with $B$.

### Example 9

**Problem.** Consider the triangle in $\mathbb{R}^3$ whose vertices are $A = (3, 2, 5)$, $B = (5, 2, 1)$, and $C = (2, 1, 3)$. Show that the angle formed by $\overrightarrow{CA}$ and $\overrightarrow{CB}$ is a right angle.

**Solution.** You want to show that the angle formed by $\overrightarrow{CA}$ and $\overrightarrow{CB}$ is a right angle. But $\overrightarrow{CA}$ is equivalent to $A - C$ and $\overrightarrow{CB}$ is equivalent to $B - C$. So, you only have to show that $A - C = (1, 1, 2)$ is orthogonal to $B - C = (3, 1, -2)$. And it is: $(A - C) \cdot (B - C) = (1, 1, 2) \cdot (3, 1, -2) = 0$.

One of the most beautiful theorems in mathematics is the Pythagorean Theorem. Does it extend to $\mathbb{R}^n$?

In $\mathbb{R}^2$, if $A$ and $B$ are nonzero vectors that aren't scalar multiples of each other, then $A+B$ and $A-B$ are the two diagonals of the parallelogram whose sides are $A$ and $B$. So the lengths of the diagonals are $\|A+B\|$ and $\|A-B\|$.

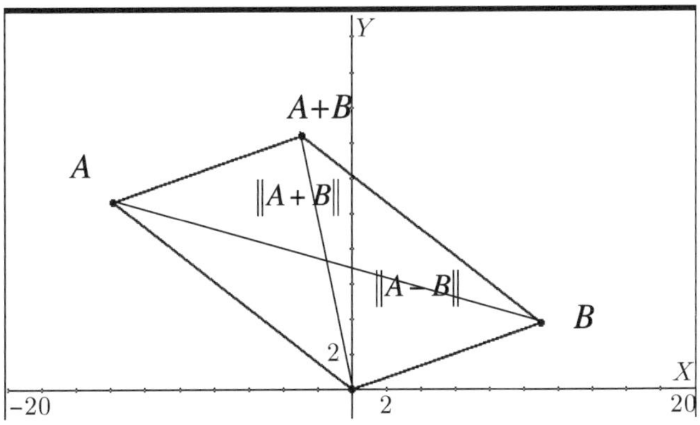


Now, from plane geometry, if the diagonals of a parallelogram have the same length, the parallelogram is a rectangle. So, if the parallelogram determined by $A$ and $B$ is a rectangle, then $A$ is perpendicular to $B$. Thus, $A$ is perpendicular to $B$ if and only if the diagonals, $A+B$ and $A-B$, have the same length. The same fact is true in $\mathbb{R}^n$, but the proof is via algebra.

**Lemma 2.3**

*If $A$ and $B$ are vectors in $\mathbb{R}^n$, $A$ is orthogonal to $B$ if and only if $\|A+B\| = \|A-B\|$.*

***Proof.*** Suppose $\|A+B\| = \|A-B\|$. By Theorem 2.2 and squaring both sides, you have $(A+B)\cdot(A+B) = (A-B)\cdot(A-B)$. This simplifies to $4(A \cdot B) = 0$, so $A \cdot B = 0$, and $A$ is orthogonal to $B$. The proof of the converse is just as simple. ■

←
A **lemma** is a result that's needed to prove another result. In this case, Lemma 2.3 is needed to prove Theorem 2.4. Usually, people discover that they need some fact when they try to prove a theorem, so they call that fact a lemma and prove it separately. The German word for lemma is *hilfsatz*—"helping statement." Note the connection between the words "lemma" and "dilemma."

In the next figure, $A$, $B$, and $A-B$ are three sides of a triangle. Since $A$ is orthogonal to $B$, it follows from Lemma 2.3 that $\|A-B\| = \|A+B\|$, so the hypotenuse of the right triangle whose legs are $A$ and $B$ has length $\|A+B\|$.

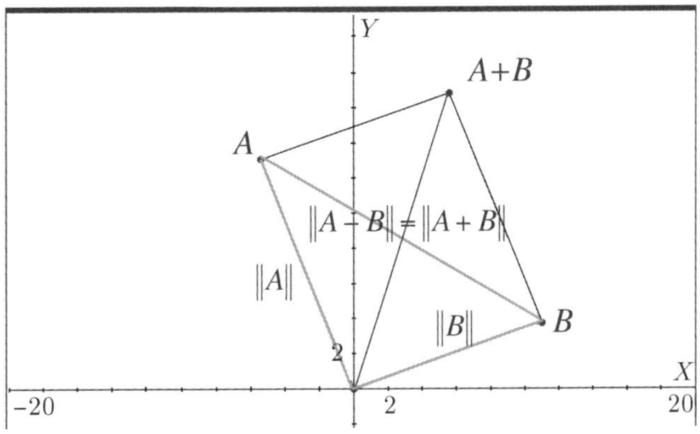

**Theorem 2.4 (The Pythagorean Theorem)**

*If $A$ and $B$ are vectors in $\mathbb{R}^n$ and $A$ is orthogonal to $B$, then $\|A+B\|^2 = \|A\|^2 + \|B\|^2$.*

$\longleftarrow$ The converse of the Pythagorean Theorem also holds in $\mathbb{R}^n$ (see Exercise 9).

***Proof.*** Since $A$ is orthogonal to $B$, $A \cdot B = 0$. So,

$$\begin{aligned} \|A+B\|^2 &= (A+B)\cdot(A+B) \\ &= A\cdot A + 2(A\cdot B) + B\cdot B \\ &= A\cdot A + B\cdot B \\ &= \|A\|^2 + \|B\|^2 \end{aligned}$$

■

**Developing Habits of Mind**

**Use algebra to extend geometric ideas.** The extension program is now fully underway. Look at the proof of the Pythagorean Theorem. It looks just like an algebraic calculation with numbers—the difference is that the letters stand for vectors and the operation is dot product, so it is carried out using a different set of basic rules.

There are two advantages to a proof like this:

1. It is extremely simple and compact.
2. It establishes the result for *any* dimension.

A disadvantage is that it doesn't seem very geometric—gone are the lovely "squares upon the hypotenuse" from plane geometry. With time and practice, you'll be able to look at a calculation like this and *see* the geometry.

## Exercises

**1.** For each $A$ and $B$, find

(i) $A \cdot B$
(ii) $(A+B)\cdot(A+B)$
(iii) $(A+B)\cdot(A-B)$
(iv) $(2A+3B)\cdot(A-B)$
(v) $(A+B)\cdot(3A-3B)$

**a.** $A = (1,4,2,1)$, $B = (-2,1,3,2)$
**b.** $A = (-2,3)$, $B = (5,1)$
**c.** $A = (-2,3,0)$, $B = (5,1,0)$
**d.** $A = (1,4,2)$, $B = (2,1,-3)$
**e.** $A = (1,5,2,3,1)$, $B = (1,4,-2,0,-3)$

**2.** If $A$ and $B$ are vectors in $\mathbb{R}^n$ and $c$ is a number, characterize each of the following by one of the words "vector" or "number."

**a.** $A\cdot(cB)$
**b.** $(A\cdot B)A$
**c.** $(A\cdot A)B + (B\cdot B)A$
**d.** $(cA+cB)\cdot A$
**e.** $((cA\cdot B)B)\cdot A$
**f.** $\dfrac{A\cdot B}{B\cdot B}B \quad (B \neq O)$

**3.** Find a nonzero vector $X$ in $\mathbb{R}^3$ orthogonal to $(1,3,2)$.

**4.** Characterize all vectors $X$ in $\mathbb{R}^3$ orthogonal to $A = (1,3,2)$ and $B = (-1,-2,1)$.

**5.** Characterize all vectors $X$ in $\mathbb{R}^3$ orthogonal to $A = (1,3,2)$, $B = (-1,-2,1)$, and $C = (0,1,3)$.

**6.** Characterize all vectors $X$ in $\mathbb{R}^3$ orthogonal to $A = (1,3,2)$, $B = (-1,-2,1)$, and $C = (0,1,4)$.

**7.** Let $A = (5,3,3)$, $B = (1,3,1)$, and $C = (2,6,-1)$. One angle of $\triangle ABC$ is a right angle. Which one is it?

**8.** In $\mathbb{R}^4$, let $A = (4,2,5,3)$, $B = (1,1,1,1)$, and $C = (0,4,2,-1)$. Show that $\triangle ABC$ is a right triangle.

**9.** Prove the converse of the Pythagorean Theorem: if $A$ and $B$ are vectors in $\mathbb{R}^n$ so that $\|A+B\|^2 = \|A\|^2 + \|B\|^2$, then $A$ is orthogonal to $B$.

**10.** Suppose $A$ and $B$ are vectors in $\mathbb{R}^n$ and $X$ is a vector orthogonal to both $A$ and $B$. Show that $X$ is orthogonal to every vector of the form $c_1A + c_2B$.

**11.** **a.** If $A_1, A_2, \ldots, A_r$ are vectors in $\mathbb{R}^n$, a **linear combination** of $A_1, A_2, \ldots, A_r$ is a vector $B$ which can be written as $c_1A_1 + c_2A_2 + \cdots + c_rA_r$ for some numbers $c_1, c_2, \ldots, c_r$. Show that $(3,9,4,7)$ is a linear combination of $(1,3,0,1)$, $(2,1,4,2)$, and $(-1,2,0,3)$, while $(3,9,4,8)$ is not.

**b.** If $A_1, A_2, \ldots, A_r$ are vectors in $\mathbb{R}^n$, and if $X$ is orthogonal to $A_i$ for each $i$, show that $X$ is orthogonal to every linear combination of $A_1, A_2, \ldots, A_r$.

**c.** If $A_1, A_2, A_3$ and $B_1, B_2$ are two sets of vectors in $\mathbb{R}^n$ so that each $B_j$ is orthogonal to all the $A_i$'s, show that any linear combination of $A_i$'s is orthogonal to any linear combination of the $B_j$'s.

**12.** Let $A$ and $B$ be nonzero vectors in $\mathbb{R}^n$ and suppose $C$ is a linear combination of $A$ and $B$. If $C$ is orthogonal to both $A$ and $B$, show that $C = O$.

**13.** Suppose $A = (2,11,10)$. Find

**a.** $\|A\|$

**b.** another vector $B$ that has the same length as $A$

**c.** a vector $B$ that has the same length as $A$ and is orthogonal to $A$

**d.** a vector $B$ that has the same length as $A$, that is orthogonal to $A$, and that has integer coordinates

$\longleftarrow$ A point whose coordinates are all integers is called a **lattice point**.

**14.** Two adjacent vertices of a square are at $O$ and $A = (-14, -2, 5)$.

**a.** How many such squares are there?
**b.** Find two vertices that will complete the square.
**c.** Find two vertices that complete the square and that are lattice points.

**15.** Show that the triangle whose vertices are $A = (4, 3, 0, 1)$, $B = (5, 4, 1, 2)$, and $C = (5, 2, 1, 0)$ is an isosceles right triangle.

**16.** If $A$ and $B$ are vectors in $\mathbb{R}^n$, show that $(A + B) \cdot (A - B) = A \cdot A - B \cdot B$.

**17.** If $A$ and $B$ are vectors in $\mathbb{R}^n$, show that $(A+B)\cdot(A+B) = A\cdot A + B\cdot B$ if and only if $A$ is orthogonal to $B$.

**18.** Show that if $A$ is orthogonal to $B$, $A$ is orthogonal to every scalar multiple of $B$.

**19.** Let $A_1, A_2, \ldots, A_r$ be mutually orthogonal nonzero vectors in $\mathbb{R}^n$. If $c_1 A_1 + c_2 A_2 + \cdots + c_r A_r = O$, show that each $c_i = 0$.

**20.** Let $A$ and $B$ be vectors so that $(A+B)\cdot(A+B) = (A-B)\cdot(A-B)$. Show that $A$ is orthogonal to $B$.

**21.** Let $A = (2, 1, 3, 2)$ and $B = (2, 1, 4, 1)$. Show that $A - \left(\frac{A\cdot B}{B\cdot B}\right) B$ is orthogonal to $B$.

**22.** True or false? If $A \cdot B = A \cdot C$ and if $A \neq O$, then $B = C$.

**23.** Find all vectors $X$ that have length 3 and that are orthogonal to both $(-1, 0, 1)$ and $(3, 2, -4)$.

**24.** Prove part (2) of Theorem 2.1.

**25.** Derman wrote down the incorrect definition of dot product. His notes say that

$$(a_1, a_2, a_3) \cdot (b_1, b_2, b_3) = a_1 b_1 + a_2 b_2$$

TONY: Derman, it's supposed to be $a_1 b_1 + a_2 b_2 + a_3 b_3$.

DERMAN: OK, but my dot product obeys the same basic rules as the ones in Theorem 2.1.

Is Derman right? Explain.

**26.** If $A$ and $B$ are vectors in $\mathbb{R}^n$, show that

$$A \cdot A + B \cdot B \geq 2(A \cdot B)$$

← When will both sides be equal?

**27.** Let $A = (a_1, a_2, a_3)$, $B = (b_1, b_2, b_3)$, and

$$C = (a_2 b_3 - a_3 b_2, a_3 b_1 - a_1 b_3, a_1 b_2 - a_2 b_1).$$

Show that $C$ is orthogonal to both $A$ and $B$.

**28.** Use Theorem 2.2 and the Basic Rules for Dot Product (Theorem 2.1) to prove parts (1) and (2) of Theorem 1.5 from Lesson 1.4.

> **Theorem 1.5.** *Let $A$ and $B$ be vectors in $\mathbb{R}^n$ and let $c$ be a real number. Then*
> (1) $\|A\| \geq 0$, *and* $\|A\| = 0$ *if and only if* $A = O$
> (2) $\|cA\| = |c|\,\|A\|$

**29.** If $A$ and $B$ are vectors, show that

**a.** $\|A+B\|^2 = \|A\|^2 + \|B\|^2 + 2(A \cdot B)$
**b.** $\|A+B\|^2 - \|A-B\|^2 = 4(A \cdot B)$
**c.** $\|A+B\|^2 + \|A-B\|^2 = 2\,\|A\|^2 + 2\,\|B\|^2$

⟵
What does this say in $\mathbb{R}^2$? Draw a picture.

**30.** If $A$ is a scalar multiple of $B$, show that

$$(A \cdot A)(B \cdot B) - (A \cdot B)^2 = 0$$

**31.** If $A, B,$ and $C$ are vectors in $\mathbb{R}^n$ so that $d(A,B) = d(C,B)$, show that

$$\frac{A \cdot A}{2} - A \cdot B = \frac{C \cdot C}{2} - C \cdot B$$

**32.** If $A$ and $B$ are orthogonal vectors in $\mathbb{R}^n$ so that $\|A\| = \|B\| = 1$, show that $d(A,B) = \sqrt{2}$.

**33.** Let $A$ and $B$ be vectors in $\mathbb{R}^n$, and let $c$ and $d$ be numbers. Prove the following identities.

⟵
The last two of these identities will be useful in later sections.

**a.** $(A+2B) \cdot (A-B) = A \cdot A + A \cdot B - 2B \cdot B$
**b.** $(cA+B) \cdot (cA+B) = c^2(A \cdot A) + 2c(A \cdot B) + B \cdot B$
**c.** $(cA+dB) \cdot (cA+dB) = c^2 A \cdot A + 2cdA \cdot B + d^2 B \cdot B$
**d.** $(\|B\|\,A + \|A\|\,B) \cdot (\|B\|\,A + \|A\|\,B)$
$= 2\,\|A\|\,\|B\|\,(\|A\|\,\|B\| + A \cdot B)$
**e.** $\left(\frac{A \cdot B}{B \cdot B} B\right) \cdot \left(\frac{A \cdot B}{B \cdot B} B\right) = \frac{(A \cdot B)^2}{B \cdot B} \qquad (B \neq O)$

**34.** Show that if $A \cdot B = A \cdot (B + C)$, then $A$ is orthogonal to $C$.

**35.** If $A$ and $B$ are vectors in $\mathbb{R}^n$, $B \neq O$, and $c$ is a number so that $A - cB$ is orthogonal to $B$, show that $c = \frac{A \cdot B}{B \cdot B}$.

**36.** If $A$ and $B$ are vectors in $\mathbb{R}^n$, $B \neq O$, and $P = \frac{A \cdot B}{B \cdot B} B$, show that

**a.** $P \cdot P = A \cdot P$
**b.** $A \cdot A = P \cdot P + (A - P) \cdot (A - P)$

**37.** If $A$ and $B$ are vectors in $\mathbb{R}^n$, $B \neq O$, and $c$ is a number so that $A - cB$ is orthogonal to $A + cB$, show that

$$c = \pm \frac{\|A\|}{\|B\|}$$

## 2.3 Projection

An important use of dot product is to determine the projection of a vector onto another vector. Projection has a number of applications throughout the study of linear algebra.

**In this lesson, you will learn how to**

- find the component of a vector along another vector
- find the projection of a vector along another vector

In $\mathbb{R}^2$, if $A$ and $B$ are vectors and $B \neq 0$, the projection of $A$ along $B$ is the vector obtained by intersecting the line through $A$ perpendicular to the line along $B$ with that line. This is illustrated for several situations in the following figure.

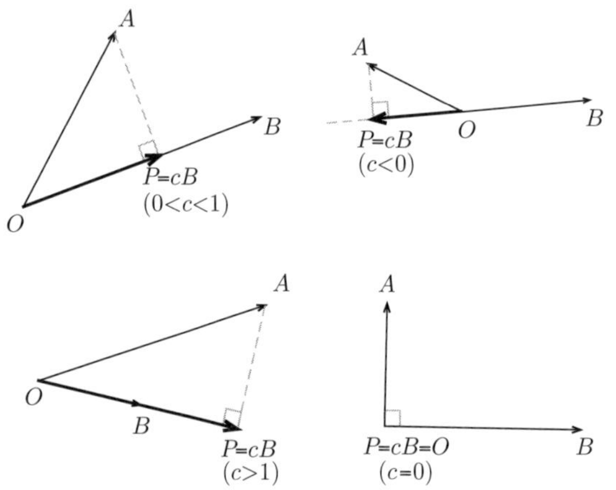


In each case, $P$ is the projection of $A$ along $B$. To extend this notion to $\mathbb{R}^n$, you need to describe $P$ with vector algebra.

(1) $P = cB$ for some scalar $c$.

(2) $\overrightarrow{PA}$ meets the line along $B$ at right angles.

Condition ((2)) can be reformulated.

2′. $A - P$ is orthogonal to $B$.

Since $P = cB$, you will have a formula for $P$ if you can determine $c$. To this end, use condition $((2)')$.

$$\begin{aligned} (A-P)\cdot B &= 0 \\ A\cdot B - P\cdot B &= 0 \\ P\cdot B &= A\cdot B \\ cB\cdot B &= A\cdot B \\ c(B\cdot B) &= A\cdot B \\ c &= \frac{A\cdot B}{B\cdot B} \qquad \text{(since } B\neq O, B\cdot B\neq 0) \end{aligned}$$

⟵
Fill in a reason for each step.

Hence $P = \dfrac{A\cdot B}{B\cdot B}B$. This formula makes sense in $\mathbb{R}^n$.

### Definition

Let $A$ and $B$ be vectors in $\mathbb{R}^n$, with $B \neq 0$.

- The **component** of $A$ along $B$, written $\text{comp}_B\, A$, is the *number*

$$\text{comp}_B\, A = \frac{A\cdot B}{B\cdot B}$$

- The **projection** of $A$ along $B$, written $\text{Proj}_B\, A$, is the *vector* defined by the formula

$$\text{Proj}_B\, A = (\text{comp}_B\, A)B = \frac{A\cdot B}{B\cdot B}B$$

**Habits of Mind**

**Find general purpose tools.** The projection ties together quite a bit of geometry into one little package. You'll see in the exercises and in the next sections that it's a very useful tool.

### Example 1

**Problem.** In $\mathbb{R}^2$, let $A = (5,1)$ and $B = (-3,0)$. Find $\text{Proj}_B\, A$.

**Solution.** You might expect that $\text{Proj}_B\, A = (5,0)$. (Why?) Use the definition to find that

$$\text{comp}_B\, A = \frac{-15}{9} = \frac{-5}{3}$$

so

$$\text{Proj}_B\, A = \frac{-5}{3}(-3,0) = (5,0)$$

### Example 2

**Problem.** In $\mathbb{R}^4$, let $A = (-3,1,-2,4)$ and $B = (1,1,2,0)$. Find $\text{Proj}_A\, B$ and $\text{Proj}_B\, A$.

**Solution.**

$$\text{comp}_B\, A = \frac{-6}{6} = -1 \qquad \text{so} \qquad \text{Proj}_B\, A = (-1,-1,-2,0)$$

Similarly, $\text{Proj}_A\, B = -\frac{1}{5}A = \left(\frac{3}{5}, \frac{-1}{5}, \frac{2}{5}, -\frac{4}{5}\right)$.

Note that if $A$ and $B$ are vectors in $\mathbb{R}^n$ and $P = \text{Proj}_B A$, then $P$ satisfies conditions ((1)) and ((2)′), which were used to characterize projections in $\mathbb{R}^2$. Clearly, $P$ is a multiple of $B$, and Example 7 from Lesson 2.2 shows that $A - P$ is orthogonal to $B$.

### Example 3

**Problem.** If $A$ and $B$ are vectors in $\mathbb{R}^n$ and $B \neq 0$, let $P = \text{Proj}_B A$ and show that

$$\|A\|^2 = \|P\|^2 + \|A - P\|^2$$

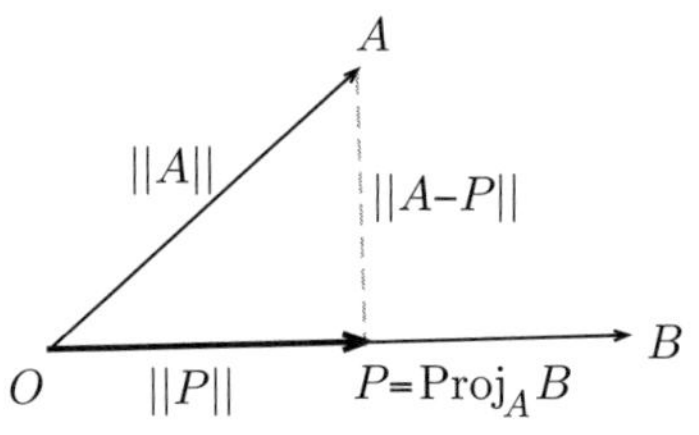


**Solution 1.** Let $c = \text{comp}_B A$ so that $P = cB$. Since $A - P$ is orthogonal to $B$, $A - P$ is also orthogonal to $P$ (Lesson 2.2, Exercise 18), so $P \cdot (A - P) = 0$. But then

$$\begin{aligned}
\|P\|^2 + \|A - P\|^2 &= P \cdot P + (A - P) \cdot (A - P) \\
&= P \cdot P + A \cdot A - 2A \cdot P + P \cdot P \\
&= 2P \cdot P - 2A \cdot P + A \cdot A \\
&= -2P \cdot (A - P) + A \cdot A \\
&= -2\,(P \cdot (A - P)) + A \cdot A \\
&= A \cdot A = \|A\|^2
\end{aligned}$$

**Solution 2.** Since $A - P$ is orthogonal to $P$, you can apply the Pythagorean Theorem to $A - P$ and $P$.

$$\|A - P\|^2 + \|P\|^2 = \|A - P + P\|^2 = \|A\|^2$$

## Exercises

1. For each of the given points $A$ and $B$, find $d(A, B)$, $\text{Proj}_B A$, and $\text{Proj}_A B$.

   **a.** $A = (3, 1), B = (4, 2)$ **b.** $A = (1, 0, 1), B = (0, 1, 0)$
   **c.** $A = (1, 3, 2), B = (4, 1, 3)$ **d.** $A = (1, 3, -1, 4), B = (2, 1, 3, 8)$

2. Suppose $A$, $B$, and $C$ are vectors. Characterize each expression with one of the words "vector," "number," or "meaningless."

   **a.** $\|A + B\|$ **b.** $A \cdot (B + C)$ **c.** $\|A \cdot B\|$
   **d.** $\text{Proj}_B A$ **e.** $\|\text{Proj}_B A\|$ **f.** $\text{Proj}_A(\text{comp}_B A)$
   **g.** $(A \cdot B) \cdot A$ **h.** $A - A \cdot B$ **i.** $\|(A \cdot B)C\|$
   **j.** $d(A, \text{Proj}_B A)$ **k.** $A \cdot \text{Proj}_A B$ **l.** $(\text{comp}_B A \ \text{comp}_A B)C$

3. If $A = (1, 0, 3)$ and $B = (-1, 2, 0)$, find

   **a.** $d(A, \text{Proj}_B A)$
   **b.** $\|\text{Proj}_A B\|$
   **c.** $\text{Proj}_A(\text{Proj}_B A)$
   **d.** $\text{comp}_B A \ \text{comp}_A B$
   **e.** $(\text{Proj}_B A - A) \cdot B$
   **f.** $A \cdot \text{Proj}_B A$

4. If $A$ and $B$ are nonzero vectors, show that $A$ is orthogonal to $B$ if and only if $\text{Proj}_B A = O$.

5. If $A$ and $B$ are nonzero vectors, show that $\text{comp}_B A$ and $\text{comp}_A B$ cannot have opposite signs.

←
Is there a geometric interpretation of this common sign?

6. If $A$ and $B$ are nonzero vectors, show that

$$\frac{\text{comp}_B A}{\text{comp}_A B} = \left(\frac{\|A\|}{\|B\|}\right)^2$$

7. If $A$ and $B$ are vectors in $\mathbb{R}^n$ $(B \neq O)$, show that $\|\text{Proj}_B A\| = \frac{|A \cdot B|}{\|B\|}$.

8. Show that if $A$ and $B$ are nonzero vectors,

$$\frac{\|A\|}{\|\text{Proj}_B A\|} = \frac{\|B\|}{\|\text{Proj}_A B\|}$$

   What is the value of this common ratio?

9. If $A$ and $B$ are vectors in $\mathbb{R}^n$ $(B \neq O)$, and $A$ is a scalar multiple of $B$, show that $\text{Proj}_B A = A$.

10. Suppose $A = (1, 4, -1)$. Find

    **a.** the projection of $A$ on the $x$–$y$ plane
    **b.** the projection of $A$ on the $x$–$z$ plane
    **c.** the projection of $A$ on the $y$–$z$ plane

←
What does it mean to project a vector on a plane? Part of this problem is for you to figure out a reasonable answer.

11. Suppose $A$ and $B$ are nonzero points in $\mathbb{R}^2$. Show that the area of the triangle whose vertices are $A$, $B$, and $O$ is $\frac{1}{2}\sqrt{(A \cdot A)(B \cdot B) - (A \cdot B)^2}$.

**Hint**: Show that the area is $\frac{1}{2}\|B\|\sqrt{\|A\|^2 - \|\text{Proj}_B A\|^2}$ and then simplify.

12. Use Exercise 11 to show that if $A$ and $B$ are vectors in $\mathbb{R}^2$, $(A \cdot A)(B \cdot B) - (A \cdot B)^2 \geq 0$.

13. Suppose $A$ and $B$ are nonzero points in $\mathbb{R}^n$ and let $P = \text{Proj}_B A$.

    **a.** Show that

$$\|A\|^2 \geq \|P\|^2$$

    with equality if and only if $A = P$.

    **b.** Use this to show that

$$(A \cdot A) \geq \frac{(A \cdot B)^2}{(B \cdot B)}$$

←
See Example 3 in this lesson.

14. Use Exercise 11 to find the area of the triangle whose vertices are

    **a.** $(0, 0), (3, 1), (7, 0)$
    **b.** $(0, 0), (4, -2), (5, 3)$
    **c.** $(0, 0), (5, 2), (-1, -3)$
    **d.** $(1, 3), (2, 1), (7, -2)$

←
For part **d**, translate to $(0, 0)$.

**15.** Suppose $A = (1, 4, -1)$ and $B = (-4, 0, 2)$. Let $\mathfrak{P}$ be the parallelogram whose vertices are $O$, $A$, $B$, and $A + B$.

**a.** Find the vertices of $\mathfrak{P}'$, the projection of $\mathfrak{P}$ on the $x$-$y$ plane.
**b.** Find the vertices of $\mathfrak{P}''$, the projection of $\mathfrak{P}$ on the $x$-$z$ plane.
**c.** Find the vertices of $\mathfrak{P}'''$, the projection of $\mathfrak{P}$ on the $y$-$z$ plane.
**d.** Find the areas of $\mathfrak{P}'$, $\mathfrak{P}''$, and $\mathfrak{P}'''$.

$\longleftarrow$ $\mathfrak{P}'$, $\mathfrak{P}''$, and $\mathfrak{P}'''$ are also parallelograms. Can you prove it?

## 2.4 Angle

The geometric image of vectors in $\mathbb{R}^2$ allows you to think about the angle between two vectors. Working with such an image even lets you measure that angle in a familiar way.

**In this lesson, you will learn how to**

- find the angle between two vectors in any dimension
- understand and use the triangle inequality in $\mathbb{R}^n$

**Minds in Action** Episode 6

*Sasha and Tony are thinking about Exercise 5 from Getting Started (Lesson 2.1).*

SASHA: Say, Tony, I was thinking about projection. I bet we can use it to find the angle between two vectors. Remember in the Getting Started, we tried to find the angle between $A = (2, 9)$ and $B = (6, 4)$?

TONY: Vaguely. How did we solve it then?

SASHA: Using the Law of Cosines. But look:

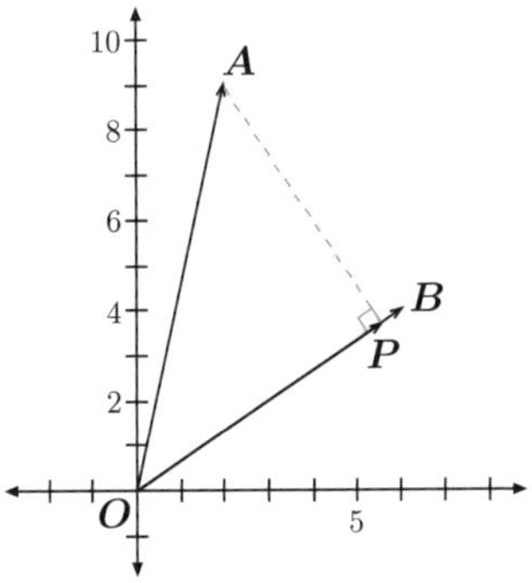


If I drop a perpendicular from $A$ to $B$, I get a right triangle. And $P$ is the projection of $A$ onto $B$. Now finding the cosine is pretty basic.

TONY: Wow, Sasha, how do you come up with these crazy things! So that new triangle has sides $\|A\|$, $\|\text{Proj}_B\, A\|$, and—

SASHA: We don't need that third side. I can use cosine with just those two:

$$\cos\theta = \frac{\|\text{Proj}_B\, A\|}{\|A\|}$$

TONY: But wait, how is this easier?

SASHA: Well, we've also seen that numerator before. Here it is, Exercise 7 from Lesson 2.3: *If $A$ and $B$ are vectors in $\mathbb{R}^n$ ($B \neq O$), show that* $\|\text{Proj}_B\, A\| = \frac{|A \cdot B|}{\|B\|}$. So, in our case,

$$\frac{|A \cdot B|}{\|B\|} = \frac{|(2,9)\cdot(6,4)|}{\sqrt{(6,4)\cdot(6,4)}} = \frac{|12+36|}{\sqrt{36+16}} = \frac{48}{\sqrt{52}}$$

TONY: And $\|A\| = \sqrt{(2,9)\cdot(2,9)} = \sqrt{85}$. So $\cos\theta = \frac{48}{\sqrt{52}} \cdot \frac{1}{\sqrt{85}}$. That makes $\theta = \cos^{-1}\left(\frac{48}{\sqrt{52}\sqrt{85}}\right)$.

SASHA: Yeah, that's what I got before. It works out to be about 43.8°.

TONY: Hmm . . . I bet we're on to something here.

---

Sasha and Tony have started using the new facts about vectors to algebraically find the measure of an angle in $\mathbb{R}^2$. The next step is to see if their discovery helps to think about angles in higher dimensions.

### Developing Habits of Mind

**Use algebra to extend geometric ideas.** You are looking for a way to measure the angle between two vectors in $\mathbb{R}^n$, but you don't want to be restricted to the geometry of $\mathbb{R}^2$. So you want to use the geometry of $\mathbb{R}^2$ to measure an angle algebraically in a way that lets you extend it to any dimension. The best tool for that job? As Sasha and Tony discovered, it's trigonometry.

---

As Sasha and Tony discovered, you can calculate the measure of an angle between vectors using trigonometry, basing your calculation on the lengths of vectors. Since you already know how to find the length of the projection of one vector onto another, cosine is the best choice. Cosine also has another benefit: on the interval $0 \le \theta \le \pi$, there is a one-to-one correspondence between the measure of an angle and its cosine.

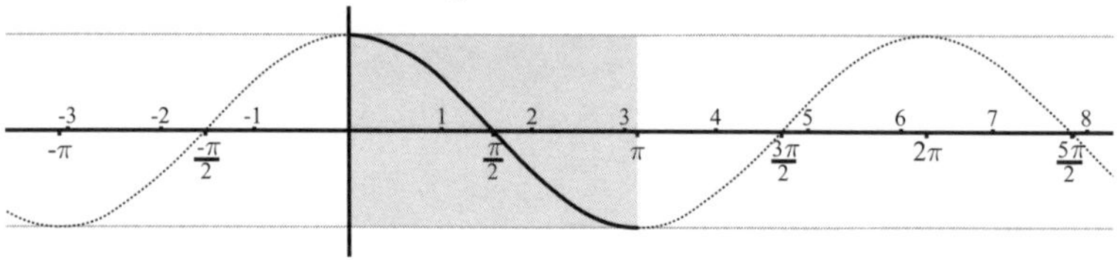


So, for angles within that interval, knowing the cosine of an angle is enough to tell you what the angle is. And, if you can find a way to compute the cosine of the angle between two vectors from what you know about the vectors, that will provide a way of defining the angle between two vectors in $\mathbb{R}^n$.

When you look at two vectors in $\mathbb{R}^2$, you might see two different angles that could be considered "between $A$ and $B$": one could be considered "the angle from $A$ to $B$" and one "the angle from $B$ to $A$." For any two vectors in $\mathbb{R}^2$, these two measurements will always add to $2\pi$. One will measure between 0 and $\pi$, and the other will measure between $\pi$ and $2\pi$. It doesn't

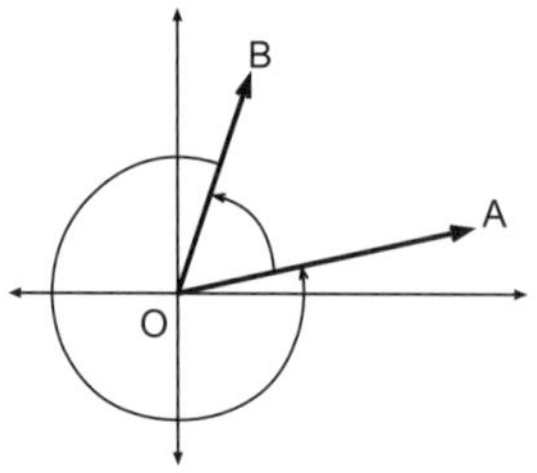


← When measuring angles, you could also "wrap around" the axis any number of times, adding (or subtracting) multiples of $2\pi$ (or, using degrees, 360°) to the angle. But your goal is to find a *unique* number, so you can ignore these equivalent measures here.

necessarily matter which you choose as long as you're consistent. So you can consider the unique angle between two nonzero vectors $A$ and $B$ in $\mathbb{R}^2$ to be the unique angle $\theta$ determined by $A$ and $B$ that satisfies the restriction $0 \le \theta \le \pi$.

$\longleftarrow$
In fact, for any angle $\theta$, $\cos(2\pi - \theta) = \cos\theta$, so the cosine would be the same whether you pick the larger or smaller angle.

So now, to invoke the extension program, you want to finalize a formula for the cosine of the angle between two vectors in $\mathbb{R}^2$ and use that formula as the *definition* for the cosine of the angle between two vectors in $\mathbb{R}^n$.

Suppose $A$ and $B$ are nonzero vectors, and $\theta$ is the angle between $A$ and $B$.

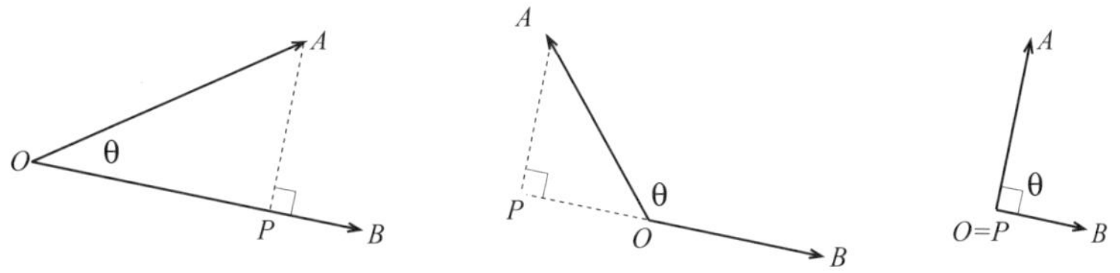


In each figure, $P = \operatorname{Proj}_B A$. In the first case, where $\theta$ is acute, you can use right triangle trigonometry, like Sasha and Tony did, to say that $\cos\theta = \frac{\|P\|}{\|A\|}$. Since $P = cB$, where $c = \operatorname{comp}_B A$, you can say that

$$\cos\theta = \frac{\|cB\|}{\|A\|} = \frac{|c|\,\|B\|}{\|A\|}$$

Now, since $\theta$ is acute, $c > 0$, so $|c| = c$. Also, from the definition of component, you know that $c = \frac{A\cdot B}{B\cdot B}$, so

$$\cos\theta = \left(\frac{A\cdot B}{B\cdot B}\right)\frac{\|B\|}{\|A\|}$$

Finally, since $\|B\| = \sqrt{B\cdot B}$, then $\|B\|^2 = B\cdot B$, and so

$$\cos\theta = \left(\frac{A\cdot B}{\|B\|^2}\right)\frac{\|B\|}{\|A\|} = \frac{A\cdot B}{\|A\|\,\|B\|}$$

### For You to Do

1. You just saw that in $\mathbb{R}^2$,
$$\cos\theta = \frac{A\cdot B}{\|A\|\,\|B\|}$$
when $0 < \theta < \frac{\pi}{2}$. Show that this formula works for any angle $\theta$ where $0 \le \theta \le \pi$.

$\longleftarrow$
Split your work into these four additional cases:
- $\frac{\pi}{2} < \theta < \pi$
- $\theta = \frac{\pi}{2}$
- $\theta = 0$
- $\theta = \pi$

For any angle $\theta$ where $0 \le \theta \le \pi$, you can calculate its cosine using the formula

$$\cos\theta = \frac{A\cdot B}{\|A\|\,\|B\|}$$

This formula uses only the length of a vector and the dot product, both of which can be calculated in $\mathbb{R}^n$ for any $n$, not just $\mathbb{R}^2$. So it appears to be a great candidate for the extension program: you can define the angle between any two vectors $A$ and $B$ in $\mathbb{R}^n$ to be the unique angle between 0 and $\pi$ that satisfies this equation. But there's one more issue to check: in $\mathbb{R}^2$, you know that the cosine of an angle ranges between $-1$ and 1. How can you be sure the formula will always produce numbers in that range in any $\mathbb{R}^n$?

In other words, to extend this formula to $\mathbb{R}^n$, you first need to show that $-1 \leq \frac{A \cdot B}{\|A\|\|B\|} \leq 1$.

**Theorem 2.5 (Cauchy-Schwarz Inequality)**
*If $A$ and $B$ are vectors in $\mathbb{R}^n$, then $\|A\|\,\|B\| \geq |A \cdot B|$.*

$\longleftarrow$
The inequality is named after Augustin-Louis Cauchy (1789–1857) and Herman Schwarz (1843–1921). It has other names as well, and it is used all over mathematics.

***Proof.*** Let $A$ and $B$ be vectors in $\mathbb{R}^n$ ($B \neq 0$) and let $P = \text{Proj}_B\, A$. From Example 3 from Lesson 2.3, you know that

$$\|A\|^2 = \|P\|^2 + \|A - P\|^2$$

Since $\|A - P\| \geq 0$, you can thus say that $\|A\|^2 \geq \|P\|^2$. Now, $P = \frac{A \cdot B}{B \cdot B} B$, so

$\longleftarrow$
Make sure you understand the reason for each step in this proof.

$$\begin{aligned}
\|A\|^2 &\geq \left\| \frac{A \cdot B}{B \cdot B} B \right\|^2 \\
\|A\|^2 &\geq \left( \frac{A \cdot B}{B \cdot B} \right)^2 \|B\|^2 \\
\|A\|^2 &\geq \left( \frac{A \cdot B}{\|B\|^2} \right)^2 \|B\|^2 \\
\|A\|^2 &\geq \frac{(A \cdot B)^2}{\|B\|^2} \\
\|A\|^2 \|B\|^2 &\geq (A \cdot B)^2 \\
(\|A\|\,\|B\|)^2 &\geq (A \cdot B)^2 \\
\|A\|\,\|B\| &\geq |A \cdot B|
\end{aligned}$$

■

### For You to Do

**2.** Use the Cauchy-Schwarz Inequality to show that

$$-1 \leq \frac{A \cdot B}{\|A\|\,\|B\|} \leq 1$$

Because $\frac{A \cdot B}{\|A\|\|B\|}$ is the cosine of a unique angle $\theta$ in $\mathbb{R}^2$ so that $0 \leq \theta \leq \pi$, and because you have shown that, even in $\mathbb{R}^n$, it never exceeds the range of the cosine function, you can feel confident using it as a definition of cosine in $\mathbb{R}^n$.

### Definition

The **angle between two nonzero vectors** in $\mathbb{R}^n$, $A$ and $B$, is the unique angle $\theta$ that satisfies $0 \le \theta \le \pi$ and $\cos\theta = \frac{A \cdot B}{\|A\|\|B\|}$.

### For You to Do

**3.** Suppose $A = (5, 5) and B = (-3, 0)$. Find the angle $\theta$ between $A$ and $B$.

### Example

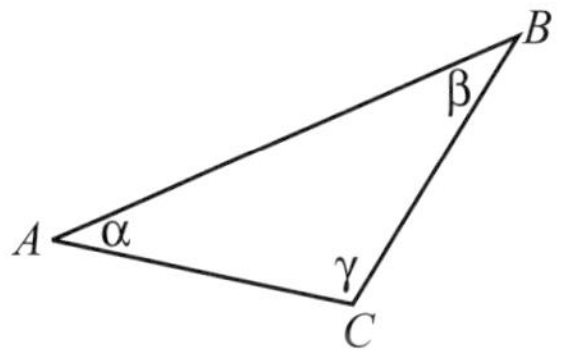


**Problem.** Find angle $\alpha$ of $\triangle ABC$, as shown in this figure, where $A = (-13, -66, 76)$, $B = (5, 60, 31)$, and $C = (27, 46, 60)$.

**Solution.** To find $\cos\alpha$, find the cosine of the angle between $B - A$ and $C - A$. So,

$$\cos\alpha = \frac{(B - A) \cdot (C - A)}{\|B - A\| \, \|C - A\|} = \frac{15552}{135 \cdot 120} = \frac{24}{25}$$

And thus, $\alpha = \cos^{-1} \frac{24}{25} \approx 16.26°$.

⟵ First move $A$ to the origin.

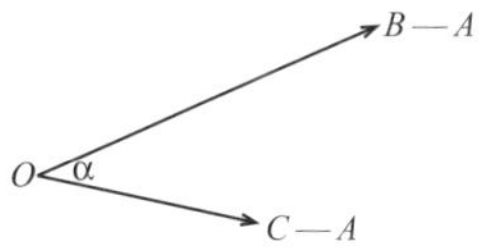


### For You to Do

**4.** Find angles $\beta$ and $\gamma$ of $\triangle ABC$ from the example above.

**5.** Find the angle between $A = (1, 3, -1, 2)$ and $B = (4, 1, -3, 0)$.

Here's another application of the Cauchy-Schwarz Inequality. You can use it to prove part (3) of Theorem 1.5 from Lesson 1.4, which is typically called the Triangle Inequality.

### Theorem (The Triangle Inequality)

*If $A$ and $B$ are vectors in $\mathbb{R}^n$, then $\|A + B\| \le \|A\| + \|B\|$.*

***Proof.***

$$\begin{aligned} \|A + B\|^2 &= (A + B) \cdot (A + B) \\ &= A \cdot A + 2(A \cdot B) + B \cdot B \\ &= \|A\|^2 + 2(A \cdot B) + \|B\|^2 \\ &\le \|A\|^2 + 2\,|A \cdot B| + \|B\|^2 \\ &\le \|A\|^2 + 2\,\|A\|\,\|B\| + \|B\|^2 \\ &= (\|A\| + \|B\|)^2 \end{aligned}$$

⟵ Give a reason for every step.

So, $\|A+B\|^2 \le (\|A\|+\|B\|)^2$; since both $\|A+B\|$ and $\|A\|+\|B\|$ are positive, you can take the square root of both sides, giving the desired result. ∎

### Developing Habits of Mind

**Use vectors to prove ideas about numbers.** The Cauchy-Schwarz Inequality makes a statement about vectors. You can restate it using coordinates: if $A = (a_1, a_2, \ldots, a_n)$ and $B = (b_1, b_2, \ldots, b_n)$, then

$$|a_1b_1 + a_2b_2 + \cdots + a_nb_n| \le \sqrt{a_1^2 + a_2^2 + \cdots + a_n^2}\sqrt{b_1^2 + b_2^2 + \cdots + b_n^2}$$

This remarkable statement is about ordinary numbers—in this form, it has nothing to do with vectors. While proving it can be quite difficult using only the algebra of numbers, as you've seen, it is pretty straightforward to do so using linear algebra.

## Exercises

**1.** Find $\cos\theta$ if $\theta$ is the angle between $A$ and $B$.

**a.** $A = (3,4)$, $B = (0,7)$
**b.** $A = (1,1,1)$, $B = (1,1,0)$
**c.** $A = (2,1,0)$, $B = (5,-3,4)$
**d.** $A = (-3,1,2,5)$, $B = (4,1,3,-4)$
**e.** $A = (-2,1,3,2), B = (1,1,1,-1)$
**f.** $A = (-5,0), B = (1,\sqrt{3})$

**2.** Find the cosine of each angle of $\triangle ABC$ where

**a.** $A = (1,5,2)$, $B = (2,6,3)$, $C = (2,5,1)$
**b.** $A = (1,-1)$, $B = (\sqrt{3},\sqrt{3})$, $C = (\sqrt{3}+1,0)$
**c.** $A = (10,68,56)$, $B = (-22,-156,136)$, $C = (-150,100,-120)$

**3.** **a.** Use trigonometry to show that the angle between $A = (1,1)$ and $B = (1,\sqrt{3})$ is $\frac{\pi}{12}$.
**b.** Show that $\cos\frac{\pi}{12} = \frac{1+\sqrt{3}}{2\sqrt{2}}$.

**4.** If $A$ and $B$ are nonzero vectors in $\mathbb{R}^n$, and $c$ and $d$ are positive scalars, show that the angle between $cA$ and $dB$ is the same as the angle between $A$ and $B$.

**5.** If $A$ and $B$ are nonzero vectors in $\mathbb{R}^n$, show that

$$\text{comp}_B\, A \ \text{comp}_A\, B = \cos^2\theta$$

where $\theta$ is the angle between $A$ and $B$.

**6.** If $A$ and $B$ are nonzero vectors in $\mathbb{R}^n$ and $\theta$ is the angle between $A$ and $B$, show that

$$\|A-B\|^2 = \|A\|^2 + \|B\|^2 - 2\,\|A\|\,\|B\|\cos\theta$$

(In $\mathbb{R}^2$, this is the Law of Cosines.)

**7.** In $\mathbb{R}^3$, let $A = (\sqrt{3}, \sqrt{3}, 1)$, $B = (-1+\sqrt{3}, 1+\sqrt{3}, 1)$, and $C = (-1, 1, 1)$. Show that the angles of the triangle whose vertices are $A$, $B$, and $C$ are 90°, 60°, and 30°. Verify that the length of the shorter leg is one half the length of the hypotenuse.

**8.** Let $A$ and $B$ be nonzero vectors in $\mathbb{R}^n$, and suppose $\|\text{Proj}_B A\| = \frac{\sqrt{3}}{2}\|A\|$. Show that the angle between $A$ and $B$ is 30°.

**9.** If $A$ and $B$ are nonzero vectors in $\mathbb{R}^n$ so that $\|A\| = \|B\|$, show that $A + B$ bisects the angle between $A$ and $B$. Draw a picture in $\mathbb{R}^2$.

**10.** Suppose $A$ and $B$ are nonzero vectors in $\mathbb{R}^n$, and $\theta$ is the angle between $A$ and $B$.

**a.** If $\cos\theta = 1$, show that $A = cB$ where $c > 0$.
**b.** If $\cos\theta = -1$, show that $A = cB$ where $c < 0$.

⟵
**Hint:** Let $c = \text{comp}_B A$. Show that $(A - cB)\cdot(A - cB) = 0$.

**11.** If $A$ and $B$ are points in $\mathbb{R}^n$ and $A = cB$ where $c > 0$, show that $\|A + B\| = \|A\| + \|B\|$.

**12.** If $A$, $B$, and $C$ are points in $\mathbb{R}^n$, show that $\|A + B + C\| \leq \|A\| + \|B\| + \|C\|$.

**13.** If $A$ and $B$ are points in $\mathbb{R}^n$, show that $\|A - B\| \geq \|A\| - \|B\|$.

**14.** Find $X$ in $\mathbb{R}^3$ so that $X$ is orthogonal to $(2, 0, -1)$, $\|X\| = 9$, and $X$ makes a 45° angle with $(0, 1, 1)$.

**15.** Find a vector $A$ in $\mathbb{R}^3$ so that $\|A\| = 9$, $A$ is orthogonal to $(4, 0, -1)$, and $\cos\theta = \frac{28}{45}$ where $\theta$ is the angle between $A$ and $(4, 3, 0)$.

**16.** Let $a_1, a_2, \ldots, a_n$ be positive real numbers. Show that

$$(a_1 + a_2 + \cdots + a_n)\left(\frac{1}{a_1} + \frac{1}{a_2} + \cdots + \frac{1}{a_n}\right) \geq n^2$$

⟵
**Hint:** Use the Cauchy-Schwarz Inequality.

**17.** **Take It Further.** In vector language, the triangle inequality says that for any vectors $A$ and $B$ in $\mathbb{R}^n$, it is true that $\|A + B\| \leq \|A\| + \|B\|$. In geometric language, the Triangle Inequality says that the sum of the lengths of any two sides of a triangle is greater than the length of the third side.

**a.** Working only in $\mathbb{R}^2$, show the *geometric* interpretation of $\|A + B\| \leq \|A\| + \|B\|$. In other words, show what $\|A + B\| \leq \|A\| + \|B\|$ has to do with triangles.
**b.** Use $\|A + B\| \leq \|A\| + \|B\|$ to prove the Cauchy-Schwartz Inequality.

## 2.5 Cross Product

In many physical applications, it is necessary to find a vector orthogonal to a given set of vectors. In $\mathbb{R}^3$, for example, one often has to find a vector orthogonal to two given vectors. In this section, you'll derive an explicit formula for a vector that is orthogonal to two given vectors in $\mathbb{R}^3$.

← In the next section, you'll need to find a vector orthogonal to two given vectors in order to find equations of planes in $\mathbb{R}^3$.

**In this lesson, you will learn how to**

- find a vector orthogonal to two given vectors in $\mathbb{R}^3$ using cross product
- determine the area of triangles and parallelograms in $\mathbb{R}^3$ using cross product
- apply the cross product to find the angle between two vectors in $\mathbb{R}^3$

### For You to Do

**1.** Find a vector orthogonal to both $(1, 5, -2)$ and $(2, 1, 0)$. Explain how you did it.

### Minds in Action Episode 7

*Tony and Sasha are trying to solve the general problem of finding a vector orthogonal to two vectors in* $\mathbb{R}^3$.

TONY: Let's just do what we did with $(1, 5, -2)$ and $(2, 1, 0)$, once and for all. Suppose $A = (a_1, a_2, a_3)$ and $B = (b_1, b_2, b_3)$ are two vectors in $\mathbb{R}^3$. If $X = (x, y, z)$ is orthogonal to both $A$ and $B$, then $A \cdot X = 0$ and $B \cdot X = 0$. Expanding this we have two equations in three unknowns.

**Habits of Mind**

Tony is exercising an important algebraic habit: mimic a numerical calculation with variables.

*Tony writes on the board.*

$$\begin{aligned} a_1x + a_2y + a_3z &= 0 \\ b_1x + b_2y + b_3z &= 0 \end{aligned}$$

Two equations in *three* unknowns. Now what?

SASHA: Well, if you think about what we are trying to find, there must be infinitely many solutions, because there are infinitely many vectors that are orthogonal to both $A$ and $B$. In fact, they'll all lie on a line. Look:

*Sasha draws a picture.*

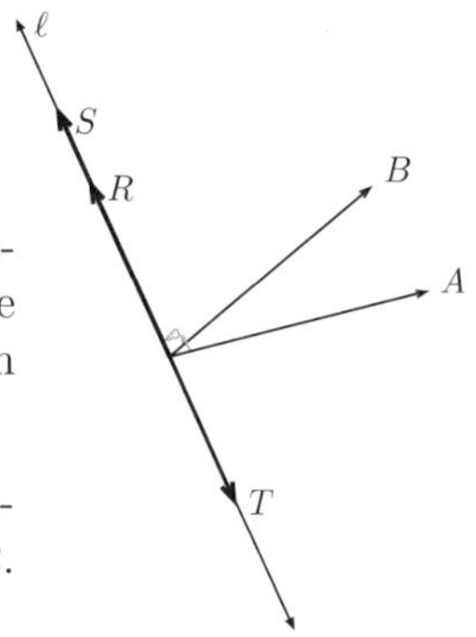


Vectors $R$, $S$, and $T$, and any other vector along $\ell$ will do the trick. So, all the solutions must be scalar multiples of each other.

TONY: Okay, well, let's just find one solution . . . I know: let's just let $z$ be, say, 12. Then we'll have only two unknowns.

SASHA: Good idea. But why not just let $z$ be $z$? Treat it like a constant and get $x$ and $y$ in terms of it.

TONY: Great. And then we can find one solution to our system by assigning any value to $z$. Here we go. Let's write it like this:

$$\begin{aligned} a_1x + a_2y &= -a_3z \\ b_1x + b_2y &= -b_3z \end{aligned}$$

making believe that the right-hand sides are constants. Multiply the first equation by $-b_1$ and the second equation by $a_1$; add and simplify.

SASHA: I get

$$(a_1b_2 - b_1a_2)y = (a_3b_1 - a_1b_3)z$$

⟵ Make sure you check Sasha's calculations.

And now I'll go back and eliminate $y$; that is, multiply the first equation by $b_2$ and the second by $-a_2$; add and simplify. I get

$$(a_1b_2 - b_1a_2)x = (a_2b_3 - a_3b_2)z$$

TONY: And now we get to let $z$ be anything we want.

SASHA: Well, I see something for $z$ that will make everything easier. Look—the coefficient of $y$ in the first equation is the same as the coefficient of $x$ in the second.

TONY: Nice catch, Sasha. If we let $z = (a_1b_2 - b_1a_2)$, we can cancel it from each side in each equation.

SASHA: What a team we are.

*Enter Derman, looking at the board.*

DERMAN: I could have told you all this yesterday. I did it in my head.

TONY: OK, Derman, can you show us that it actually works—that

$$(a_2b_3 - a_3b_2,\ a_3b_1 - a_1b_3,\ a_1b_2 - b_1a_2)$$

is actually orthogonal to $A$ and $B$?

DERMAN: It's time for lunch.

Tony and Sasha have proved the following theorem.

**Theorem 2.6**

*If* $A = (a_1, a_2, a_3)$ *and* $B = (b_1, b_2, b_3)$, *the vector*

$$X = (a_2b_3 - a_3b_2, a_3b_1 - a_1b_3, a_1b_2 - a_2b_1)$$

*is orthogonal to both* $A$ *and* $B$.

**Definition**

If $A = (a_1, a_2, a_3)$ and $B = (b_1, b_2, b_3)$, the **cross product** $A \times B$ is the vector defined by

$$A \times B = (a_2b_3 - a_3b_2, a_3b_1 - a_1b_3, a_1b_2 - a_2b_1)$$

← So, another way to state Theorem 2.6 is that $A \times B$ is orthogonal to both $A$ and $B$.

**Developing Habits of Mind**

**Generalize from numerical examples.** This is a common occurrence in mathematics: starting with a numerical example, you generalize and come up with a general solution to a problem. You give the solution a *name*, and then start investigating its properties. In this case, the process went as follows:

1. Find a vector orthogonal to both $(1, 5, -2)$ and $(2, 1, 0)$.
   - You found a solution above. One answer is $(2, -4, -9)$.
2. Find a vector orthogonal to both $(a_1, a_2, a_3)$ and $(b_1, b_2, b_3)$.
   - Tony and Sasha did this above. One answer is

$$(a_2b_3 - a_3b_2, b_1a_3 - a_1b_3, a_1b_2 - a_2b_1)$$

3. *Name* this generic solution.
   - Call it the *cross product* of $A$ and $B$ and write

$$A \times B = (a_2b_3 - a_3b_2, b_1a_3 - a_1b_3, a_1b_2 - a_2b_1)$$

4. Study the properties of the named thing.
   - In this case, there is a new operation: $\times$. You know how to study properties of operations. That comes next.

← Is $A \times B = B \times A$? Try it with numbers.

**Facts and Notation**

There is an easy way to remember the formula for $A \times B$. First, define the **determinant** of the array $\begin{pmatrix} a & b \\ c & d \end{pmatrix}$, or $\det \begin{pmatrix} a & b \\ c & d \end{pmatrix}$, to be the number $ad - bc$. Using the determinant notation, you can write

$$A \times B = \left( \det \begin{pmatrix} a_2 & a_3 \\ b_2 & b_3 \end{pmatrix}, -\det \begin{pmatrix} a_1 & a_3 \\ b_1 & b_3 \end{pmatrix}, \det \begin{pmatrix} a_1 & a_2 \\ b_1 & b_2 \end{pmatrix} \right)$$

← Determinants and their properties will be the subject of a later chapter. Sometimes, $\det \begin{pmatrix} a & b \\ c & d \end{pmatrix}$ is written as $\begin{vmatrix} a & b \\ c & d \end{vmatrix}$.

Furthermore, these arrays can all be obtained from the rectangular array $\begin{pmatrix} a_1 & a_2 & a_3 \\ b_1 & b_2 & b_3 \end{pmatrix}$ whose first row contains the coordinates of $A$ and whose second row contains the coordinates of $B$. The first coordinate of $A \times B$ is the determinant of the array obtained by crossing out the first column of $\begin{pmatrix} a_1 & a_2 & a_3 \\ b_1 & b_2 & b_3 \end{pmatrix}$, the second coordinate of $A \times B$ is the negative of the determinant of the array obtained by crossing out the second column, and the third coordinate of $A \times B$ is the determinant of the array obtained by crossing out the third column.

**Habits:** Some people cover up each column with a hand instead of crossing it out.

### Example 1

**Problem.** Find $A \times B$ if $A = (3, 1, 2)$ and $B = (-1, 4, 3)$.

**Solution.** Start with the array $\begin{pmatrix} 3 & 1 & 2 \\ -1 & 4 & 3 \end{pmatrix}$. So

$$\begin{aligned} A \times B &= \left( \begin{vmatrix} 1 & 2 \\ 4 & 3 \end{vmatrix}, -\begin{vmatrix} 3 & 2 \\ -1 & 3 \end{vmatrix}, \begin{vmatrix} 3 & 1 \\ -1 & 4 \end{vmatrix} \right) \\ &= (-5, -11, 13) \end{aligned}$$

You can check that $A \times B$ is orthogonal to both $A$ and $B$.

The next theorem gives the important geometric properties of the cross product.

**Theorem 2.7**

*If $A$ and $B$ are vectors in $\mathbb{R}^n$,*

(1) *$A \times B$ is orthogonal to both $A$ and $B$, and*

(2) $\|A \times B\|^2 = \|A\|^2 \|B\|^2 - (A \cdot B)^2$ *(Lagrange Identity)*

⟵ Compare this version of the Lagrange Identity to Exercise 11 from Lesson 2.3.

***Proof.*** Part ((1)) is just a restatement of Theorem 2.6. To prove the Lagrange Identity, let $A = (a_1, a_2, a_3)$ and $B = (b_1, b_2, b_3)$. Then

$$\|A \times B\|^2 = (a_2 b_3 - a_3 b_2)^2 + (a_3 b_1 - a_1 b_3)^2 + (a_1 b_2 - a_2 b_1)^2 \text{ while}$$
$$\|A\|^2 \|B\|^2 - (A \cdot B)^2 = (a_1^2 + a_2^2 + a_3^2)(b_1^2 + b_2^2 + b_3^2) - (a_1 b_1 + a_2 b_2 + a_3 b_3)^2$$

You can show these are equal by expanding and simplifying. ■

Next, the algebraic properties of the cross product are:

**Theorem 2.8 (The Basic Rules of Cross Product)**

*If $A$, $B$, and $C$ are vectors in $\mathbb{R}^3$ and $d$ is a scalar, then*

(1) $A \times B = -(B \times A)$

(2) a. $A \times (B \times C) = (A \cdot C)B - (A \cdot B)C$
b. $(A \times B) \times C = (A \cdot C)B - (B \cdot C)A$

⟵ Beware: Part ((2)) of Theorem 2.8 shows that the cross product is not associative; that is, in general, $A \times (B \times C) \neq (A \times B) \times C$.

(3) $A \times (B + C) = (A \times B) + (A \times C)$

(4) $(dA) \times B = d(A \times B) = A \times (dB)$

(5) $A \times O = O$

(6) $A \times A = O$

***Proof.*** The proofs use only the definition of the cross product. The proof of ((1)) is below.

←
You will prove the remaining parts in the exercises.

Suppose $A = (a_1, a_2, a_3)$ and $B = (b_1, b_2, b_3)$. Then

$$A \times B = (a_2b_3 - a_3b_2, a_3b_1 - a_1b_3, a_1b_2 - a_2b_1) \text{ and}$$
$$B \times A = (b_2a_3 - b_3a_2, b_3a_1 - b_1a_3, b_1a_2 - b_2a_1)$$

So, $A \times B = -(B \times A)$. ■

### Example 2

**Problem.** Show that if $B = cA$ for some number $c$, then $A \times B = 0$.

**Solution.** You could do this by using generic coordinates and the definition of cross product, or you could use the properties in the theorem and say something like this:

$$A \times B = A \times cA = c(A \times A) = cO = O$$

### Facts and Notation

Let $E_1 = (1, 0, 0), E_2 = (0, 1, 0), E_3 = (0, 0, 1)$. These vectors are called the **standard basis vectors** in $\mathbb{R}^3$. You should show that

←
$E_1$, $E_2$, and $E_3$ are *unit* vectors (see Lesson 2.3), each one lying along a different coordinate axis in $\mathbb{R}^3$. You'll become good friends with these vectors.

$$E_1 \times E_2 = E_3, \qquad E_2 \times E_3 = E_1, \qquad \text{and} \qquad E_3 \times E_1 = E_2$$

It follows from Theorem 2.8 that

$$E_2 \times E_1 = -E_3, \qquad E_3 \times E_2 = -E_1, \qquad \text{and} \qquad E_1 \times E_3 = -E_2$$

Here's a diagram to help you remember these facts:

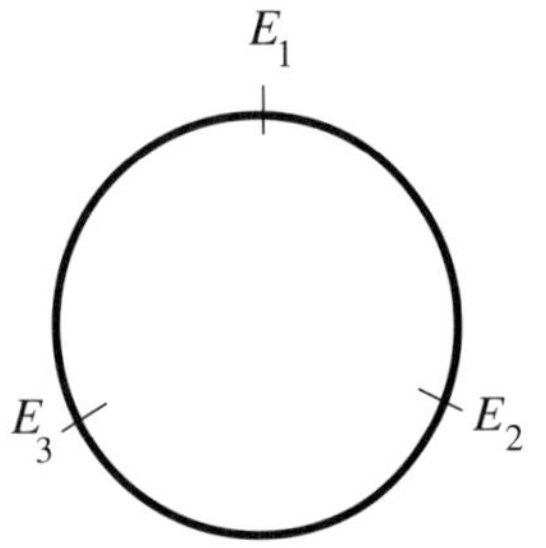


Taking the cross product of any two vectors in the clockwise direction yields the next vector. The cross product of any two vectors in the counterclockwise direction yields the negative of the next vector.

Theorem 2.7 determines the length of $A \times B$, and it also tells you that $A \times B$ lies on the line that is orthogonal to the plane determined by $A$ and $B$. However, it is not hard to see that there are *two* vectors of a given length that lie along this line.

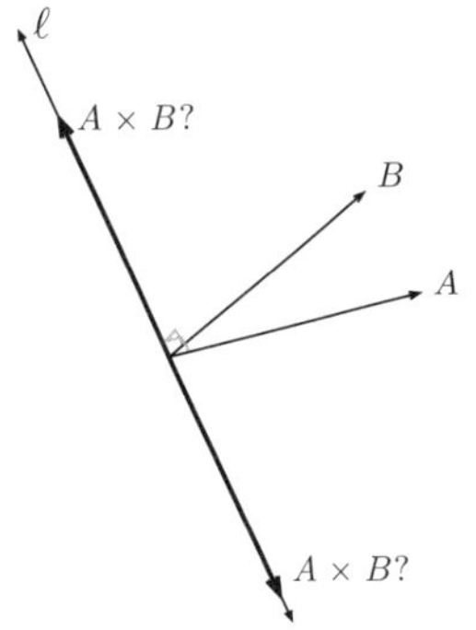


It can be shown that the direction of $A \times B$ can be determined by the following "right-hand rule": to determine the direction of $A \times B$, position your right hand as in the following figure. If you point the index finger of your right hand in the direction of $A$, and point the middle finger of your right hand in the direction of $B$, your thumb will point in the direction of $A \times B$.

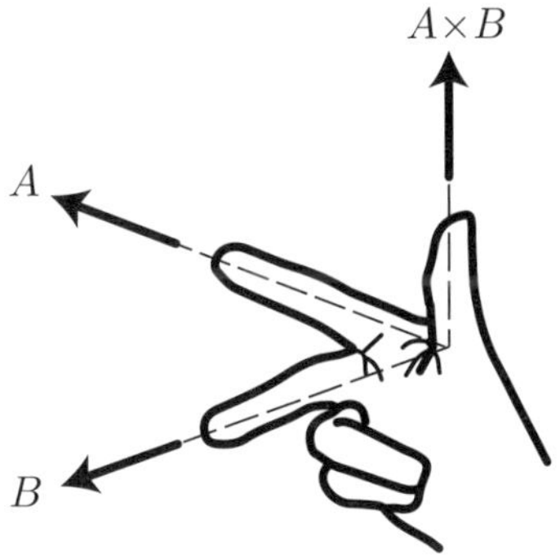


⟵
The right-hand rule works because this coordinate system is oriented to support it. For example, if you cross a vector along the positive $x$-axis (a positive multiple of $E_1$) with a vector along the positive $y$-axis (a positive multiple of $E_2$), you get a vector along the positive $z$-axis (a positive multiple of $E_3$). In this course, you will use a "right-hand" coordinate system. Some applications (computer graphics, for example), often use a left-hand coordinate system.

The formula for $\|A \times B\|$ in Theorem 2.7 can be simplified to give a more geometric formula in the case where $A$ and $B$ are nonzero vectors. Let $\theta$ be the angle between $A$ and $B$. Then $A \cdot B = \|A\|\,\|B\|\cos\theta$. So, from Theorem 2.7,

$$\begin{aligned}\|A \times B\|^2 &= \|A\|^2\,\|B\|^2 - (A \cdot B)^2 \\ &= \|A\|^2\,\|B\|^2 - \|A\|^2\,\|B\|^2\cos^2\theta \\ &= \|A\|^2\,\|B\|^2(1 - \cos^2\theta) \\ &= (\|A\|\,\|B\|\sin\theta)^2\end{aligned}$$

Since $\theta$ is between 0 and $\pi$, $\sin\theta$ is nonnegative, so you can take square roots of both sides to obtain the interesting formula

$$\|A \times B\| = \|A\|\,\|B\|\sin\theta$$

⟵
Why is $\sin\theta$ nonnegative if $\theta$ is between 0 and $\pi$?

**An Application to Area**

Suppose $A$ and $B$ are nonzero vectors that are not multiples of each other. Then the altitude of the parallelogram spanned by $A$ and $B$ is $\|B\|\sin\theta$, so that the area of the parallelogram is $\|A\|\,\|B\|\sin\theta$. But wait—this is $\|A \times B\|$. This statement would make a good theorem . . .

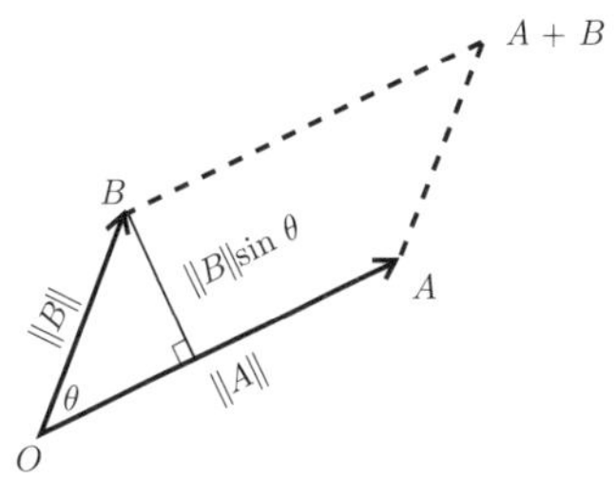


⟵
The parallelogram spanned by $A$ and $B$ has vertices $O$, $A$, $A + B$, and $B$.

**Theorem 2.9**

*The area of the parallelogram spanned by the nonzero vectors $A$ and $B$ is $\|A \times B\|$.*

**Example 3**

**Problem.** Let $A = (1,2,0), B = (-2,6,1)$, and $C = (-10,6,5)$. Find the area of $\triangle ABC$.

**Solution.** The area of the triangle is $\frac{1}{2}$ the area of the parallelogram determined by $\overrightarrow{AB}$ and $\overrightarrow{AC}$. This parallelogram has the same area as the one spanned by $B - A$ and $C - A$; that is, by $(-3,4,1)$ and $(-11,4,5)$. So, the area of the triangle is given by

$$\tfrac{1}{2}\|(-3,4,1) \times (-11,4,5)\| = \tfrac{1}{2}\|(16,4,32)\| = \tfrac{1}{2} \cdot 36 = 18$$

**In-Class Experiment**

Suppose $A = (1,4,-1)$ and $B = (-4,0,2)$. Let $\mathfrak{P}$ be the parallelogram whose vertices are $O$, $A$, $B$, and $A + B$.

← See Exercise 15 from Lesson 2.3.

1. Find the area of $\mathfrak{P}'$, the projection of $\mathfrak{P}$ on the $x$–$y$ plane.
2. Find the area of $\mathfrak{P}''$, the projection of $\mathfrak{P}$ on the $x$–$z$ plane.
3. Find the area of $\mathfrak{P}'''$, the projection of $\mathfrak{P}$ on the $y$–$z$ plane.
4. Relate these to the coordinates of $A \times B$.
5. Generalize to arbitrary pairs of vectors in $\mathbb{R}^3$.

## Exercises

**1.** Let $A = (1,3,1)$, $B = (2,-1,4)$, $C = (5,1,0)$, and $D = (4,2,5)$. Find

| | | |
|---|---|---|
| **a.** $A \times B$ | **b.** $A \times C$ | **c.** $A \times (B + C)$ |
| **d.** $(A \times B) \times C$ | **e.** $(A \times B) \times D$ | **f.** $3A \times 2B$ |
| **g.** $2A \times 2C$ | **h.** $A \cdot (B \times C)$ | **i.** $D \cdot (A \times B)$ |
| **j.** $A \cdot (A \times C)$ | **k.** $A \times (A \times B)$ | **l.** $(A \times A) \times B$ |

**2.** Find a nonzero vector orthogonal to both $(1,0,3)$ and $(2,-1,4)$.

**3.** Verify Theorem 2.8, part ((2)), for $A = (1,2,1)$, $B = (3,1,4)$, $C = (1,-1,0)$.

**4.** Verify parts ((3)) and ((4)) of Theorem 2.8 for $A = (0,1,1)$, $B = (1,3,4)$, $C = (1,2,3)$, $d = 2$.

**5.** Find the area of the triangle whose vertices are

**a.** $A = (0,0,0), B = (1,4,-1), C = (-4,0,2)$
**b.** $A = (1,3,5), B = (2,7,4), C = (-3,3,7)$
**c.** $A = (1,1,2), B = (-3,-2,5), C = (5,7,-2)$
**d.** $A = (2,1,3), B = (1,4,1), C = (3,5,9)$
**e.** $A = (1,0,0), B = (0,1,0), C = (0,0,1)$
**f.** $A = (3,4,0), B = (1,1,-1), C = (1,3,2)$
**g.** $A = (0,0,0), B = (3,1,0), C = (4,2,0)$
**h.** $A = (5,0,1), B = (9,4,-5), C = (2,-2,5)$
**i.** $A = (3,0), B = (4,3), C = (5,1)$

**6.** Use Theorem 2.8 to prove $(A+B) \times C = (A \times C) + (B \times C)$.

**7.** Find all vectors $X$ so that $(3,1,2) \times X = (-1,-1,2)$.

**8.** If $A$, $B$, and $C$ are vectors in $\mathbb{R}^3$, use Theorem 2.8, part (2), to prove: $((A \times B) \times C) + ((B \times C) \times A) = (A \times C) \times B$.

**9.** If $A$, $B$, and $C$ are vectors in $\mathbb{R}^3$, show that

$$((A \times B) \times C) - (A \times (B \times C)) = (A \times C) \times B$$

**10.** If $A$, $B$, and $C$ are nonzero vectors in $\mathbb{R}^3$ so that

$$A \times (B \times C) = (A \times B) \times C$$

show that either $A$ and $C$ are both orthogonal to $B$, or that $A$ is a multiple of $C$.

**11.** If $A$ and $B$ are vectors in $\mathbb{R}^3$, show that

$$(A \times B) \times A = A \times (B \times A)$$

**12.** In $\mathbb{R}^3$, show that $(E_1 \times E_2) \times E_3 = E_1 \times (E_2 \times E_3)$.

**13.** If $A$ and $B$ are vectors in $\mathbb{R}^3$ and $C = x_1 A + x_2 B$ for some scalars $x_1$ and $x_2$, show that $C \cdot (A \times B) = 0$. Interpret geometrically.

**14.** Use the Lagrange Identity to prove the Cauchy-Schwarz Inequality in $\mathbb{R}^3$.

**15.** Suppose that $B = (2, 2, 1)$ and $C = (1, -2, 2)$. Here is a picture of a cube that has $O$, $B$, and $C$ as vertices:

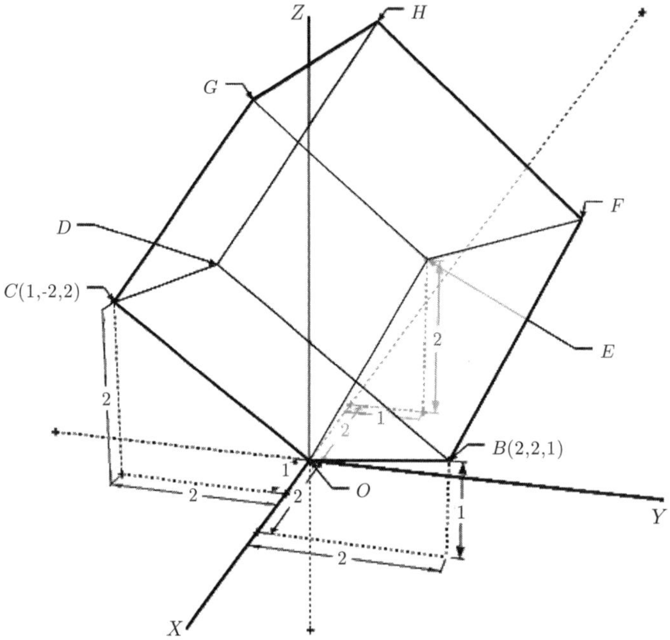


Find the coordinates of the remaining vertices.

**16.** **a.** Let $A = (1, 0, 0, 0)$, $B = (0, 1, 0, 0)$, and $C = (0, 0, 1, 0)$. Find a vector (in $\mathbb{R}^4$) perpendicular to $A$, $B$, and $C$.
**b.** Let $D = (2, -1, 2, -3)$, $E = (-1, -2, 1, 1)$, and $F = (0, -1, 1, 2)$. Find a vector perpendicular to $D$, $E$, and $F$.

**17.** **Take It Further.** Find a formula for a vector perpendicular to three vectors $A = (a_1, a_2, a_3, a_4)$, $B = (b_1, b_2, b_3, b_4)$, and $C = (c_1, c_2, c_3, c_4)$.

**18.** If $A$, $B$, and $C$ are vectors in $\mathbb{R}^3$, no two of which are collinear, use the accompanying diagram to show that the volume of the parallelepiped determined by $A$, $B$, and $C$ is $|C \cdot (A \times B)|$.

$\longleftarrow$
The operation that assigns three vectors $R$, $S$, $T$ to $R \cdot (S \times T)$ is sometimes called the **scalar triple product** of $R$, $S$, and $T$. Does it have any basic rules?

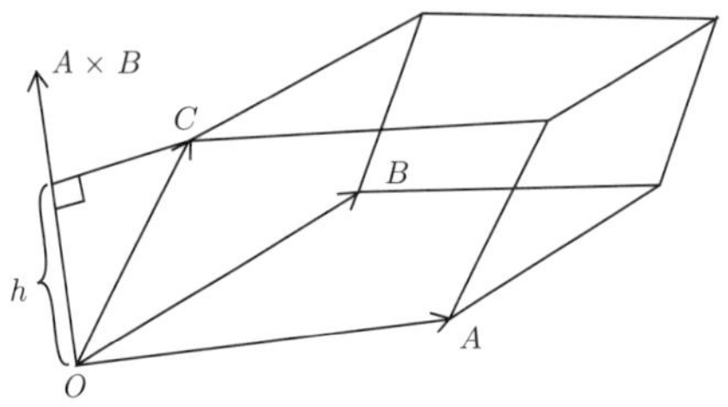


**19.** Find the volume of the parallelepiped determined by $(3, 1, 0)$, $(4, 2, 1)$, and $(1, 2, -1)$.

**20.** Prove parts ((3)) and ((4)) of Theorem 2.8.

**21.** Prove parts ((5)) and ((6)) of Theorem 2.8.

**22.** Prove part ((2)) of Theorem 2.8.

## 2.6 Lines and Planes

---

A basic problem in analytic geometry is constructing equations that define simple geometric objects. For instance, you already know how to write an equation for a line in $\mathbb{R}^2$. But the typical methods, which involve calculating slope, don't work in $\mathbb{R}^3$ and beyond. You need to find another way to characterize a line (and other geometric objects) that you can generalize to higher dimensions.

**In this lesson, you will learn how to**

- find a vector equation of a line given a point and direction vector
- find the vector equation of a plane given a point and two nonparallel direction vectors
- find an equation of a plane given a point and a direction normal to the plane
- recognize the difference between the vector and coordinate equations of a hyperplane
- find the distance from a point to a line and from a point to a plane

---

### Lines

Begin with lines in $\mathbb{R}^2$. In analytic geometry, you learned that a straight line in $\mathbb{R}^2$ has an equation of the form

$$ax + by = c$$

where $a$, $b$, and $c$ are real numbers (and at least one of the coefficients $a$ or $b$ is not zero). This means that the line consists precisely of those points $(r, s)$ so that

$$ar + bs = c$$

← The equation is the *point-tester* for the line.

For example, if $\ell$ is the line containing $(3, 0)$ and $(1, -1)$, the equation for $\ell$ is $x - 2y = 3$; $(4, 1)$ is not on $\ell$ because $4 - 2(1) \neq 3$, and $(5, 1)$ is on $\ell$ because $5 - 2(1) = 3$.

← Make sure you can find the equation for $\ell$.

To generalize the notion of straight line to $\mathbb{R}^n$, you need a vector characterization of lines in $\mathbb{R}^2$.

### Example 1

Consider, for example, the line $\ell$ whose equation is $3x + 5y = 14$. Suppose you take two points on $\ell$, say $P = (3, 1)$ and $Q = (-2, 4)$. If $A = P - Q = (5, -3)$, then another way to characterize $\ell$ is as follows:

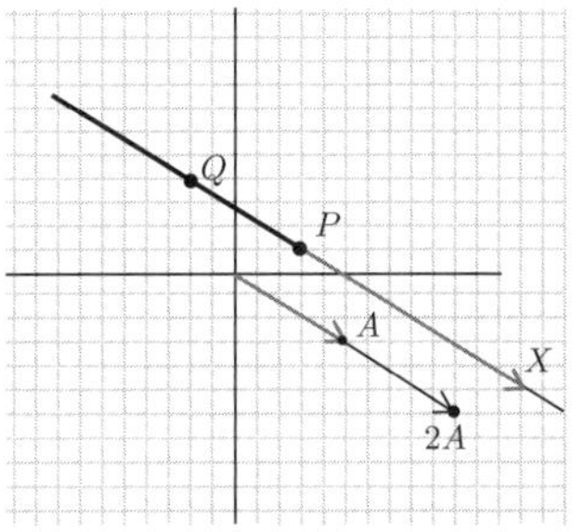


$\ell$ consists of all points $X$ so that $\overrightarrow{PX}$ is parallel to $A$. For example, if $X = (13, -5)$, then $X - P = (10, -6) = 2A$ so that $\overrightarrow{PX}$ is parallel to $A$, and hence $X$ is on $\ell$.

⟵ See Exercise 3 from Lesson 1.2.

In general, then, you can characterize a line in terms of a point $P$ on the line and a vector $A$ that sets its direction:

> The line through $P$ in the direction of $A$ is the set of all points $X$ so that $\overrightarrow{PX}$ is parallel to $A$.

But if $\overrightarrow{PX}$ is parallel to $A$, then $X - P$ is a scalar multiple of $A$. That is, $X - P = tA$ for some number $t$, or equivalently, $X = P + tA$ for some number $t$. So, a point $X$ is on $\ell$ if and only if $X$ can be written as "$P$ plus a multiple of $A$."

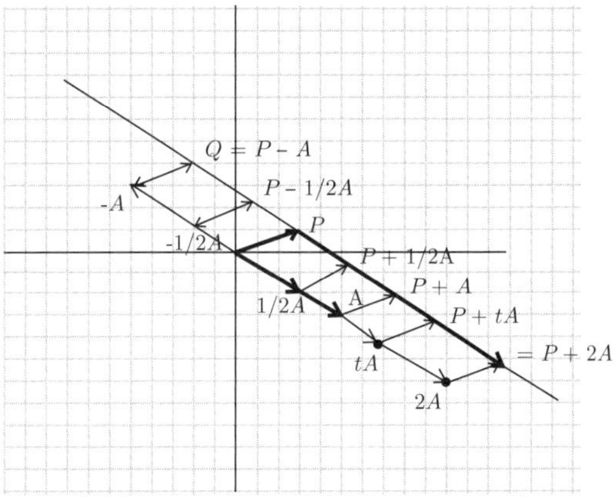


⟵ Some students call this the "ladder image" for $\ell$.

## Developing Habits of Mind

**Different forms for different purposes.** There are two kinds of equations you can use to describe the line $\ell$ from Example 1: $3x + 5y = 14$ and $X = (3, 1) + t(5, -3)$.

- The equation $3x + 5y = 14$ is a point-*tester*, since you can use it to test any point to see if it is on the graph: $(2, 1)$ is not on the graph, because $3(2) + 5(1) = 11$, not 14. But $(3, 1)$ is, since $3(3) + 5(1) = 14$.
- The equation $X = (3, 1) + t(5, -3)$ is a point-*generator*: any real value of $t$ generates a point on the line.

⟵ Equations of this form will be referred to as **coordinate equations** in this book.

⟵ Equations of this form will be referred to as **vector equations** in this book.

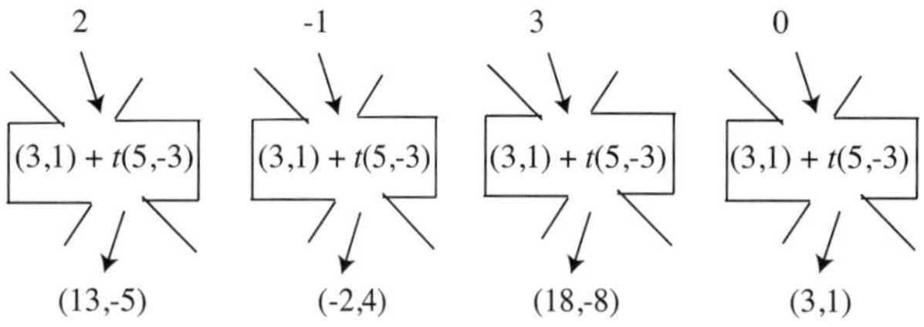


## Example 2

**Problem.** Find a vector equation for the line $m$ between $R = (5, 3)$ and $S = (7, 8)$.

**Solution.** The direction of $m$ is set by, for example,

$$A = S - R = (2, 5)$$

⟵ Any multiple of $S - R$, including $R - S$, can be used as a direction vector (why?).

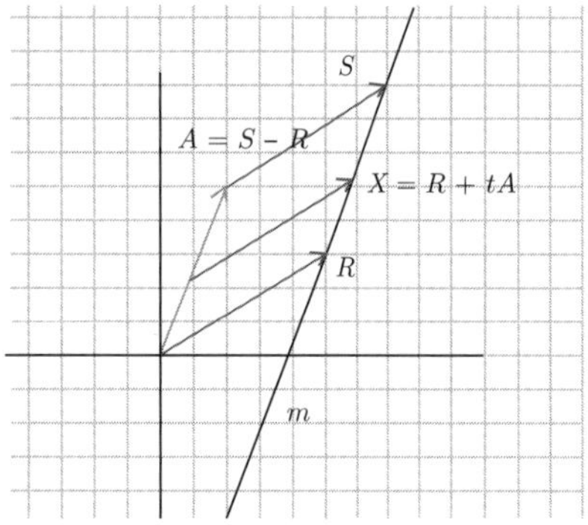


A point $X$ is on $m$ if and only if $X$ can be written as $R$ plus a multiple of $A$. In other words, the equation for $m$ is

$$X = R + tA \quad \text{or}$$
$$X = (5, 3) + t(2, 5)$$

This equation can be used to find any number of points on the line.

By picking a number for $t$, you can find points on $m$. For example, if $t = -3$,

$$(5, 3) + -3(2, 5) = (-1, -12)$$

and $(-1, -12)$ is on $m$. If $t = 1$,

$$(5, 3) + 1(2, 5) = (7, 8)$$

so $(7, 8)$ is also on $m$.

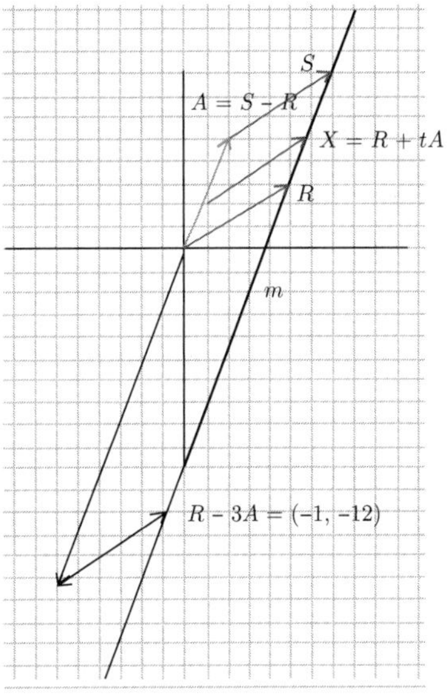


Can you use the vector equation to test points for being on $m$? Well, try it!

- Is $\left(6, \frac{11}{2}\right)$ on $m$? In other words, is there a number $t$ so that

  $$\left(6, \tfrac{11}{2}\right) \stackrel{?}{=} (5, 3) + t(2, 5)$$

  To see if such a $t$ exists, expand the right-hand side:

  $$\left(6, \tfrac{11}{2}\right) \stackrel{?}{=} (5 + 2t, 3 + 5t)$$

  This leads to two equations:

  $$\begin{aligned} 6 &= 5 + 2t \quad \text{and} \\ \tfrac{11}{2} &= 3 + 5t \end{aligned}$$

  The first equation implies that $t = \frac{1}{2}$. And $\frac{1}{2}$ works in the second equation, too. So,

  $$\left(6, \frac{11}{2}\right) = (5, 3) + \tfrac{1}{2}(2, 5)$$

  and $\left(6, \frac{11}{2}\right)$ is on $m$.

- But $(6, 6)$ is not on $m$, because if you set up the equations as above, you end up with

  $$\begin{aligned} 6 &= 5 + 2t \quad \text{and} \\ 6 &= 3 + 5t \end{aligned}$$

  The only value of $t$ that satisfies the first equation is $t = \frac{1}{2}$, and this doesn't work in the second equation.

### For You to Do

1. Find the coordinate equation (something like $ax + by = c$) for the line $m$ in Example 2. Which equation is easier to test points: the coordinate equation or the vector equation? Why?

### Developing Habits of Mind

**Different forms for different purposes.** Yes, you can use the vector equation of $m$ to test a point to see if it is on $m$, but it takes some work: you have to solve two equations (for $t$) and see whether they are consistent. The coordinate equation, on the other hand, is a true point-tester—you substitute the coordinates of the point you're testing for the variables. If the equation is true, the point passes the test: it's on the line!

This vector characterization of lines—a point plus all multiples of a direction vector—makes perfect sense in $\mathbb{R}^3$. Invoking the extension program, you can *define* a line in $\mathbb{R}^n$ by this vector equation.

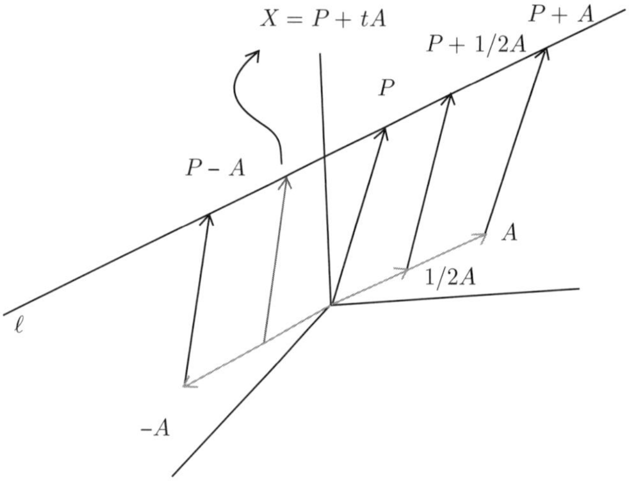


### Definition

Let $P$ and $A$ be elements of $\mathbb{R}^n$, $A \neq O$. The **line** $\ell$ **through** $P$ **in the direction of** $A$ is the set of all points $X$ so that $X = P + tA$ for some number $t$.

The equation $X = P + tA$ is called the **vector equation of** $\ell$. $P$ is an initial point of $\ell$ and $A$ is a **direction vector** for $\ell$.

← Some people call $\ell$ the **line through** $P$ **along** $A$. This book will use both ways to describe lines.

← Sometimes the vector equation is called a **parametric equation** and the variable $t$ is called the **parameter**.

### Example 3

**Problem.** Find the vector equation of the line $\ell$ in $\mathbb{R}^2$ whose coordinate equation is $3x + 2y = 6$.

**Solution.** There are many vector equations for $\ell$. First, take two points on $\ell$, say $P = (2, 0)$ and $Q = (0, 3)$.

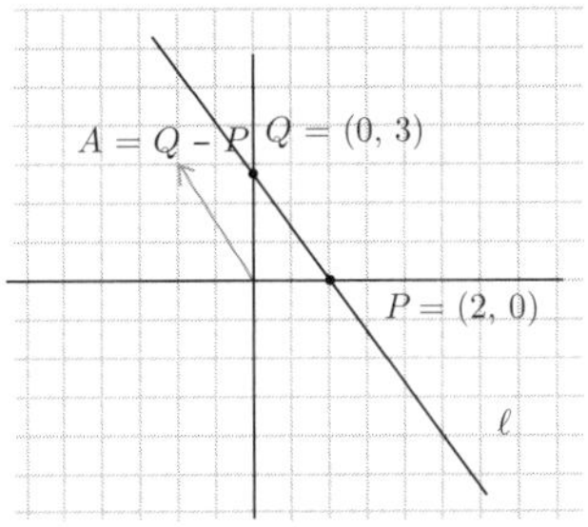


As a direction vector, take $A = Q - P = (-2, 3)$ and take $P$ as an initial point. So, one equation for $\ell$ is $X = (2, 0) + t(-2, 3)$. Other possible equations include $X = (2, 0) + t(2, -3)$ and $X = (4, -3) + t(-4, 6)$. Can you find another one?

### Example 4

**Problem.** Find an equation for the line $\ell$ in $\mathbb{R}^3$ that contains the points $P = (3, -1, 4)$ and $Q = (1, 3, -1)$.

**Solution.** Take $P$ as an initial point and $A = Q - P = (-2, 4, -5)$ as a direction vector. The equation is then $X = (3, -1, 4) + t(-2, 4, -5)$. Sometimes the equation is written in terms of coordinates

$$\begin{aligned} x &= 3 - 2t \\ y &= -1 + 4t \\ z &= 4 - 5t \end{aligned}$$

but, unlike the analogous situation in $\mathbb{R}^2$, there is no simple single equation relating $x$, $y$, and $z$.

Many people think that since a line has equation $ax + by = c$ in $\mathbb{R}^2$, a line should have equation $ax + by + cz = d$ in $\mathbb{R}^3$. That's not the case. In fact, the graph of $ax + by + cz = d$ in three dimensions is a *plane*, not a line, as you'll see shortly.

That's one real advantage for using vector equations: the vector equation of a line looks like $X = P + tA$ in *any* dimension.

### Example 5

**Problem.** In $\mathbb{R}^4$, show that $A = (3, -1, 1, 2)$, $B = (4, 0, 1, 6)$, and $C = (1, -3, 1, -6)$ are collinear.

**Solution.** Collinearity in $\mathbb{R}^4$ means, by definition, that there is some vector equation of a line that is satisfied by all three points. Now, an equation for the line containing $A$ and $B$ can be obtained by taking $A$ as an initial point and $B - A = (1, 1, 0, 4)$ as a direction vector; this gives

$$X = (3, -1, 1, 2) + t(1, 1, 0, 4)$$

To see that $C$ is on this line, you need to find a scalar $t$ so that $C = A + t(1, 1, 0, 4)$. But $C - A = (-2, -2, 0, -8) = -2(1, 1, 0, 4)$, so $t = -2$.

Note that all of the other geometric criteria for collinearity in $\mathbb{R}^2$ and $\mathbb{R}^3$ are satisfied by $A$, $B$, and $C$. For example, you should check that $B - A$, $C - A$, and $C - B$ are all scalar multiples of each other, that $d(C, A) + d(A, B) = d(C, B)$, and that $\cos\theta = 1$ or $-1$, where $\theta$ is the angle between $\overrightarrow{BA}$ and $\overrightarrow{BC}$.

← These are good things to do.

### Planes

Back in Lesson 1.3, Sasha and Tony came up with this vector equation of a plane:

$$X = k_1 A + k_2 B$$

At the time, Sasha said she thought something was missing. Seeing the vector equation of a line, she has a new thought.

**Minds in Action** Episode 8

SASHA: I know what I was missing before. Look: any plane with an equation like $X = k_1 A + k_2 B$ has to go through the origin.

TONY: I can see that. If both $k_1$ and $k_2$ are set to 0, then $X$ is going to be $O$ no matter what $A$ and $B$ are.

SASHA: Exactly. But not all planes have to go through the origin.

TONY: You're right. Well, what did we do for lines?

*Tony looks at his notes.*

We first came up with the vector equation of a line by looking at a line through the origin, then adding a fixed vector to all the points on that line—remember the ladder image? Will that work for planes?

SASHA: Well, yeah, let's just look at the way we wrote the vector equation of a line. It's a vector plus all multiples of another vector. So what if we just add another vector like this:

$$X = P + k_1 A + k_2 B$$

There. The linear combination of two vectors added to another vector.

TONY: So $P$ is the initial point—it just represents some point we know is on the plane. And $A$ and $B$ are direction vectors.

You can think of the "ladder image" for a plane as being ladder images for infinitely many lines. The figure below shows three, but there are many other lines not shown that will fill out the entire plane.

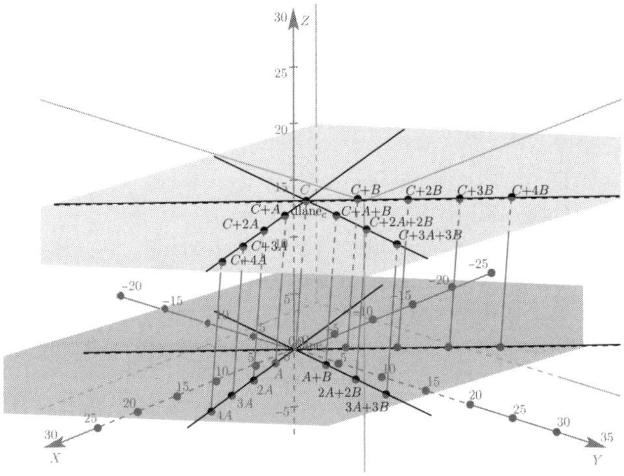

### For You to Do

**2.** **a.** Describe geometrically the graph of $X = k_1(1,0,0) + k_2(0,1,0)$.

**b.** Describe geometrically the graph of

$$X = (0,0,1) + k_1(1,0,0) + k_2(0,1,0).$$

### Minds in Action Episode 9

*Tony and Sasha are thinking about how to describe the following vector equation geometrically:*

$$X = (0,0,1,0) + k_1(1,0,0,0) + k_2(0,1,0,0)$$

TONY: (*thinking aloud*) Those vectors are in $\mathbb{R}^4$. Can we *do* geometry in $\mathbb{R}^4$?

SASHA: Well, that equation looks like the vector equation of a plane.

TONY: Yeah, but before we were looking at vectors in $\mathbb{R}^3$. This is $\mathbb{R}^4$. Can we still say it's a plane?

SASHA: Sure! It's just the extension program all over again. See, we said that equations that look like $X = A + tB$ were lines, no matter what $\mathbb{R}^n$ we were in. So if we said this was a plane in $\mathbb{R}^3$, we can just define a plane to be a fixed point plus all linear combinations of two vectors, then yeah, it's a plane, even if we don't know what a plane *looks* like in four dimensions.

⟵ This definition works as long as the two vectors don't fall on the same line.

### Definition

Let $P$, $A$, and $B$ be elements of $\mathbb{R}^n$, where $A$ and $B$ are not parallel. The **plane** $E$ **through** $P$ **spanned by** $A$ **and** $B$ is the set of all points $X$ so that $X = P + t_1A + t_2B$ for some numbers $t_1$ and $t_2$.

⟵ If $A$ and $B$ are not parallel, neither can equal $O$. Why?

The equation $X = P + sA + tB$ is called the **vector equation of a plane** $E$. $P$ is an **initial point** of $E$, and $A$ and $B$ are **direction vectors** for $E$.

**Remember**

A vector equation is sometimes called a **parametric equation** and the variables $s$ and $t$ are called the **parameters**.

### Hyperplanes

There is another way to find the equation of a plane in $\mathbb{R}^3$. Let $E$ be an arbitrary plane in $\mathbb{R}^3$, and let $P$ be a point on $E$.

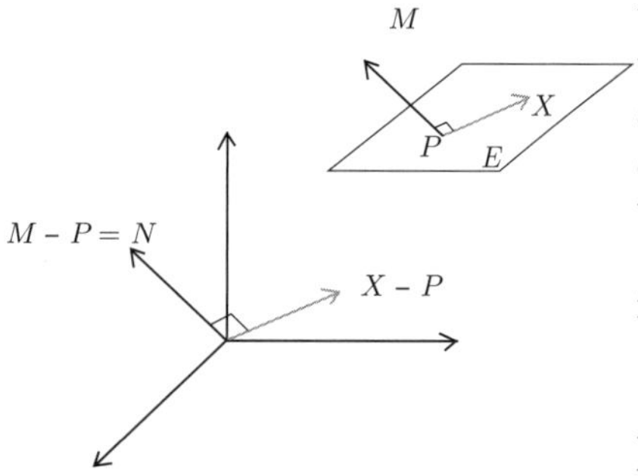


From $P$, draw a vector $\overrightarrow{PM}$ that is orthogonal to $E$. Next, let $N = M - P$, the vector equivalent to $\overrightarrow{PM}$ starting at $O$. $N$ is called a **normal** to $E$ (and all other normals to $E$ are scalar multiples of $N$). $E$ can be described as "the set of all points $X$ so that $X - P$ is orthogonal to $N$."

That is, $E$ is the set of all points $X$ so that $(X - P) \cdot N = 0$. But this equation is equivalent to $X \cdot N = P \cdot N$.

**Remember**

In geometry, it is enough to make $\overrightarrow{PM}$ orthogonal to $\overrightarrow{PX}$ and $\overrightarrow{PY}$ where $X$ and $Y$ are two points in $E$ distinct from $P$ and each other.

←
Alternatively, $E$ is the set of all points $X$ so that $\overrightarrow{PX}$ is orthogonal to $\overrightarrow{PM}$.

#### Example 6

Let $E$ be the plane containing $(3, 1, 0)$ orthogonal to $(1, 1, -1)$. Then an equation of $E$ is $X \cdot (1, 1, -1) = (3, 1, 0) \cdot (1, 1, -1)$. That is, $E$ consists of all points $X$ so that $X \cdot (1, 1, -1) = 4$. So, $(2, 2, 0)$ is on $E$, while $(2, 1, 1)$ is not. So the coordinate equation for $E$ is all points $(x, y, z)$ such that $x + y - z = 4$.

←
Just as a line is determined by a point on it and a direction vector, a plane in $\mathbb{R}^3$ can be determined by a point on *it* and a *normal* vector.

#### For You to Do

**3.** **a.** Show that the graph of $x + 2y - z = 8$ is a plane in $\mathbb{R}^3$.
**b.** Find three points on that plane.
**c.** Find a normal vector to the plane.
**d.** Find an equation for the plane in the form

$$X \cdot N = P \cdot N$$

←
How does part 3**d** help you with part 3**a**?

So, a plane in $\mathbb{R}^3$ also has an equation of the form

$$X \cdot N = P \cdot N$$

You can use the extension program and this equation to define objects in other dimensions. Those objects have special properties, and so they are given a special name.

#### Definition

Suppose $P$ and $N$ are in $\mathbb{R}^n$, $N \neq O$. The **hyperplane** containing $P$ orthogonal to $N$ is the set of all points $X$ so that $X \cdot N = P \cdot N$.

←
Think of $P$ as a point and $N$ as a vector.

$N$ is called a **normal** to $E$, $P$ is an **initial point** for $E$, and the equation $X \cdot N = P \cdot N$ is an **equation** for $E$.

This definition for hyperplane works in $\mathbb{R}^n$ for *any* $n$, but the geometric object it describes changes as $n$ changes.

- What is a hyperplane in $\mathbb{R}^2$? Suppose $E$ is a hyperplane in $\mathbb{R}^2$ containing $P = (p_1, p_2)$ orthogonal to $N = (a, b)$. Then an equation of $E$ is $(x, y) \cdot (a, b) = (p_1, p_2) \cdot (a, b)$. Letting $c$ be the number $p_1 a + p_2 b$, this equation becomes $ax + by = c$. So, hyperplanes in $\mathbb{R}^2$ are simply lines.

$\longleftarrow$
You can amaze your friends next time you talk about the line with equation $5x + 2y = 3$. Say "This is a hyperplane in $\mathbb{R}^2$ that is orthogonal to $(5, 2)$."

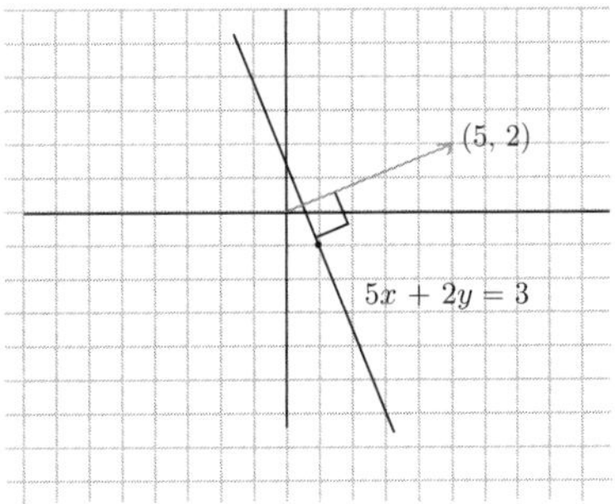


This gives a nice interpretation to equations for lines from analytic geometry. The line with equation $ax + by = c$ is orthogonal to the vector $(a, b)$.

- In $\mathbb{R}^3$, a hyperplane is simply a plane. Suppose $E$ is a plane in $\mathbb{R}^3$ containing $P$ orthogonal to $N = (a, b, c)$. Then if $Q$ is any point on $E$, $Q \cdot N = P \cdot N$. If you let $d = P \cdot N$, an equation for $E$ can be written as $X \cdot N = d$. Let $X = (x, y, z)$; the equation becomes $ax + by + cz = d$, an equation in three variables similar to the ordinary equation in two variables for a line in $\mathbb{R}^2$.

The equation in the definition of hyperplanes can be expanded to produce coordinate equations.

$$\begin{aligned} P \text{ and } N\text{:} \quad & (5, 1), (7, -1) \\ \text{Equation:} \quad & X \cdot (7, -1) = (5, 1) \cdot (7, -1) \\ \text{or} \quad & 7x - y = 34 \end{aligned}$$

$$\begin{aligned} P \text{ and } N\text{:} \quad & (5, 1, 3), (7, -1, 2) \\ \text{Equation:} \quad & X \cdot (7, -1, 2) = (5, 1, 3) \cdot (7, -1, 2) \\ \text{or} \quad & 7x - y + 2z = 40 \end{aligned}$$

$$\begin{aligned} P \text{ and } N\text{:} \quad & (5, 1, 3, 2), (7, -1, 2, 1) \\ \text{Equation:} \quad & X \cdot (7, -1, 2, 1) = (5, 1, 3, 2) \cdot (7, -1, 2, 1) \\ \text{or} \quad & 7x - y + 2z + w = 42 \end{aligned}$$

Coordinate equations are often called **linear equations**, not because they are equations of lines, but because each side of the equation is a *linear combination* of variables. You will learn more about linear combinations in Chapter 3. The following theorem formalizes the relationship between linear equations and their graphs.

← In other words, each term is a constant or the product of a constant and a variable raised to the first power. Some examples of nonlinear equations are $xy - y = 3$, $x_1^2 + 3x_2 = 5$, $\sin x + y - \cos z = 0$, and $3^x + 2y = 3$.

### Theorem 2.10

*Every hyperplane in $\mathbb{R}^n$ is the graph of a linear equation. Conversely, the graph of $a_1x_1 + a_2x_2 + \cdots + a_nx_n = d$ is a hyperplane with normal vector $(a_1, a_2, \ldots, a_n)$.*

***Proof.*** If $E$ is a hyperplane containing $P$ orthogonal to $N = (a_1, a_2, \ldots, a_n)$, $E$ is the graph of

$$a_1x_1 + a_2x_2 + \cdots + a_nx_n = d$$

where $d = P \cdot N$.

Conversely, let $a_1x_1 + a_2x_2 + \cdots + a_nx_n = d$ be a linear equation and suppose $a_1 \neq 0$. If $P = \left(\frac{d}{a_1}, 0, 0, \ldots, 0\right)$, this equation can be written as $X \cdot N = P \cdot N$, where $X = (x_1, x_2, \ldots, x_n)$ and $N = (a_1, a_2, \ldots, a_n)$. So the graph of the equation is a hyperplane containing $P$ orthogonal to $N$. ■

### Example 7

**a.** Let $E$ be the hyperplane in $\mathbb{R}^4$ containing $(3, -1, 1, 4)$ orthogonal to $(1, -1, 3, 1)$. The coordinate equation defining $E$ is $x - y + 3z + w = 11$. This comes from expanding the equation

$$X \cdot (1, -1, 3, 1) = (3, -1, 1, 4) \cdot (1, -1, 3, 1)$$

**b.** Let $E$ be the plane in $\mathbb{R}^3$ whose linear equation is $2x + y - z = 5$. Then $(2, 1, -1)$ is a vector normal to $E$; $E$ contains, for example, $(0, 0, -5)$.

### Developing Habits of Mind

**The extension program.** You used the program three times in this lesson.

- A line in $\mathbb{R}^n$ is described by a point on it, $P$, and a direction vector, $A$; it has equation
$$X = P + tA$$
- A plane in $\mathbb{R}^n$ is described by a point on it, $P$, and two direction vectors, $A$ and $B$; it has equation
$$X = P + t_1A + t_2B$$
- A hyperplane in $\mathbb{R}^n$ is described by a point on it, $P$, and a normal vector $N$; it has equation
$$X \cdot N = P \cdot N$$

Things to notice:

1. The equations can be used in $\mathbb{R}^n$ for any $n$, and the graph of a linear equation in $\mathbb{R}^n$ is always a hyperplane in $\mathbb{R}^n$.
2. So, a hyperplane in $\mathbb{R}^2$ is a line, since $n = 2$, and thus $n - 1 = 1$. Hence, in $\mathbb{R}^2$, lines can be described by either a vector or a linear equation.
3. And, as you saw already, a hyperplane in $\mathbb{R}^3$ is a plane, so the graph of a "linear" equation in $\mathbb{R}^3$ is a plane.
4. In $\mathbb{R}^n$, you've now seen that there are
   - points (0-dimensional)
   - lines (1-dimensional)
   - planes (2-dimensional)
   - hyperplanes ($(n-1)$-dimensional)
   - all of $\mathbb{R}^n$ ($n$-dimensional)

   In fact, there are similar kinds of objects of every dimension between 1 and $n - 1$.

← For instance, you can describe a 5-dimensional flat object in $\mathbb{R}^8$ if you want. Try it.

### Example 8

**Problem.** Find an equation for the plane $E$ containing $P_1 = (1, 0, 3)$, $P_2 = (-1, 1, 2)$, and $P_3 = (2, -1, 0)$.

**Solution.** Take $P_1$ as the initial point for $E$. To find a normal to $E$, you need a vector orthogonal to $P_2 - P_1 = (-2, 1, -1)$ and $P_3 - P_1 = (1, -1, -3)$. That's exactly what the cross product is for.

$$(-2, 1, -1) \times (1, -1, -3) = (-4, -7, 1)$$

So you can take $(-4, -7, 1)$ as the normal. Since $P_1 \cdot (-4, -7, 1) = -1$, the desired equation is $X \cdot (-4, -7, 1) = -1$, which you can rewrite as

$$-4x - 7y + z = -1$$

Any multiple of this coordinate equation will also define $E$.

### For You to Do

**4.** Let $F$ be a plane containing points $P_1 = (0, 2, 1)$, $P_2 = (3, -1, 4)$, and $P_3 = (5, 0, 1)$.

**a.** Find a coordinate equation for the plane $F$.

**b.** Find a vector equation for the plane $F$.

### Example 9

**Problem.** In $\mathbb{R}^3$, let $\ell$ be the line through $(-5, -2, -4)$ in the direction of $(2, 1, 3)$, and let $E$ be the plane containing $(2, 0, 0)$ orthogonal to $(2, -3, 1)$. Find the intersection of $\ell$ and $E$.

**Solution.** The equation for $\ell$ is $X = (-5, -2, -4) + t(2, 1, 3)$, and the equation for $E$ is $X \cdot (2, -3, 1) = 4$. If $X$ satisfies both equations, there is some value for $t$ so that $((-5, -2, -4) + t(2, 1, 3)) \cdot (2, -3, 1) = 4$. Using the rules for dot product, this equation becomes

$$(-5, -2, -4) \cdot (2, -3, 1) + t((2, 1, 3) \cdot (2, -3, 1)) = 4$$

or $-8 + 4t = 4$. So, $t = 3$ and $X = (-5, -2, -4) + 3(2, 1, 3) = (1, 1, 5)$ is the point where $\ell$ meets $E$.

← Check to see that $(1, 1, 5)$ is on both $\ell$ and $E$.

### Example 10

**Problem.** Let $E$ be the plane defined by the equation $4x - y + 8z = 10$. If $Q = (9, 0, 17)$, find the distance from $Q$ to $E$.

**Solution.** The strategy is illustrated below.

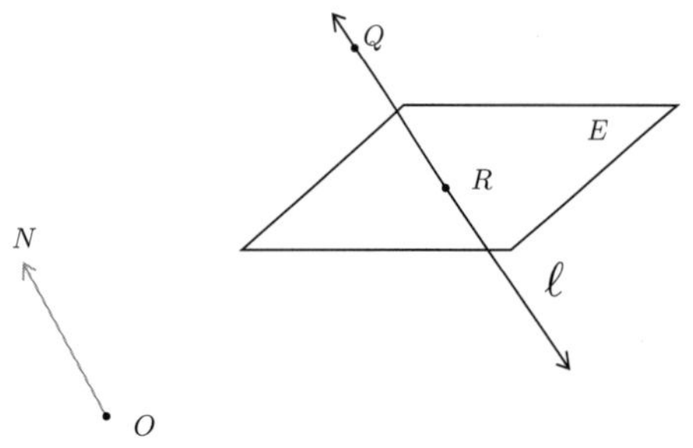


The distance from $Q$ to $E$ is $d(Q, R)$, where $R$ is the point of intersection of line $\ell$—that goes through $Q$ orthogonal to $E$—with $E$.

Now, a direction vector for $\ell$ is precisely a normal vector to $E$, that is, $(4, -1, 8)$. So an equation for $\ell$ is $X = (9, 0, 17) + t(4, -1, 8)$. Proceeding as in Example 9, find that $R = (1, 2, 1)$. The desired distance is then $d(Q, R) = 18$.

### Example 11

**Problem.** Find the equation of the intersection of the two planes whose equations are $x + y - 3z = 1$ and $2x + y + 3z = 4$.

**Solution.** The intersection of two planes will be a line.

← . . . unless the planes are parallel. How do you know that the planes are not parallel?

The intersection of the planes consists of all points that satisfy both equations at once. Solve the first equation for $x$.

$$x = 1 - y + 3z$$

Substitute this in the second equation and solve for $y$.

$$y = -2 + 9z$$

Since $x = 1 - y + 3z$,

$$x = 1 - (-2 + 9z) + 3z = 3 - 6z$$

There are no more constraints (equations) so $z$ can be anything. So, the two planes intersect on the set of all points $(x, y, z)$ so that

$$\begin{aligned} x &= 3 - 6z \\ y &= -2 + 9z \\ z &\text{ is arbitrary} \end{aligned}$$

Let $z = t$ ($t$ stands for any real number). The intersection is the set of all points of the form $(3 - 6t, -2 + 9t, t)$. But

$$\begin{aligned} (3 - 6t, -2 + 9t, t) &= (3, -2, 0) + (-6t, 9t, t) \\ &= (3, -2, 0) + t(-6, 9, 1) \end{aligned}$$

So, the intersection is the set of all points $X$ so that $X = (3, -2, 0) + t(-6, 9, 1)$. This is the equation of the intersection of the two planes.

$\longleftarrow$ This rewriting $(3 - 6t, -2 + 9t, t)$ as $(3, -2, 0) + t(-6, 9, 1)$ (a constant vector plus a "$t$" part) will be a valuable technique throughout this book.

## Exercises

**1.** For each given set of conditions, find a vector equation of the line $\ell$.

**a.** $\ell$ contains $(3, 0)$ and is parallel to $(4, 1)$
**b.** $\ell$ contains $(4, 1)$ and $(5, -2)$
**c.** $\ell$ contains $(3, 1, 2)$ and $(2, 0, -1)$
**d.** $\ell$ contains $(1, 0, 1, 0)$ and $(3, -1, 4, 2)$
**e.** $\ell$ contains $(2, -1, 4)$ and is parallel to the $z$-axis
**f.** $\ell$ contains $(3, 4)$ and is orthogonal to $(5, -3)$

**2.** For each given set of conditions, find an equation in the form $X \cdot N = P \cdot N$ of the hyperplane $E$.

**a.** $E$ contains $(3, 1, 2)$ and is normal to $(1, 2, 0)$
**b.** $E$ contains $(3, 1, 2, -1)$ and is normal to $(4, 1, -1, 2)$
**c.** $E$ contains $(3, 1, 2)$, $(2, -1, 4)$, and $(1, 0, 1)$
**d.** $E$ contains $(1, 3, 1)$, $(4, 1, 0)$, and $(1, 3, 2)$
**e.** $E$ contains $(1, 0, 1)$, $(2, 1, 3)$, and $(0, 0, 0)$
**f.** $E$ contains $(2, 0, 1)$ and is parallel to the plane whose vector equation is $X \cdot (3, 1, 2) = 4$

**3.** Find the linear equations for the hyperplanes found in Exercise 2.

**4.** In $\mathbb{R}^3$, find the equations of the $x$–$y$ plane, the $x$–$z$ plane, and the $y$–$z$ plane.

Suppose $E$ and $E'$ are hyperplanes in $\mathbb{R}^n$. $E$ and $E'$ are **parallel** if their normal vectors are scalar multiples of each other. $E$ and $E'$ are **perpendicular** if their normal vectors are orthogonal, and the **angle between $E$ and $E'$** is the angle between their normal vectors.

**5.** For each pair of equations, determine whether they define parallel hyperplanes; if not, find the angle between them.

**a.** $2x - y = 3$ and $4x - 2y = 3$
**b.** $x + y = 4$ and $x - y = 1$
**c.** $x - y + z = 3$ and $x - 2y + z = 1$
**d.** $x - 3y + z = 1$ and $x - z = 2$
**e.** $x = 0$ and $z = 0$ (in $\mathbb{R}^3$)
**f.** $y = 3$ and $z = 1$ (in $\mathbb{R}^3$)

⟵
Which pairs of equations define perpendicular planes?

**6.** **a.** Find an equation of the plane containing $(3, 1, 4)$ and parallel to the plane whose equation is $3x - y + z = 7$.
**b.** Find an equation of the plane containing $(3, 1, 4)$ and perpendicular to the planes whose equations are $3x - y + z = 7$ and $x + y - z = 4$.

**7.** Find the cosine of the angle of intersection of the hyperplanes defined by each pair of equations.

**a.** $3x - 2y + z = 4$ and $x + y - z = 1$
**b.** $x + y - z + w = 0$ and $2x + 2y - 2z + 2w = 7$
**c.** $x + 3y - z = 2$ and $z = 0$

**8.** Let $E$ be the set of all vectors in $\mathbb{R}^3$ that are orthogonal to $(3, 2, 1)$. Show that $E$ is a plane and find an equation for $E$.

**9.** Let $\ell$ be the set of all vectors in $\mathbb{R}^3$ that are orthogonal to $(2, 0, 1)$ and $(3, 1, -1)$. Show that it is a line and find a parametric equation for $\ell$.

**10.** Let $E$ be the set of all points of the form

$$(3, 2, 1) + s(2, 0, 1) + t(1, 1, 2)$$

where $s$ and $t$ are arbitrary parameters. Show that $E$ is a plane and that the linear equation for $E$ is $x + 3y - 2z = 7$.

⟵
This is a vector equation for a plane. It takes the form

$$X = A + sB + tC$$

**11.** Let $E$ be the plane in $\mathbb{R}^3$ containing $(0, 0, 2)$ and orthogonal to $(2, -3, 4)$ and suppose $\ell$ is the line containing $(2, 3, 0)$ in the direction of $(2, 1, 3)$. Find the intersection of $\ell$ and $E$.

**12.** Let $E$ have equation $x + y - 2z = 3$. If $\ell$ has equation $X = (2, 1, 3) + t(0, 1, 4)$, find the intersection of $\ell$ and $E$.

**13.** **a.** Suppose the equation of $\ell$ is $X = (2, 2, 10) + t(0, 1, 6)$ and the equation of $\ell'$ is $X = (0, -5, -4) + t(1, 3, 4)$. Find the intersection of $\ell$ and $\ell'$.
**b.** Let $\ell$ have equation $X = (2, 1, 3) + t(4, 1, 0)$ and let $\ell'$ have equation $X = (2, 4, 0) + t(1, 1, 0)$. Show that $\ell$ and $\ell'$ are not parallel but do not intersect.

**14.** Let $E$ be the plane whose equation is $5x + 2y + 14z = -306$. If $Q = (8, 3, 7)$, find the distance from $Q$ to $E$.

**15.** Let $E$ be the hyperplane in $\mathbb{R}^n$ containing $P$ orthogonal to $N$. If $Q$ is any point in $\mathbb{R}^n$, show that the distance from $Q$ to $E$ is given by

$$\frac{|(P-Q)\cdot N|}{\|N\|}$$

⟵ In $\mathbb{R}^2$, this gives you the formula for the distance from a point to a line. Check it out on Exercise 6 from Lesson 2.2.

**16.** Find the distance between the given point and the hyperplane with the given equation.

**a.** $(1,14,25)$ and $x+4y+8z=14$
**b.** $(2,0,1)$ and $3x+2y+6z=11$
**c.** $(3,1,2)$ and $x+y+z=6$
**d.** $(2,0,3)$ and $2x+3y-z=7$
**e.** $(4,1)$ and $3x+4y=5$
**f.** $(1,0,13,1)$ and $x+y+z+w=-2$

**17.** Find the equation of the intersection of the planes whose equations are $x-y+z=4$ and $2x-y+z=-1$.

**18.** If $\ell$ has equation $X=(8,2,9)+t(3,1,4)$ and $\ell'$ has equation $X=(9,1,4)+t(7,1,3)$,

**a.** find the intersection of $\ell$ and $\ell'$
**b.** find an equation of the plane containing $\ell$ and $\ell'$

**19.** Find equations of two distinct hyperplanes in $\mathbb{R}^4$ containing $(1,1,0,0)$, $(1,0,1,1)$, and $(0,1,1,1)$.

# Chapter 2 Mathematical Reflections

These problems will help you summarize what you have learned in this chapter.

1. For each given $A$ and $B$, find $A \cdot B$. Determine whether $A$ is orthogonal to $B$.
   - **a.** $A = (-2, 3, 1)$, $B = (4, 3, -1)$
   - **b.** $A = (0, 8, -6)$, $B = (9, 2, 4)$
   - **c.** $A = (-2, 3, 1, 0)$, $B = (10, 7, -1, -5)$

2. Let $A = (2, -1, -1)$ and $B = (-4, 0, 2)$. For each exercise, calculate the given expression.
   - **a.** $d(A, B)$ **b.** $\text{comp}_B A$ **c.** $\text{Proj}_B A$
   - **d.** $\text{Proj}_A B$ **e.** $\|\text{Proj}_B A\|$ **f.** $\|\text{Proj}_A B\|$

3. Find $\theta$ if $\theta$ is the angle between $A$ and $B$.
   - **a.** $A = (-2, 1)$, $B = (4, 3)$
   - **b.** $A = (2, 0, 3)$, $B = (-1, 4, 2)$
   - **c.** $A = (1, 0, 0, 2)$, $B = (-2, 1, 0, 5)$

4. Let $A = (4, -5, 0)$ and $B = (0, 3, -2)$.
   - **a.** Find $A \times B$
   - **b.** Find the area of the parallelogram whose vertices are $O$, $A$, $B$, and $A + B$.

5. Let $A = (2, -1, 3)$ and $B = (1, 1, 2)$. Find a vector equation of the line $\ell$ containing $A$ and $B$.

6. How can you determine whether two vectors (of any dimension) are orthogonal?

7. How can you find a vector orthogonal to two given vectors in $\mathbb{R}^3$?

8. Let $A = (2, -1, 3)$, $B = (1, 1, 2)$, and $C = (2, 0, 5)$. Find an equation of the hyperplane $E$ containing $A$, $B$, and $C$.

**Vocabulary**

In this chapter, you saw these terms and symbols for the first time. Make sure you understand what each one means, and how it is used.

- angle (between two vectors)
- component
- cross product ( $\times$ )
- determinant
- direction vector of a line
- dot product ( $\cdot$ )
- hyperplane
- initial point
- lemma
- linear equation
- normal
- orthogonal
- projection
- right-hand rule
- standard basis vectors
- vector equation of a line

# Chapter 2 Review

In Lesson 2.2, you learned to

- find the dot product of two vectors of any dimension
- determine whether two vectors are orthogonal
- use the basic properties of dot product to prove statements and solve problems

The following questions will help you check your understanding.

**1.** For each given $A$ and $B$,

(i) find $A \cdot B$
(ii) find $(A + B) \cdot B$
(iii) find $2A \cdot B + 2B \cdot B$
(iv) is $A$ orthogonal to $B$? Explain

**a.** $A = (-4, -6)$, $B = (3, -2)$
**b.** $A = (5, 0, -1)$, $B = (2, -3, 7)$
**c.** $A = (1, 0, -2, 3)$, $B = (2, -5, 4, 2)$

**2.** If $A$ and $B$ are vectors in $\mathbb{R}^n$ and $c$ is a number, characterize each of the following by one of the words "vector" or "number."

**a.** $A \cdot B$
**b.** $A \cdot B + A \cdot A$
**c.** $cA$
**d.** $c(A \cdot B)$
**e.** $(A \cdot B)A$
**f.** $A - cB$
**g.** $(A - cB) \cdot B$

**3.** Characterize all vectors $X$ in $\mathbb{R}^3$ orthogonal to $A = (3, -1, 2)$ and $B = (2, 1, -1)$.

**4.** Let $A = (-2, 1, 4, 3)$, $B = (3, 5, -2, 3)$, and $C = (2, -1, 6, 4)$.

**a.** Show that $\Delta ABC$ is a right triangle.
**b.** Verify Pythagoras' Theorem for $\Delta ABC$.

In Lesson 2.3, you learned to

- find the component of a vector along another vector
- find the projection of a vector along another vector

The following questions will help you check your understanding.

**5.** For the given points $A$ and $B$, find $d(A, B)$, $\text{Proj}_B A$, $\text{Proj}_A B$, and $\|\text{Proj}_B A\|$.

**a.** $A = (-2, 3)$, $B = (2, 0)$
**b.** $A = (0, 5, -10)$, $B = (1, 0, -1)$
**c.** $A = (2, -1, 3, 2)$, $B = (-3, 0, 2, -1)$

6. Let $A$, $B$, and $C$ be vectors, and let $e$ be a scalar. For each exercise, characterize the expression as "vector," "number," or "meaningless."

| | | |
|---|---|---|
| **a.** $\frac{A \cdot B}{B \cdot B}$ | **b.** $\text{Proj}_A B$ | **c.** $\text{comp}_B A$ |
| **d.** $\|\text{Proj}_B A\|$ | **e.** $A \cdot \text{comp}_B A$ | **f.** $(\text{comp}_B A)(\text{Proj}_B A)$ |
| **g.** $\|A \cdot C + \text{Proj}_C A\|$ | **h.** $\|\text{comp}_A C\|$ | **i.** $(eA \cdot \text{Proj}_B C)B$ |

7. Let $A = (2, 5)$ and $B = (6, 3)$, Find the area of the triangle whose vertices are $A$, $B$, and $O$.

In Lesson 2.4, you learned to

- find the angle between two vectors in any dimension
- understand and use the triangle inequality in $\mathbb{R}^n$

The following questions will help you check your understanding.

8. Find $\cos\theta$ if $\theta$ is the angle between $A$ and $B$.

   **a.** $A = (2, -1)$, $B = (-1, 3)$
   **b.** $A = (3, 0, -1)$, $B = (3, -1, 2)$
   **c.** $A = (4, 3, 2, -1)$, $B = (8, 4, -6, 2)$

9. Find the measure of each angle of $\triangle ABC$, where $A = (0, 0, 0)$, $B = (2, -4, 1)$, and $C = (5, -2, 3)$.

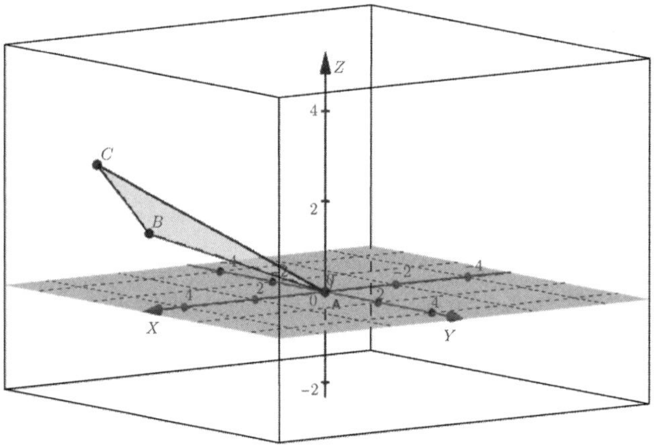


10. $A$ and $B$ are nonzero vectors in $\mathbb{R}^n$ and $\theta$ is the angle between $A$ and $B$. If $A \cdot B = 10$, $A \cdot A = 36$, and $B \cdot B = 25$, find $\theta$.

In Lesson 2.5, you learned to

- find a vector orthogonal to two given vectors in $\mathbb{R}^3$ using cross product
- determine the area of triangles and parallelograms in $\mathbb{R}^3$ using cross product
- apply the cross product to find the angle between two vectors in $\mathbb{R}^3$

The following questions will help you check your understanding.

**11.** Let $A = (-1, -2, 3)$, $B = (4, 0, 1)$, and $C = (1, 1, 1)$. Find

**a.** $A \times B$ **b.** $A \times 2B$

**c.** $2A \times 2B$ **d.** $B \times A$

**e.** $(A \times B) \times C$ **f.** $A \times (B \times C)$

**g.** $A \times A$ **h.** $(A \times A) \times B$

**12.** Find a nonzero vector orthogonal to both $(2, -1, 3)$ and $(-4, 0, 8)$.

**13.** Find the area of the triangle whose vertices are

**a.** $A = (0, 0, 0)$, $B = (0, 2, -1)$, $C = (2, 3, -4)$
**b.** $A = (1, 2, -1)$, $B = (0, 5, 3)$, $C = (2, 4, -3)$
**c.** $A = (3, -1, 2)$, $B = (4, -2, -1)$, $C = (2, 1, 0)$

In Lesson 2.6, you learned to

- find a vector equation of a line given a point and direction parallel to the line
- find an equation of a plane given a point and a direction normal to the plane
- recognize the difference between the vector and coordinate equations of a hyperplane
- find the distance from a point to a line and from a point to a plane

The following questions will help you check your understanding.

**14.** For each given set of conditions, find a vector equation of line $\ell$. Then use the vector equation to find another point on line $\ell$.

**a.** $\ell$ contains $(4, -2)$ and is parallel to $(-1, 3)$
**b.** $\ell$ contains $(1, -2, 6)$ and $(2, 0, -1)$
**c.** $\ell$ contains $(2, -2, 5, 1)$ and $(3, 2, 1, 0)$

**15.** For each given set of conditions, find

(i) an equation in the form $X \cdot N = P \cdot N$ for the hyperplane $E$
(ii) a linear equation for the hyperplane $E$

**a.** $E$ contains $(2, -1, 4)$ and is normal to $(3, 0, 2)$
**b.** $E$ contains $(2, -1, 4)$, $(3, 1, 5)$, and $(-2, 4, 6)$
**c.** $E$ contains $(2, -1, 4)$ and is parallel to the plane whose vector equation is $X \cdot (3, 6, -1) = 5$

**16.** For each pair of equations, determine whether they define parallel hyperplanes; if not, find the angle between them (that is, find the angle between their normal vectors).

**a.** $x - 2y + 3z = 4$ and $x + y - 2z = 8$
**b.** $x - 2y + 3z = 4$ and $2x - 4y + 6z = 11$
**c.** $x - 2y + 3z = 4$ and $x + 2y + z = 4$

# Chapter 2 Test

### Multiple Choice

**1.** Which of the following pairs of vectors is orthogonal?

**A.** $A = (2, -3)$, $B = (3, -2)$
**B.** $A = (1, -1, 0)$, $B = (2, 0, -2)$
**C.** $A = (1, 2, -1)$, $B = (2, -1, 0)$
**D.** $A = (4, -2, 5, -1)$, $B = (0, 2, 1, 0)$

**2.** Suppose $A = (3, -4, 1, 2)$ and $B = (0, 3, 5, -1)$. What is the value of $(A + B) \cdot B$?

**A.** 16 **B.** 26 **C.** 29 **D.** 35

**3.** Let $A = (4, 0, 5)$ and $B = (2, -1, 2)$. What is the value of $\|\text{Proj}_B A\|$?

**A.** 2 **B.** 3 **C.** 6 **D.** 18

**4.** Let $A$, $B$, and $C$ be vectors. Which expression represents a vector?

**A.** $\text{comp}_B A$ **B.** $\text{Proj}_A B$ **C.** $\|A + C\|$ **D.** $A \cdot (B + C)$

**5.** Let $E_1 = (1, 0, 0)$, $E_2 = (0, 1, 0)$, and $E_3 = (0, 0, 1)$. What is $E_2 \times E_3$?

**A.** $(0, 0, 0)$ **B.** $(1, 0, 0)$ **C.** $(0, 1, 0)$ **D.** $(0, 0, 1)$

**6.** A vector equation for line $\ell$ is $X = (5, -3) + t(1, -2)$. Which point is on line $\ell$?

**A.** $(2, -1)$ **B.** $(4, -1)$ **C.** $(6, -6)$ **D.** $(8, -4)$

### Open Response

**7.** Consider the triangle whose vertices are $A = (-2, 1, 4)$, $B = (1, -4, 3)$, and $C = (-3, -2, 5)$. Show that $\angle ACB$ is a right angle.

**8.** Let $A = (1, -1, 0)$, $B = (-4, -2, 5)$, and $P = \text{Proj}_A B$. Find $d(B, P)$.

**9.** Let $A = (1, -2, 0, 5)$ and $B = (-1, 3, -4, 2)$. Find $\theta$ if $\theta$ is the angle between $A$ and $B$.

**10.** Find a nonzero vector orthogonal to both $A = (-3, 1, 0)$ and $B = (2, 4, -1)$.

**11.** Find the area of the triangle whose vertices are $A = (-1, 5, -3)$, $B = (1, 2, -4)$, and $C = (-2, 6, -3)$.

**12.** Let $E$ be a plane containing points $P_1 = (5, -2, 3)$, $P_2 = (2, 0, -4)$, and $P_3 = (4, -6, 8)$.

**a.** Find a normal to $E$.

**b.** Find a vector equation in the form $X \cdot N = P \cdot N$ for $E$.

**c.** Find a coordinate equation for $E$.

CHAPTER

# 3 The Solution of Linear Systems

Prior to this course, the systems of equations you worked with were most likely two equations in two unknowns. You may have encountered systems with three equations in three unknowns, or where the number of equations and unknowns were different. When solving systems like these, you can easily get confused using methods such as substitution and elimination because you sometimes forget to use one equation, or you use the same equation twice by accident, or you just lose track of where you are.

⟵ For example, take a look at Example 11 from Lesson 2.6.

This chapter will focus on a way to keep track of the steps. It will use a technique called *Gaussian elimination*, a mechanical process that guarantees that, when you perform the process to a linear system, you will end up with a simpler system that has the same solutions as the original. The process is implemented using a *matrix*—a rectangular array of numbers—to keep track of the coefficients and constants in a linear system of any size.

**By the end of this chapter, you will be able to answer questions like these:**

1. How can you tell if a system of equations has a solution?
2. How can you tell if three vectors are linearly independent in $\mathbb{R}^2$? in $\mathbb{R}^3$? in $\mathbb{R}^n$?
3. Find the kernel of this matrix:

$$\begin{pmatrix} 3 & -2 & 1 & 4 \\ 1 & 2 & -1 & 0 \\ 7 & -10 & 5 & 12 \end{pmatrix}$$

   Describe this kernel geometrically.

**You will build good habits and skills for ways to**

- extend old methods to new systems
- look for similarity in structure
- reason about calculations
- use different forms for different purposes

### Vocabulary and Notation

- augmented matrix
- coefficient matrix
- dimension
- elementary row operations
- equivalent matrices
- equivalent systems
- Gaussian elimination
- kernel
- linear combination
- linearly dependent
- linearly independent
- matrix
- row-reduced echelon form
- trivial solution

## 3.1 Getting Started

**1.** Solve this system of equations:

$$\begin{aligned} x + 2y + 13z &= -4 \\ x - 5y - 8z &= 11 \\ 2x - 3y + 6z &= 7 \end{aligned}$$

**2.** Write $(-4, 11, 7)$ as a nonzero linear combination of the vectors $(1, 1, 2)$, $(2, -5, -3)$, and $(13, -8, 6)$.

**3.** Find the solution set for this system of three equations and three unknowns:

$$\begin{aligned} 3x + y - z &= 0 \\ x + y + 3z &= 0 \\ 2x + y + z &= 0 \end{aligned}$$

$\longleftarrow$ This is the same as asking for the intersection of the three planes defined by these equations.

**4.** Write $(0, 0, 0)$ as a nonzero linear combination of the vectors $(3, 1, 2)$, $(1, 1, 1)$, and $(-1, 3, 1)$.

**5.** Find all vectors $X$ orthogonal to $(3, 1, -1)$, $(1, 1, 3)$, and $(2, 1, 1)$.

**6.** Write $(-1, 3, 1)$ as a linear combination of $(3, 1, 2)$ and $(1, 1, 1)$.

**7.** Here is a system of equations:

$$\begin{aligned} x + 4y &= 15 \\ x - 2y &= 3 \end{aligned}$$

Here is a second system with the same solution set:

$$\begin{aligned} x \quad &= 7 \\ y &= 2 \end{aligned}$$

In the second system, each variable appears in only one equation, so its solution can be read immediately.

**a.** Find two more systems of equations, using $x$ and $y$ in *each* equation, with the same solution set as the two systems above.

**b.** If possible, find a system of equations in the form

$$\begin{aligned} ax + by &= 0 \\ cx + dy &= 1 \end{aligned}$$

with the same solution set.

**c.** If possible, find a system of equations in the form

$$\begin{aligned} ax + by &= 0 \\ cx + dy &= 0 \end{aligned}$$

with the same solution set.

**8.** Here is a system of equations:

$$\begin{aligned} x - 2y &= -1 \\ 3x + 4y &= 27 \end{aligned}$$

For each of the following systems, determine whether it has the same solution set as the system given above:

a. $\begin{aligned} 2x - 4y &= -2 \\ 3x + 4y &= 27 \end{aligned}$

b. $\begin{aligned} 2x - 4y &= -2 \\ 5x &= 25 \end{aligned}$

c. $\begin{aligned} 2x - 4y &= -2 \\ x &= 5 \end{aligned}$

d. $\begin{aligned} 2x - 4y &= -10 \\ x &= 5 \end{aligned}$

**9.** Lines $\ell_1$ and $\ell_2$ intersect at point $P = (4, 11, 2)$ and are parallel to $V_1 = (4, -2, 3)$ and $V_2 = (-5, 3, -6.5)$, respectively.

**a.** Find an equation of the plane containing these lines.

**b.** Find the vector equation of line $\ell_3$ through point $P$ and orthogonal to the plane containing $\ell_1$ and $\ell_2$.

**c.** Characterize all vectors $V_3$ orthogonal to vectors $V_1$ and $V_2$.

**d.** Let $V_4 = 3V_1 - 2V_2$ and $V_5 = 3V_1 - 5V_2$. Characterize all vectors $V_6$ orthogonal to vectors $V_4$ and $V_5$.

**e.** For what $c$ and $d$ are $V_3$ and $V_6$ both orthogonal to $cV_1 + dV_2$?

## 3.2 Gaussian Elimination

Many of the problems that you have encountered in this book have led to solving systems of equations. The sizes of these systems have varied: two equations in two unknowns, three equations in two unknowns, two equations in three unknowns. Each time, the algebra was essentially the same, but as these systems get bigger, it can be difficult to keep track of what you're doing. There is a more mechanical method that can solve any of these systems. The trick is to figure out how to interpret the solution.

**In this lesson, you will learn how to**

- represent a system of equations with an augmented matrix
- reduce an augmented matrix to its row-reduced echelon form
- interpret the nature of the solution set of a system given its row-reduced echelon form
- express an infinite solution set to a system as a vector equation

### Developing Habits of Mind

**Look for structural similarity.** Look back at Exercises 3–5 in the Getting Started lesson. The questions are different, but the numbers underlying them all look the same.

$$\begin{pmatrix} 3 & 1 & -1 & 0 \\ 1 & 1 & 3 & 0 \\ 2 & 1 & 1 & 0 \end{pmatrix}$$

You may have noticed you took similar steps when solving each of these exercises. This happens because so many different kinds of problems boil down to solving systems of linear equations. In this lesson, you'll learn an efficient, mechanical way to solve systems of linear equations.

Here is the system from Exercise 1 in the Getting Started lesson:

$$\begin{cases} x + 2y + 13z & = & -4 \\ x - 5y - 8z & = & 11 \\ 2x - 3y + 6z & = & 7 \end{cases}$$

This system can be rewritten in terms of dot products. For example, $x + 2y + 13z$ is the dot product of $(1, 2, 13)$ and $(x, y, z)$.

$$\begin{array}{rcl} (1, 2, 13) \cdot (x, y, z) & = & -4 \\ (1, -5, -8) \cdot (x, y, z) & = & 11 \\ (2, -3, 6) \cdot (x, y, z) & = & 7 \end{array}$$

What makes this system of equations different from other systems with three equations and three unknowns? The only information that distinguishes *this* system from others are the four numbers in each row. These numbers can be written independently in a matrix of numbers:

$$\begin{pmatrix} 1 & 2 & 13 & -4 \\ 1 & -5 & -8 & 11 \\ 2 & -3 & 6 & 7 \end{pmatrix}$$

When you solve a system of equations, it doesn't matter what variables are being used. By using matrices, you look only at the numbers involved.

← A **matrix** (plural *matrices*) is just a rectangular array of numbers.

← Look back at the two systems above. What parts of each system are represented in this 3-by-4 matrix?

### For You to Do

**1.** For each matrix, write a corresponding system of equations.

**a.** $\begin{pmatrix} 1 & -2 & 1 \\ 3 & 4 & 27 \end{pmatrix}$ **b.** $\begin{pmatrix} 1 & 0 & 3 & 5 \\ 0 & 1 & 0 & 3 \end{pmatrix}$ **c.** $\begin{pmatrix} 1 & 0 & 0 & 2 \\ 0 & 1 & 0 & 3 \\ 0 & 0 & 1 & -1 \end{pmatrix}$

Matrices like these are arrays of coefficients augmented by the constant terms of each equation. Because of this, they are called **augmented matrices**. A matrix that consisted only of the coefficients (the same matrix without the rightmost column) is called a **coefficient matrix**.

### Equivalent Systems

#### Definition

- Two systems of equations are **equivalent systems** if they have the same solution set.
- Two augmented matrices are **equivalent matrices** if their corresponding systems are equivalent.

Here are two equivalent systems and their corresponding matrices:

$$\begin{cases} x + y &= 7 \\ x - y &= 3 \end{cases} \rightarrow \begin{pmatrix} 1 & 1 & 7 \\ 1 & -1 & 3 \end{pmatrix}$$

and

$$\begin{cases} x + y &= 7 \\ 2x &= 10 \end{cases} \rightarrow \begin{pmatrix} 1 & 1 & 7 \\ 2 & 0 & 10 \end{pmatrix}$$

These systems are equivalent because they both have the solution set $\{(5, 2)\}$, and the two augmented matrices are (by definition) also equivalent. The second system and matrix are simpler, though—this system, and its corresponding matrix, make it more clear how to find any solutions to the system.

← What does 0 in the matrix signify?

← What step(s) might you take next to make an even simpler equivalent system?

### Developing Habits of Mind

**Revisit old methods.** One method you may have used in the past to solve a system of equations is the "elimination" method: you can get the equation $2x = 10$ in the second system above by adding together the two equations from the first system, $x + y = 7$ and $x - y = 3$.

It turns out, though, that if you *replace* either of the two original equations by their sum, the new system has the same solution set! So, by definition, this substitution creates an equivalent system. Here, the matrix representation of that second system has a 0 in the second row. Having lots of zeros in a matrix makes the solution to the system easier to see (as you may have noticed in the For You to Do problem 1 earlier in this lesson). So the strategy is, through a series of steps, to eliminate one variable in one equation in each step, ending up with a system (with mostly zeros) simple enough so you can read the solution set directly. Each step relies on the same process of the "elimination" method: add two equations (or some multiple of each) together so a variable will end up with a coefficient of 0. But here, you *replace* one of those equations with that sum to get a new, simpler, equivalent system, then repeat until you're done.

⟵ The two augmented matrices are also equivalent: the second row in the second matrix comes from adding the two rows in the first matrix. So adding the two rows is the same as adding the two equations, only you don't need to deal with the variables.

⟵ You should convince yourself that replacing an equation by the sum of itself and another equation produces an equivalent system. Try it with a specific system and then generalize.

### For You to Do

**2.** These seven systems of equations are all equivalent. For each given system, write its corresponding augmented matrix, and then describe the steps you could follow to transform the previous system into that system.

$$\begin{cases} x + 2y + 13z = -4 \\ x - 5y - 8z = 11 \\ 2x - 3y + 6z = 7 \end{cases} \longrightarrow \begin{pmatrix} 1 & 2 & 13 & -4 \\ 1 & -5 & -8 & 11 \\ 2 & -3 & 6 & 7 \end{pmatrix}$$

**a.** $\begin{cases} x + 2y + 13z = -4 \\ -x + 5y + 8z = -11 \\ 2x - 3y + 6z = 7 \end{cases}$ **b.** $\begin{cases} x + 2y + 13z = -4 \\ 7y + 21z = -15 \\ 2x - 3y + 6z = 7 \end{cases}$

**c.** $\begin{cases} x + 2y + 13z = -4 \\ 2x - 3y + 6z = 7 \\ 7y + 21z = -15 \end{cases}$ **d.** $\begin{cases} x + 2y + 13z = -4 \\ -7y - 20z = 15 \\ 7y + 21z = -15 \end{cases}$

**e.** $\begin{cases} x + 2y + 13z = -4 \\ z = 0 \\ 7y + 21z = -15 \end{cases}$ **f.** $\begin{cases} x + 2y + 13z = -4 \\ z = 0 \\ y + 3z = -\frac{15}{7} \end{cases}$

### Elementary Row Operations

In problem 2, there is at least one example of each of the following operations, all of which produce equivalent systems of equations.

- Change the order of the equations.
- Replace any equation by a nonzero multiple of itself.
- Replace any equation by a multiple of any other equation plus itself.

These operations on systems can be translated into statements about augmented matrices. Each of the following operations on an augmented matrix produces an equivalent matrix.

- Change the order of the rows.
- Replace any row by a nonzero multiple of itself.
- Replace any row by a multiple of any other row plus itself.

These are called the **elementary row operations** and are the key to solving systems of linear equations efficiently. By working with matrices, not only can you see the structure of linear systems more easily, but you can also more easily build an efficient algorithm for solving systems. This algorithm is convenient and simple to implement on a calculator or computer.

#### For Discussion

**3.** As stated above, the three operations on systems all produce equivalent systems. To prove this,

For each operation, show that the solution set of the original system must be the same as the solution set of the new, modified system.

#### Example 1

**Problem.** Solve this system of equations using elementary row operations:

$$\begin{cases} x + y + z &= 10 \\ -y + z &= 3 \\ -4x + 3y + 5z &= 35 \end{cases} \rightarrow \begin{pmatrix} 1 & 1 & 1 & 10 \\ 0 & -1 & 1 & 3 \\ -4 & 3 & 5 & 35 \end{pmatrix}$$

← When building an augmented matrix, 0 must be included whenever a variable is not present, and terms must be included in the same order in each row.

**Solution.** Your goal is to eliminate all but one variable from each equation. One way is to make the "$x$" entry in the third row equal to 0 by replacing that row with 4 times the first row plus itself.

$$4\begin{pmatrix}1 & 1 & 1 & 10\end{pmatrix} + \begin{pmatrix}-4 & 3 & 5 & 35\end{pmatrix} = \begin{pmatrix}0 & 7 & 9 & 75\end{pmatrix}$$

$$\begin{pmatrix} 1 & 1 & 1 & 10 \\ 0 & -1 & 1 & 3 \\ -4 & 3 & 5 & 35 \end{pmatrix} \xrightarrow{R_3 = 4r_1 + r_3} \begin{pmatrix} 1 & 1 & 1 & 10 \\ 0 & -1 & 1 & 3 \\ 0 & 7 & 9 & 75 \end{pmatrix}$$

← The notation between the matrices means "replace the third row by 4 times the first row plus the (old) third row."

Then make the "$y$" entry in the first row equal to 0 by replacing the first row with the second row plus itself.

$$\begin{pmatrix}0 & -1 & 1 & 3\end{pmatrix} + \begin{pmatrix}1 & 1 & 1 & 10\end{pmatrix} = \begin{pmatrix}1 & 0 & 2 & 13\end{pmatrix}$$

$$\begin{pmatrix}1 & 1 & 1 & 10\\ 0 & -1 & 1 & 3\\ 0 & 7 & 9 & 75\end{pmatrix} \xrightarrow{R_1=r_1+r_2} \begin{pmatrix}1 & 0 & 2 & 13\\ 0 & -1 & 1 & 3\\ 0 & 7 & 9 & 75\end{pmatrix}$$

Next, multiply the second row by $-1$ to make the $y$ entry in the second row positive.

$$-1\begin{pmatrix}0 & -1 & 1 & 3\end{pmatrix} = \begin{pmatrix}0 & 1 & -1 & -3\end{pmatrix}$$

$$\begin{pmatrix}1 & 0 & 2 & 13\\ 0 & -1 & 1 & 3\\ 0 & 7 & 9 & 75\end{pmatrix} \xrightarrow{R_2=-r_2} \begin{pmatrix}1 & 0 & 2 & 13\\ 0 & 1 & -1 & -3\\ 0 & 7 & 9 & 75\end{pmatrix}$$

Now add $-7$ times the second row to the third row to make the $y$ entry in the third row equal to 0.

$$-7\begin{pmatrix}0 & 1 & -1 & -3\end{pmatrix} + \begin{pmatrix}0 & 7 & 9 & 75\end{pmatrix} = \begin{pmatrix}0 & 0 & 16 & 96\end{pmatrix}$$

$$\begin{pmatrix}1 & 0 & 2 & 13\\ 0 & 1 & -1 & -3\\ 0 & 7 & 9 & 75\end{pmatrix} \xrightarrow{R_3=-7r_2+r_3} \begin{pmatrix}1 & 0 & 2 & 13\\ 0 & 1 & -1 & -3\\ 0 & 0 & 16 & 96\end{pmatrix}$$

Next, divide the third row by 16 (or, multiply by $\frac{1}{16}$).

$$\begin{pmatrix}1 & 0 & 2 & 13\\ 0 & 1 & -1 & -3\\ 0 & 0 & 16 & 96\end{pmatrix} \xrightarrow{R_3=\frac{r_3}{16}} \begin{pmatrix}1 & 0 & 2 & 13\\ 0 & 1 & -1 & -3\\ 0 & 0 & 1 & 6\end{pmatrix}$$

The third row now translates back to the equation "$z = 6$," so a significant part of the system is now solved. Continue to eliminate nonzero entries from the other rows.

$$\begin{pmatrix}1 & 0 & 2 & 13\\ 0 & 1 & -1 & -3\\ 0 & 0 & 1 & 6\end{pmatrix} \xrightarrow{R_2=r_2+r_3} \begin{pmatrix}1 & 0 & 2 & 13\\ 0 & 1 & 0 & 3\\ 0 & 0 & 1 & 6\end{pmatrix} \xrightarrow{R_1=r_1-2r_3} \begin{pmatrix}1 & 0 & 0 & 1\\ 0 & 1 & 0 & 3\\ 0 & 0 & 1 & 6\end{pmatrix}$$

The last augmented matrix gives the solution to the system: $x = 1$, $y = 3$, $z = 6$.

### For You to Do

4. For each of the last two steps, describe the row operation that is performed, and explain why that row operation is chosen.

### Gaussian Elimination

There is more than one set of row operations that can solve a system. The following is an algorithm for solving any system of linear equations. The **Gaussian elimination** method gives a process for choosing which row operations to perform at which time. The basic concept is to choose a nonzero "pivot" location in each column, scale the row to make its value 1, if necessary, and then use row operations to "clear out" any nonzero values in the column containing that pivot. The example below shows the process on a system of three equations and three variables.

### Example 2

**Problem.** Solve this system of equations using Gaussian elimination:

$$\begin{aligned} 3x + y + z &= 8 \\ x + 2y - z &= 9 \\ x + 3y + 2z &= 9 \end{aligned} \rightarrow \begin{pmatrix} 3 & 1 & 1 & 8 \\ 1 & 2 & -1 & 9 \\ 1 & 3 & 2 & 9 \end{pmatrix}$$

**Solution.** Choose a pivot for the first column—any row with a nonzero entry in the first column will do. Since you want to make the pivot entry 1, the simplest choice is a row that already has a 1 in that column. Both the second and third rows have a 1, so pick one—in this case, the second row. By convention, the pivot is placed as high as possible within the augmented matrix, so begin by changing the order of the first two rows.

← You could choose to divide the first row by 3, but you would then introduce fractional coefficients.

$$\begin{pmatrix} 3 & 1 & 1 & 8 \\ \mathbf{1} & 2 & -1 & 9 \\ 1 & 3 & 2 & 9 \end{pmatrix} \xrightarrow{\text{swap } r_1 \text{ and } r_2} \begin{pmatrix} \mathbf{1} & 2 & -1 & 9 \\ 3 & 1 & 1 & 8 \\ 1 & 3 & 2 & 9 \end{pmatrix}$$

Now use the pivot to "clear out" its column, turning all other entries into 0. To do this, replace the second row by $(-3 \cdot \text{row}_1 + \text{row}_2)$ and the third row by $(-\text{row}_1 + \text{row}_3)$.

$$\begin{pmatrix} \mathbf{1} & 2 & -1 & 9 \\ 3 & 1 & 1 & 8 \\ 1 & 3 & 2 & 9 \end{pmatrix} \xrightarrow{R_2 = -3r_1 + r_2} \begin{pmatrix} \mathbf{1} & 2 & -1 & 9 \\ 0 & -5 & 4 & -19 \\ 1 & 3 & 2 & 9 \end{pmatrix} \xrightarrow{R_3 = -r_1 + r_3} \begin{pmatrix} \mathbf{1} & 2 & -1 & 9 \\ 0 & -5 & 4 & -19 \\ 0 & 1 & 3 & 0 \end{pmatrix}$$

A pivot is complete when its column has a single 1 with all other entries 0, so your work on the first pivot is complete. The first row is now locked in place.

Now find a pivot in the second column, and place it as high as possible (remember, the first row is now locked in). By changing the order of the last two rows, a 1 is established as high as possible in the second column.

$$\begin{pmatrix} 1 & 2 & -1 & 9 \\ 0 & -5 & 4 & -19 \\ 0 & \mathbf{1} & 3 & 0 \end{pmatrix} \xrightarrow{\text{swap } r_2 \text{ and } r_3} \begin{pmatrix} 1 & 2 & -1 & 9 \\ 0 & \mathbf{1} & 3 & 0 \\ 0 & -5 & 4 & -19 \end{pmatrix}$$

As before, use the pivot to "clear out" its column.

$$\begin{pmatrix} 1 & 2 & -1 & 9 \\ 0 & \mathbf{1} & 3 & 0 \\ 0 & -5 & 4 & -19 \end{pmatrix} \xrightarrow{R_1 = -2r_2 + r_1} \begin{pmatrix} 1 & 0 & -7 & 9 \\ 0 & \mathbf{1} & 3 & 0 \\ 0 & -5 & 4 & -19 \end{pmatrix} \xrightarrow{R_3 = 5r_2 + r_3} \begin{pmatrix} 1 & 0 & -7 & 9 \\ 0 & \mathbf{1} & 3 & 0 \\ 0 & 0 & 19 & -19 \end{pmatrix}$$

Now find a pivot in the third column. The only remaining possible pivot comes from the third row, but there is a 19 instead of a 1. Divide the last row by 19.

$$\begin{pmatrix} 1 & 0 & -7 & 9 \\ 0 & 1 & 3 & 0 \\ 0 & 0 & \mathbf{19} & -19 \end{pmatrix} \xrightarrow{R_3 = \frac{1}{19} r_3} \begin{pmatrix} 1 & 0 & -7 & 9 \\ 0 & 1 & 3 & 0 \\ 0 & 0 & \mathbf{1} & -1 \end{pmatrix}$$

Once again, use the pivot to "clear out" its column.

$$\begin{pmatrix} 1 & 0 & -7 & 9 \\ 0 & 1 & 3 & 0 \\ 0 & 0 & \mathbf{1} & -1 \end{pmatrix} \xrightarrow{R_1=7r_3+r_1} \begin{pmatrix} 1 & 0 & 0 & 2 \\ 0 & 1 & 3 & 0 \\ 0 & 0 & \mathbf{1} & -1 \end{pmatrix} \xrightarrow{R_2=-3r_3+r_2} \begin{pmatrix} 1 & 0 & 0 & 2 \\ 0 & 1 & 0 & 3 \\ 0 & 0 & \mathbf{1} & -1 \end{pmatrix}$$

There are no more columns, so the Gaussian elimination process is complete. The final augmented matrix corresponds to the system

$$\begin{cases} x & & & = & 2 \\ & y & & = & 3 \\ & & z & = & -1 \end{cases}$$

The solution of this system and, therefore, the solution of the original system, can be determined immediately: $x = 2$, $y = 3$, $z = -1$.

← The advantage of this system is pretty clear!

### For You to Do

**5.** Solve each system using Gaussian elimination:

**a.** $\begin{cases} 5x - 3y & = & 8 \\ 3x + 2y & = & 1 \end{cases}$ **b.** $\begin{cases} 3x + 4y + z & = & 10 \\ x + 2y + 4z & = & 13 \\ 2x + 3y + 3z & = & 5 \end{cases}$ **c.** $\begin{cases} x + y - z & = & 0 \\ 2x + 3y + z & = & 0 \end{cases}$

The process of Gaussian elimination is sometimes called *reducing a matrix to echelon form*, and leads to this definition.

← In this book, the term "row-reduced echelon form" is sometimes shortened to "reduced echelon form" or simply "echelon form."

#### Definition

A matrix is in **row-reduced echelon form** if

(1) the first nonzero entry in any row is 1, and is to the right of the first nonzero entry in any row above it

(2) any column containing the first 1 from any row is, except for that 1, all 0

(3) any rows with all 0 must be at the bottom of the matrix

← On many calculators, the `rref` function takes in a matrix, and returns its row-reduced echelon form.

Roughly speaking, a matrix is in row-reduced echelon form if there are 1's running down the diagonal, and if the columns containing these 1's otherwise contain all 0's.

← There are, of course, exceptions, and the precise definition takes these into account.

### Example 3

**Problem.** Which matrices are in echelon form?

**a.** $\begin{pmatrix} 1 & 9 & 0 \\ 0 & 1 & 0 \\ 0 & 0 & 1 \end{pmatrix}$ **b.** $\begin{pmatrix} 1 & 9 & 0 \\ 0 & 0 & 1 \\ 0 & 0 & 0 \end{pmatrix}$ **c.** $\begin{pmatrix} 1 & 9 & 0 \\ 0 & 0 & 0 \\ 0 & 0 & 1 \end{pmatrix}$

**d.** $\begin{pmatrix} 1 & 0 & 0 & 0 \\ 0 & 1 & 0 & 0 \\ 0 & 0 & 1 & 0 \end{pmatrix}$ **e.** $\begin{pmatrix} 1 & 0 & 7 & 0 \\ 0 & 1 & 3 & 0 \\ 0 & 0 & 0 & 1 \end{pmatrix}$ **f.** $\begin{pmatrix} 1 & 0 & 0 & 7 \\ 0 & 1 & 0 & 6 \\ 0 & 0 & 1 & 5 \end{pmatrix}$

**Solution.**

**a.** Not in echelon form. The first 1 in the second row is in column 2), and there's a 9 above that pivotal 1, so this matrix violates part ((2)) of the definition.

**b.** In echelon form. This matrix meets all of the requirements of the definition.

**c.** Not in echelon form. The row of zeros is not below the nonzero rows, so it violates part ((3)) of the definition.

**d.** In echelon form. This matrix meets all of the requirements of the definition.

**d.** In echelon form. This matrix meets all of the requirements of the definition.

**f.** In echelon form. This matrix meets all of the requirements of the definition.

---

### Example 4

**Problem.** Solve this system of equations. Does the graph of the solution set represent a point, a line, or a plane?

$$\begin{cases} x - y + z &= 7 \\ 2x + y + z &= 5 \end{cases}$$

**Solution.** Reduce the corresponding augmented matrix to echelon form.

$\longleftarrow$ Each equation in this system has a plane as its graph. In general, what do you get if you intersect two planes?

$$\begin{pmatrix} 1 & -1 & 1 & 7 \\ 2 & 1 & 1 & 5 \end{pmatrix} \xrightarrow{R_2=-2r_1+r_2}$$

$$\begin{pmatrix} 1 & -1 & 1 & 7 \\ 0 & 3 & -1 & -9 \end{pmatrix} \xrightarrow{R_2=\frac{1}{3}r_2} \begin{pmatrix} 1 & -1 & 1 & 7 \\ 0 & 1 & -\frac{1}{3} & -3 \end{pmatrix} \xrightarrow{R_1=r_2+r_1} \begin{pmatrix} 1 & 0 & \frac{2}{3} & 4 \\ 0 & 1 & -\frac{1}{3} & -3 \end{pmatrix}$$

Write the final matrix as a system of equations to read the solution.

$$\begin{cases} x + \frac{2}{3}z &= 4 \\ y - \frac{1}{3}z &= -3 \end{cases}$$

Unlike Example 2, this solution is not quite so automatic. You can rewrite each equation to be a statement in terms of $z$: $x = 4 - \frac{2}{3}z$ and $y = -3 + \frac{1}{3}z$. The $z$ is not restricted—it can be any value at all. So the solution set is

$$\begin{aligned} X &= \left(4 - \tfrac{2}{3}z,\ -3 + \tfrac{1}{3}z,\ z\right) \text{ for any number } z \\ &= (4, -3, 0) + z\left(-\tfrac{2}{3}, \tfrac{1}{3}, 1\right) \end{aligned}$$

$\longleftarrow$ Where can the first two numbers in each of these ordered triples be found in the matrix?

As seen in Chapter 2, this is the equation of a line in $\mathbb{R}^3$.

---

### Example 5

**Problem.** Solve this system of equations:

$$\begin{cases} x + 3y + z + 2w &= 4 \\ 2x + y - 3z - w &= 3 \\ 4x + 3y - 5z - w &= 7 \end{cases}$$

**Solution.** Write the augmented matrix for the system, then reduce it to echelon form.

$$\begin{pmatrix} 1 & 3 & 1 & 2 & 4 \\ 2 & 1 & -3 & -1 & 3 \\ 4 & 3 & -5 & -1 & 7 \end{pmatrix} \xrightarrow[R_3=-4r_1+r_3]{R_2=-2r_1+r_2} \begin{pmatrix} 1 & 3 & 1 & 2 & 4 \\ 0 & -5 & -5 & -5 & -5 \\ 0 & -9 & -9 & -9 & -9 \end{pmatrix}$$

$$\xrightarrow{R_2=-\frac{1}{5}r_2} \begin{pmatrix} 1 & 0 & -2 & -1 & 1 \\ 0 & 1 & 1 & 1 & 1 \\ 0 & -9 & -9 & -9 & -9 \end{pmatrix} \xrightarrow{R_3=9r_2+r_3} \begin{pmatrix} 1 & 0 & -2 & -1 & 1 \\ 0 & 1 & 1 & 1 & 1 \\ 0 & 0 & 0 & 0 & 0 \end{pmatrix}$$

Write the final matrix as a system of equations.

$$\begin{cases} x - 2z - w &= 1 \\ y + z + w &= 1 \end{cases}$$

← The system has only two equations because all the entries in the third row of the augmented matrix became 0. Strictly speaking, that row represents the equation $0x + 0y + 0z + 0w = 0$. Since any values of $x$, $y$, $z$, and $w$ satisfy that equation, it gives you no additional information as to the final solution, so it can be ignored.

As in Example 4, each equation has more than one variable. You can write the variables $x$ and $y$ each in terms of $z$ and $w$.

$$\begin{cases} x &= 1 + 2z + w \\ y &= 1 - z - w \end{cases}$$

Variables $z$ and $w$ are not restricted, and can be any value at all. The solution set is

$$\begin{aligned} X &= (1 + 2z + w,\quad 1 - z - w,\quad z,\quad w) \text{ for any } z \text{ and } w \\ &= (1, 1, 0, 0) + z\,(2, -1, 1, 0) + w\,(1, -1, 0, 1) \end{aligned}$$

---

There is also more than one way to write the solution set to this example. The given solution solved two equations for $x$ and $y$ in terms of $z$ and $w$, but this is not the only possibility. Here are the same equations solved for $z$ and $w$ in terms of $x$ and $y$.

$$\begin{cases} z &= -2 + x + y \\ w &= 3 - x - 2y \end{cases}$$

One way to find this is to build the initial augmented matrix with a different order of variables. The variables placed first are the ones most easily solved for when using Gaussian elimination.

The equations above led to the solution set

$$\begin{aligned} X &= (x,\quad y,\quad -2 + x + y,\quad 3 - x - 2y) \text{ for any } x \text{ and } y \\ &= (0, 0, -2, 3) + x\,(1, 0, 1, -1) + y\,(0, 1, 1, -2) \end{aligned}$$

This doesn't look like the same solution set, but it has to be, since it comes from the same set of equations.

---

### For You to Do

**6.** **a.** Show that $(0, 0, -2, 3)$ is in the solution set of both of these equations:

$$\begin{aligned} X &= (1, 1, 0, 0) + s(2, -1, 1, 0) + t(1, -1, 0, 1) \\ X &= (0, 0, -2, 3) + s(1, 0, 1, -1) + t(0, 1, 1, -2) \end{aligned}$$

**b.** Find two more points $(x, y, z, w)$ that are solutions, and check them against both solution sets.

### Example 6

**Problem.** Solve this system of equations. Describe the graph of the solution set.

$$\begin{cases} x - y + z &= 3 \\ 4x + y + z &= 5 \\ 5x + 2z &= 12 \end{cases}$$

**Solution.** Write the augmented matrix, then reduce to echelon form.

$$\begin{pmatrix} 1 & -1 & 1 & 3 \\ 4 & 1 & 1 & 5 \\ 5 & 0 & 2 & 12 \end{pmatrix} \rightarrow \begin{pmatrix} 1 & 0 & \frac{2}{5} & \frac{12}{5} \\ 0 & 1 & -\frac{3}{5} & -\frac{7}{5} \\ 0 & 0 & 0 & 4 \end{pmatrix}$$

Write the final matrix as a system of equations.

$$\begin{cases} x + \frac{2}{5}z &= 0 \\ y - \frac{3}{5}z &= 0 \\ 0 &= 4 \end{cases}$$

The last equation, $0 = 4$, is a false statement—no matter what values of $x$, $y$, and $z$, it will always be false. When one equation is always false, there is *no solution* to the entire system of equations. No values of $x$, $y$, and $z$ can satisfy all three equations simultaneously. The graph is empty, since no points make all the equations true.

## Exercises

**1.** Find all vectors $X$ that are orthogonal to $(1, 2, 3)$, $(4, 5, 6)$, and $(7, 8, 9)$ by solving this system of equations:

$$\begin{cases} X \cdot (1,2,3) &= 0 \\ X \cdot (4,5,6) &= 0 \\ X \cdot (7,8,9) &= 0 \end{cases} \longrightarrow \begin{cases} x + 2y + 3z &= 0 \\ 4x + 5y + 6z &= 0 \\ 7x + 8y + 9z &= 0 \end{cases}$$

⟵ When all the constants are 0, the system is called a **homogeneous system**.

**2.** The solution to Exercise 1 says that any multiple of $(1, -2, 1)$ is perpendicular to the three given vectors. How is this possible geometrically?

**3.** Find the linear combination of $(1, 2, 3)$, $(4, 5, 6)$, and $(7, 8, 10)$ that produces $(17, 19, 12)$ by solving this system of equations:

$$\begin{cases} a + 4b + 7c &= 17 \\ 2a + 5b + 8c &= 19 \\ 3a + 6b + 10c &= 12 \end{cases}$$

**4.** Let $A = (1, -3, 2)$, $B = (5, 0, -4)$, and $C = (2, 1, -1)$.

**a.** Write $(13, 6, -16)$ as a linear combination of $A$, $B$, and $C$.
**b.** Find all vectors that are orthogonal to $A$, $B$, and $C$.

**5.** Let $A = (1, -1, 2)$, $B = (2, 1, -1)$, and $C = (-1, 1, -3)$. Write the following vectors as linear combinations of $A$, $B$, and $C$:

**a.** $(1, 0, 0)$ **b.** $(0, 1, 0)$ **c.** $(0, 0, 1)$
**d.** $(1, 1, 1)$ **e.** $(17, -13, 19)$

**6.** Reduce this matrix to echelon form and compare this result to the results you got in Exercise 5.

$$\begin{pmatrix} 1 & 2 & -1 & 1 & 0 & 0 & 1 & 17 \\ -1 & 1 & 1 & 0 & 1 & 0 & 1 & -13 \\ 2 & -1 & -3 & 0 & 0 & 1 & 1 & 19 \end{pmatrix}$$

**7.** Solve each system by reducing an augmented matrix to echelon form.

**a.** $\begin{cases} 2x + 3y = 11 \\ 3x + y = -1 \end{cases}$ **b.** $\begin{cases} 2a - 3b = 19 \\ a + 5b = -10 \end{cases}$

**c.** $\begin{cases} x + y + z = -2 \\ 2x - 3y + z = -11 \\ -x + 2y - z = 8 \end{cases}$ **d.** $\begin{cases} x + y + z = 6 \\ 2x - 3y + 4z = 3 \\ 4x - 8y + 4z = 12 \end{cases}$

**e.** $\begin{cases} x + 2y + z = 24 \\ 2x - 3y + z = -1 \\ x - 2y + 2z = 7 \end{cases}$ **f.** $\begin{cases} a + b + c = 0 \\ 3a - 2b + 5c = 1 \\ 2a + b + 2c = -1 \end{cases}$

**g.** $\begin{cases} a - 2b - 4c = -3 \\ 2a + 3b + 7c = 13 \\ 3a - 2b + 5c = -15 \end{cases}$ **h.** $\begin{cases} x + y + z + w = 10 \\ 2x - y + 3z - w = 5 \\ 3x + y + z + w = 12 \\ -x - y + z + w = 4 \end{cases}$

**8.** Solve each system by reducing an augmented matrix to echelon form. Write solution sets as fixed vectors plus linear combinations of other vectors when needed, as seen in Examples 4 and 5.

**a.** $\begin{cases} x - y + z = 2 \\ 3x + y - z = 3 \\ x + y + z = 3 \end{cases}$ **b.** $\begin{cases} x + 3y = 7 \\ x - y = -1 \end{cases}$

**c.** $\begin{cases} x - y + 3z = 5 \\ 5x + y - z = 9 \\ 2x + y + 3z = 7 \end{cases}$ **d.** $\begin{cases} x - y + 3z = 2 \\ 2x - y + z = 3 \end{cases}$

**e.** $\begin{cases} x - y + 3z = 2 \\ 2x - y + z = 3 \\ x + y + z = 4 \end{cases}$ **f.** $\begin{cases} x - y + 3z = 2 \\ 2x - y + z = 3 \\ 3x - 2y + 4z = 5 \end{cases}$

**g.** $\begin{cases} x - y + 3z = 2 \\ 2x - y + z = 3 \\ 3x - 2y + 4z = 6 \end{cases}$ **h.** $\begin{cases} x - y + z - w = 2 \\ 2x + 3y + z - w = 5 \end{cases}$

**9.** **a.** Suppose $A = (a, b, c, d)$ and $B = (e, f, g, h)$ are solutions to the system in Exercise 8**h**. Is $A + B$ a solution to the system? Explain.

**b.** Find a system of two equations in three unknowns so that if $A$ and $B$ are solutions, so is $A + B$.

**10.** Solve each homogeneous system.

**a.** $\begin{cases} x - y + z &= 0 \\ 3x + y - z &= 0 \\ x + y + z &= 0 \end{cases}$ **b.** $\begin{cases} x + 3y &= 0 \\ x - y &= 0 \end{cases}$

**11.** For each pair of homogeneous systems, solve them separately, then compare the two systems and their solution sets.

**a.** (i) $\begin{cases} x - y + 3z &= 0 \\ 5x + y - z &= 0 \\ 2x + y + 3z &= 0 \end{cases}$ (ii) $\begin{cases} x - y + 3z &= 0 \\ 2x + y + 3z &= 0 \end{cases}$

**b.** (i) $\begin{cases} x - y + 3z &= 0 \\ 2x - y + z &= 0 \\ x + y + z &= 0 \end{cases}$ (ii) $\begin{cases} x - y + 3z &= 0 \\ 2x - y + z &= 0 \end{cases}$

**c.** (i) $\begin{cases} x - y + 3z + w &= 0 \\ 2x - y + z + w &= 0 \\ 3x - 2y + 4z + w &= 0 \end{cases}$ (ii) $\begin{cases} x - y + 3z + w &= 0 \\ 2x - y + z + w &= 0 \\ 3x - 2y + 4z + w &= 0 \\ 3x + 2y + 4z + w &= 0 \end{cases}$

**12.** Solve each homogeneous system.

**a.** $\begin{cases} x - y + z - w &= 0 \\ 2x + 3y + z - w &= 0 \end{cases}$ **b.** $\begin{cases} x - y + 3z &= 0 \\ 2x - y + z &= 0 \\ 3x - 2y + 4z &= 0 \end{cases}$

**13.** On the basis of what you've seen in Exercises 10–12, under what circumstances *must* there be more than one solution to a system of equations? You've also seen that there *can* be more than one solution under other circumstances. Try to describe such a circumstance.

**Remember**

A *homogeneous system* is one where all the constants are 0.
The zero vector $O$ is a solution to any homogeneous system—why? For each system, determine if there are additional nonzero solutions.

$\longleftarrow$
Why is there more than one solution for part **a**?

## 3.3 Linear Combinations

In Chapter 1, you saw a definition for linear combination. It was not always easy to find out if one vector was a linear combination of two other vectors. The row-reduced echelon form of a matrix made up of those vectors is a quick way to find out.

**In this lesson, you will learn how to**

- determine whether a vector is a linear combination of other given vectors
- find the linear combination of one or more vectors that results in a given vector

Here is the definition you saw in Chapter 1:

**Definition**

Let $A_1, A_2, \ldots, A_k$ be vectors in $\mathbb{R}^n$ and let $c_1, c_2, \ldots, c_k$ be real numbers. A vector $B$ that can be written as

$$c_1A_1 + c_2A_2 + \cdots + c_kA_k = B$$

is a **linear combination** of $A_1$ through $A_k$.

← Informally, a *linear combination* is any sum of scalar multiples of a given set of vectors.

One way to find out *if* a vector $B$ can be expressed as a linear combination of other vectors also tells you *how* to make that combination (if it is possible).

**Example 1**

**Problem.** Determine whether $(17, 19, 12)$ can be written as a linear combination of $(1, 2, 3)$, $(4, 5, 6)$, and $(7, 8, 10)$. If so, how? If not, explain.

← This question uses the same vectors as Exercise 3 from the last lesson. Pay close attention to the process.

**Solution.** The vector $(17, 19, 12)$ is a linear combination of the given vectors if and only if there exist $a$, $b$, and $c$ so that

$$a(1, 2, 3) + b(4, 5, 6) + c(7, 8, 10) = (17, 19, 12)$$

This leads to three equations that must be solved simultaneously. Here are the three equations and the corresponding augmented matrix.

$$\begin{cases} a + 4b + 7c &= 17 \\ 2a + 5b + 8c &= 19 \\ 3a + 6b + 10c &= 12 \end{cases} \quad \rightarrow \quad \begin{pmatrix} 1 & 4 & 7 & 17 \\ 2 & 5 & 8 & 19 \\ 3 & 6 & 10 & 12 \end{pmatrix}$$

Note that the four vectors you started with—the three given vectors and the target vector—have become the four *columns* of the augmented matrix. Now, reduce the

← This makes it possible to move directly from the problem statement to the augmented matrix.

augmented matrix to echelon form, and write the corresponding system. If the system has a solution, then a linear combination exists. If not, it doesn't.

$$\begin{pmatrix} 1 & 4 & 7 & 17 \\ 2 & 5 & 8 & 19 \\ 3 & 6 & 10 & 12 \end{pmatrix} \rightarrow \begin{pmatrix} 1 & 0 & 0 & -12 \\ 0 & 1 & 0 & 23 \\ 0 & 0 & 1 & -9 \end{pmatrix} \rightarrow \begin{cases} a & = & -12 \\ b & = & 23 \\ c & = & -9 \end{cases}$$

So a linear combination can be made using the values $a = -12$, $b = 23$, and $c = -9$.

Check your solution to make sure it works.

$$\begin{aligned} -12(1,2,3) + 23(4,5,6) - 9(7,8,10) \\ &= (-12 + 92 - 63, -24 + 115 - 72, -36 + 138 - 90) \\ &= (17, 19, 12) \end{aligned}$$

The solution to the system also indicates that this is the *only* possible linear combination that produces $(17, 19, 12)$.

$\longleftarrow$ You will soon see cases where there are more than one possible linear combinations.

---

### For You to Do

**1.** Determine whether $(7, 8, 9)$ can be written as a linear combination of $(1, 2, 3)$ and $(4, 5, 6)$. If so, how? If not, explain.

---

### Example 2

**Problem.** Determine whether $(11, 13, 15)$ can be written as a linear combination of $(1, 2, 3)$, $(4, 5, 6)$, and $(7, 8, 9)$. If so, how? If not, explain.

**Solution.** Set up the matrix, reduce to echelon form, and translate.

$$\begin{pmatrix} 1 & 4 & 7 & 11 \\ 2 & 5 & 8 & 13 \\ 3 & 6 & 9 & 15 \end{pmatrix} \rightarrow \begin{pmatrix} 1 & 0 & -1 & -1 \\ 0 & 1 & 2 & 3 \\ 0 & 0 & 0 & 0 \end{pmatrix} \rightarrow \begin{cases} a - c & = & -1 \\ b + 2c & = & 3 \end{cases}$$

The final system says that $a \cdot (1, 2, 3) + b \cdot (4, 5, 6) + c \cdot (7, 8, 9) = (11, 13, 15)$ as long as $a - c = -1$ and $b + 2c = 3$. There is a linear combination for any value of $c$, since the equations can be rewritten as

$$\begin{cases} a & = & c - 1 \\ b & = & -2c + 3 \end{cases}$$

Find any particular linear combination by picking a value of $c$. For example, $c = 2$ leads to the solution $a = 1$, $b = -1$, $c = 2$ and the linear combination

$$1(1, 2, 3) - 1(4, 5, 6) + 2(7, 8, 9) = (11, 13, 15)$$

$\longleftarrow$ So, unlike the previous two examples, there are infinitely many linear combinations.

---

### Developing Habits of Mind

**Seek structural similarity.** The first three columns of the matrix in Example 2—both the original and the echelon form—are identical to the first three columns in For You to Do problem 1. In that problem, you found that $(7,8,9)$ was a linear combination of $(1,2,3)$ and $(4,5,6)$.

$$(7,8,9) = -1(1,2,3) + 2(4,5,6)$$

Example 2 asks to find out whether $(11,13,15)$ is a linear combination of $(1,2,3)$, $(4,5,6)$, and $(7,8,9)$. Since $(7,8,9)$ is itself a linear combination of the other two, this means that the question is identical to finding out whether $(11,13,15)$ is a linear combination of just $(1,2,3)$ and $(4,5,6)$.

For example, take the solution just found in Example 2:

$$1(1,2,3) - 1(4,5,6) + 2(7,8,9) = (11,13,15)$$

Replace $(7,8,9)$ by a combination of $(1,2,3)$ and $(4,5,6)$, then recollect terms.

$$1(1,2,3) - 1(4,5,6) + 2[-1(1,2,3) + 2(4,5,6)] = -1(1,2,3) + 3(4,5,6)$$

This means that if you can find a solution to $(11,13,15) = a\cdot(1,2,3) + b\cdot(4,5,6) + c\cdot(7,8,9)$, you can also find a solution to $(11,13,15) = p\cdot(1,2,3) + q\cdot(4,5,6)$.

It is even possible to solve the problem by ignoring the $(7,8,9)$ column altogether. Here is Example 3 resolved without the vector $(7,8,9)$:

$$\begin{pmatrix}1 & 4 & 11\\ 2 & 5 & 13\\ 3 & 6 & 15\end{pmatrix} \rightarrow \begin{pmatrix}1 & 0 & -1\\ 0 & 1 & 3\\ 0 & 0 & 0\end{pmatrix} \rightarrow \begin{cases} a = -1\\ b = 3\end{cases}$$

The result states that $(11,13,15)$ can be written just as a linear combination of $(1,2,3)$ and $(4,5,6)$.

⟵ Be careful! This works because $(7,8,9)$ is a linear combination of the other two vectors. Only if you are sure in advance that a vector is a linear combination of the other vectors should you use this technique.

### Example 3

**Problem.** Determine whether $(17,19,12)$ can be written as a linear combination of $(1,2,3)$, $(4,5,6)$, and $(7,8,9)$. If so, how? If not, explain.

**Solution.** As before, set up the augmented matrix and reduce it to echelon form.

$$\begin{pmatrix}1 & 4 & 7 & 17\\ 2 & 5 & 8 & 19\\ 3 & 6 & 9 & 12\end{pmatrix} \rightarrow \begin{pmatrix}1 & 0 & -1 & 0\\ 0 & 1 & 2 & 0\\ 0 & 0 & 0 & 1\end{pmatrix} \rightarrow \begin{cases} a - c = 0\\ b + 2c = 0\\ 0 = 1\end{cases}$$

The third row says $0a+0b+0c = 1$, and that's impossible for any choice of the variables. It is impossible to solve the system of equations needed here, and that means that $(17,19,12)$ *cannot* be written as a linear combination of $(1,2,3)$, $(4,5,6)$, and $(7,8,9)$.

⟵ You have seen that the vectors $(1,2,3)$, $(4,5,6)$, and $(7,8,9)$ all lie on the same plane in $\mathbb{R}^3$. Is $(17,19,12)$ on that plane?

### Developing Habits of Mind

**Reason about calculations.** If you arrive at $0 = 1$, where is the error in logic? All the algebra (represented by the elementary row operations that reduced the matrix to echelon form) was correct! The only place the error *can* be is in the initial statement of the linear combination

$$a(1,2,3) + b(4,5,6) + c(7,8,9) = (17,19,12)$$

that led to this set of algebraic steps. The original statement must be invalid, so there can be no numbers $a$, $b$, $c$ for which the equation is true.

---

The examples in this lesson are perfectly general: they suggest a method for determining whether a given vector $B$ is a linear combination of other vectors $A_1, \ldots, A_k$, and, if it is, the method shows how to find the coefficients. These facts are summarized in the following lemma.

**Lemma 3.1**

*To test if $B$ is a linear combination of $k$ vectors $A_1$ to $A_k$, build a matrix whose first $k$ columns are $A_1$ through $A_k$, with $B$ in column $(k+1)$. Then reduce to echelon form.*

*If any row of the echelon form has a nonzero entry in column $k+1$ and zeros everywhere else, the linear combination cannot be made. Otherwise, it can be made, and translating the final matrix back to a system of equations shows how.*

## Exercises

**1.** When asked to write $D$ as a linear combination of three vectors $A$, $B$, and $C$, Derman ended up with this echelon form:

$$\begin{pmatrix} 1 & 0 & 0 & 3 \\ 0 & 1 & 0 & -2 \\ 0 & 0 & 1 & 1 \end{pmatrix}$$

Write $D$ as a linear combination of $A$, $B$, and $C$.

**2.** Can $D = (13, 17, 3)$ be written as a linear combination of $A = (1, 2, 3)$, $B = (4, 5, 6)$, and $C = (7, 8, 0)$? If so, how?

**3.** Can $D = (13, 17, 3)$ be written as a linear combination of $A = (1, 2, 3), B = (4, 5, 6)$, and $C = (7, 8, 9)$? If so, how?

**4.** Can $D = (13, 17, 21)$ be written as a linear combination of $A = (1, 2, 3), B = (4, 5, 6)$, and $C = (7, 8, 9)$? If so, how?

**5.** To see if vector $D$ can be written as a linear combination of $A$, $B$, and $C$, you create a matrix and reduce to echelon form. For each of the following echelon forms, determine whether $D$ is a linear combination of $A$, $B$, and $C$, and if so, how.

**a.** $\begin{pmatrix} 1 & 0 & 0 & -\frac{30}{7} \\ 0 & 1 & 0 & -\frac{115}{7} \\ 0 & 0 & 1 & -60 \end{pmatrix}$ **b.** $\begin{pmatrix} 1 & 0 & 0 & 0 \\ 0 & 1 & 0 & 5 \\ 0 & 0 & 1 & 7 \end{pmatrix}$

**c.** $\begin{pmatrix} 1 & \frac{1}{2} & -\frac{3}{2} & 0 \\ 0 & 0 & 0 & 1 \\ 0 & 0 & 0 & 0 \end{pmatrix}$ **d.** $\begin{pmatrix} 1 & \frac{1}{2} & 0 & 0 \\ 0 & 0 & 1 & 0 \\ 0 & 0 & 0 & 1 \end{pmatrix}$

**e.** $\begin{pmatrix} 1 & \frac{1}{2} & 0 & \frac{5}{2} \\ 0 & 0 & 1 & 7 \\ 0 & 0 & 0 & 0 \end{pmatrix}$ **f.** $\begin{pmatrix} 1 & \frac{1}{2} & -\frac{3}{2} & \frac{5}{2} \\ 0 & 0 & 0 & 0 \\ 0 & 0 & 0 & 0 \end{pmatrix}$

**6.** Find the intersection of the two planes whose equations are $X \cdot (1, -1, 7) = -2$ and $X \cdot (2, -1, 12) = -1$ by setting up an augmented matrix and reducing to echelon form.

**7.** Find the intersection of the two lines with equations $X = (4, -2, 2) + t(-1, 2, 1)$ and $X = (1, 2, -1) + s(2, -3, 1)$.

$\longleftarrow$ First find the values of $t$ and $s$. Then what?

**8.** Find the intersection of the three planes whose equations are

$$\begin{aligned} x + 3y - z &= -2 \\ 2x + 7y - 3z &= -6 \\ x + y + 2z &= -3 \end{aligned}$$

**9.** Determine if $(13, 6, -16)$ can be written as a linear combination of $(1, -3, 2)$, $(5, 0, -4)$, and $(2, 1, -1)$. If so, show how.

**10.** **a.** In $\mathbb{R}^4$, find a vector orthogonal to all of these vectors:

$$(22, 16, 3, -2) \quad (22, 22, 9, -1) \quad (17, 16, 6, -1)$$

**b.** **Take It Further.** Show that there is a nonzero vector orthogonal to any three vectors in $\mathbb{R}^4$.

**11.** In $\mathbb{R}^4$, let

$$X = y(22, 16, 3, -2) + z(22, 22, 9, -1) + w(17, 16, 6, -1)$$

**a.** Find a coordinate equation in the form $X \cdot N = 0$ with the same solution set as the equation above. Compare your answer to the result in Exercise 10.

**b.** **Take It Further.** Show that every point that satisfies your solution to part **a** also satisfies the original equation.

$\longleftarrow$ So, your answer to part **a** defines a hyperplane in $\mathbb{R}^4$.

**12.** Suppose $P$, $A$, $B$, and $C$ are vectors in $\mathbb{R}^4$ and you have this equation:

$$X = P + yA + zB + wC$$

Explain why the solution also satisfies an equation of the form $X \cdot N = d$.

$\longleftarrow$ If $N$ is orthogonal to $A$, $B$, and $C$, dot both sides with $N$. Does such an $N$ exist?

**13.** Let $E$ be the set of points $X$ in $\mathbb{R}^4$ such that

$$X \cdot (1, 1, 1, 1) = 10$$

Find vectors $P$, $A$, $B$, and $C$ so that $E$ is the solution set to

$$X = P + yA + zB + wC$$

## 3.4 Linear Dependence and Independence

You can determine special relationships among a set of vectors by determining whether a linear combination of them can produce the zero vector.

**In this lesson, you will learn how to**

- recognize the identity matrix by notation and by form
- use row-reduced echelon form to find a linear combination of a vector in terms of a group of other vectors (if it exists)
- interpret linear dependency in $\mathbb{R}^3$ geometrically

**Example 1**

**Problem.** Write $(0,0,0)$ as a linear combination of the given vectors.

**a.** $A = (1,2,3), B = (4,5,6), C = (7,8,10)$

**b.** $A = (1,2,3), B = (4,5,6), C = (7,8,9)$

**Solution.** Using Lemma 3.1 from Lesson 3.3, construct an augmented matrix whose first three columns are the three given vectors, and whose fourth column is the zero vector. Then reduce the matrix to echelon form.

**a.** $$\begin{pmatrix} 1 & 4 & 7 & 0 \\ 2 & 5 & 8 & 0 \\ 3 & 6 & 10 & 0 \end{pmatrix} \rightarrow \begin{pmatrix} 1 & 0 & 0 & 0 \\ 0 & 1 & 0 & 0 \\ 0 & 0 & 1 & 0 \end{pmatrix} \rightarrow \begin{aligned} a &= 0 \\ b &= 0 \\ c &= 0 \end{aligned}$$

According to the echelon form, the only possible solution is to set all the scalars to 0.

$\longleftarrow$ The "all zero" solution is typically referred to as the **trivial solution**.

**b.** $$\begin{pmatrix} 1 & 4 & 7 & 0 \\ 2 & 5 & 8 & 0 \\ 3 & 6 & 9 & 0 \end{pmatrix} \rightarrow \begin{pmatrix} 1 & 0 & -1 & 0 \\ 0 & 1 & 2 & 0 \\ 0 & 0 & 0 & 0 \end{pmatrix} \rightarrow \begin{aligned} a - c &= 0 \\ b + 2c &= 0 \end{aligned}$$

For these vectors, there are nontrivial solutions: any combination of $a$, $b$, and $c$, where $a = c$ and $b = -2c$. One such possibility is

$$2(1,2,3) - 4(4,5,6) + 2(7,8,9) = (0,0,0)$$

which you can get by letting $c = 2$.

You can always find a linear combination of any set of vectors that will produce $O$: simply let all the scalars equal zero. But special relationships can be revealed if you can find nonzero combinations.

### Definition

- Vectors $A_1, A_2, \dots, A_k$ are **linearly dependent** if there are numbers $c_1, c_2, \dots, c_k$ that are *not all zero* so that

$$c_1 A_1 + c_2 A_2 + \cdots + c_k A_k = O$$

  where $O = (0, 0, \dots, 0)$.

- On the other hand, the vectors are **linearly independent** if the only solution to $c_1 A_1 + c_2 A_2 + \cdots + c_k A_k = O$ is $c_1 = c_2 = \cdots = c_k = 0$.

← The key phrase in the definition is *not all zero*, otherwise every set of vectors would be linearly dependent.

The definitions of linearly dependent and linearly independent are stated in terms of algebra. But there is a geometric interpretation in $\mathbb{R}^3$: three vectors in $\mathbb{R}^3$ are linearly dependent if and only if the three vectors lie in the same plane. So, for example, $(7, 8, 9)$ is in the plane spanned by $(1, 2, 3)$ and $(4, 5, 6)$ but $(7, 8, 10)$ is not. The next theorem states this in general.

← You can extend this idea to any dimension, even if you cannot visualize it: "$n$ vectors in $\mathbb{R}^n$ are linearly dependent if and only if they all lie in the same hyperplane." What does this mean in $\mathbb{R}^2$?

### Theorem 3.2

*Vectors $A_1, A_2, \dots, A_k$ are linearly dependent if and only if one of the vectors is a linear combination of the others.*

### For Discussion

1. Prove the statement "If vectors $A_1, A_2, \dots, A_k$ are linearly dependent, then one of the vectors $A_i$ is a linear combination of the other vectors."

   To begin, you can say that if the vectors are linearly dependent, then by the definition of linear dependence,

$$c_1 A_1 + c_2 A_2 + \cdots + c_k A_k = O$$

   for some set of numbers $c_1, c_2, \dots, c_k$ where at least one scalar is not zero. Let $c_i$ be the first nonzero among the scalars.

   Complete this half of the proof by showing that $A_i$ can be written as a linear combination of the other vectors.

← Since the statement in Theorem 3.2 is "if and only if," the proof comes in two parts to prove each direction. This For Discussion problem proves one direction. You will prove the other direction in Exercise 9.

### Example 2

**Problem.** Are the vectors $(1, 1, -1), (-1, 1, 0)$, and $(2, 1, 1)$ linearly dependent or independent?

**Solution.** The vectors are linearly dependent if $a(1, 1, -1) + b(-1, 1, 0) + c(2, 1, 1) = (0, 0, 0)$ has a nonzero solution. Set up an augmented matrix and reduce to echelon form.

$$\begin{cases} a - b + 2c &= 0 \\ a + b + c &= 0 \\ -a + c &= 0 \end{cases} \quad \rightarrow \quad \begin{pmatrix} 1 & -1 & 2 & 0 \\ 1 & 1 & 1 & 0 \\ -1 & 0 & 1 & 0 \end{pmatrix}$$

Reducing that matrix to echelon form gives

$$\begin{pmatrix} 1 & 0 & 0 & 0 \\ 0 & 1 & 0 & 0 \\ 0 & 0 & 1 & 0 \end{pmatrix}$$

So the only solution is $a = 0$, $b = 0$, $c = 0$. Thus, these three vectors are linearly independent.

**Remember**

You can construct the augmented matrix directly by writing the given vectors as the first columns, and then the desired linear combination vector as the last column.

### Example 3

**Problem.** Are the vectors $(1, 2, 3), (4, 5, 6)$, and $(7, 8, 9)$ linearly dependent or independent?

**Solution.** These vectors should be old friends by now. You could set up a matrix and reduce it to echelon form, but you've seen in Example 1 earlier in this lesson that

$$(7, 8, 9) = 2(4, 5, 6) - (1, 2, 3)$$

So, the vectors are linearly dependent by Theorem 3.2.

⟵ Note that $(1, 2, 3) - 2(4, 5, 6) + (7, 8, 9) = (0, 0, 0)$

Here is a graph of the three linearly independent vectors from Example 2.

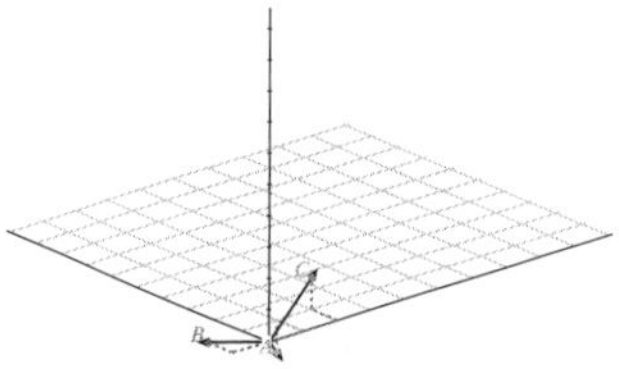

The vectors $(1, 1, -1), (-1, 1, 0)$, and $(2, 1, 1)$ do not lie in the same plane.

Here are two views of the three linearly dependent vectors from Example 3.

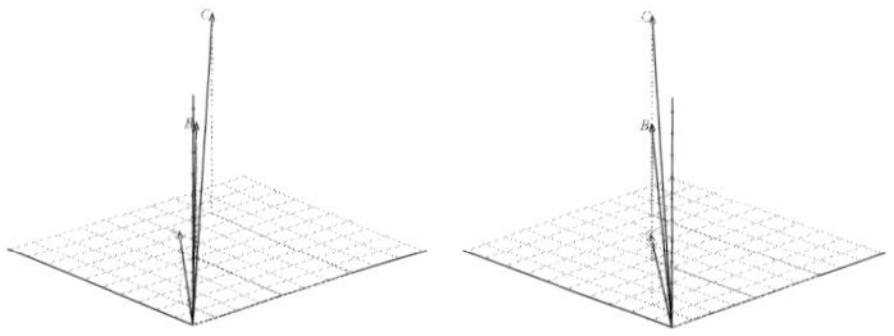

These vectors lie in the same plane.

### Facts and Notation

The elementary row operations will leave any column in a matrix consisting entirely of zeros unchanged. So when testing linear dependence or independence, adding that final column of zeros provides no additional information—it will not change no matter what elementary row operations you take to reduce the matrix.

⟵ Why?

When a set of vectors in $\mathbb{R}^3$ is linearly independent, the echelon form of the augmented matrix, which represents the only possible solution, is

$$\begin{pmatrix} 1 & 0 & 0 & 0 \\ 0 & 1 & 0 & 0 \\ 0 & 0 & 1 & 0 \end{pmatrix}$$

which translates to the equations $a = 0$, $b = 0$, $c = 0$. If you remove that extraneous final column of zeros, you get

$$\begin{pmatrix} 1 & 0 & 0 \\ 0 & 1 & 0 \\ 0 & 0 & 1 \end{pmatrix}$$

This form, a matrix with 1's on the diagonal and 0 otherwise, is an example in $\mathbb{R}^3$ of the result you get when you reduce a matrix of $n$ linearly independent vectors in $\mathbb{R}^n$. This matrix is called the **identity matrix**, with the shorthand $I$, for reasons that will become clear in the next chapter.

---

The statement about linearly independent vectors made in the Facts and Notation above can be expressed as the following theorem.

**Theorem 3.3**

*A set of $n$ vectors $A_1, \ldots, A_n$ in $\mathbb{R}^n$ is linearly independent if and only if the echelon form of the matrix whose columns are $A_1$ through $A_n$ is the identity matrix $I$.*

$\longleftarrow$ This theorem will grow into a really big theorem over the coming chapters.

## Exercises

1. Solve each problem by writing an augmented matrix and reducing it to echelon form.
   - **a.** Show that $(3, 1, 0)$, $(2, 4, 3)$, and $(0, -10, -9)$ are linearly dependent.
   - **b.** Are $(2, 1, 4)$, $(3, 0, 1)$, $(7, 1, 2)$, and $(8, -1, 0)$ linearly dependent?
   - **c.** Are $(4, 1, 2)$, $(3, 0, 1)$, and $(7, 1, 4)$ linearly dependent?
   - **d.** Write $(9, 8, 17)$ as a linear combination of $(2, 1, 3)$, $(4, 1, 2)$, and $(7, 5, 6)$.
   - **e.** Show that $(9, 8, 17)$, $(2, 1, 3)$, $(4, 1, 2)$, and $(7, 5, 6)$ are linearly dependent.
   - **f.** Show that $(10, 5, 10, 6)$ is a linear combination of $(4, 1, 2, 0)$ and $(3, 2, 4, 3)$.
   - **g.** Show that $(10, 5, 10, 6)$, $(4, 1, 2, 0)$, and $(3, 2, 4, 3)$ are linearly dependent.

2. 
   - **a.** Show that $(6, 9, 12)$ is a linear combination of the vectors $(1, 2, 3)$, $(4, 5, 6)$, and $(7, 8, 9)$.
   - **b.** Show that $(6, 9, 10)$ is *not* a linear combination of the vectors $(1, 2, 3)$, $(4, 5, 6)$, and $(7, 8, 9)$.

3. **a.** Describe all vectors that are orthogonal to $(1,0,3)$ and $(7,1,2)$.
   **b.** Describe all vectors that are orthogonal to $(1,0,3)$, $(7,1,2)$, and $(9,1,9)$.
   **c.** Describe all vectors that are orthogonal to $(1,0,3)$, $(7,1,2)$, and $(9,1,8)$.
   **d.** Which of the three sets of vectors (from parts **a**, **b**, and **c** above) are linearly independent?
   **e.** Compare your answers to parts **a**, **b**, and **c**. How did your answer to part **d** relate to those answers?

4. Find the intersection of the graphs of $X \cdot (2,1,4) = 8$ and $X \cdot (1,1,5) = 6$. Describe that intersection geometrically.

← That is, is it a point? a line? a plane? something else?

5. Find the intersection of the graphs of

$$X = (3,1,2) + t(4,1,6) \text{ and}$$
$$X = (-2,5,5) + s(3,-1,1)$$

6. Write $(1,2,5)$ as a linear combination of $(4,1,6)$ and $(3,-1,1)$.

7. Are the lines with equations $X = (3,1,4) + t(0,1,6)$ and $X = (1,1,7) + s(1,2,4)$ parallel, intersecting, or skew?

Lines are **skew** in $\mathbb{R}^3$ if they are neither parallel nor intersecting. This "third" possibility can't exist in $\mathbb{R}^2$, right?

8. Show that any set of vectors that contains $O$ is linearly dependent.

9. Complete the second half of the proof of Theorem 3.2 by proving the following statement:

   "Given a set of vectors $A_1, A_2, \ldots, A_k$, if one of the vectors $A_i$ is a linear combination of the other vectors in the set, then the vectors are linearly dependent."

10. Find numbers $a$, $b$, and $c$, not all zero, so that

$$a(7,1,3) + b(2,1,4) + c(8,-1,-6) = (0,0,0)$$

← Will any other values for $a$, $b$, and $c$ work?

11. Show that if $a(1,4,7)+b(2,5,8)+c(3,6,0) = (0,0,0)$, then $a = b = c = 0$.

12. Find all vectors orthogonal to the rows of

$$\begin{pmatrix} 3 & 1 & 2 & 0 \\ 4 & 1 & 6 & 1 \\ 1 & 3 & 2 & 0 \end{pmatrix}$$

13. Show that any vector orthogonal to the rows of

$$\begin{pmatrix} 1 & 2 & 3 \\ 4 & 5 & 6 \\ 7 & 8 & 0 \end{pmatrix}$$

must be $(0,0,0)$.

**14.** When asked to write $D$ as a linear combination of three vectors $A$, $B$, and $C$, Sasha ended up with this echelon form:

$$\begin{pmatrix} 1 & 0 & 2 & 0 \\ 0 & 1 & -1 & 0 \\ 0 & 0 & 0 & 0 \end{pmatrix}$$

**a.** Is $D$ a linear combination of $A$, $B$, and $C$? Explain.

**b.** Are $A$, $B$, $C$, and $D$ linearly dependent? Explain how you know.

## 3.5 The Kernel of a Matrix

In a number of exercises and examples, you found the set of vectors orthogonal to the rows of a given matrix. This set of vectors is called the **kernel** of a matrix.

⟵ For two examples, see Exercises 12 and 13 from Lesson 3.4.

**In this lesson, you will learn how to**

- find the kernel of a matrix
- understand the connection between the kernel of a matrix and the linear dependence/independence of the columns

### Definition

If $A$ is a matrix, the **kernel** of $A$, written $\ker(A)$, is the set of all vectors orthogonal to the rows of $A$.

⟵ Throughout the rest of this course, you'll see generalizations, refinements, and equivalent formulations of this definition. But this is a good place to start.

### Example 1

**Problem.** Find the kernel of $\begin{pmatrix} 3 & 1 & 2 \\ 4 & 0 & 1 \end{pmatrix}$.

**Solution.** Finding the kernel of this matrix is equivalent to finding the set of vectors orthogonal to both $(3, 1, 2)$ and $(4, 0, 1)$. If $X = (x, y, z)$ is a vector in the kernel, then $X$ satisfies

$$\begin{aligned} X \cdot (3,1,2) &= 0 \\ X \cdot (4,0,1) &= 0 \end{aligned} \rightarrow \begin{cases} 3x + y + 2z = 0 \\ 4x + z = 0 \end{cases}$$

This produces a homogeneous system of equations. Solve by writing the augmented matrix, then reducing to echelon form:

**Remember**

A system of equations is *homogeneous* if all its constants are zero.

$$\begin{pmatrix} 3 & 1 & 2 & 0 \\ 4 & 0 & 1 & 0 \end{pmatrix} \rightarrow \begin{pmatrix} 3 & 1 & 2 & 0 \\ 1 & -1 & -1 & 0 \end{pmatrix} \rightarrow \cdots \rightarrow \begin{pmatrix} 1 & -1 & -1 & 0 \\ 0 & 4 & 5 & 0 \end{pmatrix} \rightarrow \begin{pmatrix} 1 & 0 & \frac{1}{4} & 0 \\ 0 & 1 & \frac{5}{4} & 0 \end{pmatrix}$$

The kernel is the set of solutions generated by the equations

$$\begin{aligned} x &= -\tfrac{1}{4}z \\ y &= -\tfrac{5}{4}z \end{aligned}$$

The kernel of this matrix is thus

$$X = (-\tfrac{1}{4}z, -\tfrac{5}{4}z, z) \text{ or } X = t(-1, -5, 4)$$

Note that the zero vector $(0, 0, 0)$ is in the kernel (when $t = 0$), but there are other solutions, so the kernel includes other vectors as well.

### For You to Do

**1.** Find the kernel of each matrix:

**a.** $\begin{pmatrix} 1 & 2 & 3 \\ 1 & 4 & 7 \end{pmatrix}$

**b.** $\begin{pmatrix} 1 & 2 & 3 \\ 1 & 4 & 7 \\ 1 & 6 & 0 \end{pmatrix}$

**c.** $\begin{pmatrix} 1 & 2 & 3 \\ 1 & 4 & 7 \\ 1 & 6 & 11 \end{pmatrix}$

**d.** $\begin{pmatrix} 1 & 2 & 3 & 4 \\ 1 & 4 & 7 & 0 \\ 1 & 6 & 11 & 0 \\ 1 & 0 & 3 & 0 \end{pmatrix}$

### For Discussion

**2.** Will the zero vector $O$ *always* be in the kernel of any matrix?

### Developing Habits of Mind

**Look for shortcuts.** You may have noticed in the preceding For You to Do that when finding the kernel of a matrix, you could jump directly to the augmented matrix for the system of equations. The augmented matrix is built by tacking on a column of zeros to the right of the original.

But you can even skip that part! As stated in the last lesson, any row operations performed on a matrix leaves a column of zeros unaffected. That means you can take the original matrix, reduce it directly to echelon form, then tack on that column of zeros at the end. Or, you can just picture it being there, and not bother inserting it at all!

### Minds in Action Episode 10

TONY: I feel like we are doing the same thing over and over in this chapter.

DERMAN: Yeah, I keep typing "rref" a lot.

TONY: No, I mean we keep using the echelon form to answer different questions. Sometimes, the echelon form is really simple, and it makes the answer simple.

SASHA: So what have you found?

TONY: One thing that's been nice in this lesson is when the constants are all zero, you never run into that sticky situation where there is no solution to a system.

DERMAN: Oh, the $0 = 1$ thing. That always confuses me.

TONY: There always has to be at least one solution, the zero vector. That happens with these kernels, too. $O$ is always in the kernel.

SASHA: I noticed that a lot of the time $O$ is the *only* thing in the kernel.

TONY: Right. In the last lesson, zero was the only solution when the vectors were linearly independent.

SASHA: And in this lesson . . . hey, that's a great idea!

TONY: What's a great idea?

SASHA: I was wondering when the kernel would be just the zero vector. And you just figured it out!

TONY: Ohh. Well, I'll take credit if you want. Very smooth, Sasha.

DERMAN: You guys are confusing. I need a worked-out example.

---

### Example 2

**Problem.**

**a.** Show that the vectors $(1, 1, 1)$, $(2, 4, 6)$, and $(3, 7, 2)$ are linearly independent.

**b.** Find the kernel of $\begin{pmatrix} 1 & 2 & 3 \\ 1 & 4 & 7 \\ 1 & 6 & 2 \end{pmatrix}$.

**Solution.**

**a.** Write these vectors as columns of a matrix. As stated in Theorem 3.3, the vectors are independent if and only if the echelon form of that matrix is the identity matrix. The matrix reduced to echelon form is

$$\begin{pmatrix} 1 & 2 & 3 \\ 1 & 4 & 7 \\ 1 & 6 & 2 \end{pmatrix} \rightarrow \begin{pmatrix} 1 & 0 & 0 \\ 0 & 1 & 0 \\ 0 & 0 & 1 \end{pmatrix}$$

which is the identity. So the vectors are linearly independent.

**Remember**

The identity matrix has 1's on its diagonal and zeros everywhere else.

**b.** Note that this is the same matrix from part **a**. To find vectors $(x, y, z)$ orthogonal to the rows of this matrix, look for solutions to the system

$$\begin{cases} x + 2y + 3z &= 0 \\ x + 4y + 7z &= 0 \\ x + 6y + 2z &= 0 \end{cases} \rightarrow \begin{pmatrix} 1 & 2 & 3 & 0 \\ 1 & 4 & 7 & 0 \\ 1 & 6 & 2 & 0 \end{pmatrix}$$

What is the echelon form of this augmented matrix? Look back at part **a**, which uses the same matrix without a column of zeros. The column of zeros has no effect so the echelon form must be

$$\begin{pmatrix} 1 & 0 & 0 & 0 \\ 0 & 1 & 0 & 0 \\ 0 & 0 & 1 & 0 \end{pmatrix}$$

This states that the zero vector $(0, 0, 0)$ is the only vector orthogonal to the three-row vectors. The kernel is only the zero vector.

---

This example suggests the following two theorems.

**Theorem 3.4**

*The kernel of a matrix is $O$ if and only if the columns of the matrix are linearly independent.*

**Theorem 3.5**

*Given a square matrix $A$,* $\ker(A) = O$ *if and only if* $\text{rref}(A) = I$.

⟵
A **square matrix** is a matrix that has the same number of rows and columns.

**For Discussion**

**3.** Sketch proofs for these two theorems using the examples as templates, and using the theorems from Lesson 3.4.

**Developing Habits of Mind**

**Look for connections.** The theorems in this chapter say that these conditions on a square matrix $A$ are all connected:

1. The columns of $A$ are linearly independent.
2. $\text{rref}(A) = I$.
3. $\ker(A) = O$.

In later chapters, you'll add to this list of connections. It turns out these statements are *equivalent*: for a given matrix $A$, if one is true, they are all true, and if one is false, they all fail.

This means that if you are looking for any one of these properties to be true for a matrix, you can choose to establish whichever property is most convenient. In some situations, this is a very good reason to construct a matrix in the first place.

## Exercises

**1.** In $\mathbb{R}^2$, characterize

**a.** all linear combinations of $(1, 3)$
**b.** all linear combinations of $(1, 3)$ and $(3, 6)$
**c.** all linear combinations of $(1, 3)$ and $(-3, -9)$

⟵
Draw a picture, if you think it will help.

**2.** In $\mathbb{R}^3$, give a geometric description of

**a.** all linear combinations of $(1, 3, 7)$
**b.** all linear combinations of $(1, 3, 7)$ and $(3, 6, -1)$
**c.** all linear combinations of $(1, 3, 7)$, $(3, 6, -1)$, and $(5, 6, 1)$
**d.** all linear combinations of $(1, 3, 7)$, $(3, 6, -1)$, and $(4, 9, 6)$

⟵
The set of all linear combinations of $(1, 3, 7)$, $(3, 6, -1)$, and $(5, 6, 1)$ is called the **linear span** of these vectors.

**3.** In $\mathbb{R}^3$, give a geometric description of

**a.** all vectors orthogonal to $(1, 3, 7)$

**b.** the kernel of $\begin{pmatrix} 1 & 3 \\ 3 & 6 \\ 7 & -1 \end{pmatrix}$

**c.** the kernel of $\begin{pmatrix} 1 & 3 & 5 \\ 3 & 6 & 6 \\ 7 & -1 & 1 \end{pmatrix}$

**d.** the kernel of $\begin{pmatrix} 1 & 3 & 4 \\ 3 & 6 & 9 \\ 7 & -1 & 6 \end{pmatrix}$

**4.** For each matrix, given in echelon form, find its kernel:

**a.** $\begin{pmatrix} 1 & 0 & 3 \\ 0 & 1 & -2 \\ 0 & 0 & 0 \end{pmatrix}$ **b.** $\begin{pmatrix} 1 & 0 & -4 & 3 \\ 0 & 1 & 2 & -5 \\ 0 & 0 & 0 & 0 \end{pmatrix}$ **c.** $\begin{pmatrix} 1 & 0 & 0 \\ 0 & 1 & 0 \\ 0 & 0 & 1 \end{pmatrix}$

← You may prefer to write an additional column of zeros to the right of the matrices given in these exercises. Remember, the zero vector is in the kernel of any matrix.

**5.** For each given matrix, find its kernel.

**a.** $\begin{pmatrix} 3 & 1 & 2 \\ 1 & 2 & 4 \end{pmatrix}$ **b.** $\begin{pmatrix} 3 & 1 & 0 & 4 \\ 1 & 4 & 3 & 1 \end{pmatrix}$

**c.** $\begin{pmatrix} 3 & 1 & 0 & 4 \\ 1 & 4 & 3 & 1 \\ 2 & -3 & -3 & 3 \end{pmatrix}$ **d.** $\begin{pmatrix} 1 & 2 & 3 \\ 4 & 5 & 6 \\ 7 & 8 & 9 \end{pmatrix}$

**e.** $\begin{pmatrix} 1 & 2 & 3 \\ 4 & 5 & 6 \\ 7 & 8 & 0 \end{pmatrix}$ **f.** $\begin{pmatrix} 1 & 3 & 0 \\ 2 & 1 & 3 \\ 3 & 4 & 3 \\ 5 & 5 & 6 \end{pmatrix}$

**6.** **a.** Suppose $A$ is an $m \times n$ matrix and vectors $X$ and $Y$ are in $\ker(A)$. Show that every linear combination of $X$ and $Y$ is also in $\ker(A)$.

**b.** Show that the kernel of a matrix is either just the zero vector or contains infinitely many vectors.

**7.** **Take It Further.** If $A$ is a $6 \times 11$ matrix, show that $\ker(A)$ is infinite.

**8.** Without a calculator, compute the row-reduced echelon form of

$$A = \begin{pmatrix} 1 & 0 & -2 & 8 & 4 \\ 1 & 0 & -1 & 5 & 2 \\ 2 & 0 & -3 & 18 & 6 \end{pmatrix}$$

**9.** Find the solution set of the system

$$\begin{cases} 2x + 3y - z + w = 17 \\ x + y + z - 2w = -6 \\ x + 2z + 4w = 18 \\ 2y + z - 2w = 5 \end{cases}$$

**10.** When asked to solve a system of equations, Derman ended up with this echelon form:

$$\begin{pmatrix} 1 & 0 & 1 & 0 \\ 0 & 1 & -2 & 1 \end{pmatrix}$$

Find a solution for the system of equations.

**11.** Find $\ker(A)$ if $A = \begin{pmatrix} 1 & 1 & 1 & 2 \\ 2 & 1 & -1 & 1 \\ 3 & 2 & 0 & 3 \end{pmatrix}$.

**12.** Find the intersection of the two lines with equations $X = (3, 0, 1) + t(1, 1, 2)$ and $X = (7, 3, 4) + s(2, 1, -1)$.

**13.** Find a set of vectors in $\mathbb{R}^3$ so that all of the following are true:

- $(1, 1, 0)$ is in your set.
- Every vector $(x, y, z)$ in $\mathbb{R}^3$ is a linear combination of the vectors in your set.
- The vectors in your set are linearly independent.

**14.** Find a set of vectors in $\mathbb{R}^4$ so that all of the following are true:

- $(1, 0, 1, 0)$ is in your set.
- Every vector $(x, y, z, w)$ in $\mathbb{R}^4$ is a linear combination of the vectors in your set.
- The vectors in your set are linearly independent.

**15.** Let $A$ be an $n \times n$ matrix. Show that if any one of these statements is true, the other three are also true.

**a.** $\text{rref}(A) = I$.
**b.** The columns of A are linearly independent.
**c.** $\ker(A) = O$.
**d.** Every vector in $\mathbb{R}^n$ is a linear combination of the columns of $A$.

$\longleftarrow$
Another way to read Exercise 15**d**: "The linear span of the columns of $A$ is all of $\mathbb{R}^n$."

**16.** Show that if any one of the statements in Exercise 15 are false, they all are.

# Chapter 3 Mathematical Reflections

These problems will help you summarize what you have learned in this chapter.

1. Solve each system by reducing an augmented matrix to echelon form. Write solution sets as fixed vectors plus linear combinations of other vectors when needed.

   **a.** $\begin{cases} x + 4y + 3z &= -10 \\ 2x + 3y - z &= 5 \\ 5x - 5y - 8z &= 15 \end{cases}$ **b.** $\begin{cases} x + y + z &= 4 \\ 2x - y + 5z &= -1 \\ 3x + 2y + 4z &= 9 \end{cases}$

2. Let $A = (2, 1, -2)$, $B = (3, 4, 1)$, and $C = (2, 1, 3)$.
   Determine whether $(4, 12, 2)$ can be written as a linear combination of $A$, $B$, and $C$. If so, how? If not, explain.

3. Are the vectors $(2, 1, -2)$, $(3, 4, 1)$, and $(2, 1, 3)$ linearly dependent or independent? Do these vectors lie in the same plane?

4. Find the kernel of each matrix:

   **a.** $\begin{pmatrix} 1 & -3 & 2 \\ 1 & 5 & -1 \\ 1 & 6 & 3 \end{pmatrix}$ **b.** $\begin{pmatrix} 1 & -3 & 2 \\ 1 & 5 & -1 \\ 1 & -19 & 8 \end{pmatrix}$ **c.** $\begin{pmatrix} 1 & -3 & 2 & 1 \\ 1 & 5 & -1 & 3 \\ 1 & 6 & 3 & -2 \end{pmatrix}$

5. In $\mathbb{R}^3$, give a geometric description of

   **a.** all linear combinations of $(4, -1, 2)$, $(3, 7, -5)$, and $(5, -9, 10)$
   **b.** all linear combinations of $(4, -1, 2)$, $(3, 7, -5)$, and $(5, -9, 9)$
   **c.** the kernel of $\begin{pmatrix} 4 & -1 & 2 \\ 3 & 7 & -5 \\ 5 & -9 & 9 \end{pmatrix}$

6. How can you tell if a system of equations has a solution?

7. How can you tell if three vectors are linearly independent in $\mathbb{R}^2$? in $\mathbb{R}^3$? in $\mathbb{R}^n$?

8. Find the kernel of this matrix:

   $$\begin{pmatrix} 3 & -2 & 1 & 4 \\ 1 & 2 & -1 & 0 \\ 7 & -10 & 5 & 12 \end{pmatrix}$$

   Describe this kernel geometrically.

**Vocabulary**

In this chapter, you saw these terms and symbols for the first time. Make sure you understand what each one means, and how it is used.

- augmented matrix
- coefficient matrix
- dimension
- elementary row operations
- equivalent matrices
- equivalent systems
- Gaussian elimination
- kernel
- linear combination
- linearly dependent
- linearly independent
- matrix
- row-reduced echelon form
- trivial solution

# Chapter 3 Review

In Lesson 3.2, you learned to

- represent a system of equations with an augmented matrix
- reduce an augmented matrix to its row-reduced echelon form
- interpret the nature of the solution of a system given its row-reduced echelon form
- express an infinite solution set to a system as a vector equation

The following problems will help you check your understanding.

**1.** Solve each system by reducing an augmented matrix to echelon form. Write solution sets as fixed vectors plus linear combinations of other vectors when needed, as seen in Examples 4 and 5.

**a.** $\begin{cases} x - 3y - 2z &= -15 \\ 2x - 5y + 3z &= 1 \\ 4x - y + 3z &= 17 \end{cases}$ **b.** $\begin{cases} x - y + 2z &= 8 \\ 3x + y - z &= 1 \\ 4x + z &= -2 \end{cases}$

**c.** $\begin{cases} x - 3y - 5z &= 18 \\ 3x + 2y - 4z &= 21 \\ 2x + 5y + z &= 3 \end{cases}$ **d.** $\begin{cases} 2x - 3y - 6z &= 5 \\ x - 2y - 8z &= 7 \end{cases}$

**2.** Find the linear combination of $(-1, -5, 3)$, $(4, 2, -5)$, and $(7, -1, 2)$ that produces $(22, -16, 19)$ by solving this system of equations:

$$\begin{cases} -a + 4b + 7c &= 22 \\ -5a + 2b - c &= -16 \\ 3a - 5b + 2c &= 19 \end{cases}$$

**3.** Find all vectors that are orthogonal to $(1, 3, 1)$, $(2, 1, 1)$, and $(3, 4, 2)$ by solving this system of equations:

$$\begin{cases} X \cdot (1,3,1) &= 0 \\ X \cdot (2,1,1) &= 0 \\ X \cdot (3,4,2) &= 0 \end{cases} \longrightarrow \begin{cases} 1x + 3y + 1z &= 0 \\ 2x + 1y + 1z &= 0 \\ 3x + 4y + 2z &= 0 \end{cases}$$

In Lesson 3.3, you learned to

- determine whether a vector is a linear combination of other given vectors
- find the linear combination of one or more vectors that results in a given vector

The following problems will help you check your understanding.

**4.** To see if vector $D$ can be written as a linear combination of $A$, $B$, and $C$, you create a matrix whose columns are $A$, $B$, and $C$, and $D$, and reduce to echelon form. For each of the following echelon forms, determine whether $D$ is a linear combination of $A$, $B$, and $C$, and if so, how.

**a.** $\begin{pmatrix} 1 & 0 & 0 & 3 \\ 0 & 1 & 0 & -4 \\ 0 & 0 & 1 & -2 \end{pmatrix}$ **b.** $\begin{pmatrix} 1 & 0 & 0 & -5 \\ 0 & 1 & 0 & \frac{4}{5} \\ 0 & 0 & 1 & \frac{1}{5} \end{pmatrix}$

**c.** $\begin{pmatrix} 1 & 0 & -2 & 3 \\ 0 & 1 & -1 & 1 \\ 0 & 0 & 0 & 0 \end{pmatrix}$ **d.** $\begin{pmatrix} 1 & 0 & 0 & -\frac{2}{9} \\ 0 & 1 & 0 & \frac{4}{9} \\ 0 & 0 & 0 & 2 \end{pmatrix}$

**5.** Determine if $D = (-15, 12, 4)$ can be written as a linear combination of $A = (1, 1, 2)$, $B = (4, 3, 1)$, and $C = (-2, 3, 0)$? If so, show how.

**6.** Find the intersection of the two planes whose equations are $X \cdot (2, -3, 5) = 4$ and $X \cdot (1, -1, 1) = -3$ by setting up an augmented matrix and reducing to echelon form.

In Lesson 3.4, you learned to

- recognize the identity matrix by notation and by form
- use row-reduced echelon form to find a linear combination of a vector in terms of a group of other vectors (if it exists)
- interpret linear dependency in $\mathbb{R}^3$ geometrically

The following questions will help you check your understanding.

**7.** Write $(0, 0, 0)$ as a linear combination of the given vectors. Are the vectors linearly dependent or independent?

**a.** $A = (2, -3, 4)$, $B = (-5, 1, 0)$, $C = (1, -8, 12)$
**b.** $A = (2, -3, 4)$, $B = (-5, 1, 0)$, $C = (1, -8, 11)$

**8.** Let $A = (2, 1, -2)$, $B = (3, 4, 1)$, and $C = (2, 1, 3)$.

**a.** Describe all vectors that are orthogonal to $A$, $B$, and $C$.
**b.** Are these vectors linearly dependent or independent? How do you know?

**9.** Let $A = (-3, 4, 5)$, $B = (2, 1, -1)$, and $C = (-8, 7, 11)$.

**a.** Describe all vectors that are orthogonal to $A$, $B$, and $C$.
**b.** Are these vectors linearly dependent or independent? How do you know?

In Lesson 3.5, you learned to

- find the kernel of a matrix
- understand the connection between the kernel of a matrix and the linear dependence/independence of the columns
- show that the vectors in the kernel of a matrix are orthogonal to the row vectors of the matrix

The following problems will help you check your understanding.

**10.** Find the kernel of each matrix.

**a.** $\begin{pmatrix} 1 & 3 & 5 \\ 1 & 2 & 4 \end{pmatrix}$

**b.** $\begin{pmatrix} 1 & 3 & 5 \\ 1 & 2 & 4 \\ 1 & 5 & 6 \end{pmatrix}$

**c.** $\begin{pmatrix} 1 & 3 & 5 \\ 1 & 2 & 4 \\ 1 & 5 & 7 \end{pmatrix}$

**d.** $\begin{pmatrix} 1 & 3 & 5 & 11 \\ 1 & 2 & 4 & 10 \\ 1 & 5 & 7 & 13 \end{pmatrix}$

**11.** In $\mathbb{R}^3$, give a geometric description of

**a.** all linear combinations of $(2, 4, 7)$ and $(3, 5, 9)$
**b.** all linear combinations of $(2, 4, 7)$, $(3, 5, 9)$, and $(1, 1, 3)$
**c.** all linear combinations of $(2, 4, 7)$, $(3, 5, 9)$, and $(1, 1, 2)$

**12.** In $\mathbb{R}^3$, give a geometric description of

**a.** all vectors orthogonal to $(2, 4, 7)$

**b.** the kernel of $\begin{pmatrix} 2 & 3 & 1 \\ 4 & 5 & 1 \\ 7 & 9 & 2 \end{pmatrix}$

**c.** the kernel of $\begin{pmatrix} 2 & 3 & 1 \\ 4 & 5 & 1 \\ 7 & 9 & 3 \end{pmatrix}$

# Chapter 3 Test

## Multiple Choice

**1.** To solve a system of equations, Tanya finds the reduced echelon form of

$$\begin{pmatrix} 1 & 0 & 1 & -2 \\ 0 & 1 & -2 & 3 \\ 0 & 0 & 0 & 0 \end{pmatrix}$$

What is the solution to the system?

**A.** $X = (0, 0, 0)$ **B.** $X = (-2, 3, 0)$
**C.** $X = (-2, 3, 0) + t(-1, 2, 1)$ **D.** No solution

**2.** Let $A = (2, -3, 1)$, $B = (4, 0, 5)$, $C = (-2, 1, -1)$, and $D = (-6, -7, -9)$. Which equation shows that $D$ is a linear combination of $A$, $B$, and $C$?

**A.** $D = -3A + 2B - 2C$ **B.** $D = -2A + 2B + 3C$
**C.** $D = 2A - 2B + 3C$ **D.** $D = 3A - 2B + 2C$

**3.** Which equation describes all vectors that are orthogonal to $(-2, 3, 2)$ and $(3, -4, 1)$?

**A.** $X = (0, 0, 0)$ **B.** $X = (11, 8, -1)$
**C.** $X = t(-11, -8, 1)$ **D.** $X = t(8, 11, -1)$

**4.** In $\mathbb{R}^3$, let $A = (2, -1, 5)$ and $B = (1, 0, -3)$. Which of the following describes all linear combinations of $A$ and $B$?

**A.** the point $O$
**B.** the line $X = t(2, -1, 5)$
**C.** the plane $X = s(2, -1, 5) + t(1, 0, -3)$
**D.** all of $\mathbb{R}^3$

**5.** A matrix in echelon form is

$$\begin{pmatrix} 1 & 0 & -3 & 2 \\ 0 & 1 & 2 & -4 \\ 0 & 0 & 0 & 0 \end{pmatrix}$$

What is the kernel of this matrix?

**A.** $O$
**B.** $X = s(3, -2, 0) + t(-2, 4, 0)$
**C.** $X = s(3, -2, 0, 1) + t(-2, 4, 1, 0)$
**D.** $X = s(3, -2, 1, 0) + t(-2, 4, 0, 1)$

**6.** Suppose that $A$ is a $3 \times 3$ matrix and $\text{rref}(A) = I$. Which of the following statements is *not* true?

**A.** The kernel of $A$ is $O$.
**B.** The column vectors lie in the same plane.
**C.** The zero vector $(0, 0, 0)$ is the only vector orthogonal to the row vectors.
**D.** The column vectors of $A$ are linearly independent.

**Open Response**

**7.** Solve the system using Gaussian elimination:

$$\begin{cases} x + 2y + 5z &= 15 \\ 3x - 3y + 2z &= 20 \\ 2x - 5y + 4z &= 33 \end{cases}$$

**8.** Determine if $D = (4, 16, -22)$ can be written as a linear combination of $A = (1, -2, 5)$, $B = (3, 4, -8)$, and $C = (5, -2, 0)$. If so, show how.

**9.** Find the intersection of the two lines with equations $X = (1, -2, 4) + t(3, -1, 6)$ and $X = (2, 1, 6) + s(2, -2, 4)$.

**10.** Let $A = (2, 7, 5)$, $B = (-3, 3, 6)$, and $C = (2, 1, -1)$.

**a.** Write $(0, 0, 0)$ as a linear combination of $A$, $B$, and $C$.
**b.** Are $A$, $B$, and $C$ linearly dependent or independent? Explain.

**11.** Let $A = (1, -3, 2)$, $B = (2, 0, -4)$, and $C = (3, 5, -1)$.

**a.** Describe all vectors that are orthogonal to $A$, $B$, and $C$.
**b.** Are $A$, $B$, and $C$ linearly dependent or independent? Explain.

**12.** Find the kernel of each matrix.

**a.** $\begin{pmatrix} 1 & 4 & 3 \\ -2 & 1 & 6 \\ 5 & -4 & 1 \end{pmatrix}$ **b.** $\begin{pmatrix} 1 & -1 & 0 & 3 \\ 1 & -2 & 1 & 4 \\ 3 & -2 & 4 & 2 \end{pmatrix}$

CHAPTER

# 4 Matrix Algebra

In the last chapter, you used matrices as a bookkeeping organizer: the matrix kept track of the coefficients in a linear system. By operating on the matrix in certain ways (using the elementary row operations), you transformed the matrix without changing the solution set of the underlying linear system. But matrices can be objects in their own right. They have their own algebra, their own basic rules, and their own operations beyond the elementary row operations.

If you think of a vector as a matrix, you can start extending vector operations, such as addition and scalar multiplication, to matrices of any dimension. This chapter defines three operations—addition, scaling, and multiplication—and develops an algebra of matrices that allows you to perform complicated calculations and solve seemingly difficult problems very efficiently. So, think of this chapter as an expansion of your algebra toolbox.

**By the end of this chapter, you will be able to answer questions like these:**

1. When can you multiply two matrices?
2. How can you tell if a matrix equation has a unique solution?
3. Let $A = \begin{pmatrix} 1 & 4 & 3 \\ -1 & 1 & 2 \\ 5 & 4 & 2 \end{pmatrix}$. What is $A^{-1}$.

**You will build good habits and skills for ways to**

- look for similarity in structure
- reason about calculations
- create a process
- seek general results
- look for connections

### Vocabulary and Notation

- $A_{ij}$, $A_{*j}$, $A_{i*}$
- diagonal matrix
- entry
- equal matrices
- identity matrix
- inverse
- invertible matrix, nonsingular matrix
- kernel
- lower triangular matrix
- $m \times n$ matrix
- matrix multiplication, matrix product
- multiplication by a scalar
- scalar matrix
- singular matrix
- skew-symmetric matrix
- square matrix
- sum of matrices
- symmetric matrix
- transpose
- upper triangular matrix

## 4.1 Getting Started

A **matrix** is a rectangular array of numbers. Here are some matrices:

$$H = \begin{pmatrix} 1 & 2 \\ 3 & 4 \end{pmatrix}, \qquad I = \begin{pmatrix} 1 & 0 & 0 \\ 0 & 1 & 0 \\ 0 & 0 & 1 \end{pmatrix},$$

← You've seen matrices before.

$$J = \begin{pmatrix} 0 & 1 & 1 \\ 1 & 0 & 1 \\ 1 & 1 & 0 \end{pmatrix}, \qquad K = \begin{pmatrix} 3 & -1 & 12 & 5 \\ 0 & 6 & 3 & 1 \\ 0 & 0 & -2 & 10 \end{pmatrix},$$

$$L = \begin{pmatrix} 2 & 1 \\ 3 & 4 \end{pmatrix}, \qquad M = \begin{pmatrix} 2 & 3 & 4 \\ 5 & 6 & 7 \end{pmatrix},$$

$$N = \begin{pmatrix} 2 & 5 \\ 3 & 6 \\ 4 & 7 \end{pmatrix}, \qquad O = \begin{pmatrix} 0 & 0 \\ 0 & 0 \\ 0 & 0 \end{pmatrix}$$

**1.** Describe conditions that you think would make two matrices equal.

← Make your definition so clear that you could read it over the phone and the person at the other end would understand.

**2.** **a.** Find three pairs of matrices above that you think can be added together. Calculate the sum of each pair.
**b.** Find three pairs of matrices above that you think *cannot* be added together, and explain why.

**3.** **a.** Cal says that matrix $K$ is basically a pile of four vectors. What do you think he means by this? What dimension would these vectors have?
**b.** Rose says that matrix $K$ is actually a pile of three vectors. What do you think she means by this? What dimension would these vectors have?

**4.** **a.** If $A$ is a vector and $c$ is a real number, what is $cA$ and how is it calculated?
**b.** If $A$ is a matrix and $c$ is a real number, how do you think $cA$ should be defined?

← What would Cal and Rose say about this?

The map below shows the number of daily nonstop flights between certain U.S. cities on an airline.

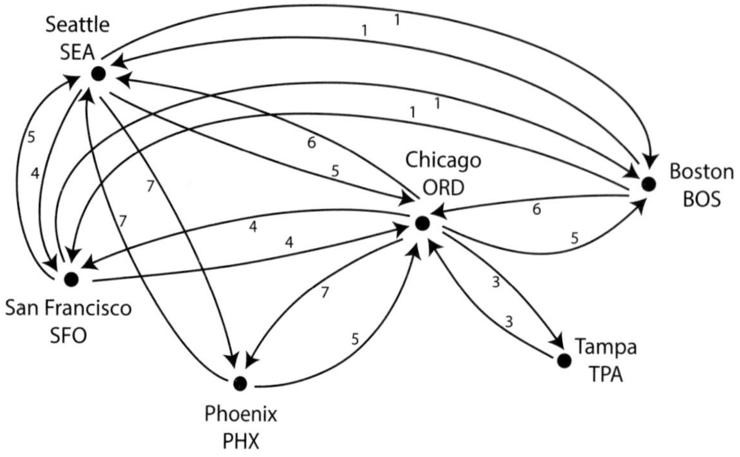


**5.** There is no direct flight from Boston to Phoenix.

**a.** If you had a one-stop flight from Boston to Phoenix, in what cities could the stop be?

**b.** How many possible one-stop flights are there from Boston to Phoenix?

**c.** How many possible one-stop flights are there from Phoenix to Boston?

**6.** Carrie lives in Chicago.

*Carrie:* I had some great flights, nonstop both ways.

Assuming she went to one of the five other cities on the map above, how many possible round trips could she have taken?

**7.** A six-by-six matrix can be used to contain all the information in the map. Complete the matrix below; some entries have been given.

| TO→ FROM↓ | BOS | ORD | SEA | SFO | TPA | PHX |
|---|---|---|---|---|---|---|
| BOS | | 6 | | | | 0 |
| ORD | 5 | | | | | |
| SEA | | | 0 | | | |
| SFO | 1 | | | | 0 | |
| TPA | | | | | | |
| PHX | | | 7 | | | |

⟵ The 6 says there are six direct flights from Boston to Chicago, while the 5 says there are five direct flights from Chicago to Boston. What should the entry be for Boston to Tampa?

**8.** Use *only the matrix* to answer the following questions:

**a.** How many one-stop flights are there from Boston to Phoenix?

**b.** How many one-stop flights are there from Chicago to Seattle?

**c.** How many one-stop flights are there from Chicago to Chicago?

**d.** How many one-stop flights are there from San Francisco to Tampa?

⟵ Part **c** is just another way to state the trip Carrie takes in Exercise 6.

**9.** Describe how to use the matrix to find the number of one-stop flights from one given city to another given city.

**10.** Write $a$ and $b$ in terms of $x$ and $y$.

$$a = 3m + 2n + 4p$$
$$b = m - 4n + 6p$$

$$m = 2x + 3y$$
$$n = 4x - 5y$$
$$p = x + 10y$$

**11.** Write $a$ and $b$ in terms of $x$ and $y$.

$$a = 2m + 4n + p$$
$$b = 3m - 5n + 10p$$

$$m = 3x + y$$
$$n = 2x - 4y$$
$$p = 4x + 6y$$

**12.** Compare the steps you followed to solve Exercises 8 and 10. How are they similar?

**Habits of Mind**

Look for common structure in different problems.

## 4.2 Adding and Scaling Matrices

---

Vectors can be added and scaled just like numbers. Those operations can be extended, using the extension program, to matrices of any dimension.

**In this lesson, you will learn how to**

- use clear notation for the different entries in a matrix
- determine when two matrices can be added
- multiply a matrix by a scalar
- apply the properties of vector addition and scalar multiplication to matrices
- find the transpose of a matrix

---

As you saw in Chapter 3, a matrix is just a rectangular array of numbers. You can classify matrices according to size: an $m \times n$ matrix is a matrix with $m$ rows and $n$ columns. So, if $A = \begin{pmatrix} 3 & 1 & 2 & 4 \\ 0 & 1 & 7 & 8 \\ 1 & 1 & 3 & 2 \end{pmatrix}$, $A$ is a $3 \times 4$ matrix; it has three rows and four columns.

⟵ When you see "$m \times n$," say "m by n."

If $A$ is $m \times n$, the notation $A_{ij}$ means the **entry** in the $i^{\text{th}}$ row and $j^{\text{th}}$ column of $A$. In the above matrix, $A_{32} = 1$ and $A_{23} = 7$. If $A$ is $m \times n$, $A$ can be written as

⟵ In some books, the entries of the generic $m \times n$ matrix are marked using lower case letters like $a_{ij}$.

$$A = \begin{pmatrix} A_{11} & A_{12} & A_{13} & A_{14} & \cdots & A_{1n} \\ A_{21} & A_{22} & A_{23} & A_{24} & \cdots & A_{2n} \\ A_{31} & A_{32} & A_{33} & A_{34} & \cdots & A_{3n} \\ \vdots & \vdots & \vdots & \vdots & \ddots & \vdots \\ A_{m1} & A_{m2} & A_{m3} & A_{m4} & \cdots & A_{mn} \end{pmatrix}$$

### Facts and Notation

You can think of a matrix as a list of rows or a list of columns. The notation $A_{i*}$ means the $i^{\text{th}}$ row of $A$. It is a row vector whose entries are

$$\begin{pmatrix} A_{i1} & A_{i2} & \cdots & A_{in} \end{pmatrix}$$

The vector $A_{i*}$ is in $\mathbb{R}^n$ for all $i$ between 1 and $m$.

Similarly, the notation $A_{*j}$ means the $j^{\text{th}}$ column of $A$. It is a column vector whose entries are

$$\begin{pmatrix} A_{1j} \\ A_{2j} \\ \vdots \\ A_{mj} \end{pmatrix}$$

The vector $A_{*j}$ is in $\mathbb{R}^m$ for all $j$ between 1 and $n$.

⟵ Think of the star as meaning "all." $A_{3*}$ means the third row and all the columns of $A$. Similarly, $A_{*2}$ means the entire second column of $A$. Any row or column of $A$ is a vector with dimension equal to the other dimension of the matrix. This notation is used in some computer languages.

### For You to Do

1. Let $B = \begin{pmatrix} 2 & -1 & 2 & 4 \\ 3 & 1 & 8 & 8 \end{pmatrix}$. If $B$ is an $m \times n$ matrix, what are $m$ and $n$?

2. Determine the following:
   a. $B_{23} + B_{11}$ b. $B_{1*}$ c. $B_{*1} + B_{*2}$

The best place to start building an algebra of matrices is to decide what it means for two matrices to be equal. You described conditions that would make two matrices equal in Exercise 1 from Lesson 4.1. You might have said something along the lines of "two matrices are equal if they have the same size and if all their entries are equal." The following definition expands on that idea.

**Habits of Mind**

**The extension program.** You have already extended the algebra of numbers to vectors. Now, the task is to extend the algebra of vectors to matrices.

### Definition

Two matrices $A$ and $B$ are said to be **equal** if they have the same size and if any of these equivalent conditions is met:

- $A_{ij} = B_{ij}$ for all $i$ and $j$
- $A_{i*} = B_{i*}$ for all $i$
- $A_{*j} = B_{*j}$ for all $j$

The three conditions stand for different ideas. If $A$ and $B$ are $m \times n$ matrices, the second condition is equality in $\mathbb{R}^n$ among the $m$ rows of $A$ and $B$. The third condition is equality in $\mathbb{R}^m$ among the $n$ columns of $A$ and $B$. The equal sign means different things here! The first time, it means equality of *numbers*, while the second and third times, it is equality of *vectors*.

← Why do all three methods give you the same result?

### For Discussion

3. Let $M = \begin{pmatrix} 2 & 3 & 4 \\ 5 & 6 & 7 \end{pmatrix}$ and $N = \begin{pmatrix} 2 & 5 \\ 3 & 6 \\ 4 & 7 \end{pmatrix}$. Are $M$ and $N$ equal? If so, explain why. If not, describe the relationship between $M$ and $N$.

If you think of a matrix as a collection of vectors—as the definition of equality suggests—then you can use that idea to extend operations of vectors to matrices. So, in order to add two matrices, they must be exactly the same size, and the sum is calculated by adding the corresponding entries of the two matrices.

**Definition**

If $A$ and $B$ are $m \times n$ matrices, then the **sum of matrices** $A$ and $B$, written $A + B$, is an $m \times n$ matrix defined by any of these:

- $(A+B)_{ij} = A_{ij} + B_{ij}$
- $(A+B)_{i*} = A_{i*} + B_{i*}$
- $(A+B)_{*j} = A_{*j} + B_{*j}$

As with the definition of matrix equality, the definition of addition has three different options. Addition may be carried out by adding *entries*, but it can also be carried out by adding row or column *vectors*.

**For You to Do**

**4.** Let $A = \begin{pmatrix} 2 & 4 & -3 \\ -4 & 3 & 1 \end{pmatrix}$ and $B = \begin{pmatrix} 4 & 8 & -6 \\ 4 & -3 & -1 \end{pmatrix}$. Calculate $A + B$ using each of the three methods given in the definition.

Scalar multiplication is defined for vectors, and that definition carries over to matrices. This is one place where thinking of a matrix as a list of vectors comes in handy.

**Definition**

If $A$ is an $m \times n$ matrix and $c$ is a scalar, then the **multiplication of a matrix by a scalar**, written $cA$, is an $m \times n$ matrix defined by any of these:

- $(cA)_{ij} = cA_{ij}$
- $(cA)_{i*} = cA_{i*}$
- $(cA)_{*j} = cA_{*i}$

Scalar multiplication can be carried out by multiplying through each *entry*, but it can also be carried out by performing scalar multiplication on the row or column *vectors*.

With these two operations, you can start building the list of basic rules of matrix algebra.

**Theorem 4.1 (The Basic Rules of Matrix Algebra)**

*Suppose that $A$, $B$, and $C$ are matrices of the same size, and that $d$ and $e$ are scalars. Then*

(1) $A + B = B + A$

(2) $A + (B + C) = (A + B) + C$

(3) $A + O = A$

(4) $A + (-1)A = O$

← What do you think $O$ represents here?

(5) $(d+e)A = dA + eA$

(6) $d(A+B) = dA + dB$

(7) $d(eA) = (de)A$

(8) $1A = A$

These properties should look familiar. They were seen in Chapter 1 applied to vectors. The notation $\begin{pmatrix} 1 & 2 & 3 \end{pmatrix}$ can now stand for a vector, a point, or a *matrix* (with one row), and the properties in Theorem 4.1 apply to all three.

### Developing Habits of Mind

**Look for structural similarity.** Because of part ((4)) of Theorem 4.1, the matrix $-1A$ is called either the *opposite* of $A$ or the *negative* of $A$. The shorthand $-A$ can be used, but it really means $-1A$.

← Some call $-A$ the "additive inverse" of $A$.

You may have noticed there was no definition for subtraction of matrices. For numbers, subtraction is defined in terms of adding the opposite: $x - y = x + (-y)$. The same definition can now be used for subtraction of matrices: $A - B = A + (-B)$, where $-B$ means $-1B$.

What other properties of real numbers might apply to matrices? For the time being, only addition and subtraction are defined. A kind of multiplication of matrices will be defined later, and it will be useful to see what properties of real numbers (commutativity, associativity, identity, inverses) carry over into the system of matrices.

### Definition

If $A$ is an $m \times n$ matrix, the **transpose** of $A$, written $A^\top$ is the $n \times m$ matrix defined by any of these:

← Other books may write ${}^tA, A^{tr}$, or $A'$ for transpose.

- $A^\top_{ij} = A_{ji}$
- $A^\top_{i*} = A_{*i}$
- $A^\top_{*j} = A_{j*}$

← Note that $A^\top_{i*}$ is a *row* vector, while $A_{*i}$ is a *column* vector.

For example, if $A = \begin{pmatrix} 1 & 3 & 2 \\ -1 & 4 & 7 \end{pmatrix}$, $A^\top = \begin{pmatrix} 1 & -1 \\ 3 & 4 \\ 2 & 7 \end{pmatrix}$.

### For Discussion

**5.** Give a short explanation for why each of the following facts is true about the transpose operator:

**a.** $(A+B)^\top = A^\top + B^\top$
**b.** $(cA)^\top = c(A^\top)$
**c.** $(A^\top)^\top = A$

← Say "the transpose of the sum is the sum of the transposes."

## Exercises

1. **a.** Prove one of the statements in Theorem 4.1 using the entries $A_{ij}, B_{ij}$ of the matrices involved.
   **b.** Prove one of the statements in Theorem 4.1 using the rows $A_{i*}, B_{i*}$ of the matrices involved.
   **c.** Prove one of the statements in Theorem 4.1 using the columns $A_{*j}, B_{*j}$ of the matrices involved.

2. Some of these statements are always true. Some aren't always true. And some don't make sense. Decide which is which.

   **a.** $A + cB = cA + B$ **b.** $c(d + A) = cd + cA$
   **c.** $c(A - B) = cA - cB$ **d.** $A - A^\top = O$
   **e.** $0A = O$ **f.** $c(A - B) = c(A + B) - cB$

⟵ Capital letters represent matrices and lowercase letters represent scalars.

3. Let $A = \begin{pmatrix} 1 & -1 & -3 \\ 2 & -1 & 0 \end{pmatrix}$ and $B = \begin{pmatrix} 3 & 1 & 4 \\ 1 & 2 & -3 \end{pmatrix}$.

   Calculate the following:

   **a.** $2A + 3B$ **b.** $A - 3B$ **c.** $\frac{1}{2}(A + B) + \frac{1}{2}(A - B)$
   **d.** $3B + \operatorname{rref}(A)$ **e.** $2A + 2B$ **f.** $2(A + B)$
   **g.** $A + O$ **h.** $3B + 2A$ **i.** $5A + 2A$

4. If $A = \begin{pmatrix} 1 & -7 \\ 1 & 6 \end{pmatrix}$, find $B$ if $A + 4B = 3B - 2A$.

5. Let $A = \begin{pmatrix} 1 & 3 \\ -1 & 4 \\ 7 & 2 \end{pmatrix}$, $B = \begin{pmatrix} 0 & 3 \\ 1 & 2 \\ 3 & -1 \end{pmatrix}$, and $C = \begin{pmatrix} -3 & 1 \\ 1 & 2 \\ -2 & 3 \end{pmatrix}$.

   Calculate the following:

   **a.** $A + (B + C)$ **b.** $(A + B) + C$ **c.** $A^\top$
   **d.** $B^\top$ **e.** $A^\top + B^\top$ **f.** $(A + B)^\top$
   **g.** $(3A)^\top$ **h.** $(-2A)^\top$ **i.** $((A + B) + C)^\top$

6. Let $A = \begin{pmatrix} 3 & 1 & 7 & -1 \\ 3 & 2 & 5 & 2 \\ 1 & 3 & 1 & 7 \\ 3 & 4 & -1 & 0 \end{pmatrix}$.

   Calculate the following:

   **a.** $A + A^\top$ **b.** $A - A^\top$ **c.** $\frac{1}{2}(A + A^\top) + \frac{1}{2}(A - A^\top)$

7. Let $A = \begin{pmatrix} 1 & 3 & -1 & 4 \\ 2 & 0 & 10 & 6 \\ 5 & 7 & 8 & 9 \end{pmatrix}$.

   Determine the following:

   **a.** $A_{14}$ **b.** $A_{21}$ **c.** $A_{33}$
   **d.** $A_{2*}$ **e.** $A_{*3}$ **f.** $A_{1*} + A_{3*}$

**8.** Let $A$ be an $m \times n$ matrix with generic entry $A_{ij}$.

**a.** What are the entries in the $i^{\text{th}}$ row of $A$?
**b.** What are the entries in the $j^{\text{th}}$ column of $A$?
**c.** What are the diagonal entries of $A$?
**d.** What are the entries in the last row of $A$?
**e.** What are the entries in the next to last column of $A$?

**9.** Let $A$ be a $3 \times 5$ matrix whose generic entry is $A_{ij}$. Write the entries of $A$ given each of the following conditions:

**a.** $A_{ij} = i + j$ **b.** $A_{pq} = p - 2q$
**c.** $A_{pk} = p$ **d.** $A_{rs} = rs$
**e.** $A_{rs} = r^2 + s^2$ **f.** $A_{ij} = i^2 - j^2$
**g.** $A_{ij} = \max\{i, j\}$ **h.** $A_{vw} = \min\{v, w\}$

**10.** Let $A = \begin{pmatrix} 1 & 0 & 4 \\ 3 & 1 & 2 \end{pmatrix}$. Find the matrix $B$ if $B_{ij} = A_{ji}$.

**11.** Let $A = \begin{pmatrix} 1 & 3 & -1 \\ 2 & 0 & 4 \end{pmatrix}$ and $B = \begin{pmatrix} 2 & 7 & 4 \\ 1 & -1 & 0 \\ 3 & 2 & 1 \end{pmatrix}$.

Calculate the following:

**a.** $A_{1*} \cdot A_{2*}$
**b.** $A_{1*} \cdot B_{*1}$
**c.** $A_{2*} \cdot B_{*3}$
**d.** $(A_{1*} + A_{2*}) \cdot (B_{*1} - B_{*2} + B_{*3})$
**e.** $\text{Proj}_{A_{1*}} B_{*2}$
**f.** $A_{11}B_{*1} + A_{12}B_{*2} + A_{13}B_{*3}$

**12.** Find $x, y, z$, and $w$ if

$$\begin{pmatrix} x + 2y - z & 2x + 3y \\ x + z - 5 & 3x + z + 6 \end{pmatrix} = \begin{pmatrix} 3 - w & 5 + z - 2w \\ w - y & 4y - 2w \end{pmatrix}$$

## 4.3 Different Types of Square Matrices

---

You have already encountered many types of square matrices. Some of them have special uses or properties, so it is helpful to refer to them by special names.

**In this lesson, you will learn how to**

- recognize the different types of square matrices
- decompose any square matrix into its symmetric and skew-symmetric parts

---

A **square matrix** is an $n \times n$ matrix: a matrix with the same number of rows and columns. Here is a generic square matrix:

$$A = \begin{pmatrix} A_{11} & A_{12} & A_{13} & \cdots & A_{1n} \\ A_{21} & A_{22} & A_{23} & \cdots & A_{2n} \\ A_{31} & A_{32} & A_{33} & \cdots & A_{3n} \\ \vdots & \vdots & \vdots & \ddots & \vdots \\ A_{n1} & A_{n2} & A_{n3} & \cdots & A_{nn} \end{pmatrix}$$

Square matrices are common enough that they warrant further classification. Here are some special kinds of square matrices. The examples are all $3 \times 3$ or $4 \times 4$, but these special types of square matrices can occur for any $n \times n$.

| Special type | Example | Description |
|---|---|---|
| Identity matrix | $I = \begin{pmatrix} 1 & 0 & 0 \\ 0 & 1 & 0 \\ 0 & 0 & 1 \end{pmatrix}$ | $I_{ij} = \begin{cases} 1 & \text{if } i = j \\ 0 & \text{otherwise} \end{cases}$ |
| Diagonal matrix | $\begin{pmatrix} 3 & 0 & 0 \\ 0 & 5 & 0 \\ 0 & 0 & \pi \end{pmatrix}$ | $A_{ij} = 0$ when $i \neq j$ |
| Scalar matrix | $\begin{pmatrix} 3 & 0 & 0 \\ 0 & 3 & 0 \\ 0 & 0 & 3 \end{pmatrix}$ | $A_{ij} = 0$ when $i \neq j$, and<br>$c = A_{11} = A_{22} = A_{33} = \cdots$<br>Note: $A = cI$ |

| Special type | Example | Description |
|---|---|---|
| Upper triangular matrix | $\begin{pmatrix} 1 & 2 & 3 \\ 0 & 4 & 5 \\ 0 & 0 & 6 \end{pmatrix}$ | $A_{ij} = 0$ when $i > j$ |
| Lower triangular matrix | $\begin{pmatrix} 1 & 0 & 0 & 0 \\ 2 & 3 & 0 & 0 \\ 4 & 5 & 6 & 0 \\ 7 & 8 & 9 & 10 \end{pmatrix}$ | $A_{ij} = 0$ when $i < j$ |
| Symmetric matrix | $\begin{pmatrix} 1 & 2 & 3 & 4 \\ 2 & 5 & 6 & 7 \\ 3 & 6 & 8 & 9 \\ 4 & 7 & 9 & 10 \end{pmatrix}$ | $A_{ij} = A_{ji}$ or $A_{k*} = A_{k*}^{\top}$ or $A^{\top} = A$ |
| Skew-symmetric matrix | $\begin{pmatrix} 0 & 1 & 2 & 3 \\ -1 & 0 & 4 & 5 \\ -2 & -4 & 0 & 6 \\ -3 & -5 & -6 & 0 \end{pmatrix}$ | $A_{ij} = -A_{ji}$ or $A_{k*} = -A_{k*}^{\top}$ or $A^{\top} = -A$ |

**For You to Do**

**1.** Let $A = \begin{pmatrix} 1 & 2 & 3 \\ 4 & 6 & 9 \\ 5 & -2 & 0 \end{pmatrix}$. Calculate $A + A^{\top}$ and $A - A^{\top}$. What kind of matrices are these?

Your work on problem 1 suggests the following lemma.

**Lemma 4.2**

*If $A$ is a square matrix, then*

- $A + A^{\top}$ *is symmetric, and*
- $A - A^{\top}$ *is skew-symmetric*

There are multiple possible proofs of the lemma, including ones that track $A$ through its entries, or its rows and columns. One simple proof uses only the definitions of symmetric and skew-symmetric, along with some properties of the transpose operator you explored in For Discussion problem 5 from Lesson 4.2.

***Proof.*** $A + A^\top$ is symmetric if and only if its transpose equals itself, and it does:

$$(A + A^\top)^\top = A^\top + (A^\top)^\top = A^\top + A = A + A^\top$$

The proof that $A - A^\top$ is skew-symmetric is almost identical. A matrix is skew-symmetric if and only if its transpose is also its negative.

$$(A - A^\top)^\top = A^\top - (A^\top)^\top = A^\top - A = -A + A^\top = -(A - A^\top)$$

■

Symmetry and skew-symmetry are preserved under scalar multiplication, as the following lemma states.

**Lemma 4.3**

*If $A$ is symmetric, then $cA$ is also symmetric. If $A$ is skew-symmetric, then $cA$ is also skew-symmetric.*

You will be asked to prove this lemma in Exercise 2.

Look back at Exercise 6**c** from Lesson 4.2. There, you were asked to calculate

$$\tfrac{1}{2}(A + A^\top) + \tfrac{1}{2}(A - A^\top)$$

This expression turned out to equal the original matrix $A$. But this expression is also the sum of a symmetric matrix and a skew-symmetric matrix, according to the lemmas above. The algebra is general, leading to the following theorem.

**Theorem 4.4**

*Every square matrix is the sum of a symmetric matrix and a skew-symmetric matrix.*

***Proof.*** If the square matrix is $A$, the symmetric matrix is $\frac{1}{2}(A + A^\top)$, and the skew-symmetric matrix is $\frac{1}{2}(A - A^\top)$.

$$\begin{aligned}
\underbrace{\tfrac{1}{2}(A + A^\top)}_{\text{symmetric}} + \underbrace{\tfrac{1}{2}(A - A^\top)}_{\text{skew-symmetric}} &= \tfrac{1}{2}A + \tfrac{1}{2}A^\top + \tfrac{1}{2}A - \tfrac{1}{2}A^\top \\
&= (\tfrac{1}{2}A + \tfrac{1}{2}A) + (\tfrac{1}{2}A^\top - \tfrac{1}{2}A^\top) \\
&= A + O \\
&= A
\end{aligned}$$

■

**For Discussion**

**2.** **a.** Let $A = \begin{pmatrix} 3 & 4 & 2 \\ -2 & 7 & 0 \\ 2 & -1 & 5 \end{pmatrix}$. Write $A$ as the sum of a symmetric and a skew-symmetric matrix.

**b.** Explain why $B = \begin{pmatrix} 3 & 4 & 2 \\ -2 & 7 & 0 \end{pmatrix}$ cannot be written as the sum of a symmetric and a skew-symmetric matrix.

## Exercises

**1.** Prove Lemma 4.2.

**a.** Use the entry-by-entry definitions of symmetric and skew-symmetric.
**b.** Use the row-by-row definitions of symmetric and skew-symmetric.

**2.** Prove the two parts of Lemma 4.3.

**a.** Use any definition of symmetric to show that if $A$ is symmetric and $c$ is a scalar, then $cA$ is symmetric.
**b.** Use any definition of skew-symmetric to show that if $A$ is skew-symmetric and $c$ is a scalar, then $cA$ is skew-symmetric.

**3.** If $A = \begin{pmatrix} a & b \\ c & d \end{pmatrix}$, write $A$ as the sum of a symmetric matrix and a skew-symmetric matrix.

**4.** Theorem 4.4 gives a method to write a square matrix $A$ as the sum of a symmetric matrix and a skew-symmetric matrix. Prove that this pair of matrices is unique, or find some other pairs of matrices that can also be used.

**5.** For each given matrix $A$, find an expression in terms of $i$ and $j$ that defines the entries $A_{ij}$ of the matrix. All matrices can be defined as

$$A = \begin{pmatrix} A_{11} & A_{12} & A_{13} & A_{14} \\ A_{21} & A_{22} & A_{23} & A_{24} \\ A_{31} & A_{32} & A_{33} & A_{34} \end{pmatrix}$$

**a.** $A = \begin{pmatrix} 0 & -1 & -2 & -3 \\ 1 & 0 & -1 & -2 \\ 2 & 1 & 0 & -1 \end{pmatrix}$ **b.** $A = \begin{pmatrix} 4 & 6 & 8 & 10 \\ 6 & 8 & 10 & 12 \\ 8 & 10 & 12 & 14 \end{pmatrix}$

**c.** $A = \begin{pmatrix} 2 & 2 & 2 & 2 \\ 4 & 4 & 4 & 4 \\ 6 & 6 & 6 & 6 \end{pmatrix}$ **d.** $A = \begin{pmatrix} 3 & 5 & 7 & 9 \\ 4 & 6 & 8 & 10 \\ 5 & 7 & 9 & 11 \end{pmatrix}$

**e.** $A = \begin{pmatrix} 2 & 3 & 4 & 5 \\ 5 & 6 & 7 & 8 \\ 10 & 11 & 12 & 13 \end{pmatrix}$

**6.** Classify each given matrix as scalar, diagonal, upper or lower triangular, symmetric, skew-symmetric, or none of the above.

**a.** $A$ is $3 \times 3$ and $A_{ij} = i^2 + j^2$
**b.** $A$ is $4 \times 4$ and $A_{ij} = i^2 - j^2$
**c.** $A$ is $3 \times 3$ and $A_{pq} = p^2 + pq + q^2$
**d.** $A$ is $2 \times 2$ and $A_{pq} = p - 2q$
**e.** $A$ is $2 \times 2$ and $A_{rs} = 2s - r$

**f.** $A = \operatorname{rref}\begin{pmatrix} 1 & 2 & 3 \\ 4 & 5 & 6 \\ 7 & 8 & 9 \end{pmatrix}$ **g.** $A = \operatorname{rref}\begin{pmatrix} 1 & 2 & 3 \\ 4 & 5 & 6 \\ 7 & 8 & 0 \end{pmatrix}$

**h.** $A$ is $3 \times 3$ and $A_{ij} = \begin{cases} i+j & \text{if } i > j \\ 0 & \text{if } i \leq j \end{cases}$

**i.** $A$ is $4 \times 4$ and $A_{ij} = \begin{cases} i^2 & \text{if } i = j \\ 0 & \text{if } i \neq j \end{cases}$

**7.** Write $a$ in terms of $x$.

$$a = 2m - 3n + 5p$$

$$\begin{aligned} m &= 4x \\ n &= 5x \\ p &= -x \end{aligned}$$

**8.** Write $a$ and $b$ in terms of $x$.

$$\begin{aligned} a &= 2m - 3n + 5p \\ b &= -2m + 6n + 10p \end{aligned}$$

$$\begin{aligned} m &= 4x \\ n &= 5x \\ p &= -x \end{aligned}$$

**9.** Write $a$ in terms of $x$ and $y$.

$$a = 2m - 3n + 5p$$

$$\begin{aligned} m &= 4x - 2y \\ n &= 5x + 3y \\ p &= -x + 5y \end{aligned}$$

**10.** Write $a$ and $b$ in terms of $x, y, z$, and $w$.

$$\begin{aligned} a &= 2m - 3n + 5p \\ b &= -2m + 6n + 10p \end{aligned}$$

$$\begin{aligned} m &= 4x - 2y + 7z + w \\ n &= 5x + 3y - 3z + w \\ p &= -x + 5y + 2z + w \end{aligned}$$

**11.** As in Exercise 7 from Lesson 4.1, this $5 \times 5$ matrix gives the number of one-way, nonstop flights between four different cities served by an airline.

| | PDX | LAS | DEN | CLT | PHL |
|---|---|---|---|---|---|
| PDX | 0 | 0 | 5 | 0 | 1 |
| LAS | 0 | 0 | 0 | 5 | 5 |
| DEN | 5 | 0 | 0 | 4 | 3 |
| CLT | 0 | 5 | 4 | 0 | 10 |
| PHL | 1 | 5 | 3 | 8 | 0 |

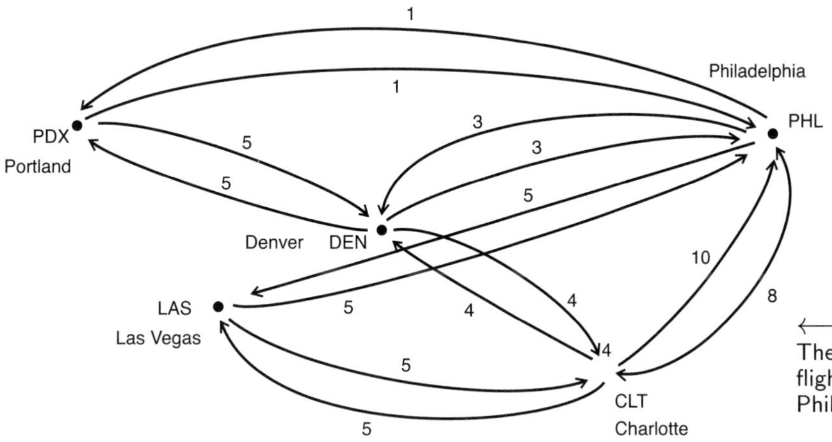


← There are 10 daily nonstop flights from Charlotte to Philadelphia.

**a.** How many different one-stop flights are there from Portland to Las Vegas?

**b.** How many different one-stop flights are there from Portland to Denver?

**c.** How many different one-stop flights are there from Portland to Charlotte?

**d.** How many different one-stop flights are there from Portland to Philadelphia?

**e.** How many different one-stop flights are there from Philadelphia to Denver?

**f.** How many different *two-stop* flights are there from Philadelphia to Denver?

← Yes, one of those stops can be Denver or Philadelphia.

## 4.4 Matrix Multiplication

You have already seen how to add two matrices. You may wonder if you can multiply two matrices, and if it works like adding and multiplying two numbers. It turns out that the operation for multiplication of matrices is a little more complicated than addition, but the results end up being very helpful.

**In this lesson, you will learn how to**

- calculate the product of two matrices
- determine whether you can multiply two matrices
- find the transpose of the product of two matrices

**Minds in Action** Episode 11

*Sasha and Derman are working on Exercise* 10 *from the previous lesson.*

DERMAN: This problem is all about matrices. It reminds me of Chapter 3. Look, here are the two matrices involved:

$$\begin{pmatrix} 2 & -3 & 5 \\ -2 & 6 & 10 \end{pmatrix} \text{ and } \begin{pmatrix} 4 & -2 & 7 & 1 \\ 5 & 3 & -3 & 1 \\ -1 & 5 & 2 & 1 \end{pmatrix}$$

DERMAN: All I do is copy the numbers. I guess I might need to put a zero sometimes like we did in Chapter 3, but that didn't happen here.

SASHA: Hey, that was a pretty good idea. But what do you *do* with those matrices? You can't add them; they're not the same size.

DERMAN: Beats me. But the matrices have all the numeric information from the problem.

SASHA: Hmm, okay. Let's try to solve Exercise 10 using just your matrices.

DERMAN: Well, we know $a$ is going to be some number of $x$'s.

SASHA: Right: $a = 2m - 3n + 5p$. I can read that from the first row of your first matrix.

DERMAN: Then I read the first row of the other matrix?

SASHA: That's not going to work. Look at the original problem. Where are all the $x$'s?

DERMAN: Oh! It's the first *column*. We use the row of the first matrix, and the column of the second matrix.

SASHA: I noticed in those exercises we kept performing the same calculation, and it was exactly like doing a dot product. You know, with your matrices, we can really *do* a dot product. Let's calculate the dot product of the first row of the first matrix and the first column of the second matrix.

DERMAN: All right, this will take me a while.

*Derman carefully counts out a row of the first matrix and a column of the second matrix.*

$$\begin{pmatrix} \mathbf{2} & \mathbf{-3} & \mathbf{5} \\ -2 & 6 & 10 \end{pmatrix} \text{ and } \begin{pmatrix} \mathbf{4} & -2 & 7 & 1 \\ \mathbf{5} & 3 & -3 & 1 \\ \mathbf{-1} & 5 & 2 & 1 \end{pmatrix}$$

$$\begin{pmatrix} 2 & -3 & 5 \end{pmatrix} \cdot \begin{pmatrix} 4 \\ 5 \\ -1 \end{pmatrix} =$$

$$2 \cdot 4 + (-3) \cdot 5 + 5 \cdot (-1) = -12$$

DERMAN: Is that okay? Can I do a dot product with vectors that are pointed differently?

SASHA: As long as they have the same number of elements. You got $-12$, so the answer to the exercise should include $a = -12x$ ... hey, it does!

DERMAN: Seriously? Wow. And if I want to do the other stuff with $a$, I keep using the first row and switch the column ... the last column will be pretty easy to use, huh? Then to do $b$, I use the second row, I think.

SASHA: Very smooth. We should check to make sure we're getting the same answers with or without matrices. I think your method would work pretty well on those airline problems, too.

---

### For You to Do

**1.** Use Derman and Sasha's method to verify the rest of Exercise 10.

---

The operation Derman and Sasha use in the above dialogue comes up very frequently in problems that involve matrices. Today, this operation is known as **matrix multiplication**. This operation isn't as simple to describe as matrix addition, but has many applications, including the solution of systems of equations like the ones from Chapter 3.

To find the product $AB$ of two matrices $A$ and $B$, calculate all the dot products of the *rows* of $A$ with the *columns* of $B$.

← Derman's calculation above gives the entry in the first row and first column of $AB$, since he used the first row of $A$ and the first column of $B$.

### Definition

Let $A$ and $B$ be matrices, where $A$ is $m \times n$ and $B$ is $n \times p$. The **matrix product** of $A$ and $B$, written as $AB$, is an $m \times p$ matrix given by

$$(AB)_{ij} = A_{i*} \cdot B_{*j}$$

← Read this carefully. The $ij^{\text{th}}$ entry of $AB$ is the dot product of ... what, specifically?

For example, suppose

$$A = \begin{pmatrix} A_{11} & A_{12} \\ A_{21} & A_{22} \\ A_{31} & A_{32} \end{pmatrix}, \qquad B = \begin{pmatrix} B_{11} & B_{12} & B_{13} & B_{14} \\ B_{21} & B_{22} & B_{23} & B_{24} \end{pmatrix}$$

Then

$$AB = \begin{pmatrix} (AB)_{11} & (AB)_{12} & (AB)_{13} & (AB)_{14} \\ (AB)_{21} & (AB)_{22} & (AB)_{23} & (AB)_{24} \\ (AB)_{31} & (AB)_{32} & (AB)_{33} & (AB)_{34} \end{pmatrix}$$
$$= \begin{pmatrix} A_{1*} \cdot B_{*1} & A_{1*} \cdot B_{*2} & A_{1*} \cdot B_{*3} & A_{1*} \cdot B_{*4} \\ A_{2*} \cdot B_{*1} & A_{2*} \cdot B_{*2} & A_{2*} \cdot B_{*3} & A_{2*} \cdot B_{*4} \\ A_{3*} \cdot B_{*1} & A_{3*} \cdot B_{*2} & A_{3*} \cdot B_{*3} & A_{3*} \cdot B_{*4} \end{pmatrix}$$

This process is followed when a calculator is asked to multiply two matrices.

### For Discussion

**2.** **a.** What would it mean to "square" a matrix, and what kinds of matrices could be squared?

**b.** If possible, square the matrix from Exercise 11 from the last lesson. What happens?

### Developing Habits of Mind

**Reason about calculations.** Each entry $(AB)_{ij}$ is the dot product of the $i^{\text{th}}$ row of $A$ and the $j^{\text{th}}$ column of $B$. For the matrix product $AB$ to exist, the dot products must exist. That means that the number of columns in $A$ must equal the number of rows in $B$.

If, as above, $A$ is $3 \times 2$ and $B$ is $2 \times 4$, each dot product is in $\mathbb{R}^2$, and the resulting matrix $AB$ is $3 \times 4$.

The order in which you perform a matrix multiplication is important. If you tried to compute $BA$, you'd find it won't work; each row of $B$ has four terms, while each column of $A$ has three. The dot products are undefined, and so is the matrix product.

Some people visualize the matrix product $AB$ as shown below. Consider any entry in $AB$ (the dot pictured below). Its value is the dot product of the row in $A$ to its left and the column in $B$ above it.

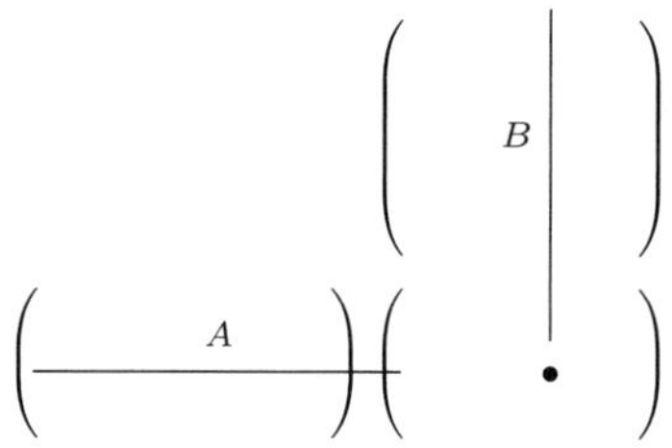


⟵ The size of the row vectors in $A$ must be the same as the size of the column vectors in $B$.

⟵ Even if $AB$ and $BA$ are defined, it's not necessarily true that $AB = BA$ or even that they have the same size. Try it with an example, say, with $A$ $2 \times 3$ and $B$ $3 \times 2$.

⟵ This way of remembering how multiplication works is often called the *over and down* method.

### For You to Do

**3.** Which of these matrices can be multiplied together? Find all pairs, the order in which they can be multiplied, and the size of the matrix product that will result.

$A$ is $3 \times 2$ $B$ is $2 \times 4$ $C$ is $3 \times 1$ $D$ is $4 \times 2$ $E$ is $4 \times 4$

### Example

**Problem.** Calculate the product of $A = \begin{pmatrix} 1 & -1 \\ 3 & 1 \\ 7 & 2 \end{pmatrix}$ and $B = \begin{pmatrix} 3 & 2 & 1 \\ 2 & -1 & 5 \end{pmatrix}$.

**Solution.** Use the format described above.

$$\begin{array}{cc} & \begin{pmatrix} 3 & 2 & 1 \\ 2 & -1 & 5 \end{pmatrix} \\ \begin{pmatrix} 1 & -1 \\ 3 & 1 \\ 7 & 2 \end{pmatrix} & \begin{pmatrix} 1 & 3 & -4 \\ 11 & 5 & 8 \\ 25 & 12 & 17 \end{pmatrix} \end{array} \qquad \text{So } AB = \begin{pmatrix} 1 & 3 & -4 \\ 11 & 5 & 8 \\ 25 & 12 & 17 \end{pmatrix}$$

Each dot product is a row of $A$ and a column of $B$. Some examples:

$$AB_{23} = A_{2*} \cdot B_{*3} = \begin{pmatrix} 3 & 1 \end{pmatrix} \cdot \begin{pmatrix} 1 \\ 5 \end{pmatrix} = 8$$

$$AB_{12} = A_{1*} \cdot B_{*2} = \begin{pmatrix} 1 & -1 \end{pmatrix} \cdot \begin{pmatrix} 2 \\ -1 \end{pmatrix} = 3$$

$$AB_{31} = A_{3*} \cdot B_{*1} = \begin{pmatrix} 7 & 2 \end{pmatrix} \cdot \begin{pmatrix} 3 \\ 2 \end{pmatrix} = 25$$

### For You to Do

**4.** Use $A$ and $B$ from the previous example. Does the matrix product $BA$ exist? If so, calculate it. If not, explain why it can't be done.

### Minds in Action Episode 12

DERMAN: I saw matrices in a movie recently.

SASHA: I've had enough of your jokes about *The Matrix*, okay?

DERMAN: That wasn't the one. It was *Good Will Hunting*. There was a map, and then he made a matrix.

SASHA: I'll bet it's on YouTube.

*Derman and Sasha go to http://www.youtube.com/watch?v=l0y7HeDqrV8*

DERMAN: See, matrices. There's a map in the upper left, it looks like this:

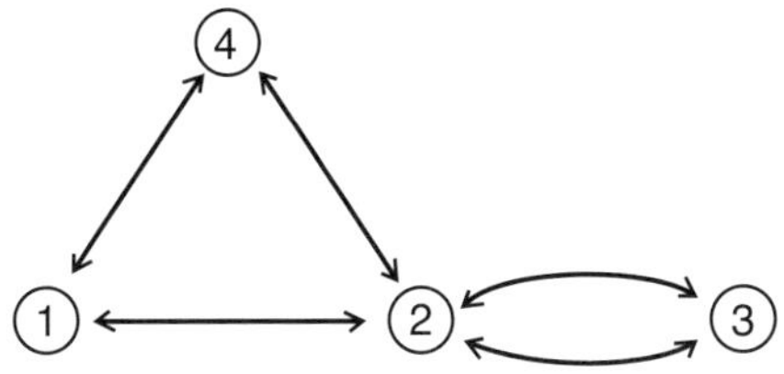


SASHA: He answers the first question with a $4 \times 4$ matrix called $A$. I'll bet it's the matrix of paths, like the airline problems. There are two paths from 2 to 3 . . . hey, this is what's in the movie!

DERMAN: I don't understand why the MIT people in this movie are so impressed if we can do it. The second question looks harder: "Find the matrix giving the number of three-step walks."

SASHA: Well, $A$ by itself gives the number of *one*-step walks.

DERMAN: Maybe we should make a matrix of two-step walks first. If you wanted to do a two-step walk from 1 to 1, you could go from 1 to 2, then 2 to 1 . . . or from 1 to 3 . . . no you can't do that . . . or from 1 to 4, then 4 to 1. So, there are two ways to get from 1 to 1 in a two-step walk. This is going to take a while.

SASHA: Derman, this is matrix multiplication at work! If we take $A$ and multiply it with itself, that'll be our matrix of two-step walks.

DERMAN: How do you like them apples?

SASHA: Huh? Anyway, let's get to work. It's interesting that in the movie, he writes $A^3$ next to his answer for the three-step problem.

### For You to Do

**5.** **a.** Use the map to construct the $4 \times 4$ matrix $A$.
**b.** Use matrix multiplication to construct a second $4 \times 4$ matrix giving the number of two-step walks from any point to any other point.
**c.** Find the matrix giving the number of three-step walks. Is the answer given in the movie correct?

⟵ To check, there should be six two-step paths from point 2 to itself, two two-step paths from point 1 to point 3, and one two-step path from point 4 to point 2.

### In-Class Experiment

Carefully write out the result of each matrix multiplication. What do you notice?

1. $\begin{pmatrix} g & h \end{pmatrix} \begin{pmatrix} i & j & k \\ l & m & n \end{pmatrix}$

2. $\begin{pmatrix} c & d \end{pmatrix} \begin{pmatrix} i & j & k \\ l & m & n \end{pmatrix}$

3. $\begin{pmatrix} a & b \\ c & d \\ e & f \\ g & h \end{pmatrix} \begin{pmatrix} i & j & k \\ l & m & n \end{pmatrix}$

4. $\begin{pmatrix} a & b \\ c & d \\ e & f \\ g & h \end{pmatrix} \begin{pmatrix} j \\ m \end{pmatrix}$

5. $\begin{pmatrix} a & b \\ c & d \\ e & f \\ g & h \end{pmatrix} \begin{pmatrix} k \\ n \end{pmatrix}$

6. Here is a large matrix multiplication problem. Use a group of six people to determine the matrix product $AB$ without using a calculator.

$$A = \begin{pmatrix} 1 & 2 & -1 & 3 & 0 & 4 & -2 & 1 & 1 & 4 \\ 0 & 0 & 0 & 3 & 4 & 3 & 1 & 10 & -1 & 2 \\ 1 & 1 & 1 & 1 & 1 & 1 & 1 & 1 & 1 & 1 \\ 3 & 6 & -3 & 9 & 0 & 12 & -6 & 3 & 3 & 12 \\ 9 & 8 & 7 & 6 & 5 & 4 & 3 & 2 & 1 & 0 \\ 10 & 10 & 6 & 9 & 5 & 8 & 1 & 3 & 2 & 4 \end{pmatrix}, \quad B = \begin{pmatrix} 2 & 0 & -1 & -2 & 1 & 3 \\ 1 & 0 & -2 & -1 & 2 & 3 \\ -1 & 0 & 0 & 1 & 3 & 2 \\ 3 & 1 & 0 & -3 & 4 & 7 \\ 0 & 1 & 2 & 0 & 5 & 5 \\ 4 & 1 & 3 & -4 & 5 & 9 \\ -2 & 1 & 1 & 2 & 4 & 2 \\ 1 & 1 & 5 & -1 & 3 & 4 \\ 1 & 1 & 0 & -1 & 2 & 3 \\ 4 & 1 & 0 & -4 & 100 & 104 \end{pmatrix}$$

When you multiply two matrices $A$ and $B$, the first row of the product $AB$ uses *only* the first row of $A$. Similarly, the third column of $AB$ uses *only* the third column of $B$. This means you can think of a matrix multiplication as operating on each row of the first matrix, or each column of the second matrix.

**Theorem 4.5**

*If the matrix product* $AB$ *exists, then*

- $(AB)_{i*} = A_{i*} \cdot B$, *and*
- $(AB)_{*j} = A \cdot B_{*j}$

← Say, "The $i^{\text{th}}$ column of $AB$ is $A$ times the $i^{\text{th}}$ column of $B$." What would you say for rows?

**For You to Do**

**6.** Calculate each matrix product.

**a.** $\begin{pmatrix} 1 & -1 \\ 3 & 1 \\ 7 & 2 \end{pmatrix} \cdot \begin{pmatrix} 3 & 1 \\ 2 & 5 \end{pmatrix}$ **b.** $\begin{pmatrix} 1 & 3 & 7 \\ -1 & 1 & 2 \end{pmatrix} \cdot \begin{pmatrix} 3 & 2 \\ 1 & 5 \end{pmatrix}$ **c.** $\begin{pmatrix} 3 & 2 \\ 1 & 5 \end{pmatrix} \cdot \begin{pmatrix} 1 & 3 & 7 \\ -1 & 1 & 2 \end{pmatrix}$

The first two exercises above show that $AB$ and $A^\top B^\top$ are unrelated; in general, $(AB)^\top \neq A^\top B^\top$. But the third exercise reveals the following theorem.

← Some say that transpose doesn't "distribute" over matrix multiplication.

**Theorem 4.6**

*If the matrix product* $AB$ *exists, then*

$$(AB)^\top = B^\top A^\top$$

***Proof.*** The proof relies on the definitions of transpose and matrix multiplication, and the fact that the dot product, the operation underlying matrix multiplication, is commutative.

← This proof was conceived by a high school student.

$$\begin{aligned} (AB)^\top_{ij} &= (AB)_{ji} && \text{by definition of transpose} \\ &= A_{j*} \cdot B_{*i} && \text{by definition of matrix multiplication} \\ &= B_{*i} \cdot A_{j*} && \text{since dot product is commutative} \\ &= B^\top_{i*} \cdot A^\top_{*j} && \text{by definition of transpose} \\ &= (B^\top A^\top)_{ij} && \text{by definition of matrix multiplication} \end{aligned}$$

This proves that $(AB)^\top$ and $B^\top A^\top$ are equal for any entry, so they are equal matrices. ■

## Exercises

**1.** Calculate each matrix product without using a calculator.

**a.** $\begin{pmatrix}1 & 3 & -1\\ 2 & 4 & 0\end{pmatrix}\begin{pmatrix}1\\ 3\\ 7\end{pmatrix}$ **b.** $\begin{pmatrix}1 & 3 & -1\\ 2 & 4 & 0\end{pmatrix}\begin{pmatrix}-1\\ 1\\ 2\end{pmatrix}$

**c.** $\begin{pmatrix}1 & 3 & -1\\ 2 & 4 & 0\end{pmatrix}\begin{pmatrix}1 & -1\\ 3 & 1\\ 7 & 2\end{pmatrix}$ **d.** $\begin{pmatrix}2 & 4 & 0\end{pmatrix}\begin{pmatrix}1 & -1\\ 3 & 1\\ 7 & 2\end{pmatrix}$

**e.** $\begin{pmatrix}4 & 8 & 0\end{pmatrix}\begin{pmatrix}1 & -1\\ 3 & 1\\ 7 & 2\end{pmatrix}$ **f.** $\begin{pmatrix}3 & 7 & -1\end{pmatrix}\begin{pmatrix}1 & -1\\ 3 & 1\\ 7 & 2\end{pmatrix}$

**2.** Calculate this matrix product: $\begin{pmatrix}3 & 1\\ 5 & 2\end{pmatrix}\begin{pmatrix}a & b\\ c & d\end{pmatrix}$.

**3.** Use the result from Exercise 2 to find a $2 \times 2$ matrix $M$ so that $\begin{pmatrix}3 & 1\\ 5 & 2\end{pmatrix}M = \begin{pmatrix}1 & 0\\ 0 & 1\end{pmatrix}$.

**4.** Find two matrices $A$ and $B$ so that

**a.** $AB$ is defined but $BA$ isn't.
**b.** $AB$ and $BA$ are both defined but have different sizes.
**c.** $AB$ and $BA$ are both defined and have the same size, but $AB \neq BA$.
**d.** $AB = BA$.

**5.** Let

$$A = \begin{pmatrix}1 & 3 & -1\\ 2 & 4 & 0\end{pmatrix} \quad B = \begin{pmatrix}1 & -1\\ 3 & 1\\ 7 & 2\end{pmatrix} \quad C = \begin{pmatrix}3 & -1\\ 1 & 4\end{pmatrix}$$

$$D = \begin{pmatrix}-1 & 1 & 5\\ 1 & 0 & 2\end{pmatrix} \quad E = \begin{pmatrix}-1 & 1\\ 5 & 2\end{pmatrix}$$

Calculate each of the following:

**a.** $AB$ **b.** $A_{2*} \cdot B$ **c.** $A \cdot B_{*1}$ **d.** $DB$

**e.** $C^2$ **f.** $CE$ **g.** $EC$
**h.** $(A + D)B$ **i.** $AB + DB$ **j.** $(AB)C$
**k.** $A(BC)$ **l.** $A(3B)$ **m.** $(4A)B$

**n.** Find a $2 \times 2$ matrix $F$ so that $EF = \begin{pmatrix}1 & 0\\ 0 & 1\end{pmatrix}$

**6.** Compute the matrix product $AB$ given the two matrices below. Try to organize your algebraic work to make the result as "clean" as possible.

$$A = \begin{pmatrix}a & b & c\\ d & e & f\end{pmatrix} \quad B = \begin{pmatrix}u & v\\ w & x\\ y & z\end{pmatrix}$$

**7.** Using the same matrices from Exercise 6, compute the matrix product $BA$.

**8.** Describe how the results of the last two exercises can be used to explain Theorem 4.5.

**9.** Let $A = \begin{pmatrix} 5 & 3 \\ 2 & 1 \end{pmatrix}$ and $B = \begin{pmatrix} 5 & 4 \\ 3 & 2 \end{pmatrix}$.

Show that $(AB)^\top \neq A^\top B^\top$.

**10.** For each given set of square matrices $A$ and $B$, determine whether $AB = BA$.

**a.** $A = \begin{pmatrix} 1 & 0 \\ 1 & 3 \end{pmatrix}$ and $B = \begin{pmatrix} -2 & 4 \\ 1 & 6 \end{pmatrix}$

**b.** $A = \begin{pmatrix} 1 & -1 & 1 \\ 3 & 1 & 2 \\ 4 & -1 & 1 \end{pmatrix}$ and $B = \begin{pmatrix} 2 & 1 & 3 \\ -1 & 4 & 2 \\ 0 & 1 & 7 \end{pmatrix}$

**c.** $A = \begin{pmatrix} 1 & -1 & 1 \\ 4 & 2 & 3 \\ 1 & 0 & 1 \end{pmatrix}$ and $B = \begin{pmatrix} 1 & 0 & 0 \\ 0 & 1 & 0 \\ 0 & 0 & 1 \end{pmatrix}$

**d.** $A = \begin{pmatrix} 1 & -1 & 1 \\ 4 & 2 & 3 \\ 1 & 0 & 1 \end{pmatrix}$ and $B = \begin{pmatrix} 2 & 0 & 0 \\ 0 & 2 & 0 \\ 0 & 0 & 2 \end{pmatrix}$

**e.** $A = \begin{pmatrix} 1 & -1 & 1 \\ 4 & 2 & 3 \\ 1 & 0 & 1 \end{pmatrix}$ and $B = \begin{pmatrix} 2 & 0 & 0 \\ 0 & 3 & 0 \\ 0 & 0 & 2 \end{pmatrix}$

**f.** $A = \begin{pmatrix} 1 & -2 \\ 2 & 1 \end{pmatrix}$ and $B = \begin{pmatrix} -2 & 4 \\ -4 & -2 \end{pmatrix}$

**11.** Given $A = \begin{pmatrix} 1 & 3 & -1 \\ 2 & 0 & 1 \end{pmatrix}$, $B = \begin{pmatrix} 1 & -1 & 1 \\ 3 & 4 & 0 \\ 2 & 6 & 1 \end{pmatrix}$, and $C = \begin{pmatrix} 3 & -1 \\ 1 & 4 \\ 2 & 7 \end{pmatrix}$,

calculate each of the following:

**a.** $A(BC)$

**b.** $(AB)C$

**12.** Does matrix multiplication distribute over addition? In other words, if $A$, $B$, and $C$ are matrices and all of the operations are defined, are either or both of these true?

$$A(B + C) = AB + AC$$
$$(B + C)A = AB + AC$$

Use examples or counterexamples to illustrate or disprove the statements.

**13.** Suppose $A$ and $B$ are matrices that can be multiplied and $c$ is a number. Show that $A(cB) = (cA)B = c(AB)$.

**14.** Theorem 4.1 gives some properties of matrix addition and scalar multiplication. Find and prove some properties of matrix multiplication.

**15.** True or false: If $X$ and $Y$ are square matrices of the same size and $XY = O$, then either $X = O$ or $Y = O$.

If true, prove it. If not, find a counterexample.

**16.** Find all $2 \times 2$ matrices $A$ so that

**a.** $A^2 = \begin{pmatrix} 25 & 0 \\ 0 & 16 \end{pmatrix}$ **b.** $A^2 = \begin{pmatrix} -1 & 0 \\ 0 & -1 \end{pmatrix}$

**c.** $A^2 = \begin{pmatrix} 0 & -1 \\ 0 & 0 \end{pmatrix}$ **d.** $A^2 = \begin{pmatrix} 0 & -1 \\ 1 & 0 \end{pmatrix}$

**17.** Suppose that $y \neq 0$ and that

$$A = \begin{pmatrix} x & y \\ \frac{1-x^2}{y} & -x \end{pmatrix}$$

Show that $A^2 = I$.

← So, there are infinitely many $2 \times 2$ matrices whose square is the identity matrix.

**18.** Suppose $A = \begin{pmatrix} 2 & 3 \\ 4 & 5 \end{pmatrix}$. Show that $A^2 - 7A - 2I = O$.

**19.** Suppose $A = \begin{pmatrix} a & b \\ c & d \end{pmatrix}$. Show that

$$A^2 - (a + d)A + (ad - bc)I = O$$

← $a+d$ is called the **trace** of $A$ and $ad - bc$ is called the **determinant** of $A$.

**20.** If $X$ and $Y$ are square matrices of the same size, is it true that

$$(X + Y)(X - Y) = X^2 - Y^2\,?$$

If so, prove it. If not, correct the statement so it becomes true.

**21.** Here is a graph of connections on a map.

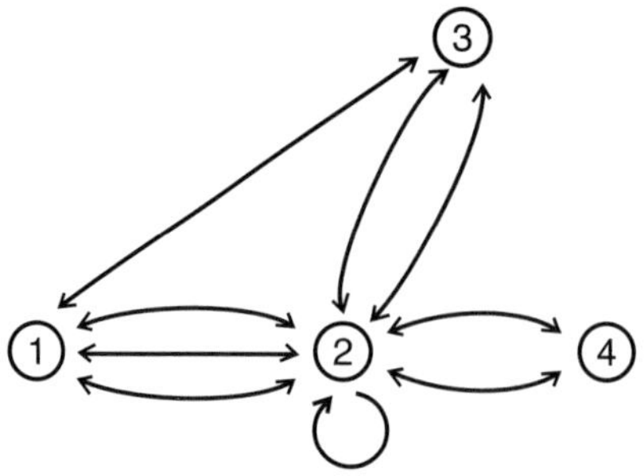


**a.** Use the map to construct a $4 \times 4$ matrix $A$ to represent the situation.

**b.** Find the total number of ways to get from 1 to 2 in one step, in two steps, in three steps.

**c.** Compute $A^2$ and $A^3$.

**d.** Determine, using matrices, the total number of ways to get from 1 to 4 in five or fewer steps.

← Note that 2 has a path to itself . . .

**22.** Solve this system of linear equations:

$$\begin{cases} 3x + 5y + z &= 16 \\ x + 3y + 4z &= 25 \\ 4x - 2y - 3z &= -9 \end{cases}$$

**23.** Calculate the result of this matrix multiplication:

$$\begin{pmatrix} 3 & 5 & 1 \\ 1 & 3 & 4 \\ 4 & -2 & -3 \end{pmatrix} \begin{pmatrix} 2 \\ 1 \\ 5 \end{pmatrix}$$

**24.** Write $\begin{pmatrix} 11 \\ 11 \\ 6 \end{pmatrix}$ as a linear combination of the columns of the matrix $A$ where

$$A = \begin{pmatrix} 3 & 5 & 1 \\ 1 & 3 & 4 \\ 4 & -2 & -3 \end{pmatrix}$$

**25.** Show that

$$\begin{pmatrix} 3 & 5 & 1 \\ 1 & 3 & 4 \\ 4 & -2 & -3 \end{pmatrix} \begin{pmatrix} x \\ y \\ z \end{pmatrix} = x \begin{pmatrix} 3 \\ 1 \\ 4 \end{pmatrix} + y \begin{pmatrix} 5 \\ 3 \\ -2 \end{pmatrix} + z \begin{pmatrix} 1 \\ 4 \\ -3 \end{pmatrix}$$

**26.** Find a vector $\begin{pmatrix} x \\ y \\ z \end{pmatrix}$ so that

$$\begin{pmatrix} 3 & 5 & 1 \\ 1 & 3 & 4 \\ 4 & -2 & -3 \end{pmatrix} \begin{pmatrix} x \\ y \\ z \end{pmatrix} = \begin{pmatrix} 11 \\ 11 \\ 6 \end{pmatrix}$$

**27.** If $A$ is an $n \times n$ matrix, the trace of $A$, written $\mathrm{Tr}(A)$, is the sum of the elements on the "main diagonal" of $A$.

$$\mathrm{Tr}(A) = A_{11} + A_{22} + \cdots + A_{nn}$$

True or false? If true, prove it. If false, give a counterexample.

**a.** $\mathrm{Tr}(A + B) = \mathrm{Tr}(A) + \mathrm{Tr}(B)$
**b.** $\mathrm{Tr}(cA) = c\,\mathrm{Tr}(A)$
**c.** $\mathrm{Tr}(AB) = \mathrm{Tr}(A)\,\mathrm{Tr}(B)$
**d.** $\mathrm{Tr}(AB) = \mathrm{Tr}(BA)$
**e.** $\mathrm{Tr}(A^2) = (\mathrm{Tr}(A))^2$

## 4.5 Operations, Identity, and Inverse

In the last lesson, matrix multiplication was formally defined as a generalized dot product. In this lesson, you will explore another way to think about matrix multiplication: a generalized linear combination. This representation of matrix multiplication will prove to be very useful.

**In this lesson, you will learn how to**

- set up a matrix multiplication as a linear combination of vectors
- understand the relationship between matrix multiplication, dot product, and linear combination
- find the inverse of a matrix, if it exists
- use the inverse to solve a matrix equation

Consider Exercises 22–25 from the previous lesson. Two of the exercises are types you explored in Chapter 3. The others align closely with them but take an approach through matrix multiplication.

Exercise 25 gives the result

$$\begin{pmatrix} 3 & 5 & 1 \\ 1 & 3 & 4 \\ 4 & -2 & -3 \end{pmatrix} \begin{pmatrix} x \\ y \\ z \end{pmatrix} = \begin{pmatrix} 3x + 5y + z \\ x + 3y + 4z \\ 4x - 2y - 3z \end{pmatrix}$$

If you let $X = \begin{pmatrix} x \\ y \\ z \end{pmatrix}$, $AX$ is the left side of a system of linear equations

$$\begin{cases} 3x + 5y + z &= b_1 \\ x + 3y + 4z &= b_2 \\ 4x - 2y - 3z &= b_3 \end{cases}$$

The system can be described using the matrix equation $AX = B$, where $A$ is a matrix and $X$ is a column vector. Any possible solution to the system can be tested by calculating the matrix product $AX$ and determining whether it solves $AX = B$.

### For You to Do

**1.** Calculate each matrix product.

**a.** $\begin{pmatrix} 3 & 5 & 1 \\ 1 & 3 & 4 \\ 4 & -2 & -3 \end{pmatrix} \begin{pmatrix} 3 \\ 0 \\ 2 \end{pmatrix}$ **b.** $\begin{pmatrix} 3 & 5 & 1 \\ 1 & 3 & 4 \\ 4 & -2 & -3 \end{pmatrix} \begin{pmatrix} 0 \\ 1 \\ 0 \end{pmatrix}$ **c.** $\begin{pmatrix} 3 & 5 & 1 \\ 1 & 3 & 4 \\ 4 & -2 & -3 \end{pmatrix} \begin{pmatrix} 3 & 0 \\ 0 & 1 \\ 2 & 0 \end{pmatrix}$

**d.** Find $X$ if $\begin{pmatrix} 3 & 5 & 1 \\ 1 & 3 & 4 \\ 4 & -2 & -3 \end{pmatrix} X = \begin{pmatrix} 6 \\ 8 \\ 8 \end{pmatrix}$.

As seen in part **c** from the For You to Do problem 1, you can test many possible columns $X_1, X_2, \ldots$ at once by creating a matrix with each possible $X$ as a column, and read the results from the columns of the matrix product. This follows directly from Theorem 4.5, since matrix multiplication operates on each column of the second matrix.

In Chapter 3, you learned that solutions to systems are closely tied to linear combinations. Matrix multiplication gives you a good way to express these ties. For example:

$$\begin{aligned}\begin{pmatrix} 3 & 5 & 1 \\ 1 & 3 & 4 \\ 4 & -2 & -3 \end{pmatrix}\begin{pmatrix} x \\ y \\ z \end{pmatrix} &= \begin{pmatrix} 3x+5y+z \\ x+3y+4z \\ 4x-2y-3z \end{pmatrix} \\ &= x\begin{pmatrix} 3 \\ 1 \\ 4 \end{pmatrix} + y\begin{pmatrix} 5 \\ 3 \\ -2 \end{pmatrix} + z\begin{pmatrix} 1 \\ 4 \\ -3 \end{pmatrix}\end{aligned}$$

There's nothing special about the numbers in this calculation. You could do the same thing for any matrix and column. In other words, you have the essence of the proof of a theorem that will be very useful for the rest of this book.

**Theorem 4.7**

*The product $AX$, where $A$ is a matrix and $X$ is a column vector, is a linear combination of the columns of $A$. More precisely, if $A$ is $m \times n$,*

$$A\begin{pmatrix} x_1 \\ x_2 \\ \vdots \\ x_n \end{pmatrix} = x_1 A_{*1} + x_2 A_{*2} + \cdots + x_n A_{*n}$$

⟵ You might try proving this theorem for a general matrix $A$. The proof is just a generic version of the above calculation.

**Facts and Notation**

A similar result is true for rows: a row vector times a matrix is a linear combination of the rows of the matrix. More precisely, if $A$ is $m \times n$

$$(x_1, x_2, \ldots, x_m)A = x_1 A_{1*} + x_2 A_{2*} + \cdots + x_m A_{m*}$$

Work out a few examples to see how the proof would go.

Combining Theorems 4.5 and 4.7, you can pick apart a matrix calculation: any column of $AB$ is $A$ times the corresponding column of $B$. A similar calculation shows that the same thing is true for the rows of $AB$.

**Theorem 4.8 (The Pick-Apart Theorem)**
*Suppose $A$ is $m \times n$ and $B$ is $n \times p$. Then*

(1) *the $j^{th}$ column of $AB$ is $A$ times the $j^{th}$ column of $B$, and this is a linear combination of the columns of $A$:*

$$(AB)_{*j} = A\,B_{*j} = A \begin{pmatrix} B_{1j} \\ B_{2j} \\ \vdots \\ B_{nj} \end{pmatrix} = B_{1j}A_{*1} + B_{2j}A_{*2} + \cdots + B_{nj}A_{*n}$$

(2) *the $i^{th}$ row of $AB$ is the $i^{th}$ row of $A$ times $B$, and this is a linear combination of the rows of $B$:*

$$\begin{aligned} (AB)_{i*} &= A_{i*}\,B \\ &= (A_{i1}, A_{i2}, \ldots, A_{im})\,B \\ &= A_{i1}\,B_{1*} + A_{i2}B_{2*} + \cdots + A_{im}B_{m*} \end{aligned}$$

### Some Properties of Matrix Multiplication

Theorem 4.6 leads to the following theorem about symmetric matrices.

**Theorem 4.9**
*If $A$ and $B$ are symmetric matrices, then $(AB)^\top = BA$.*

***Proof.*** A matrix $A$ is symmetric if $A = A^\top$. Now use Theorem 4.6:

$$\begin{aligned} (AB)^\top &= B^\top A^\top && \text{by Theorem 4.6} \\ &= BA && B \text{ and } A \text{ are symmetric} \end{aligned}$$

■

Matrix multiplication may not be commutative, but it is associative.

**Theorem 4.10**
*If $A$ is $m \times n$, $B$ is $n \times p$, and $C$ is $p \times q$, then*

$$(AB)C = A(BC)$$

The associative property of matrix multiplication can be proved right from the definition of matrix multiplication, but the proof is extremely messy. See Exercises 15 and 16 for examples in the case of $2 \times 2$ matrices. In Chapter 5, you'll see a simple proof of associativity that depends on using matrices to represent certain kinds of functions.

### For You to Do

**2.** In Theorem 4.10, show that $(AB)C$ and $A(BC)$ have the same size.

Because matrix multiplication is not commutative, care must be taken whenever multiplying both sides by a matrix. If $AX = B$ and you want to multiply both sides by $C$, it is *not* true that $C(AX) = BC$. Instead, a correct step is $C(AX) = CB$ ("left multiplication") or $(AX)C = BC$ ("right multiplication").

**Habits of Mind**

Theorem 4.10 lets you write $ABC$ for a product of three matrices. How you do the multiplication doesn't matter: $(AB)C$ or $A(BC)$. But switching the order (like $(AC)B$) will change the answer.

### For You to Do

**3.** Let $A = \begin{pmatrix} 3 & 5 & 1 \\ 1 & 3 & 4 \\ 4 & -2 & -3 \end{pmatrix}$. Find a $3 \times 3$ matrix $I$ such that $AI = A$.

One of the special matrices introduced in Lesson 4.3 was the identity matrix, a square matrix with 1 on the diagonal and 0 everywhere else. The use of the term *identity* matches its use in other mathematical systems: it is the identity for matrix multiplication.

⟵ So, the $3 \times 3$ identity matrix is $I = \begin{pmatrix} 1 & 0 & 0 \\ 0 & 1 & 0 \\ 0 & 0 & 1 \end{pmatrix}$. When the size needs to be made clear, you can write $I_3$ for the $3 \times 3$ identity matrix.

⟵ The identity for real-number addition is 0, and the identity for real-number multiplication is 1. Is there an identity for matrix addition?

**Theorem 4.11**

*If $A$ is an $n \times n$ matrix and $I$ is the $n \times n$ identity matrix, then*

$$AI = IA = A$$

### In-Class Experiment

**4.** For each matrix $A$, find a matrix $B$ so that $AB = I$, or show that no such $B$ exists.

a. $\begin{pmatrix} 1 & 2 \\ 3 & 4 \end{pmatrix}$ b. $\begin{pmatrix} 1 & 2 \\ 2 & 4 \end{pmatrix}$ c. $\begin{pmatrix} 3 & 0 \\ 0 & 4 \end{pmatrix}$ d. $\begin{pmatrix} 1 & 2 & 3 \\ 4 & 5 & 6 \\ 7 & 8 & 9 \end{pmatrix}$ e. $\begin{pmatrix} 1 & 2 & 3 \\ 4 & 5 & 6 \\ 7 & 8 & 0 \end{pmatrix}$

Come up with some more $2 \times 2$ and $3 \times 3$ matrices. Determine if a matrix $B$ always exists, and if not, some conditions that guarantee $B$ exists (or doesn't).

⟵ The numbers used in these last two might seem familiar from Chapter 3.

In other mathematical systems, two elements are *inverses* for an operation if they produce the identity as output under that operation, and that definition carries over to matrices.

**Definition**

A square matrix $A$ is **invertible** if there exists a square matrix $B$ such that $AB = I = BA$. This matrix $B$ is the **matrix inverse** of $A$, denoted $A^{-1}$.

⟵ This definition is a little redundant. If $A$ is square and $AB = I$, you can prove that $BA = I$. See Exercise 32.

Invertible matrices are sometimes called **nonsingular**; matrices that do not have inverses are therefore **singular**.

As you saw in the In-Class Experiment above, not all square matrices $A$ are invertible. One important question that will be answered soon is how to determine whether or not a square matrix $A$ is invertible. If $A^{-1}$ exists, it becomes very helpful in solving the matrix equation $AX = B$, as the following theorem shows.

← Calculators can find the inverse of a matrix when it exists, either by typing $A^{-1}$ or using the `inv` function. The notation $A^{-1}$ is the same notation used for inverse functions (their composite is the identity function), and for inverses under real-number multiplication (their product is 1).

**Theorem 4.12**

*If an $n \times n$ matrix $A$ is invertible and $B$ is* any *column vector in $\mathbb{R}^n$, the system $AX = B$ has a unique column vector $X$ as its solution.*

***Proof.*** Since $A$ is invertible, $A^{-1}$ exists. If $AX = B$ were an "algebra" problem, it would be solved by multiplying by the multiplicative inverse of $A$ on both sides. The equivalent step here is to perform left multiplication by $A^{-1}$.

← All of the observations made in Chapter 3 about linear combinations, especially those about linear dependence and independence, can be stated in terms of the matrix equation $AX = B$.

$$\begin{aligned} AX &= B \\ A^{-1}(AX) &= A^{-1}B \\ (A^{-1}A)X &= A^{-1}B \\ IX &= A^{-1}B \\ X &= A^{-1}B \end{aligned}$$

■

## Developing Habits of Mind

**Create a process.** As with some of the proofs in Chapters 1 and 2, the proof of Theorem 4.12 not only proves the theorem, but *gives the solution.*

As you work with inverse matrices, you will find that multiplying by $A^{-1}$ serves many of the same purposes as multiplying by $a^{-1} = \frac{1}{a}$ in an ordinary linear equation from Algebra 1 (both "undo" an operation). However, be careful: for matrices, $A^{-1}$ may not exist.

← In fact, not all real numbers have multiplicative inverses: $0^{-1}$ doesn't exist for numbers.

The next lesson will focus more on the existence of inverses and solving the matrix equation $AX = B$. Some facts from Chapter 3 will prove very helpful.

## For You to Do

**5.** Prove the converse of Theorem 4.12: if $A$ is $n \times n$ and $AX = B$ has a unique solution for every column vector $B$ in $\mathbb{R}^n$, then $A$ is invertible.

Hint: Let $B$ be the matrix whose first column is the unique solution to $AX = \begin{pmatrix} 1 \\ 0 \\ 0 \\ \vdots \\ 0 \end{pmatrix}$, second column is the unique solution to $AX = \begin{pmatrix} 0 \\ 1 \\ 0 \\ \vdots \\ 0 \end{pmatrix}$, and so on. What is $AB$?

## Exercises

**1.** Find a solution to this system of linear equations.

$$\begin{cases} 2x + 4y + 5z &= 57 \\ 3x + 2y + 5z &= 48 \\ -x + 2y &= 9 \end{cases}$$

**2.** Calculate the result of each matrix multiplication.

**a.** $\begin{pmatrix} 2 & 4 & 5 \\ 3 & 2 & 5 \\ -1 & 2 & 0 \end{pmatrix} \begin{pmatrix} 1 \\ 5 \\ 7 \end{pmatrix}$ **b.** $\begin{pmatrix} 2 & 4 & 5 \\ 3 & 2 & 5 \\ -1 & 2 & 0 \end{pmatrix} \begin{pmatrix} -5 \\ 2 \\ 10 \end{pmatrix}$

**c.** $\begin{pmatrix} 2 & 4 & 5 \\ 3 & 2 & 5 \\ -1 & 2 & 0 \end{pmatrix} \begin{pmatrix} 7 \\ 8 \\ 4 \end{pmatrix}$ **d.** $\begin{pmatrix} 2 & 4 & 5 \\ 3 & 2 & 5 \\ -1 & 2 & 0 \end{pmatrix} \begin{pmatrix} 1 \\ 0 \\ 0 \end{pmatrix}$

**e.** $\begin{pmatrix} 2 & 4 & 5 \\ 3 & 2 & 5 \\ -1 & 2 & 0 \end{pmatrix} \begin{pmatrix} 1 & -5 & 7 & 1 \\ 5 & 2 & 8 & 0 \\ 7 & 10 & 4 & 0 \end{pmatrix}$

**3.** Write $\begin{pmatrix} 3 \\ 4 \\ 5 \end{pmatrix}$ as a linear combination of the columns of the matrix

$$\begin{pmatrix} 2 & 4 & 5 \\ 3 & 2 & 5 \\ -1 & 2 & 0 \end{pmatrix}$$

or prove that it is impossible.

**4.** Suppose $A = \begin{pmatrix} 2 & 4 & 5 \\ 3 & 2 & 5 \\ -1 & 2 & 0 \end{pmatrix}$. Write the result of the following matrix multiplication as a linear combination of the columns of $A$.

$$\begin{pmatrix} 2 & 4 & 5 \\ 3 & 2 & 5 \\ -1 & 2 & 0 \end{pmatrix} \begin{pmatrix} x_1 & y_1 \\ x_2 & y_2 \\ x_3 & y_3 \end{pmatrix}$$

**5.** Find two different vectors that make this matrix multiplication true.

$$\begin{pmatrix} 2 & 4 & 5 \\ 3 & 2 & 5 \\ -1 & 2 & 0 \end{pmatrix} \begin{pmatrix} x \\ y \\ z \end{pmatrix} = \begin{pmatrix} 4 \\ 6 \\ -2 \end{pmatrix}$$

**6.** For each given matrix, find its inverse without using a calculator, or show that the inverse does not exist.

**a.** $\begin{pmatrix} 2 & 4 & 5 \\ 3 & 2 & 5 \\ -1 & 2 & 0 \end{pmatrix}$ **b.** $\begin{pmatrix} 2 & 4 & 5 \\ 3 & 2 & 5 \\ -1 & 2 & 1 \end{pmatrix}$ **c.** $\begin{pmatrix} 2 & 0 & 0 \\ 0 & 5 & 0 \\ 0 & 0 & -10 \end{pmatrix}$

**7.** **a.** If $A$ is symmetric, prove that $A^2$ is symmetric.
**b.** If $A$ is skew-symmetric, what can be said about $A^2$? Prove it.

**8.** Theorem 4.9 states that if $A$ and $B$ are symmetric, then $(AB)^\top = BA$. Is this true . . .

**a.** if $A$ and $B$ are both skew-symmetric?
**b.** if $A$ is symmetric and $B$ is skew-symmetric?

**9.** If $A$ is a square matrix and $AI = A$, prove that $IA = A$.

**10.** Justify each step in the proof of Theorem 4.12.

**11.** **What's Wrong Here?** Derman claims that right-multiplication could have been used in the proof of Theorem 4.12. Here are his steps:

$$\begin{aligned} AX &= B \\ (AX)A^{-1} &= BA^{-1} \\ (AA^{-1})X &= BA^{-1} \\ X &= BA^{-1} \end{aligned}$$

He concludes this means that $A^{-1}B = BA^{-1}$ for any invertible matrix $A$ and any matrix $B$. What is wrong with Derman's reasoning?

**12.** Given $M = \begin{pmatrix} a & b \\ b & a+b \end{pmatrix}$, calculate each of the following:

**a.** $M^2$ **b.** $M^3$ **c.** $M^4$ **d.** $M^5$ **e.** $M^6$ **f.** $M^7$

**g.** What is happening in general? Can you explain it?

**13.** For $A = \begin{pmatrix} 1 & 2 & 3 \\ 3 & 2 & 1 \\ 2 & 3 & 1 \end{pmatrix}$, find $X$ if

**a.** $AX = \begin{pmatrix} 1 \\ 0 \\ 0 \end{pmatrix}$ **b.** $AX = \begin{pmatrix} 0 \\ 1 \\ 0 \end{pmatrix}$ **c.** $AX = \begin{pmatrix} 0 \\ 0 \\ 1 \end{pmatrix}$

**14.** For $A = \begin{pmatrix} 1 & 2 & 3 \\ 3 & 2 & 1 \\ 2 & 3 & 1 \end{pmatrix}$, determine $A^{-1}$.

**15.** Let $A = \begin{pmatrix} 2 & 1 \\ 1 & 3 \end{pmatrix}$, $B = \begin{pmatrix} 4 & -1 \\ 5 & 1 \end{pmatrix}$, and $C = \begin{pmatrix} 1 & -1 \\ 5 & 0 \end{pmatrix}$. Calculate each of the following:

**a.** $A(BC)$ **b.** $(AB)C$

**16.** Let $A = \begin{pmatrix} a & b \\ c & d \end{pmatrix}$, $B = \begin{pmatrix} e & f \\ g & h \end{pmatrix}$, and $C = \begin{pmatrix} w & x \\ y & z \end{pmatrix}$. Calculate each of the following:

**a.** $A(BC)$ **b.** $(AB)C$

**17.** For each $2 \times 2$ matrix $A$, calculate $A^{-1}$. Use fractions when needed.

**a.** $A = \begin{pmatrix} 10 & -1 \\ 3 & 10 \end{pmatrix}$ **b.** $A = \begin{pmatrix} 10 & -1 \\ 3 & 20 \end{pmatrix}$ **c.** $A = \begin{pmatrix} 10 & 1 \\ 3 & 20 \end{pmatrix}$

**d.** $A = \begin{pmatrix} 5 & 8 \\ 8 & 13 \end{pmatrix}$ **e.** $A = \begin{pmatrix} a & b \\ c & d \end{pmatrix}$

**18.** Under what circumstances will a $2 \times 2$ matrix be invertible? Find some examples that are invertible and some that are not.

**19.** Find a $3 \times 3$ matrix $B$ so that $\begin{pmatrix} 3 & 2 & 1 \\ 2 & 0 & -1 \\ 0 & 1 & 1 \end{pmatrix} B = \begin{pmatrix} 1 & 0 & 0 \\ 0 & 1 & 0 \\ 0 & 0 & 1 \end{pmatrix}$.

**20.** Under what circumstances will a $3 \times 3$ matrix be invertible? Find some examples that are invertible and some that are not.

**21.** **a.** Pick any $3 \times 3$ diagonal matrix and square it. What do you notice?

**b.** If $A = \begin{pmatrix} 0.9 & 0 & 0 \\ 0 & 1 & 0 \\ 0 & 0 & 0.5 \end{pmatrix}$, compute $A^3$.

**c.** For matrix $A$, give a good estimate for $A^{100}$ without computing it.

**22.** Determine whether each statement is true or false. Justify your answers. Assume the given matrices are square and the same size.

**a.** The product of two scalar matrices is a scalar matrix.
**b.** The product of two diagonal matrices is a diagonal matrix.
**c.** The sum of two upper triangular matrices is an upper triangular matrix.
**d.** The product of two upper triangular matrices is an upper triangular matrix.
**e.** The product of an upper triangular matrix and a lower triangular matrix is a diagonal matrix.
**f.** The product of two symmetric matrices is a symmetric matrix.
**g.** The sum of two skew-symmetric matrices is a skew-symmetric matrix.
**h.** The product of two skew-symmetric matrices is a skew-symmetric matrix.

**23.** Find all matrices $A$ so that $A^2 = \begin{pmatrix} 1 & 0 \\ 0 & 1 \end{pmatrix}$.

**24.** Find all matrices $B$ so that $B^3 = \begin{pmatrix} 1 & 0 \\ 0 & 1 \end{pmatrix}$.

**25.** Find all matrices $C$ so that $C^4 = \begin{pmatrix} 1 & 0 \\ 0 & 1 \end{pmatrix}$.

**26.** Let $A = \begin{pmatrix} 2 & 3 \\ 1 & -1 \end{pmatrix}$ and $P = \begin{pmatrix} 5 & 7 \\ 3 & 4 \end{pmatrix}$. Compute the following:

**a.** $P^{-1}$ **b.** $M = PAP^{-1}$

**c.** $M^2$ **d.** $PA^2P^{-1}$

← Since multiplying matrices is associative, you can read part **b** as either $(PA)P^{-1}$ or $P(AP^{-1})$. But it is *not* the same as $A$!

**27.** Using the same matrices as Exercise 26, use a calculator to compute each of the following:

**a.** $M^{10} = \left(PAP^{-1}\right)^{10}$

**b.** $PA^{10}P^{-1}$

**28.** Prove that if $P$ is invertible and $n$ is any positive integer, then

$$\left(PAP^{-1}\right)^n = PA^nP^{-1}$$

**29.** Let $A = \begin{pmatrix} 0.9 & 0.3 \\ 0.1 & 0.7 \end{pmatrix}$ and $P = \begin{pmatrix} 1 & 1 \\ -1 & 3 \end{pmatrix}$.

**a.** Compute $P^{-1}$, and then compute $M = PAP^{-1}$.

**b.** Compute $M^{10}$ to four decimal places.

**c.** Explain why $M^{100}$ is very close to $\begin{pmatrix} 1 & 0 \\ 0 & 0 \end{pmatrix}$.

**d.** Use Exercise 28 to explain why $A^{100} = P^{-1}M^{100}P$.

**e.** Give a very good approximation to $A^{100}$ without a calculator.

**30.** Suppose $A$ and $B$ are $n \times n$ matrices so that $AB = I$. Show that the system $(x_1, x_2, \ldots, x_n)A = (c_1, c_2, \ldots, c_n)$ has a unique solution for every $(c_1, c_2, \ldots, c_n)$ in $\mathbb{R}^n$.

**31.** Suppose $A$ and $B$ are $n \times n$ matrices so that $AB = I$. Show that there is an $n \times n$ matrix $C$ so that $CA = I$.

← Use Exercise 30 to find vectors $C_i$ so that $C_iA = (0, 0, \ldots, 0, 1, 0, \ldots, 0)$, where the 1 is in the $i^{\text{th}}$ place. Let $C$ be the matrix whose rows are the $C_i$. What is $CA$?

**32.** Suppose $A$ and $B$ are $n \times n$ matrices so that $AB = I$. Use Exercise 31 to show that $BA = I$.

## 4.6 Applications of Matrix Algebra

You can use matrices to solve systems of equations. There is a useful relationship between the solutions to a system and the kernel of its corresponding matrix. Finding the kernel will often give you a faster way to find the whole set of solutions.

**In this lesson, you will learn how to**

- determine when a matrix has an inverse
- find all solutions to a matrix equation given one solution and the kernel of the matrix
- find the set of vectors left invariant or just simply scaled after multiplication by a matrix

In Chapter 3, you learned this definition of the **kernel** of a matrix:

If $A$ is a matrix, the kernel of $A$, written $\ker(A)$, is the set of all vectors that are orthogonal to the rows of $A$.

Consider the matrix

$$A = \begin{pmatrix} 2 & 2 & 3 \\ 7 & 2 & 10 \\ 3 & -2 & 4 \end{pmatrix}$$

A vector $X = \begin{pmatrix} x \\ y \\ z \end{pmatrix}$ is in the kernel if it is orthogonal to all the rows of $A$. The three dot products involved produce this homogeneous system of equations:

$$\begin{cases} 2x + 2y + 3z &= 0 \\ 7x + 2y + 10z &= 0 \\ 3x - 2y + 4z &= 0 \end{cases}$$

**Remember**

A system of equations is *homogeneous* if all its constant terms are zero.

Oh, but this is familiar from the last lesson. The written system of equations can be written as one equation in the form $AX = B$, where $A$ is a matrix of coefficients, $X$ is a column vector of unknowns, and $B$ is a column vector of constants.

$$\begin{pmatrix} 2 & 2 & 3 \\ 7 & 2 & 10 \\ 3 & -2 & 4 \end{pmatrix} \begin{pmatrix} x \\ y \\ z \end{pmatrix} = \begin{pmatrix} 0 \\ 0 \\ 0 \end{pmatrix}$$

The right side is the zero vector, so the system can be written as $AX = O$.

So, $X$ is in the kernel if it solves the homogeneous system of equations, and a second equivalent definition of kernel is the following:

### Definition

The **kernel** $\ker(A)$ of a matrix $A$ is the set of vectors $X$ that make $AX = O$.

### For Discussion

**1.** Suppose vector $S$ solves $AX = B$ and nonzero vector $Y$ is in the kernel of $A$. Find three more vectors $Z$ that each solve $AZ = B$.

$\longleftarrow$ What does the new kernel definition say about $AY$, and how can this be used to find more solutions?

The zero vector will always be in the kernel: $AO = O$. Some matrices have other vectors in their kernels while others do not.

### For You to Do

**2.** **a.** Given matrix $A = \begin{pmatrix} 2 & 2 & 3 \\ 7 & 2 & 10 \\ 3 & -2 & 4 \end{pmatrix}$, determine whether there is a nonzero vector $X$ in the kernel of $A$. You may wish to use methods from Chapter 3 or from Lesson 4.5.

**b.** If matrix $A$ has a nonzero kernel, change it slightly to make a new matrix that does not. If matrix $A$ does not have a nonzero kernel, change it slightly to make a new one that does.

In the last lesson, you learned that *when it exists*, a matrix's inverse can be used to solve the system $AX = B$. The system $AX = O$ is no exception, which leads to this corollary to Theorem 4.12.

**Corollary 4.13**

*If a matrix $A$ is invertible,* $\ker(A)$ *is only the zero vector.*

$\longleftarrow$ The converse of this theorem is also true. You'll prove it later in this lesson.

You'll be asked to prove this corollary in the exercises.

### For You to Do

**3.** Let $A = \begin{pmatrix} 4 & 1 & 0 \\ 2 & 5 & -1 \\ 0 & 9 & -2 \end{pmatrix}$, $X = \begin{pmatrix} 1 \\ 1 \\ 1 \end{pmatrix}$, and $Y = \begin{pmatrix} 0 \\ 5 \\ 19 \end{pmatrix}$. Calculate the following:

**a.** $AX$ **b.** $AY$ **c.** $AX - AY$ **d.** $A(X - Y)$ **e.** $A(3X - 3Y)$

The results from these problems suggest the following theorem.

**Theorem 4.14**

*Given a matrix $A$, if there exist vectors $X$ and $Y$ with $AX = AY$, then $k(X - Y)$ is in* $\ker(A)$ *for any real number $k$.*

***Proof.*** If $AX = AY$, then $AX - AY = O$, a zero vector. Since matrix multiplication is distributive, then $A(X - Y) = O$ as well, and $X - Y$ is in the kernel of $A$.

Now consider $k(X-Y)$ for some real $k$. These steps prove that $k(X-Y)$ is in $\ker(A)$.

$$\begin{aligned} A(k(X-Y)) &= k(A(X-Y)) \\ &= k \cdot O \\ &= O \end{aligned}$$

■

**Corollary 4.15**

*For a matrix $A$, if $\ker(A) = O$, then $AX = AY$ if and only if $X = Y$.*

***Proof.***

- Clearly if $X = Y$, then $AX = AY$.
- If $\ker(A) = O$, then by Theorem 4.14,

  $$AX = AY \Longrightarrow k(X-Y) \text{ is in } \ker(A)$$

  specifically, $X - Y$ is in $\ker(A)$. Hence $X - Y = O$, and $X = Y$. ■

**Minds in Action** Episode 13

*Tony and Sasha are looking at a linear system.*

TONY: We want to solve $AX = B$: $\begin{pmatrix} 1 & 1 & 5 \\ 1 & 2 & 7 \\ 3 & 4 & 17 \end{pmatrix} \begin{pmatrix} x \\ y \\ z \end{pmatrix} = \begin{pmatrix} 3 \\ 4 \\ 10 \end{pmatrix}$

SASHA: I see one solution: $\begin{pmatrix} 3 \\ 4 \\ 10 \end{pmatrix}$ is the last column minus the sum of the first two.

TONY: How'd you see that?

SASHA: Practice. Anyway, this means that $S = \begin{pmatrix} -1 \\ -1 \\ 1 \end{pmatrix}$ is a solution by Theorem 4.5.

TONY: Do you see any others?

SASHA: No. But I have an idea.

TONY: Of course you do. What is it?

SASHA: Well, suppose we solve an easier system: $AX = O$.

TONY: That will give you the kernel of $A$. What do you intend to do with that?

SASHA: Well, if $AS = B$ and $AT = O$, then $A(S+T)$ will also be $B$.

TONY: Sure, $A(S+T) = AS + AT = B + O = B$. Smooth.

*Enter Derman*

TONY: Hey, Derman. Go find rref $\begin{pmatrix} 1 & 1 & 5 \\ 1 & 2 & 7 \\ 3 & 4 & 17 \end{pmatrix}$.

DERMAN: Sure. I did it in my head. It's $\begin{pmatrix} 1 & 0 & 3 \\ 0 & 1 & 2 \\ 0 & 0 & 0 \end{pmatrix}$.

TONY: Ah. That means that the columns of $A$ are linearly dependent. It also means that

$$\ker(A) = z(-3, -2, 1)$$

SASHA: And it also means that every vector of the form

$$(-1, -1, 1) + z(-3, -2, 1)$$

is a solution to our original system. Hey . . . that's a line.

TONY: So is the kernel—it's a line through the origin. And it's parallel to our solution line.

DERMAN: I did it in my head.

SASHA: I wonder if we've found *all* the solutions . . . .

---

The next result shows that Tony, Sasha, and Derman are onto something.

**Theorem 4.16**

*If $AS = B$, then* every *solution to $AX = B$ is found by letting $X = S + K$, where $K$ is in* $\ker(A)$.

***Proof.*** There are two parts to the proof. First, show that if $AS = B$, then $A(S + K) = B$. Since $K$ is in $\ker(A)$, $AK = O$.

$$A(S + K) = AS + AK = B + O = B$$

Second, show that if $AS = B$ and $AT = B$, then $T = S + K$ with $K$ in the kernel. This follows directly from Theorem 4.14. ∎

### For You to Do

**4.** Tony, Sasha, and Derman found that the set of solutions to their "inhomogeneous" system $AX = B$ was a *translate* (by $(-1, -1, 1)$) of the kernel of $A$. Since $\ker(A)$ was a line through the origin, the solution set was a parallel line through $(-1, -1, 1)$.

Come up with a matrix $A$ that has a *plane* through the origin as its kernel. Show that the solution set to $AX = B$ for some vector $B$ is a translate of this plane by $B$:

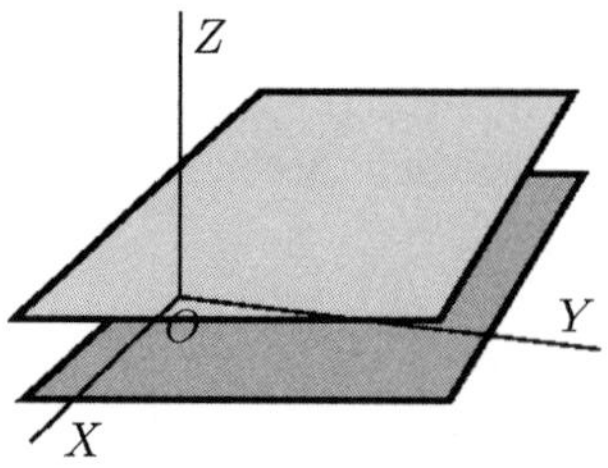

### Developing Habits of Mind

**Seek general results.** While the theorems above cover many cases, they miss an important one: the converse to Corollary 4.13. It would be nice to have some conditions that *guarantee* that a square matrix $A$ has an inverse.

You may suspect that if $A$ is a square matrix and $\ker(A)$ includes only the zero vector, then $A^{-1}$ exists. This is true, but proving it requires some work from Chapter 3. Why? Because much of the work comes down to solving equations in the form $AX = B$, which correspond directly with systems of equations.

← Nonsquare matrices cannot have inverses. More about nonsquare matrices in Chapter 5.

Suppose $A$ is $3 \times 3$. If it has an inverse $A^{-1}$, then $AA^{-1} = \begin{pmatrix} 1 & 0 & 0 \\ 0 & 1 & 0 \\ 0 & 0 & 1 \end{pmatrix}$. Now ask: what would the first column of $A^{-1}$ have to be?

Theorem 4.5 says that this first column determines the first column of the product completely. If this column is called $X_1$, then $AX_1 = \begin{pmatrix} 1 \\ 0 \\ 0 \end{pmatrix}$. This corresponds to a system of three equations and three unknowns, which may or may not have a solution.

Similarly, solve $AX_2 = \begin{pmatrix} 0 \\ 1 \\ 0 \end{pmatrix}$ and $AX_3 = \begin{pmatrix} 0 \\ 0 \\ 1 \end{pmatrix}$ and one of two things will happen:

- If all three systems were solvable, then the columns $X_1, X_2, X_3$ are, in order, the columns of the inverse $A^{-1}$.
- If at least one system was unsolvable, then $A^{-1}$ does not exist, since there is no way to make that column of the identity matrix when multiplying.

As you learned in Chapter 3, you can solve multiple systems that have the same coefficient matrix at once. The method below shows how you can use Gaussian elimination to find all the columns of $A^{-1}$ at once. What would it mean if Gaussian elimination breaks down at any step along the way?

### Example

**Problem.** Let $A = \begin{pmatrix} 2 & -1 & 0 \\ 2 & 1 & -3 \\ 0 & 1 & 1 \end{pmatrix}$. Find $A^{-1}$ or show that it does not exist.

**Solution.** The goal is to find a matrix such that

$$\begin{pmatrix} 2 & -1 & 0 \\ 2 & 1 & -3 \\ 0 & 1 & 1 \end{pmatrix} A^{-1} = \begin{pmatrix} 1 & 0 & 0 \\ 0 & 1 & 0 \\ 0 & 0 & 1 \end{pmatrix}$$

To find the first column of $A^{-1}$, you can solve the system $AX = \begin{pmatrix}1\\0\\0\end{pmatrix}$ by Gaussian elimination. This method, introduced in Chapter 3, starts by setting up an *augmented matrix*.

$$\begin{pmatrix} 2 & -1 & 0 & 1 \\ 2 & 1 & -3 & 0 \\ 0 & 1 & 1 & 0 \end{pmatrix}$$

When Gaussian elimination is complete, the rightmost column will be the first column of $A^{-1}$.

You can repeat this process to find all three columns of $A^{-1}$, or do all three at once! In Chapter 3 you learned that Gaussian Elimination can solve multiple versions of the same system of equations, by writing more columns into the augmented matrix. This means that $A^{-1}$ can be found by following Gaussian Elimination on this matrix:

$\longleftarrow$
For an example of this, see Exercise 6 in Lesson 3.2.

$$\begin{pmatrix} 2 & -1 & 0 & 1 & 0 & 0 \\ 2 & 1 & -3 & 0 & 1 & 0 \\ 0 & 1 & 1 & 0 & 0 & 1 \end{pmatrix}$$

For example, after subtracting the second row by the first row, you get

$$\begin{pmatrix} 2 & -1 & 0 & 1 & 0 & 0 \\ 0 & 2 & -3 & -1 & 1 & 0 \\ 0 & 1 & 1 & 0 & 0 & 1 \end{pmatrix}$$

### For You to Do

**5.** **a.** Find the reduced echelon form of this augmented matrix.
**b.** What is $A^{-1}$?
**c.** Verify that $AA^{-1} = I$.

The end result of this Gaussian elimination puts the identity matrix into the first three columns.

$$\begin{pmatrix} 1 & 0 & 0 & & & \\ 0 & 1 & 0 & . & . & . \\ 0 & 0 & 1 & & & \end{pmatrix}$$

The process will succeed if Gaussian elimination produces an identity matrix in its first $n$ columns. The inverse matrix can then be read from the remaining columns. This can be stated as a theorem.

**Theorem 4.17**

*For a square matrix $A$, if $\text{rref}(A) = I$, then $A^{-1}$ exists.*

In Chapter 3, you learned that $\mathrm{rref}(A) = I$ if and only if the columns of $A$ were linearly independent. This leads to a very nice corollary.

**Corollary 4.18**

*For a square matrix $A$, if its columns are linearly independent, then $A^{-1}$ exists.*

### Developing Habits of Mind

**Look for connections.** You've now learned that these conditions for a square matrix $A$ are all connected:

(1) the columns of $A$ are linearly independent

(2) $\mathrm{rref}(A) = I$

(3) $A^{-1}$ exists

(4) $AX = B$ has a unique solution for any $B$

(5) $\ker(A) = O$

⟵ You'll add to this list of connections later in the book.

These statements are, in fact, *equivalent*: for a given matrix $A$, if one is true, they all are true and, if one is false, they all fail.

How do you prove that these five statements are equivalent? You could prove that each one implies the other. But a common technique is to set up a "chain of implications" like this:

⟵ How many proofs would that be?

$$\begin{array}{ccccc} \text{(1) columns are linearly independent} & & \longrightarrow & & \text{(2) } \mathrm{rref}(A) = I \\ \uparrow & & & & \downarrow \\ \text{(5) } \ker(A) = O & \longleftarrow & \text{(4) } AX = B \text{ has a unique solution} & \longleftarrow & \text{(3) } A^{-1} \text{ exists} \end{array}$$

The next theorem formalizes this equivalence. Its proof is just a summary of theorems and corollaries you have already proved.

**Theorem 4.19 (The TFAE Theorem)**

*The following statements are all equivalent for an $n \times n$ matrix $A$:*

⟵ *TFAE* stands for "The Following Are Equivalent."

(1) *The columns of $A$ are linearly independent*

(2) $\mathrm{rref}(A) = I$

(3) $A^{-1}$ *exists*

(4) $AX = B$ *has a unique solution for any $B$*

(5) $\ker(A) = O$

***Proof.*** One chain of implications is to show that

$$(1) \Rightarrow (2) \Rightarrow (3) \Rightarrow (4) \Rightarrow (5) \Rightarrow (1)$$

You have done all but the last of these already. To show the last, suppose that $\ker(A) = O$ and suppose that

$$x_1 A_{*1} + x_2 A_{*2} + \cdots + x_n A_{*n} = O$$

← You need to show that each $x_i = 0$. Why?

Then

$$A \begin{pmatrix} x_1 \\ x_2 \\ \vdots \\ x_n \end{pmatrix} = O$$

← Why?

so $(x_1, x_2, \ldots, x_n)$ is in $\ker(A)$. Hence, each $x_i = 0$ and the columns of $A$ are linearly independent. ■

**For You to Do**

**6.** Look back over the previous chapters and find the theorems and corollaries that establish the other implications.

## Exercises

**1.** Consider $A = \begin{pmatrix} 1 & 2 & 3 \\ 4 & 5 & 6 \\ 7 & 8 & 9 \end{pmatrix}$ and $B = \begin{pmatrix} 5 \\ 11 \\ 17 \end{pmatrix}$.

**a.** Show that if $X_1 = \begin{pmatrix} -1 \\ 3 \\ 0 \end{pmatrix}$, $AX_1 = B$.

**b.** Show that if $X_2 = \begin{pmatrix} 1 \\ -1 \\ 2 \end{pmatrix}$, $AX_2 = B$.

**c.** Find some other vectors that also solve $AX = B$.

**2.** Suppose $AX_1 = B$ and $AX_2 = B$. Which of the following vectors $X$ *must* also solve $AX = B$?

**a.** $X = 2X_1$ **b.** $X = X_1 + X_2$
**c.** $X = X_1 - X_2$ **d.** $X = \frac{1}{2}X_1 + \frac{1}{2}X_2$
**e.** $X = 2X_1 - X_2$ **f.** $X = aX_1 + bX_2$ for any real $a, b$

**3.** Suppose $X_1$ and $X_2$ are each in $\ker(A)$. Which of the following vectors $X$ *must* also be in $\ker(A)$?

**a.** $X = 2X_1$ **b.** $X = X_1 + X_2$
**c.** $X = X_1 - X_2$ **d.** $X = \frac{1}{2}X_1 + \frac{1}{2}X_2$
**e.** $X = 2X_1 - X_2$ **f.** $X = aX_1 + bX_2$ for any real $a, b$

**4.** Prove that if $AX_1 = B$ and $AX_2 = B$ for nonzero $B$, then $aX_1 + bX_2$ is a solution to $AX = B$ if and only if $a + b = 1$.

**5.** Prove Corollary 4.13 using Theorem 4.12.

**6.** Use the results from this lesson to prove that a system of linear equations cannot have exactly two solutions.

**7.** Calculate each of the following, where

$$A = \begin{pmatrix} 4 & 1 & 3 \\ 2 & 1 & 0 \end{pmatrix}, B = \begin{pmatrix} 3 & 1 & 2 \\ 1 & 4 & 3 \end{pmatrix}, C = \begin{pmatrix} 1 & 2 & 3 \\ -1 & 1 & 4 \\ 1 & 3 & 2 \end{pmatrix}, D = \begin{pmatrix} 4 & 3 & 1 \\ 1 & 1 & 0 \\ 2 & 2 & 1 \end{pmatrix}$$

**a.** $AC$ **b.** $CA$ **c.** $BC$
**d.** $(A+B)C$ **e.** $A^\top C^\top$ **f.** $D^{-1}$
**g.** $(AD)_{*3}$ **h.** $(BD)_{2*}$ **i.** $(D^\top)^{-1}$

**8.** Use Gaussian elimination to determine the inverse of this matrix:

$$\begin{pmatrix} 0.4 & 0.1 & 0.3 \\ -0.2 & 0.2 & 0.6 \\ 0.2 & -0.2 & 0.4 \end{pmatrix}$$

**9.** **a.** Give some conditions that will guarantee that a square matrix $A$ has an inverse.
**b.** Give some conditions that guarantee that a square matrix $A$ will *not* have an inverse.

**10.** Here is a graph of connections on a map:

**a.** Use the map to construct a $4 \times 4$ matrix $M$ to represent the situation.

**b.** Find the total number of ways to get from 1 to 4 in one step, in two steps, in three steps.

**c.** Compute $M^2$ and $M^3$.

**d.** Determine, using matrices, the total number of ways to get from 1 to 4 in five or fewer steps.

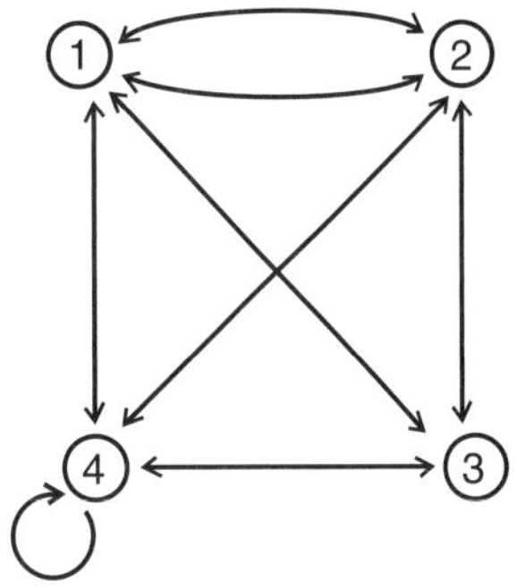


⟵ Note that 4 has a path to itself.

**11.** If $A = \begin{pmatrix} 3 & 1 & 2 \\ 1 & 7 & -1 \\ 4 & 2 & 3 \end{pmatrix}$, write $A$ as the sum of a symmetric and skew-symmetric matrix.

**12.** Given $A = \begin{pmatrix} 3 & 1 & 2 \\ 1 & 7 & -1 \\ 4 & 2 & 3 \end{pmatrix}$ and $X = \begin{pmatrix} 2 \\ -1 \\ 5 \end{pmatrix}$.

**a.** Compute $B = AX$.

**b.** Write $B$ as a linear combination of the columns of $A$.

$\longleftarrow$ In other words, find $c_1$, $c_2$, and $c_3$ so that

$$B = c_1\begin{pmatrix} 3 \\ 1 \\ 4 \end{pmatrix} + c_2\begin{pmatrix} 1 \\ 7 \\ 2 \end{pmatrix} + c_3\begin{pmatrix} 2 \\ -1 \\ 3 \end{pmatrix}$$

**13.** Solve each system by multiplying both sides by the inverse of the coefficient matrix.

**a.** $\begin{pmatrix} 7 & 5 \\ 3 & 2 \end{pmatrix}\begin{pmatrix} x \\ y \end{pmatrix} = \begin{pmatrix} -21 \\ -10 \end{pmatrix}$

**b.** $\begin{pmatrix} 2 & 1 & 1 \\ 1 & 1 & 0 \\ 2 & 0 & 3 \end{pmatrix}\begin{pmatrix} x \\ y \\ z \end{pmatrix} = \begin{pmatrix} -3 \\ 5 \\ -19 \end{pmatrix}$

**c.** $\begin{pmatrix} 2 & 3 & 1 \\ 2 & 3 & 0 \\ 1 & 2 & 1 \end{pmatrix}\begin{pmatrix} x \\ y \\ z \end{pmatrix} = \begin{pmatrix} -10 \\ -13 \\ -8 \end{pmatrix}$

**d.** $\begin{pmatrix} 6 & 1 & 1 \\ 7 & 2 & 1 \\ 2 & 1 & 0 \end{pmatrix}\begin{pmatrix} x \\ y \\ z \end{pmatrix} = \begin{pmatrix} -11 \\ -2 \\ 6 \end{pmatrix}$

**14.** For each given matrix $A$ and column vector $B$, find *all* $X$ so that $AX = B$.

$\longleftarrow$ There may be more than one solution! Find them *all*.

**a.** $A = \begin{pmatrix} 1 & 2 & 3 \\ 4 & 5 & 6 \\ 7 & 8 & 9 \end{pmatrix}$, $B = \begin{pmatrix} 5 \\ 11 \\ 17 \end{pmatrix}$

**b.** $A = \begin{pmatrix} 1 & 2 & 3 \\ 4 & 5 & 6 \\ 7 & 8 & 9 \end{pmatrix}$, $B = \begin{pmatrix} 5 \\ 11 \\ 8 \end{pmatrix}$

**c.** $A = \begin{pmatrix} 1 & 2 & 3 \\ 4 & 5 & 6 \\ 7 & 8 & 0 \end{pmatrix}$, $B = \begin{pmatrix} 5 \\ 11 \\ 8 \end{pmatrix}$

**d.** $A = \begin{pmatrix} 1 & 3 & -1 & 2 \\ 0 & 1 & 4 & 1 \\ 3 & 10 & 1 & 7 \end{pmatrix}$, $B = \begin{pmatrix} 5 \\ 2 \\ 17 \end{pmatrix}$

**e.** $A = \begin{pmatrix} 1 & 3 & -1 & 2 \\ 0 & 1 & 4 & 1 \\ 3 & 10 & 1 & 7 \end{pmatrix}$, $B = \begin{pmatrix} 4 \\ 1 \\ 8 \end{pmatrix}$

**15.** Let $A = \begin{pmatrix} 1 & 3 & 0 \\ 2 & 5 & 1 \end{pmatrix}$.

**a.** Find $\ker(A)$.

**b.** Find *all* $3 \times 1$ matrices $X$ so that $AX = \begin{pmatrix} 4 \\ 8 \end{pmatrix}$.

**16.** For each system, rewrite the system in the form $AX = B$. Then, find a solution to each system by writing $B$ as a linear combination of the columns of $A$.

⟵ There may be more than one solution, but you only have to find *one*!

**a.** $\begin{cases} x-y+z &= 0 \\ 2x+y+z &= 3 \\ x-y+2z &= 0 \\ 3x-3y-z &= 0 \end{cases}$ **b.** $\begin{cases} x-y+z &= 0 \\ 2x+y+z &= 2 \\ x-y+2z &= 1 \\ 3x+y-z &= 0 \end{cases}$

**c.** $\begin{cases} x+y+4z &= 3 \\ x-y+2z &= -1 \end{cases}$ **d.** $\begin{cases} x+2y+3z &= 1 \\ 4x+5y+6z &= 1 \\ 7x+8y+9z &= 1 \end{cases}$

**e.** $\begin{cases} x+y-z &= 2 \\ 3x+y-2z &= 5 \\ x+y+z &= 4 \end{cases}$

**17.** For each system, find one solution using the method of Exercise 16, and then find all solutions by using the kernel.

**a.** $\begin{pmatrix} 1 & 3 & 7 \\ 2 & 7 & 1 \end{pmatrix}\begin{pmatrix} x \\ y \\ z \end{pmatrix} = \begin{pmatrix} 2 \\ 5 \end{pmatrix}$ **b.** $\begin{pmatrix} 1 & 2 & 3 \\ 4 & 5 & 6 \\ 7 & 8 & 9 \end{pmatrix}\begin{pmatrix} x \\ y \\ z \end{pmatrix} = \begin{pmatrix} 1 \\ 1 \\ 1 \end{pmatrix}$

**c.** $\begin{pmatrix} 1 & 2 & 3 \\ 4 & 5 & 6 \\ 7 & 8 & 0 \end{pmatrix}\begin{pmatrix} x \\ y \\ z \end{pmatrix} = \begin{pmatrix} 1 \\ 1 \\ 1 \end{pmatrix}$ **d.** $\begin{pmatrix} 2 & 1 \\ 3 & 4 \\ 7 & 2 \end{pmatrix}\begin{pmatrix} x \\ y \end{pmatrix} = \begin{pmatrix} 4 \\ 11 \\ 11 \end{pmatrix}$

**18.** Suppose $A = \begin{pmatrix} 1 & 1 & -1 & 2 \\ 3 & 1 & -2 & 5 \\ 1 & 1 & 1 & 4 \end{pmatrix}$. Left-multiply $A$ by each of the following matrices, and describe how the result relates to $A$.

⟵ Matrix $A$ should be familiar. See Exercise 16**e**.

**a.** $\begin{pmatrix} 0 & 1 & 0 \\ 1 & 0 & 0 \\ 0 & 0 & 1 \end{pmatrix}$ **b.** $\begin{pmatrix} 2 & 0 & 0 \\ 0 & 1 & 0 \\ 0 & 0 & 1 \end{pmatrix}$ **c.** $\begin{pmatrix} 1 & 0 & 0 \\ 0 & 1 & 0 \\ -1 & 0 & 1 \end{pmatrix}$ **d.** $\begin{pmatrix} 1 & 0 & 0 \\ -3 & 1 & 0 \\ 0 & 0 & 1 \end{pmatrix}$

**19.** Suppose that $E = \begin{pmatrix} -\frac{3}{4} & \frac{1}{2} & \frac{1}{4} \\ \frac{5}{4} & -\frac{1}{2} & \frac{1}{4} \\ -\frac{1}{2} & 0 & \frac{1}{2} \end{pmatrix}$ and $A = \begin{pmatrix} 1 & 1 & -1 & 2 \\ 3 & 1 & -2 & 5 \\ 1 & 1 & 1 & 4 \end{pmatrix}$. Show that $EA$ is the echelon form of $A$.

**20.** Given $A = \begin{pmatrix} 3 & -1 & 2 & 4 & 6 \\ 1 & 7 & 1 & 0 & 5 \\ 0 & 1 & 3 & -2 & 4 \\ 1 & 5 & 1 & 0 & 6 \end{pmatrix}$ and $B = \begin{pmatrix} 1 & 3 & 7 & -1 \\ 2 & 1 & 1 & 4 \\ -1 & 2 & 6 & 3 \\ 4 & 0 & 3 & 1 \\ 2 & 8 & 1 & 7 \end{pmatrix}$, find each of these without performing the complete matrix multiplication:

**a.** $(AB)_{3*}$ **b.** $(AB)_{*2}$ **c.** $(BA)_{1*}$ **d.** $(BA)_{*2}$

⟵ Think about where the row or column of the product would come from, and find a simpler way to calculate it.

**21.** Given $A = \begin{pmatrix} a & b \\ 1-a & 1-b \end{pmatrix}$.

**a.** Compute $A^2$ and show that the sum of the entries in each column is 1.

**b.** Compute $A^3$ and show that the sum of the entries in each column is 1.

**22.** Given $A$, a $2 \times 2$ matrix where the sum of the entries in each column is $k$, show that the sum of the entries in each column of $A^2$ is $k^2$.

**23.** Given $A$, a $3 \times 3$ matrix where the sum of the entries in each column is $k$, show that the sum of the entries in each column of $A^2$ is $k^2$.

**24.** Given $A$, where the sum of the entries in each column is $a$, and $B$, where the sum of the entries in each column is $b$, show that the sum of the entries in each column of $AB$ is $ab$.

$\longleftarrow$ Be as specific or general as you like here, but the statement is even true for nonsquare $A$ and $B$!

**25.** For each matrix, determine whether it has an inverse. Compute the inverse if it exists, but not before determining whether it exists.

**a.** $\begin{pmatrix} 5 & 3 \\ 2 & 1 \end{pmatrix}$ **b.** $\begin{pmatrix} 1 & 0 & 2 \\ 2 & 1 & 1 \\ 4 & 2 & 3 \end{pmatrix}$ **c.** $\begin{pmatrix} 1 & 2 & 3 \\ 4 & 5 & 6 \\ 7 & 8 & 9 \end{pmatrix}$

**d.** $\begin{pmatrix} 1 & 2 & 3 \\ 4 & 5 & 6 \\ 7 & 8 & 0 \end{pmatrix}$ **e.** $\begin{pmatrix} 1 & 0 & 4 & 3 \\ 2 & 1 & 4 & 1 \\ 3 & 1 & 2 & 0 \\ 5 & 2 & 6 & 1 \end{pmatrix}$

**26.** Let $A = \begin{pmatrix} 5 & 3 \\ 2 & 1 \end{pmatrix}$ and $B = \begin{pmatrix} 5 & 3 \\ 3 & 2 \end{pmatrix}$.

**a.** Show that $(AB)^{-1} \neq A^{-1}B^{-1}$.

**b.** Experiment to find a different, but simple, expression for $(AB)^{-1}$ in terms of $A^{-1}$ and $B^{-1}$.

**27.** For each system $AX = B$, find one solution $X_0$ using the method of Exercise 16, and then write an expression for *all* solutions in terms of the kernel of $A$.

**a.** $\begin{pmatrix} 1 & 3 & -1 \\ 2 & 1 & 4 \end{pmatrix} \begin{pmatrix} x \\ y \\ z \end{pmatrix} = \begin{pmatrix} 2 \\ 5 \end{pmatrix}$ **b.** $\begin{pmatrix} 1 & 2 \\ 3 & -1 \\ 1 & 7 \end{pmatrix} \begin{pmatrix} x \\ y \end{pmatrix} = \begin{pmatrix} 4 \\ 5 \\ 9 \end{pmatrix}$

**c.** $\begin{pmatrix} 1 & 2 & 3 \\ 4 & 5 & 6 \\ 7 & 8 & 9 \end{pmatrix} \begin{pmatrix} x \\ y \\ z \end{pmatrix} = \begin{pmatrix} 1 \\ 1 \\ 1 \end{pmatrix}$ **d.** $\begin{pmatrix} 1 & 2 & 3 \\ 4 & 5 & 6 \\ 7 & 8 & 9 \end{pmatrix} \begin{pmatrix} x \\ y \\ z \end{pmatrix} = \begin{pmatrix} 1 \\ 1 \\ 2 \end{pmatrix}$

**e.** $\begin{pmatrix} 1 & 2 & 3 \\ 4 & 5 & 6 \\ 7 & 8 & 0 \end{pmatrix} \begin{pmatrix} x \\ y \\ z \end{pmatrix} = \begin{pmatrix} 2 \\ 2 \\ -7 \end{pmatrix}$

**28.** True or False: For $n \times n$ matrices $A$ and $B$,

if $A$ is invertible and $AB = O$, then $B = O$.

If true, prove it. If false, find a counterexample.

**29.** Sometimes, the product $AX$ determines a vector that is a multiple of the original. Let $A = \begin{pmatrix} 0.9 & 0.3 \\ 0.1 & 0.7 \end{pmatrix}$ and $X = \begin{pmatrix} x_1 \\ x_2 \end{pmatrix}$.

**a.** If $AX = kX$, write the system of equations that corresponds to the matrix equation.

**b.** Find all possible values of $k$ that could produce a possible nonzero vector $X$.

**c.** For each value of $k$ you found, determine a nonzero vector $X$ that solves $AX = kX$.

**d.** The matrix $A$ used here is the same one from Exercise 29 from Lesson 4.5. What do you notice about the results here compared with that exercise?

**30.** Pick several $2 \times 2$ matrices $A$, and follow the process of Exercise 29.

**a.** Find a matrix where $k = 0$ produces a possible nonzero vector $X$.

**b.** Find a matrix where no real number $k$ can produce a possible nonzero vector $X$.

**31.** Prove this statement:

> The values of $k$ that produce nonzero vectors $X$ solving $AX = kX$ are precisely those where $\ker(A - kI)$ has nonzero vectors in its kernel, and these vectors solve $AX = kX$.

**32.** A **block diagonal matrix** is a square matrix made of small square matrices along the diagonal of a large matrix, with zeros otherwise. One example is

$$A = \begin{pmatrix} 1 & 2 & 0 & 0 & 0 & 0 \\ 3 & 4 & 0 & 0 & 0 & 0 \\ 0 & 0 & 5 & 0 & 0 & 0 \\ 0 & 0 & 0 & 6 & 7 & 8 \\ 0 & 0 & 0 & 9 & 10 & 11 \\ 0 & 0 & 0 & 12 & 13 & 14 \end{pmatrix}$$

← Note that the smaller square matrices can be $1 \times 1$, which is an allowable matrix size.

**a.** Compute $A^2$.

**b.** What can you say about the powers of $A$?

**c.** This notation is sometimes used for block diagonal matrices:

$$A = \begin{pmatrix} X & O & O \\ O & Y & O \\ O & O & Z \end{pmatrix}$$

What does this notation mean?

**33.** A **block triangular matrix** is a matrix that, when looked at as a set of "blocks" (smaller matrices inside the larger one), becomes a triangular matrix. One example is

$$A = \begin{pmatrix} 1 & 2 & 0 & 0 & 0 & 0 \\ 3 & 4 & 0 & 0 & 0 & 0 \\ -1 & -1 & 5 & 0 & 0 & 0 \\ -2 & -2 & -3 & 6 & 7 & 8 \\ -2 & -2 & -3 & 9 & 10 & 11 \\ -2 & -2 & -3 & 12 & 13 & 14 \end{pmatrix}$$

$\longleftarrow$ The sets of $-1$, $-2$, and $-3$ show the lower-triangular blocks, though it is the fact that zeros appear above the "diagonal" that is important.

**a.** Compute $A^2$.
**b.** What can you say about the powers of $A$?
**c.** This notation is used for block triangular matrices. What does it mean?

$$A = \begin{pmatrix} X & O & O \\ M & Y & O \\ N & P & Z \end{pmatrix}$$

# Chapter 4 Mathematical Reflections

These problems will help you summarize what you have learned in this chapter.

**1.** Let

$$A = \begin{pmatrix} 1 & 2 & -1 \\ 0 & 3 & 1 \end{pmatrix}, \quad B = \begin{pmatrix} -2 & 1 & -4 \\ 3 & -1 & 5 \end{pmatrix}$$
$$C = \begin{pmatrix} 4 & 2 & 6 \\ 3 & -3 & 7 \\ 1 & 0 & -2 \end{pmatrix}, \quad D = \begin{pmatrix} 5 & -1 & 3 \\ 6 & -2 & 4 \\ 8 & 10 & -5 \end{pmatrix}$$

Calculate each of the following, if possible. If it is not possible, explain why.

**a.** $A + B$ **b.** $C + D$ **c.** $A^\top$
**d.** $C^\top$ **e.** $A^\top + B^\top$ **f.** $(A + B)^\top$
**g.** $2D$ **h.** $A + 3B$ **i.** $D_{12} + C_{32}$

**2.** Let $A = \begin{pmatrix} 3 & 6 & -2 \\ -2 & 1 & 2 \\ 12 & 10 & 7 \end{pmatrix}$.
Write $A$ as the sum of a symmetric and skew-symmetric matrix.

**3.** Let

$$A = \begin{pmatrix} 2 & -3 \\ 0 & 4 \\ -1 & 2 \end{pmatrix}, \quad B = \begin{pmatrix} 1 & -5 & 3 \\ -1 & 6 & -3 \end{pmatrix}$$
$$C = \begin{pmatrix} 5 & 3 \\ 7 & 4 \end{pmatrix}, \quad D = \begin{pmatrix} -1 & 3 \\ 2 & 1 \end{pmatrix}$$

Calculate each of the following, if possible. If it is not possible, explain why.

**a.** $AB$ **b.** $CB$ **c.** $C_{1*} \cdot B$
**d.** $D^2$ **e.** $DC$ **f.** $BC$

**g.** Find a $2 \times 2$ matrix $F$ so that $CF = \begin{pmatrix} 1 & 0 \\ 0 & 1 \end{pmatrix}$.

**4.** For each matrix, find its inverse without using a calculator, or show that the inverse does not exist.

**a.** $\begin{pmatrix} 4 & -3 \\ 10 & -5 \end{pmatrix}$ **b.** $\begin{pmatrix} -1 & 2 \\ -4 & 8 \end{pmatrix}$ **c.** $\begin{pmatrix} 3 & -1 & 0 \\ 2 & 1 & -5 \\ 3 & 0 & -4 \end{pmatrix}$

**5.** Let $A = \begin{pmatrix} 3 & -1 & 0 \\ 4 & 1 & 2 \\ -2 & 5 & 3 \end{pmatrix}$ and $X = \begin{pmatrix} 3 \\ 4 \\ -1 \end{pmatrix}$.

**a.** Compute $B = AX$.

**Vocabulary**

In this chapter, you saw these terms and symbols for the first time. Make sure you understand what each one means, and how it is used.

- $A_{ij}$, $A_{*j}$, $A_{i*}$
- diagonal matrix
- entry
- equal matrices
- identity matrix
- inverse
- invertible matrix, nonsingular matrix
- kernel
- lower triangular matrix
- $m \times n$ matrix
- matrix multiplication, matrix product
- multiplication by a scalar
- scalar matrix
- singular matrix
- skew-symmetric matrix
- square matrix
- sum of matrices
- symmetric matrix
- transpose
- upper triangular matrix

**b.** Write $B$ as a linear combination of the columns of $A$. In other words, find $c_1$, $c_2$, and $c_3$ so that

$$B = c_1 \begin{pmatrix} 3 \\ 4 \\ -2 \end{pmatrix} + c_2 \begin{pmatrix} -1 \\ 1 \\ 5 \end{pmatrix} + c_3 \begin{pmatrix} 0 \\ 2 \\ 3 \end{pmatrix}$$

**6.** Given two matrices $A$ and $B$, how can you tell if the product $AB$ exists?

**7.** How can you tell if a matrix equation has a unique solution?

**8.** Let $A = \begin{pmatrix} 1 & 4 & 3 \\ -1 & 1 & 2 \\ 5 & 4 & 2 \end{pmatrix}$. Find $A^{-1}$.

# Chapter 4 Review

In Lesson 4.2, you learned to

- use clear notation for the different entries of a matrix
- determine when two matrices can be added to each other
- multiply a matrix by a scalar
- apply the properties of vector addition and scalar multiplication to matrices
- find the transpose of a matrix

The following exercises will help you check your understanding.

**1.** Let

$$A = \begin{pmatrix} -1 & 2 \\ 4 & -5 \\ 3 & 0 \\ -6 & 1 \end{pmatrix}, B = \begin{pmatrix} 2 & 4 \\ -6 & 3 \\ 2 & 1 \\ 7 & 10 \end{pmatrix}, \text{ and } C = \begin{pmatrix} -4 & -1 \\ 3 & 8 \\ -5 & 2 \\ 1 & 0 \end{pmatrix}$$

Calculate each of the following.

**a.** $A + B$ **b.** $(A + B) + C$ **c.** $A^\top$
**d.** $A^\top + C^\top$ **e.** $(A + C)^\top$ **f.** $(2A)^\top$
**g.** $(-A)^\top$ **h.** $((A + B) + C)^\top$

**2.** Let $A = \begin{pmatrix} 2 & -1 & -3 \\ 1 & 3 & -7 \\ 4 & -9 & 2 \\ 8 & 0 & -1 \end{pmatrix}$.

Determine the following.

**a.** $A_{13}$ **b.** $A_{31}$ **c.** $A_{23} + A_{33}$
**d.** $A_{3*}$ **e.** $A_{*3}$ **f.** $A_{2*} + A_{4*}$

**3.** Let $A$ be a $2 \times 4$ matrix whose generic entry is $A_{ij}$. Write the entries of $A$ given each of the following conditions:

**a.** $A_{ij} = i$
**b.** $A_{pq} = p + q$
**c.** $A_{rs} = r + 3s$

In Lesson 4.3, you learned to

- be familiar with the different types of square matrices
- decompose any square matrix into its symmetric and skew-symmetric parts

The following exercises will help you check your understanding.

**4.** If $A = \begin{pmatrix} 1 & 4 & -3 \\ 2 & 9 & 4 \\ -4 & 5 & -7 \end{pmatrix}$, write $A$ as the sum of a symmetric matrix and skew-symmetric matrix.

**5.** Classify each given matrix as scalar, diagonal, upper or lower triangular, strictly upper or lower triangular, symmetric, skew-symmetric, or none of the above.

**a.** $A$ is $3 \times 3$ and $A_{ij} = i + j$
**b.** $A$ is $4 \times 4$ and $A_{ij} = 2i - 2j$
**c.** $A$ is $2 \times 2$ and $A_{rs} = r + 2s$
**d.** $A = \text{rref}\begin{pmatrix} 2 & 1 & 0 \\ 3 & 2 & 1 \\ 5 & 2 & 4 \end{pmatrix}$

**e.** $A$ is $3 \times 3$ and $A_{ij} = \begin{cases} i - j & \text{if } i < j \\ 0 & \text{if } i \geq j \end{cases}$

**f.** $A$ is $4 \times 4$ and $A_{ij} = \begin{cases} j & \text{if } i = j \\ 0 & \text{if } i \neq j \end{cases}$

**6.** Write $a$ and $b$ in terms of $x$ and $y$.

$$\begin{aligned} a &= 3m - 4n + 2p \\ b &= 5m + 2n - 3p \end{aligned}$$

$$\begin{aligned} m &= 2x - 3y \\ n &= -4x + y \\ p &= 3x + 5y \end{aligned}$$

In Lesson 4.4, you learned to

- determine whether the product of two matrices exists
- find the product of two matrices
- understand the operation as the dot product of a group of vectors
- find the transpose of the product of two matrices

The following exercises will help you check your understanding.

**7.** Let

$$A = \begin{pmatrix} 1 & 2 & -4 \\ 3 & 0 & 5 \\ -1 & 2 & -3 \end{pmatrix}, B = \begin{pmatrix} 2 & -6 \\ 1 & 3 \\ 4 & 0 \end{pmatrix}, \text{ and } C = \begin{pmatrix} 5 & -2 \\ 1 & 8 \end{pmatrix}$$

Calculate each of the following. If the calculation is not possible, explain why.

**a.** $AB$ **b.** $BA$ **c.** $BC$
**d.** $C^2$ **e.** $B_{2*} \cdot C$ **f.** $B(2C)$
**g.** $(AB)^\top$ **h.** $A^\top \cdot B^\top$ **i.** $B^\top \cdot A^\top$

**8.** For each set of square matrices $A$ and $B$, determine whether $AB = BA$.

**a.** $A = \begin{pmatrix} 0 & 3 \\ 6 & 3 \end{pmatrix}$ and $B = \begin{pmatrix} -1 & 2 \\ 4 & 1 \end{pmatrix}$

**b.** $A = \begin{pmatrix} -3 & -1 \\ 2 & 5 \end{pmatrix}$ and $B = \begin{pmatrix} 2 & 3 \\ 4 & 0 \end{pmatrix}$

**c.** $A = \begin{pmatrix} 1 & -3 & 1 \\ 2 & 1 & -2 \\ 5 & -1 & -1 \end{pmatrix}$ and $B = \begin{pmatrix} 2 & -1 & 3 \\ -4 & 2 & 1 \\ 1 & -3 & -10 \end{pmatrix}$

**9.** Four points on a map are labeled $A$, $B$, $C$, and $D$. This $4 \times 4$ matrix $M$ shows the number of ways to get from one point to another in one step.

$$\begin{array}{c} \\ A \\ B \\ C \\ D \end{array} \begin{array}{c} \begin{matrix} A & B & C & D \end{matrix} \\ \begin{pmatrix} 0 & 1 & 0 & 1 \\ 1 & 0 & 1 & 1 \\ 0 & 1 & 0 & 1 \\ 1 & 1 & 1 & 0 \end{pmatrix} \end{array}$$

**a.** Find the total number of ways to get from point $A$ to point $C$ in one step.
**b.** Compute $M^2$ and $M^3$.
**c.** Find the total number of ways to get from point $A$ to point $C$ in three steps.
**d.** Find the total number of ways to get from point $A$ to point $C$ in three or fewer steps.

In Lesson 4.5, you learned to

- understand the relationship between matrix multiplication, dot product, and linear combination
- set up a matrix multiplication as a linear combination of vectors
- find the inverse of a square matrix, if it exists
- use the inverse to solve a matrix equation

The following exercises will help you check your understanding.

**10.** Find two different vectors that makes this matrix multiplication true.

$$\begin{pmatrix} 1 & -2 & 5 \\ -2 & 3 & -4 \\ 0 & -1 & 6 \end{pmatrix} \begin{pmatrix} x \\ y \\ z \end{pmatrix} = \begin{pmatrix} -2 \\ 5 \\ 1 \end{pmatrix}$$

**11.** For each matrix, find its inverse without using a calculator, or show that the inverse does not exist.

**a.** $\begin{pmatrix} -1 & 3 & 5 \\ 2 & 4 & 5 \\ -1 & 0 & 1 \end{pmatrix}$ **b.** $\begin{pmatrix} 3 & 0 & 0 \\ 1 & 3 & 0 \\ -1 & 4 & 3 \end{pmatrix}$ **c.** $\begin{pmatrix} 0 & 0 & 0 \\ 1 & 0 & 0 \\ -1 & 4 & 0 \end{pmatrix}$

**12.** For each matrix $A$,

(i) find $A^{-1}$, and
(ii) verify that $A \cdot A^{-1} = I$.

**a.** $A = \begin{pmatrix} 3 & 2 \\ 0 & -1 \end{pmatrix}$

**b.** $A = \begin{pmatrix} 1 & 2 & -1 \\ 0 & 4 & 3 \\ 1 & -2 & 0 \end{pmatrix}$

In Lesson 4.6, you learned to

- determine when a square matrix has an inverse
- find the inverse of a matrix by augmenting it with the identity matrix
- find all solutions to a matrix equation given one solution and the kernel of the matrix
- understand the relationship between linear independence of the columns of a matrix and its invertibility
- find the set of vectors whose direction is left invariant after multiplication by a matrix

The following exercises will help you check your understanding.

**13.** Consider $A = \begin{pmatrix} 2 & 3 & 6 \\ 1 & 4 & 5 \\ 0 & 5 & 4 \end{pmatrix}$ and $B = \begin{pmatrix} 6 \\ -2 \\ -10 \end{pmatrix}$.

**a.** Show that if $X_1 = \begin{pmatrix} 24 \\ 6 \\ -10 \end{pmatrix}$, $AX_1 = B$.

**b.** Show that if $X_2 = \begin{pmatrix} 15 \\ 2 \\ -5 \end{pmatrix}$, $AX_2 = B$.

**c.** Find some other vectors that also solve $AX = B$.

**14.** For each system,

(i) rewrite the system in the form $AX = B$,
(ii) find a solution to each system by writing $B$ as a linear combination of the columns of $A$, and
(iii) find all solutions to the system by finding the kernel.

⟵ There may be more than one solution, but you only have to find *one*!

**a.** $\begin{cases} 2x - y + 3z = 3 \\ x + 3y + 2z = 5 \\ 3x - 5y + 4z = 1 \end{cases}$

**b.** $\begin{cases} 2x - y + 3z = 1 \\ x + 3y + 2z = 8 \\ 3x - 5y = -10 \end{cases}$

**c.** $\begin{cases} x + 2y - 4z = -5 \\ x - 3y + z = 5 \end{cases}$

**15.** Solve each system by multiplying both sides by the inverse of the coefficient matrix.

**a.** $\begin{pmatrix} 2 & 5 \\ 1 & 3 \end{pmatrix} \begin{pmatrix} x \\ y \end{pmatrix} = \begin{pmatrix} 1 \\ 5 \end{pmatrix}$

**b.** $\begin{pmatrix} 1 & 1 & 3 \\ -1 & 2 & 6 \\ 4 & 2 & -1 \end{pmatrix} \begin{pmatrix} x \\ y \\ z \end{pmatrix} = \begin{pmatrix} 4 \\ -1 \\ 0 \end{pmatrix}$

# Chapter 4 Test

### Multiple Choice

**1.** Let $A = \begin{pmatrix} 3 & 6 & -9 & 2 \\ -2 & 8 & 1 & -9 \\ -4 & 10 & -5 & 5 \\ 0 & 11 & -6 & 1 \end{pmatrix}$. What is the value of $A_{23} + A_{31}$?

**A.** $-6$ **B.** $-3$ **C.** $5$ **D.** $6$

**2.** Suppose $A$ is a $3 \times 3$ matrix whose generic entry is $A_{ij} = i - 2j$. What is the value of $A_{13}$?

**A.** $-5$ **B.** $-3$ **C.** $5$ **D.** $7$

**3.** Let $A = \begin{pmatrix} 2 & 1 & -2 \\ 4 & 0 & 3 \\ -3 & 5 & 1 \end{pmatrix}$ and $B = \begin{pmatrix} -5 & 1 & 0 \\ 2 & -3 & 4 \\ 5 & 1 & -6 \end{pmatrix}$. Which is $(AB)_{3*}$?

**A.** $\begin{pmatrix} 13 \\ -9 \\ -13 \end{pmatrix}$ **B.** $\begin{pmatrix} 16 \\ -18 \\ 14 \end{pmatrix}$ **C.** $\begin{pmatrix} 32 & -25 & -13 \end{pmatrix}$ **D.** $\begin{pmatrix} 30 & -17 & 14 \end{pmatrix}$

**4.** Let $A = \begin{pmatrix} 2 & -2 \\ 4 & 3 \end{pmatrix}$ and $B = \begin{pmatrix} -1 & 3 \\ -5 & 4 \end{pmatrix}$. Which of the following is equivalent to $(AB)^\top$?

**A.** $\begin{pmatrix} 2 & -2 \\ 4 & 3 \end{pmatrix}\begin{pmatrix} -1 & 3 \\ -5 & 4 \end{pmatrix}$ **B.** $\begin{pmatrix} -1 & 3 \\ -5 & 4 \end{pmatrix}\begin{pmatrix} 2 & -2 \\ 4 & 3 \end{pmatrix}$

**C.** $\begin{pmatrix} 2 & 4 \\ -2 & 3 \end{pmatrix}\begin{pmatrix} -1 & -5 \\ 3 & 4 \end{pmatrix}$ **D.** $\begin{pmatrix} -1 & -5 \\ 3 & 4 \end{pmatrix}\begin{pmatrix} 2 & 4 \\ -2 & 3 \end{pmatrix}$

**5.** Let $A = \begin{pmatrix} 2 & 3 & -1 \\ -1 & 2 & -4 \\ 5 & 11 & -7 \end{pmatrix}$ and $B = \begin{pmatrix} 7 \\ -4 \\ 17 \end{pmatrix}$. If $X_1 = (-2, 5, 4)$ is one solution to $AX = B$, which equation gives all solutions to $AX = B$?

**A.** $X = (-10, 9, 1) + t(-2, 5, 4)$
**B.** $X = (-10, 9, 7) + t(-2, 5, 4)$
**C.** $X = (-2, 5, 4) + t(-10, 9, 1)$
**D.** $X = (-2, 5, 4) + t(-10, 9, 7)$

**6.** Let $\begin{pmatrix} -2 & 0 & 5 \\ 3 & 1 & 3 \\ -2 & 1 & 0 \end{pmatrix}\begin{pmatrix} x \\ y \\ z \end{pmatrix} = \begin{pmatrix} 15 \\ -10 \\ 12 \end{pmatrix}$. What is the solution to this equation?

**A.** $\begin{pmatrix} x \\ y \\ z \end{pmatrix} = \begin{pmatrix} -26 \\ 7 \\ 4 \end{pmatrix}$ **B.** $\begin{pmatrix} x \\ y \\ z \end{pmatrix} = \begin{pmatrix} -5 \\ 2 \\ 1 \end{pmatrix}$

**C.** $\begin{pmatrix} x \\ y \\ z \end{pmatrix} = \begin{pmatrix} 9 \\ 70 \\ 4 \end{pmatrix}$ **D.** $\begin{pmatrix} x \\ y \\ z \end{pmatrix} = \begin{pmatrix} 30 \\ 71 \\ -40 \end{pmatrix}$

**Open Response**

**7.** Let $A = \begin{pmatrix} 1 & -6 & 2 \\ 3 & 0 & -1 \\ 2 & 5 & -4 \end{pmatrix}$ and $B = \begin{pmatrix} 3 & -2 & 7 \\ 1 & 6 & -2 \\ 3 & -4 & -8 \end{pmatrix}$. Calculate the following:

**a.** $A + 2B$

**b.** $A_{2*} + B_{3*}$

**c.** $A^\top$

**d.** $(A + B)^\top$

**8.** Let $A = \begin{pmatrix} -2 & 4 & -1 \\ 3 & 1 & 5 \\ 0 & -3 & 2 \end{pmatrix}$. Write $A$ as the sum of a symmetric and a skew-symmetric matrix.

**9.** Let

$$A = \begin{pmatrix} -2 & 4 & 1 \\ 3 & 8 & -2 \end{pmatrix}, B = \begin{pmatrix} 2 & 1 \\ -3 & 7 \\ 0 & 4 \end{pmatrix}, \text{ and } C = \begin{pmatrix} -2 & 5 \\ 3 & -1 \end{pmatrix}$$

Calculate each of the following. If the calculation is not possible, explain why.

**a.** $AB$

**b.** $BA$

**c.** $BC$

**d.** $CB$

**10.** Let $A = \begin{pmatrix} 2 & 0 & -2 \\ 1 & -3 & 1 \\ -4 & 1 & 4 \end{pmatrix}$. Determine $A^{-1}$.

**11.** Consider this system.

$$\begin{cases} 3x - y + z &= 3 \\ x + 3y - 2z &= 2 \\ 2x - 4y + 3z &= 1 \end{cases}$$

**a.** Rewrite the system in the form $AX = B$.

**b.** Find a solution to the system by writing $B$ as a linear combination of the columns of $A$.

**c.** Find all solutions to the system by finding the kernel.

⟵ There may be more than one solution, but you only have to find *one*!

**12.** Solve the system by multiplying both sides by the inverse of the coefficient matrix.

$$\begin{pmatrix} 3 & -2 \\ 3 & -1 \end{pmatrix} \begin{pmatrix} x \\ y \end{pmatrix} = \begin{pmatrix} 18 \\ 15 \end{pmatrix}$$

CHAPTER

# 5 Matrices as Functions

Up to this point, you have used matrices as a tool to interact on a set of vectors all at once, or to hold coefficients of a system of equations. Matrices can also act like functions. When you think about a matrix this way, you can examine the effect it has geometrically when applied to vectors. For instance, video game programmers and computer animators use matrices to move objects around while preserving their shape.

In this chapter, you will find methods for determining what effect a particular matrix will have on a shape in $\mathbb{R}^2$: will it preserve the shape, or stretch it, or flip it over? All of these transformations can be defined using matrices.

**By the end of this chapter, you will be able to answer questions like these:**

1. How can you find a matrix that rotates points about the origin?
2. What is the area of the parallelogram spanned by two given vectors?
3. Let $A = \begin{pmatrix} 1 & 3 & 5 \\ 4 & 6 & 2 \end{pmatrix}$.
   a. What is the image under $A$ for $\begin{pmatrix} 2 \\ 3 \\ -1 \end{pmatrix}$?
   b. What is the set of pullbacks under $A$ for $\begin{pmatrix} 7 \\ 16 \end{pmatrix}$?

**You will build good habits and skills for ways to**

- look for counterexamples
- extend using linearity
- use algebra to extend geometric ideas
- look for proofs that reveal hidden meaning
- use the extension program

- look for structure
- find alternative methods that you can generalize

### Vocabulary and Notation

- angle of rotation
- center of rotation
- conjugation
- fixed point
- image of a matrix
- linear map
- linear transformation of $\mathbb{R}^n$
- preimage
- pullback
- rotation

## 5.1 Getting Started

### Exercises

1. Given three points $T = \begin{pmatrix} 2 \\ 2 \end{pmatrix}$, $U = \begin{pmatrix} 3 \\ 0 \end{pmatrix}$, and $V = \begin{pmatrix} 9 \\ 3 \end{pmatrix}$, draw $\triangle TUV$. What kind of triangle is $TUV$? Justify your answer.

⟵ As seen before, each point corresponds to a column vector. These points are in $\mathbb{R}^2$.

2. For each matrix $J$, compute $T' = JT$, $U' = JU$, and $V' = JV$. Describe how $\triangle T'U'V'$ is obtained from $\triangle TUV$.

**a.** $J = \begin{pmatrix} 0 & 1 \\ 1 & 0 \end{pmatrix}$ **b.** $J = \begin{pmatrix} 0 & 1 \\ -1 & 0 \end{pmatrix}$ **c.** $J = \begin{pmatrix} 0.8 & -0.6 \\ 0.6 & 0.8 \end{pmatrix}$

**d.** $J = \begin{pmatrix} 2 & 0 \\ 0 & 2 \end{pmatrix}$ **e.** $J = \begin{pmatrix} 1 & 0 \\ 0 & 0 \end{pmatrix}$ **f.** $J = \begin{pmatrix} \frac{2}{5} & \frac{3}{5} \\ \frac{3}{5} & \frac{2}{5} \end{pmatrix}$

For each matrix given in Exercises 3–9, sketch and describe the effect of the matrix on each of the following:

⟵ For shapes, describe the effect by stating how the shape changed (is it similar or congruent to the original?) and the change in area.

**a.** $O$ **b.** $P = (2, 3)$ **c.** $Q = (-1, 4)$

**d.** $\angle POQ$ **e.** $\triangle POQ$

**f.** the unit square (the square with vertices at $(0, 0), (1, 0), (1, 1), (0, 1)$)

**g.** the unit circle (the circle with center $O$ with radius 1)

**h.** the line $X = t(3, 1)$

**i.** the line $X = (1, 0) + t(3, 1)$

⟵ In each case, is the result still a line?

3. $A = \begin{pmatrix} 4 & 0 \\ 0 & 4 \end{pmatrix}$ 4. $B = \begin{pmatrix} \frac{24}{25} & \frac{-7}{25} \\ \frac{7}{25} & \frac{24}{25} \end{pmatrix}$ 5. $C = \begin{pmatrix} 1 & 2 \\ 0 & 1 \end{pmatrix}$

6. $D = \begin{pmatrix} 3 & -4 \\ 4 & 3 \end{pmatrix}$ 7. $E = \begin{pmatrix} \frac{-3}{5} & \frac{-4}{5} \\ \frac{-4}{5} & \frac{3}{5} \end{pmatrix}$

8. $F = \begin{pmatrix} \frac{5}{2} & \frac{-3}{2} \\ \frac{-1}{2} & \frac{7}{2} \end{pmatrix}$ 9. $G = \begin{pmatrix} \frac{1}{10} & \frac{-3}{10} \\ \frac{-3}{10} & \frac{9}{10} \end{pmatrix}$

10. **a.** Find a $2 \times 2$ matrix $M$ so that $M \begin{pmatrix} x \\ y \end{pmatrix}$ is the reflection of $(x, y)$ over the $x$-axis.

**b.** Find a matrix $R$ so that $R \begin{pmatrix} x \\ y \end{pmatrix} = \begin{pmatrix} 2y \\ 2x \end{pmatrix}$.

⟵ What point is the reflection of $(2, 3)$ over the $x$-axis? of $(-1, -2)$? of $(x, y)$?

⟵ The transformation in 10**b** maps $(x, y) \mapsto (2y, 2x)$.

11. **a.** Find a matrix $M$ that projects any point $(x, y)$ onto the $x$-axis.

**b.** Find a matrix $N$ that projects any point $(x, y)$ onto the $y$-axis.

**c.** What are the coordinates of $(M + N) \begin{pmatrix} x \\ y \end{pmatrix}$?

12. Suppose $A$ and $B$ are $m \times n$ matrices and that $AX = BX$ for all vectors $X$ in $\mathbb{R}^n$. Show that $A = B$.

⟵ **Hint**: Choose specific vectors $X$ to look at, and try to prove that all the pieces of $A$ must equal all the pieces of $B$.

13. Suppose $A = \begin{pmatrix} 1 & 2 & 3 \\ 4 & 5 & 6 \end{pmatrix}$.

**a.** Find $\ker(A)$ and describe it geometrically.

**b.** Find all vectors $X = \begin{pmatrix} x \\ y \\ z \end{pmatrix}$ so that $AX = \begin{pmatrix} 3 \\ 9 \end{pmatrix}$. Describe your answer geometrically.

**c.** Show that *every* vector $B$ in $\mathbb{R}^2$ can be written as $AX$ for some vector $X$, or find one that cannot.

**14.** Suppose $A = \begin{pmatrix} 1 & 2 & 3 \\ 4 & 5 & 6 \\ 7 & 8 & 9 \end{pmatrix}$.

**a.** Find $\ker(A)$ and describe it geometrically.

**b.** Find all vectors $X = \begin{pmatrix} x \\ y \\ z \end{pmatrix}$ so that $AX = \begin{pmatrix} 3 \\ 9 \\ 15 \end{pmatrix}$. Describe your answer geometrically.

**c.** Show that *every* vector $B$ in $\mathbb{R}^3$ can be written as $AX$ for some vector $X$, or find one that cannot.

**15.** Suppose $A = \begin{pmatrix} 1 & 2 & 3 \\ 4 & 5 & 6 \\ 7 & 8 & 0 \end{pmatrix}$.

**a.** Find $\ker(A)$ and describe it geometrically.

**b.** Find all vectors $X = \begin{pmatrix} x \\ y \\ z \end{pmatrix}$ so that $AX = \begin{pmatrix} 3 \\ 9 \\ 15 \end{pmatrix}$. Describe your answer geometrically.

**c.** Show that *every* vector $B$ in $\mathbb{R}^3$ can be written as $AX$ for some vector $X$, or find one that cannot.

## 5.2 Geometric Transformations

---

You likely have studied geometric transformations of the plane in a previous course. Special kinds of transformations can be represented with matrices.

**In this lesson, you will learn how to**

- describe what $2 \times 2$ and $3 \times 3$ linear transformations do to a triangle, the unit square, and the unit circle
- use the properties of linear transformations to show that not all transformations are linear
- find the matrices that produce scalings, reflections, and $90°$ rotations

---

An $m \times n$ matrix can define a function from $\mathbb{R}^n$ to $\mathbb{R}^m$. When $n = m$ the matrix is square, and it defines a linear mapping from $\mathbb{R}^n$ to *itself*, called a **linear transformation** of $\mathbb{R}^n$. "Transform" means "change in form or appearance," and this is what linear transformations do to objects in $\mathbb{R}^n$.

⟵ Right now, all that "linear" means is "defined by a matrix." You will generalize this definition and make it more precise in Chapter 8.

In $\mathbb{R}^2$ and $\mathbb{R}^3$, you can think of linear transformations in terms of how the resulting points or other shapes relate to the original points in the coordinate plane or in space. Here are two matrices and their corresponding linear transformations:

- In the plane, applying the $2 \times 2$ matrix $M = \begin{pmatrix} 2 & 0 \\ 0 & 2 \end{pmatrix}$ results in points that are scaled by a factor of 2, because

$$\begin{pmatrix} 2 & 0 \\ 0 & 2 \end{pmatrix} \begin{pmatrix} x \\ y \end{pmatrix} = \begin{pmatrix} 2x \\ 2y \end{pmatrix} = 2 \begin{pmatrix} x \\ y \end{pmatrix}$$

⟵ Pick some points to test: What happens to $(3, 4)$? to $(-2, -1)$?

- In space, applying the matrix $\begin{pmatrix} 1 & 0 & 0 \\ 0 & 1 & 0 \\ 0 & 0 & -1 \end{pmatrix}$ results in a point reflected through the $xy$-plane, because

$$\begin{pmatrix} 1 & 0 & 0 \\ 0 & 1 & 0 \\ 0 & 0 & -1 \end{pmatrix} \begin{pmatrix} x \\ y \\ z \end{pmatrix} = \begin{pmatrix} x \\ y \\ -z \end{pmatrix}$$

This picture illustrates the reflection across the plane:

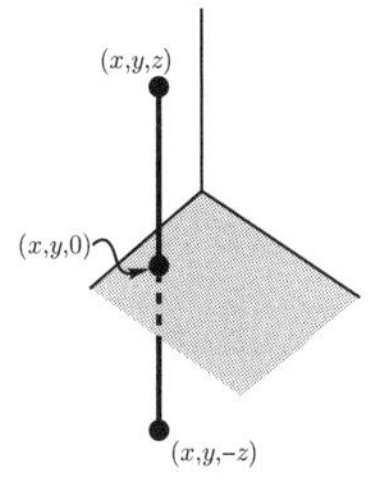


For most matrices, it is more difficult to describe the corresponding linear transformation. For example, how do points in the plane transformed by the matrix $M = \begin{pmatrix} 8 & -1 \\ 1 & 3 \end{pmatrix}$ relate to the original points? One way to analyze this question is to look at the effect of a matrix on a simple object, such as a square or circle.

### Example

**Problem.** The matrix $M = \begin{pmatrix} 8 & -1 \\ 1 & 3 \end{pmatrix}$ is applied to each of the vertices of the unit square whose vertices are $\{O(0,0), A(1,0), B(1,1), C(0,1)\}$. What is the resulting figure?

**Solution.** Calculate the image of each of the vertices:

$$MO = M\begin{pmatrix} 0 \\ 0 \end{pmatrix} = \begin{pmatrix} 0 \\ 0 \end{pmatrix} = O'$$

$$MA = M\begin{pmatrix} 1 \\ 0 \end{pmatrix} = \begin{pmatrix} 8 \\ 1 \end{pmatrix} = A'$$

$$MB = M\begin{pmatrix} 1 \\ 1 \end{pmatrix} = \begin{pmatrix} 7 \\ 4 \end{pmatrix} = B'$$

$$MC = M\begin{pmatrix} 0 \\ 1 \end{pmatrix} = \begin{pmatrix} -1 \\ 3 \end{pmatrix} = C'$$

Then, plot the new figure $O'A'B'C'$ in the plane.

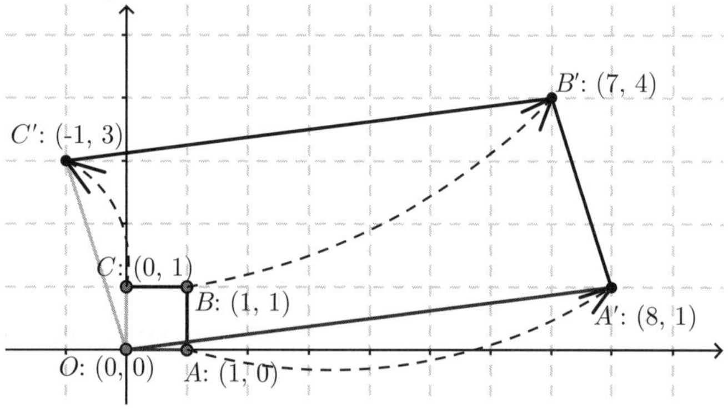


Connect consecutive vertices with segments. The unit square appears to be transformed to a parallelogram.

### Minds in Action Episode 14

*Tony, Sasha, and Derman are talking about this example.*

DERMAN: Why is it a parallelogram?

TONY: I think I see. In the original square, two vertices add to the third one, like this:

$$\begin{pmatrix} 1 \\ 1 \end{pmatrix} = \begin{pmatrix} 1 \\ 0 \end{pmatrix} + \begin{pmatrix} 0 \\ 1 \end{pmatrix}$$

So, $B = A + C$. Multiply both sides by $M$ and use the basic rules from Chapter 4:

$$M\begin{pmatrix}1\\1\end{pmatrix} = M\left(\begin{pmatrix}1\\0\end{pmatrix} + \begin{pmatrix}0\\1\end{pmatrix}\right)$$
$$= M\begin{pmatrix}1\\0\end{pmatrix} + M\begin{pmatrix}0\\1\end{pmatrix}$$

And $B' = A' + C'$. That's the parallelogram rule for adding vectors.

DERMAN: I wonder if that always works. I think it will.

TONY: I'm still not convinced of something else. We only plotted the vertices, and then connected them with segments. How do we know that the segment between $O$ and $(1, 0)$ transforms to the segment between $O$ and $(5, 2)$?

*They all think about it . . .*

SASHA: I've got it! It relies on the same basic rules. All the points along $\overrightarrow{OA}$ transform to points along $\overrightarrow{OA'}$. Consider the midpoint of $\overrightarrow{OA}$; it's $\frac{1}{2}A$, and

$$M\left(\tfrac{1}{2}A\right) = \tfrac{1}{2}M(A) = \tfrac{1}{2}A'$$

TONY: Nice. And you can do this with any point along $\overrightarrow{OA}$, since it can be written as $kA$ for some number $k$ between 0 and 1. Then

$$M(kA) = kM(A) = kA'$$

And that proves that $\overrightarrow{OA}$ transforms to $\overrightarrow{OA'}$.

DERMAN: Sounds good, but what about $\overrightarrow{CB}$?

---

**For Discussion**

1. Convince Derman that points along $\overrightarrow{CB}$ get mapped to points along $\overrightarrow{C'B'}$.
2. How does the area of the parallelogram compare with the area of the unit square?

←
Does the ratio of the areas have any connection to the matrix $M$?

---

The properties that Derman, Tony, and Sasha use in the dialogue above follow from the basic rules for matrix multiplication. Some important ones are listed in the following theorem.

**Theorem 5.1**

*Suppose $M$ is a linear transformation of $\mathbb{R}^n$. Then*

(1) *$M$ fixes the origin: $MO = O$*

(2) *The image of the sum of two vectors is the sum of their images: $M(A + B) = MA + MB$*

(3) *Scalars come out: $M(cA) = c(MA)$*

**Remember**

"Linear transformation" means "defined by a matrix." You'll generalize this notion later, but matrices will still play a central role.

The three properties of Theorem 5.1 hold for any mapping $\mathbb{R}^n \to \mathbb{R}^m$ that is defined by matrix multiplication. So, the theorem applies to more general maps.

### For You to Do

**3.** **a.** Illustrate each of the properties from Theorem 5.1 with an example from $\mathbb{R}^2$ and an example from $\mathbb{R}^3$.

**b.** Using the notation of Theorem 5.1, show that

$$M(cA + dB) = cM(A) + dM(B)$$

← Use pictures in your illustrations.

### Developing Habits of Mind

**Look for counterexamples.** One way to use Theorem 5.1 is as a tool to show that a transformation is *not* linear. For example, the mapping that adds (4,7) to every point in $\mathbb{R}^2$ isn't a linear transformation because it doesn't fix the origin. For the same reason, no $2 \times 2$ matrix can represent any translation in the plane.

### Minds in Action Episode 15

*Sasha is still thinking about the transformation matrix* $M = \begin{pmatrix} 8 & -1 \\ 1 & 3 \end{pmatrix}$

SASHA: I wonder what $M$ does to other figures? Like what about the unit circle with equation $x^2 + y^2 = 1$?

DERMAN: That's not in the book.

TONY: It is now . . .

DERMAN: I think you get another circle.

SASHA: I think you get a conic, and it might be a circle.

DERMAN: What's a conic?

TONY: It's the graph of a quadratic equation—a circle, ellipse, parabola, or hyperbola.

DERMAN: How could it be a parabola? That makes no sense. If you stretch a circle it might look like a football or something.

SASHA: $M$ took the square and stretched and squished it in a kind of uniform way. Now, to do that to a circle . . . I'm not sure what happens. So let's run the algebra.

*Sasha writes on the board*

$$M\begin{pmatrix} x \\ y \end{pmatrix} = \begin{pmatrix} 8 & -1 \\ 1 & 3 \end{pmatrix}\begin{pmatrix} x \\ y \end{pmatrix} = \begin{pmatrix} 8x - y \\ x + 3y \end{pmatrix}$$

$\begin{pmatrix} x \\ y \end{pmatrix}$ is on the circle, so $x^2 + y^2 = 1$. We need to find an equation satisfied by

$$\begin{pmatrix} x' \\ y' \end{pmatrix} = \begin{pmatrix} 8x - y \\ x + 3y \end{pmatrix}$$

TONY: Just solve for $x$ and $y$ as expressions in $x'$ and $y'$. You know an equation satisfied by $x$ and $y$, so you know an equation that will be satisfied by these expressions. Right?

DERMAN: Yeah, right. And how do you intend to solve for $x$ and $y$ when we've got

$$\begin{aligned} x' &= 8x - y \\ y' &= x + 3y \end{aligned}$$

SASHA: Look at the big picture. We have

$$\begin{pmatrix} x' \\ y' \end{pmatrix} = M \begin{pmatrix} x \\ y \end{pmatrix} \quad \text{so} \quad M^{-1} \begin{pmatrix} x' \\ y' \end{pmatrix} = \begin{pmatrix} x \\ y \end{pmatrix}$$

And we can find $M^{-1}$ in a lot of ways. If $M = \begin{pmatrix} 8 & -1 \\ 1 & 3 \end{pmatrix}$, then

$$M^{-1} = \tfrac{1}{25} \begin{pmatrix} 3 & 1 \\ -1 & 8 \end{pmatrix}$$

← Exercise 17 from Lesson 4.5 gives a formula for the inverse of a $2 \times 2$ matrix.

TONY: So,

$$\begin{pmatrix} x \\ y \end{pmatrix} = M^{-1} \begin{pmatrix} x' \\ y' \end{pmatrix} = \tfrac{1}{25} \begin{pmatrix} 3 & 1 \\ -1 & 8 \end{pmatrix} \begin{pmatrix} x' \\ y' \end{pmatrix} = \tfrac{1}{25} \begin{pmatrix} 3x' + y' \\ -x' + 8y' \end{pmatrix}$$

And

$$\begin{aligned} x &= \frac{3x' + y'}{25} \\ y &= \frac{-x' + 8y'}{25} \end{aligned}$$

SASHA: Things are heating up. If $x^2 + y^2 = 1$, then

$$\left( \frac{3x' + y'}{25} \right)^2 + \left( \frac{-x' + 8y'}{25} \right)^2 = 1$$

So $(3x' + y')^2 + (-x' + 8y')^2 = 625$. All we need to do is expand this and see what we get.

*They calculate . . .*

TONY: I get $2(x')^2 - 2x'y' + 13(y')^2 = 125$.

SASHA: The graph of that is . . . um, yeah . . . what is the graph of that?

DERMAN: And things are cooling down . . . .

### For Discussion

4. Using any tools available, graph

$$2x^2 - 2xy + 13y^2 = 125$$

in the coordinate plane, and then graph $x^2 + y^2 = 1$ on the same axes.

← You know some points, like $(1, 0)$, on the unit circle. Their images are on the graph of $2x^2 - 2xy + 13y^2 = 125$, right?

### For You to Do

5. In Exercise 18 from Lesson 4.6, you saw that each step in the process of reducing a matrix to echelon form can be accomplished by multiplying on the left by an *elementary row matrix*. For $2 \times 2$ matrices, the elementary row matrices are of three types, corresponding to the three elementary row operations.
   - $\begin{pmatrix} 0 & 1 \\ 1 & 0 \end{pmatrix}$ switches the rows.
   - $\begin{pmatrix} k & 0 \\ 0 & 1 \end{pmatrix}$ or $\begin{pmatrix} 1 & 0 \\ 0 & k \end{pmatrix}$ replaces a row by $k$ times itself.
   - $\begin{pmatrix} 1 & 0 \\ k & 1 \end{pmatrix}$ or $\begin{pmatrix} 1 & k \\ 0 & 1 \end{pmatrix}$ replaces a row by the sum of itself and $k$ times the other row.

   **a.** What does each of these matrices do to square $OACB$ in the example earlier in this lesson?

   **b. Take It Further.** What does each of these matrices do to the unit circle—the graph of $x^2 + y^2 = 1$?

## Exercises

1. For each linear transformation matrix, describe the effect it has in geometric language.

   **a.** $\begin{pmatrix} 0 & 0 \\ 0 & 1 \end{pmatrix}$ **b.** $\begin{pmatrix} 1 & 0 \\ 0 & 3 \end{pmatrix}$ **c.** $\begin{pmatrix} 0 & 2 \\ 2 & 0 \end{pmatrix}$ **d.** $\begin{pmatrix} 0 & -2 \\ -2 & 0 \end{pmatrix}$

   **e.** $\begin{pmatrix} 0 & 0 & 0 \\ 0 & 1 & 0 \\ 0 & 0 & 1 \end{pmatrix}$ **f.** $\begin{pmatrix} 2 & 0 & 0 \\ 0 & 2 & 0 \\ 0 & 0 & 2 \end{pmatrix}$

   **g.** $\begin{pmatrix} -1 & 0 & 0 \\ 0 & -1 & 0 \\ 0 & 0 & -1 \end{pmatrix}$ **h.** $\begin{pmatrix} 0 & 0 & 0 \\ 0 & 1 & 0 \\ 0 & 0 & 0 \end{pmatrix}$

← If you're not sure, test some points and see what happens.

2. For each given transformation of $\mathbb{R}^2$, find a corresponding $2 \times 2$ matrix.

   **a.** Reflect all points over the $x$-axis.
   **b.** Reflect all points over the $y$-axis.
   **c.** Scale all points by a factor of 3.
   **d.** Reflect all points through the origin.
   **e.** Reflect all points over the graph of $y = x$.

**f.** Rotate all points 90° counterclockwise around the origin.
**g.** Rotate all points 90° clockwise around the origin.
**h.** Reflect all points over the graph of $y = x$, then scale this image by 3, and then reflect *this* image over the $y$-axis.

**3.** Let $M$ be the matrix from part **f** of Exercise 2. Calculate each of the following, and describe the effect of each new matrix in geometric language.

**a.** $M^2$ **b.** $M^3$ **c.** $M^4$ **d.** $M^5$ **e.** $M^{102}$

**4.** If $B$, $C$, and $E$ are the matrices that solve parts **b**, **c**, and **e** of Exercise 2, show that $B \cdot C \cdot E$ is the matrix that solves part **h**.

⟵
If you can show that $B{\cdot}C{\cdot}E$ is the desired matrix without actually *computing* its value and comparing it to the matrix you found in part **h**, then you can probably also explain why $E \cdot C \cdot B$ does *not* solve part **h**.

**5.** Suppose $T : \mathbb{R}^2 \to \mathbb{R}^2$ is defined by

$$T(X) = X + \begin{pmatrix} 2 \\ 3 \end{pmatrix}$$

**a.** What does $T$ do to the unit square?
**b.** What does $T$ do to the unit circle?
**c.** Is $T$ a linear transformation? If so, find the matrix that represents $T$. If not, explain how you know.

**6.** **a.** Write $M = \begin{pmatrix} 8 & -1 \\ 1 & 3 \end{pmatrix}$ as a product of elementary row matrices.
**b.** Use part **a** to describe what $M$ does to points on the plane.

**7.** Show that $M = \begin{pmatrix} \frac{1}{5} & \frac{2}{5} \\ \frac{2}{5} & \frac{4}{5} \end{pmatrix}$ projects $(x, y)$ along $(1, 2)$.

**8.** Show that $\frac{1}{a^2+b^2} \begin{pmatrix} a^2 & ab \\ ab & b^2 \end{pmatrix}$ transforms $\mathbb{R}^2$ by projecting $(x, y)$ along $(a, b)$.

⟵
What is $M \begin{pmatrix} x \\ y \end{pmatrix}$?

**9.** Let $M = \begin{pmatrix} \frac{3}{5} & -\frac{4}{5} \\ \frac{4}{5} & \frac{3}{5} \end{pmatrix}$ and let $\begin{pmatrix} x' \\ y' \end{pmatrix} = M \begin{pmatrix} x \\ y \end{pmatrix}$. Find $\|(x, y)\|$ and $\|(x', y')\|$ if

**a.** $(x, y) = (2, 0)$ **b.** $(x, y) = (0, 2)$ **c.** $(x, y)$ is any vector

**10.** Let $M = \begin{pmatrix} \frac{3}{5} & -\frac{4}{5} \\ \frac{4}{5} & \frac{3}{5} \end{pmatrix}$ and let $\begin{pmatrix} x' \\ y' \end{pmatrix} = M \begin{pmatrix} x \\ y \end{pmatrix}$. Find the angle between $(x, y)$ and $(x', y')$ if

**a.** $(x, y) = (2, 0)$ **b.** $(x, y) = (0, 2)$ **c.** $(x, y)$ is any vector

**11.** Let $M = \begin{pmatrix} \frac{1}{2} & -\frac{\sqrt{3}}{2} \\ \frac{\sqrt{3}}{2} & \frac{1}{2} \end{pmatrix}$ and let $\begin{pmatrix} x' \\ y' \end{pmatrix} = M \begin{pmatrix} x \\ y \end{pmatrix}$. Find $\|(x, y)\|$ and $\|(x', y')\|$ if

**a.** $(x, y) = (2, 0)$ **b.** $(x, y) = (0, 2)$ **c.** $(x, y)$ is any vector

**12.** Let $M = \begin{pmatrix} \frac{1}{2} & -\frac{\sqrt{3}}{2} \\ \frac{\sqrt{3}}{2} & \frac{1}{2} \end{pmatrix}$ and let $\begin{pmatrix} x' \\ y' \end{pmatrix} = M\begin{pmatrix} x \\ y \end{pmatrix}$. Find the angle between $(x, y)$ and $(x', y')$ if

**a.** $(x, y) = (2, 0)$ **b.** $(x, y) = (0, 2)$ **c.** $(x, y)$ is any vector

**13.** Let $M = \begin{pmatrix} -27 & 70 \\ -12 & 31 \end{pmatrix}$.

**a.** Show what $M$ does to the unit square.
**b.** Theorem 5.1 says that $M$ fixes the origin. Does it fix any other vectors?
**c.** Does $M$ fix any *lines*? That is, are there any vectors $X$ so that $MX = kX$ for some number $k$?
**d.** Show what $M$ does to the unit circle.

**14.** Let $M = \begin{pmatrix} 35 & -12 \\ 17 & -6 \end{pmatrix}$.

**a.** Show what $M$ does to the unit square.
**b.** Theorem 5.1 says that $M$ fixes the origin. Does it fix any other vectors?
**c.** Does $M$ fix any *lines*? That is, are there any vectors $X$ so that $MX = kX$ for some number $k$?
**d.** Show what $M$ does to the unit circle.

**15.** Let $M = \begin{pmatrix} \frac{-3}{7} & \frac{-6}{7} & \frac{-2}{7} \\ \frac{-2}{7} & \frac{3}{7} & \frac{-6}{7} \\ \frac{6}{7} & \frac{-2}{7} & \frac{-3}{7} \end{pmatrix}$.

**a.** Theorem 5.1 says that $M$ fixes the origin. Does it fix any other vectors?
**b.** Show that $M$ preserves length. That is, show that for any vector $X$, $\|MX\| = \|X\|$.
**c.** **Take It Further.** Does $M$ preserve angle measure?

**16.** Let $M = \begin{pmatrix} -3 & -6 & -2 \\ -2 & 3 & -6 \\ 6 & -2 & -3 \end{pmatrix}$ and let $\ell$ be the line whose equation is $X = (4, 1, 2) + t(1, -1, 1)$. Show that the image of $\ell$ under $M$ is a line, and find a vector equation of this line.

**17.** Suppose $N$ is an $n \times n$ matrix and $\ell$ is a line in $\mathbb{R}^n$ whose equation is $X = P + tA$. Show that the image of $\ell$ under $N$ is a line, and find a vector equation of this line.

**18.** Let $R = \frac{\sqrt{2}}{2}\begin{pmatrix} 1 & 1 \\ -1 & 1 \end{pmatrix}$.

**a.** Show what $R$ does to the unit square.
**b.** Show what $R$ does to the unit circle.
**c.** Show what $R$ does to the graph of $x^2 - xy + y^2 = 1$.
**d.** Use part **c** to draw the graph of $x^2 - xy + y^2 = 1$.

**19.** Suppose that $A$ is a square matrix. Show that $(A - I)$ and $(A + I)$ commute under multiplication.

$\longleftarrow$ In other words, show that $(A - I)(A + I) = (A + I)(A - I)$.

**20.** Suppose that $A$ is a square matrix. A **fixed vector** for $A$ is a nonzero vector $X$ so that $AX = X$. Find all fixed vectors for the following matrices or show that there are none:

**a.** $\begin{pmatrix} 0.9 & 0.3 \\ 0.1 & 0.7 \end{pmatrix}$ **b.** $\begin{pmatrix} 0 & 0.6 & 0 \\ 1 & 0 & 1 \\ 0 & 0.4 & 0 \end{pmatrix}$ **c.** $\begin{pmatrix} 0 & 1 & 3 \\ -1 & 0 & 1 \\ -3 & -1 & 0 \end{pmatrix}$ **d.** $\begin{pmatrix} 0.9 & 0 & 0.3 \\ 0 & 1 & 0 \\ 0.1 & 0 & 0.7 \end{pmatrix}$

**21.** Suppose $A$ is a square matrix. Show that $A$ has a fixed vector if and only if $\ker(A - I) \neq O$.

**22.** Suppose $A$ is a square matrix without a fixed vector. Show that $A - I$ has an inverse.

**23.** Suppose $A$ is a square matrix without a fixed vector. Show that $(A - I)^{-1}$ commutes with $A + I$.

**24.** An **orthogonal matrix** is a matrix whose transpose is equal to its inverse. Show that $\begin{pmatrix} .6 & -.8 \\ .8 & .6 \end{pmatrix}$ is an orthogonal matrix.

**25.** **a.** Show that if $N$ is an orthogonal matrix, then

$$N_{*i} \cdot N_{*j} = \begin{cases} 1 & \text{if } i = j \\ 0 & \text{if } i \neq j \end{cases}$$

**b.** Use part **a** to show that each column of $N$ has unit length and is orthogonal to the other columns.

$\longleftarrow$ Hence, the name "orthogonal."

**26.** Find a $3 \times 3$ orthogonal matrix.

**27.** Show that if $N$ is an orthogonal matrix, and $A$ and $B$ are vectors,

**a.** $N$ preserves length: $\|NA\| = \|A\|$

**b.** the angle between $NA$ and $NB$ is the same as the angle between $A$ and $B$:

$$\frac{NA \cdot NB}{\|NA\| \, \|NB\|} = \frac{A \cdot B}{\|A\| \, \|B\|}$$

**28.** **Take It Further.** Prove the following theorem.

**Theorem (Cayley, 1846)**
*Consider the function $r$ defined on $n \times n$ matrices $S$ by the rule*

$$r(S) = (S - I)^{-1}(S + I)$$

*Then if $S$ is skew-symmetric with integer entries, $r(S)$ is orthogonal with entries that are rational numbers.*

**29.** Use the result of Exercise 28 to find four $3 \times 3$ orthogonal matrices with rational entries.

**30.** Find four $3 \times 3$ matrices $A$ with integer entries and with the property that $A^\top = A^{-1}$.

**31.** Suppose $A$ is a $3 \times 3$ matrix with the property that $A^\top = A^{-1}$.

**a.** Show that the columns of $A$, together with $O$, are four vertices of a cube.

**b.** How would you find the other four vertices?

**32.** Find four cubes in $\mathbb{R}^3$ whose vertices are lattice points and whose faces are not parallel to the coordinate planes.

## 5.3 Rotations

In Lesson 5.2, you saw that certain geometric transformations, such as reflecting over an axis, can be represented by a matrix, while others, such as translations, cannot. What about rotations?

**In this lesson, you will learn how to**

- find the matrix that defines a given rotation
- find one matrix for a set of successive rotations
- given a point, find its image after rotation
- find a point given its image and angle of rotation
- use conjugation to find the matrix that represents a reflection over a given line or a rotation about a given point

In $\mathbb{R}^2$, a **rotation** is determined by a point, the **center** of the rotation, and an **angle of rotation**. For example, consider the rotation $T : \mathbb{R}^2 \to \mathbb{R}^2$ defined by rotating all points 90° counterclockwise around $(5, 12)$:

← In $\mathbb{R}^3$, a rotation is determined by an **axis** and a magnitude. This lesson concentrates on rotations in the plane.

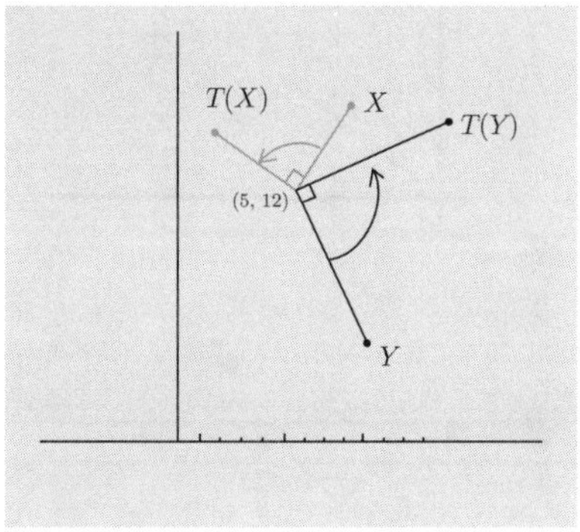


### For You to Do

1. **a.** Find and draw the image of $(8, 16)$ under $T$.
   **b.** Find and draw the image of $O$ under $T$.
   **c.** Draw the image of the unit circle under $T$.
   **d.** Find all fixed points for $T$. A fixed point satisfies $T(X) = X$.
   **e.** Is $T$ a linear transformation? Explain.

← In other words, can $T$ be represented by a matrix?

### Minds in Action    Episode 16

*Tony, Derman, and Sasha are thinking about rotations.*

SASHA: A rotation has only one fixed point—its center. Agreed?

DERMAN: Seems so. I don't think any rotation can be a linear transformation. Theorem 5.1 says a linear transformation has to fix the origin.

TONY: Pretty solid reasoning, Derman, but what if its center *is* the origin?

DERMAN: Oh, right. In that case, it has to be, so let's find the matrix for one of those.

⟵ See Exercise 2f from Lesson 5.2.

SASHA: Hold on, Derman. There were other rules in Theorem 5.1. Just because it fixes the origin doesn't mean it's a linear transformation. What about $(x, y) \mapsto (xy, x)$? It fixes the origin but violates another part of the theorem. Look, in my mapping,

$$(2,3) \mapsto (6,2) \quad \text{and} \quad (4,2) \mapsto (8,4), \quad \text{but}$$

$$(2,3) + (4,2) = (6,5) \mapsto (30,6) \neq (6,2) + (8,4)$$

It can't be linear, it violates part ((2)) of Theorem 5.1.

DERMAN: OK, but what if rotations about the origin satisfy *all* the parts of Theorem 5.1? Will rotations around the origin be linear *then*?

SASHA: Maybe, but all those things definitely have to be true. I'm guessing that if all the parts of Theorem 5.1 are satisfied, there is a matrix that does the job. But that's not what the theorem says.

TONY: All right, let's see if rotations about the origin satisfy that theorem. We know they fix the origin. What about part ((2))? It says, "the image of the sum of two vectors is the sum of their images."

*Tony draws this picture*

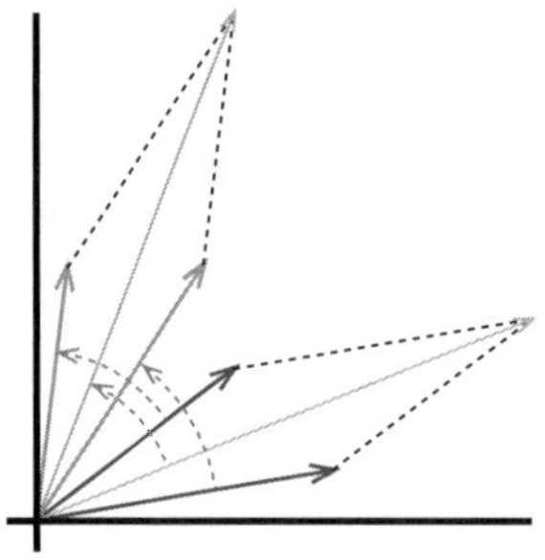

I think it works.

SASHA: Looks good to me.

DERMAN: Me too. Wait, what did we do?

### For You to Do

Tony, Sasha, and Derman are convinced that rotating the sum of two vectors through a specified angle gives the same result as rotating the vectors through that angle and then finding their sum.

⟵ In other words, rotations around the origin satisfy part ((2)) of Theorem 5.1.

**2.** Write an argument to go with the picture Tony drew, or give some other convincing argument that rotations around the origin satisfy this property.

**3.** Give an argument, geometric or otherwise, to show that rotations around the origin also satisfy part ((3)) of Theorem 5.1. If $T$ is a rotation around the origin, then $T(cA) = c(T(A))$.

### Facts and Notation

The rotation with center $O$ and magnitude $\theta$ is denoted by $R_\theta$. Starting with point $X$, $R_\theta(X)$ is defined as the point you get when you rotate $X$ through an angle of $\theta$ around $(0, 0)$.

⟵ The convention is that the rotation is in the counterclockwise direction and that angles measure between 0°and 360°.

The following theorem summarizes the properties of Theorem 5.1 for rotations around the origin.

### Theorem 5.2

*Suppose $\theta$ is an angle. The function $R_\theta$ satisfies all the properties of Theorem 5.1:*

(1) $R_\theta(O) = O$.

(2) $R_\theta(X + Y) = R_\theta(X) + R_\theta(Y)$ *for all vectors* $X$ *and* $Y$.

(3) $R_\theta(kX) = kR_\theta(X)$ *for all vectors* $X$ *and all numbers* $k$.

### Developing Habits of Mind

**Use linearity.** It turns out that Theorem 5.2 lets you find a formula for the image of any point under $R_\theta$. It comes down to the idea that any point in $\mathbb{R}^2$ can be written as

$$(x, y) = x(1, 0) + y(0, 1)$$

Parts (2) and (3) of Theorem 5.2 state that sums and scalars "pass through," so if you can say exactly what happens to $(1, 0)$ and $(0, 1)$, you can determine what happens to any point in the plane.

Given $R_\theta(1, 0)$ and $R_\theta(0, 1)$, you can determine $R_\theta(x, y)$.

$$\begin{aligned} R_\theta(x, y) &= R_\theta\left(x(1, 0) + y(0, 1)\right) \\ &= R_\theta\left(x(1, 0)\right) + R_\theta\left(y(0, 1)\right) \\ &= xR_\theta(1, 0) + yR_\theta(0, 1) \end{aligned}$$

This process, going from what a linear map does to vectors along an axis to what it does in general, is called *extension by linearity* and is a common mathematical habit.

### For Discussion

**4.** Show that $R_\theta(1, 0) = (\cos\theta, \sin\theta)$ and $R_\theta(0, 1) = (-\sin\theta, \cos\theta)$.

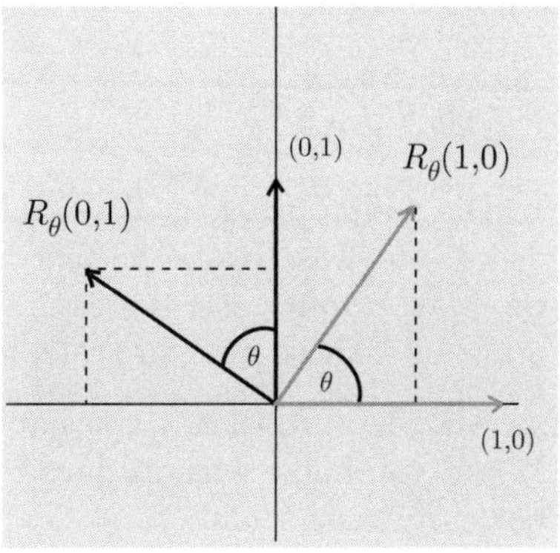


← What if $\theta$ is obtuse? greater than 180°?

The result of this For Discussion problem leads to a theorem.

**Theorem 5.3**

*For any vector* $X = (x, y)$,

$$R_\theta(X) = (x\cos\theta - y\sin\theta, x\sin\theta + y\cos\theta)$$

***Proof.***

$$\begin{aligned} R_\theta(x,y) &= R_\theta\left(x(1,0) + y(0,1)\right) \\ &= xR_\theta(1,0) + yR_\theta(0,1) \\ &= x(\cos\theta, \sin\theta) + y(-\sin\theta, \cos\theta) \\ &= (x\cos\theta - y\sin\theta, x\sin\theta + y\cos\theta) \end{aligned}$$

■

### For You to Do

**5.** Use Theorem 5.3 to find the exact $R_{30^\circ}(X)$ for each vector $X$.
**a.** $(1,0)$ **b.** $(12,0)$ **c.** $(-3,4)$ **d.** $(-9,12)$ **e.** $(9,4)$

### Minds in Action Episode 17

*Tony, Sasha, and Derman are looking at Theorem* 5.3.

DERMAN: So, now we have a formula for $R_\theta$. It looks messy. Is there a matrix that does the job?

SASHA: There sure is. Look at the next to last line of the proof:

$$x(\cos\theta, \sin\theta) + y(-\sin\theta, \cos\theta)$$

Write it as

$$x\begin{pmatrix}\cos\theta \\ \sin\theta\end{pmatrix} + y\begin{pmatrix}-\sin\theta \\ \cos\theta\end{pmatrix}$$

and think of the Pick-Apart Theorem.

$$x\begin{pmatrix}\cos\theta \\ \sin\theta\end{pmatrix} + y\begin{pmatrix}-\sin\theta \\ \cos\theta\end{pmatrix} = \begin{pmatrix}\cos\theta & -\sin\theta \\ \sin\theta & \cos\theta\end{pmatrix}\begin{pmatrix}x \\ y\end{pmatrix}$$

TONY: And those columns are the vectors from the For Discussion problem. Wow, there *is* a matrix that represents $R_\theta$. And all we needed to do was to use linearity to find what $R_\theta$ did to *two* vectors, one along each axis. Maybe your general idea about Theorem 5.1 was right after all, and there's always a matrix if those parts are satisfied.

SASHA: Yes, I think so. In fact, if you look at what we did, I think that if you have a function that satisfies all the parts of Theorem 5.1, and you know what it does to a set of vectors that "generates" somehow . . . I'm not sure what I want to say, but I think there's a more general idea here.

**Habits of Mind**

Sasha is reasoning about the calculations.

⟵ You saw the Pick-Apart Theorem—Theorem 4.8 from Lesson 4.5.

⟵ Sasha's general idea, from Episode 16, was that if all parts of Theorem 5.1 are satisfied, there must be a matrix that does the job.

Sasha's matrix for rotations deserves to be stated as a theorem.

**Theorem 5.4**

*In $\mathbb{R}^2$, rotations about the origin are linear transformations defined by the matrix*

$$R_\theta \begin{pmatrix} x \\ y \end{pmatrix} = \begin{pmatrix} \cos\theta & -\sin\theta \\ \sin\theta & \cos\theta \end{pmatrix} \begin{pmatrix} x \\ y \end{pmatrix}$$

**Example 1**

**Problem.** Find the image when the point $(\sqrt{3}, 1)$ is rotated $60°$ about the origin.

← Draw a picture. Remember, rotations are counterclockwise.

**Solution.** Use the values of $\cos 60° = \frac{1}{2}$ and $\sin 60° = \frac{\sqrt{3}}{2}$.

$$R_{60°} \begin{pmatrix} \sqrt{3} \\ 1 \end{pmatrix} = \begin{pmatrix} \frac{1}{2} & -\frac{\sqrt{3}}{2} \\ \frac{\sqrt{3}}{2} & \frac{1}{2} \end{pmatrix} \begin{pmatrix} \sqrt{3} \\ 1 \end{pmatrix} = \begin{pmatrix} 0 \\ 2 \end{pmatrix}$$

**Example 2**

**Problem.** Let $A = (4, 0)$, $B = (0, 2)$, and $C = (0, 0)$. Find the image of $\triangle ABC$ under a rotation of $45°$ about $(0, 0)$.

← Draw a picture.

**Solution.** Use the values of $\cos 45° = \frac{\sqrt{2}}{2}$ and $\sin 45° = \frac{\sqrt{2}}{2}$.

$$R_{45°} \begin{pmatrix} 4 \\ 0 \end{pmatrix} = \begin{pmatrix} \frac{\sqrt{2}}{2} & -\frac{\sqrt{2}}{2} \\ \frac{\sqrt{2}}{2} & \frac{\sqrt{2}}{2} \end{pmatrix} \begin{pmatrix} 4 \\ 0 \end{pmatrix} = \begin{pmatrix} 2\sqrt{2} \\ 2\sqrt{2} \end{pmatrix}$$

$$R_{45°} \begin{pmatrix} 0 \\ 2 \end{pmatrix} = \begin{pmatrix} \frac{\sqrt{2}}{2} & -\frac{\sqrt{2}}{2} \\ \frac{\sqrt{2}}{2} & \frac{\sqrt{2}}{2} \end{pmatrix} \begin{pmatrix} 0 \\ 2 \end{pmatrix} = \begin{pmatrix} -\sqrt{2} \\ \sqrt{2} \end{pmatrix}$$

$$R_{45°} \begin{pmatrix} 0 \\ 0 \end{pmatrix} = \begin{pmatrix} \frac{\sqrt{2}}{2} & -\frac{\sqrt{2}}{2} \\ \frac{\sqrt{2}}{2} & \frac{\sqrt{2}}{2} \end{pmatrix} \begin{pmatrix} 0 \\ 0 \end{pmatrix} = \begin{pmatrix} 0 \\ 0 \end{pmatrix}$$

The rotation of $(0, 0)$ could have been found more directly using knowledge of rotations. Since $(0, 0)$ is the center of rotation, the image must remain $(0, 0)$. Alternatively, knowing that $R_{45°}$ can be represented by a transformation matrix, it must satisfy the properties of linear transformations, including fixing the origin.

**Example 3**

**Problem.** The image of $P = (-2\sqrt{3}, 2)$ under a rotation of the plane about $(0, 0)$ through an angle of $\theta$ is $P' = (-2\sqrt{3}, -2)$. Find $\theta$.

**Solution.**

$$\cos\theta = \frac{P \cdot P'}{\|P\|\,\|P'\|} = \frac{8}{4 \cdot 4} = \frac{1}{2}$$

There is only one angle with $0° \leq \theta < 180°$ with $\cos\theta = \frac{1}{2}$, which is 60°.

**Remember**

There are two angles between 0° and 360° with cosine equal to $\frac{1}{2}$, but remember that this formula for angle restricts $\theta$ to be between 0° and 180°.

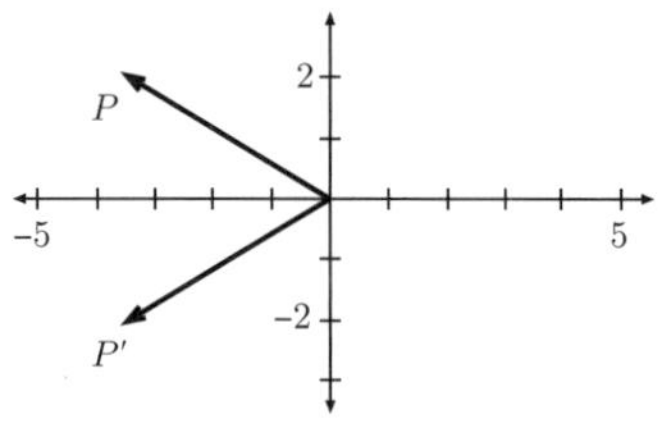


As with other linear transformations, $R_\theta$ is often identified by its matrix representation, called a **rotation matrix**:

$$R_\theta = \begin{pmatrix} \cos\theta & -\sin\theta \\ \sin\theta & \cos\theta \end{pmatrix}$$

Keep in mind that there are subtle differences between $R_\theta$, which is a transformation, and its matrix. Since the effect of $R_\theta$ is identical to multiplying by its rotation matrix, it's often said that the transformation "is" the matrix.

Some new facts about rotation matrices can be determined by looking at both sides of this coin.

**For Discussion**

**6.** **a.** What is the effect of rotating by 30°, and then rotating by 45°?
**b.** What is the effect of rotating by 45°, and then rotating by 30°?
**c.** What is the effect of rotating by 30° five times in a row?
**d.** What is the effect of rotating by 30°, and then rotating by −30°?

If you think about it geometrically, rotating a point by 30° and then rotating it by 45° produces a single rotation of 75°. But these rotations all have corresponding matrices. For example, the 30° rotation can be carried out by multiplying a vector by the matrix

$$\begin{pmatrix} \cos 30° & -\sin 30° \\ \sin 30° & \cos 30° \end{pmatrix}$$

So, for every point $\begin{pmatrix} x \\ y \end{pmatrix}$,

$$\begin{pmatrix} \cos 45° & -\sin 45° \\ \sin 45° & \cos 45° \end{pmatrix} \left( \begin{pmatrix} \cos 30° & -\sin 30° \\ \sin 30° & \cos 30° \end{pmatrix} \begin{pmatrix} x \\ y \end{pmatrix} \right) = \begin{pmatrix} \cos 75° & -\sin 75° \\ \sin 75° & \cos 75° \end{pmatrix} \begin{pmatrix} x \\ y \end{pmatrix}$$

You developed this statement based on the geometry, but it is also a statement about matrix multiplication. Since matrix multiplication is associative, then for every point $\begin{pmatrix} x \\ y \end{pmatrix}$,

$$\left(\begin{pmatrix} \cos 45^\circ & -\sin 45^\circ \\ \sin 45^\circ & \cos 45^\circ \end{pmatrix} \begin{pmatrix} \cos 30^\circ & -\sin 30^\circ \\ \sin 30^\circ & \cos 30^\circ \end{pmatrix}\right) \begin{pmatrix} x \\ y \end{pmatrix} = \begin{pmatrix} \cos 75^\circ & -\sin 75^\circ \\ \sin 75^\circ & \cos 75^\circ \end{pmatrix} \begin{pmatrix} x \\ y \end{pmatrix}$$

It follows from Exercise 12 from Lesson 5.1 that

$$\begin{pmatrix} \cos 45^\circ & -\sin 45^\circ \\ \sin 45^\circ & \cos 45^\circ \end{pmatrix} \begin{pmatrix} \cos 30^\circ & -\sin 30^\circ \\ \sin 30^\circ & \cos 30^\circ \end{pmatrix} = \begin{pmatrix} \cos 75^\circ & -\sin 75^\circ \\ \sin 75^\circ & \cos 75^\circ \end{pmatrix}$$

⟵ This is also a good way to find the exact values of $\cos 75^\circ$ and $\sin 75^\circ$.

You can use geometry to prove additional facts about rotation matrices as described in the following theorem.

### Theorem 5.5

*Let $\alpha$, $\beta$, and $\theta$ be angles.*

(1) $R_\alpha R_\beta = R_{\alpha+\beta}$

(2) $R_\alpha R_\beta = R_\beta R_\alpha$

(3) *For any nonnegative integer $n$,* $(R_\theta)^n = R_{n\theta}$

(4) $(R_\theta)^{-1} = R_{-\theta} = (R_\theta)^\top$

⟵ In this theorem, $R_\theta$ is a rotation matrix.

### For Discussion

**7.** Prove Theorem 5.5.

### Example 4

**Problem.** The image of $P$ under a rotation of $90^\circ$ about $(0, 0)$ is $(5, 3)$. Find $P$.

⟵ Draw a picture.

**Solution.** Take the equation $R_{90^\circ} \begin{pmatrix} x \\ y \end{pmatrix} = \begin{pmatrix} 5 \\ 3 \end{pmatrix}$. To solve for the point, multiply both sides by $(R_{90^\circ})^{-1}$.

$$\begin{aligned} R_{90^\circ} \begin{pmatrix} x \\ y \end{pmatrix} &= \begin{pmatrix} 5 \\ 3 \end{pmatrix} \\ (R_{90^\circ})^{-1} R_{90^\circ} \begin{pmatrix} x \\ y \end{pmatrix} &= (R_{90^\circ})^{-1} \begin{pmatrix} 5 \\ 3 \end{pmatrix} \\ \begin{pmatrix} x \\ y \end{pmatrix} &= (R_{90^\circ})^\top \begin{pmatrix} 5 \\ 3 \end{pmatrix} \\ \begin{pmatrix} x \\ y \end{pmatrix} &= \begin{pmatrix} 0 & 1 \\ -1 & 0 \end{pmatrix} \begin{pmatrix} 5 \\ 3 \end{pmatrix} \\ &= \begin{pmatrix} 3 \\ -5 \end{pmatrix} \end{aligned}$$

The point is $P = (3, -5)$.

You can describe more complex transformations using matrix multiplication. Suppose $T$ is the transformation that reflects a point over the line $\ell$

whose equation is $X = t(3,4)$. How would you find a matrix that represents $T$?

Your goal is to reflect a point $X$ over the line $\ell$. In Exercise **2a** from Lesson 5.2, you found a $2 \times 2$ matrix that corresponded to reflection of the $x$-axis: $\begin{pmatrix} 1 & 0 \\ 0 & -1 \end{pmatrix}$. You can use this reflection to reflect over any line through the origin.

A good strategy employs the algebraic habit known as *conjugation*. In this case, you want to use reflection over the $x$-axis to find reflection over $\ell$. You can do this in three steps:

1. Rotate $\ell$ so that the result lies on the $x$-axis. Transform $X$ with it.
2. Reflect the point over $\ell'$, which is now the $x$-axis. Note that reflecting $\ell'$ over the $x$-axis doesn't change $\ell'$, since they are now the same line.
3. Finally, rotate $\ell'$ back so that the result lies back on the original line $\ell$. Again, transform $X$ with it!

Try it with a sample point first, say $X(1,5)$. Here's a sketch of $X$ and $\ell$:

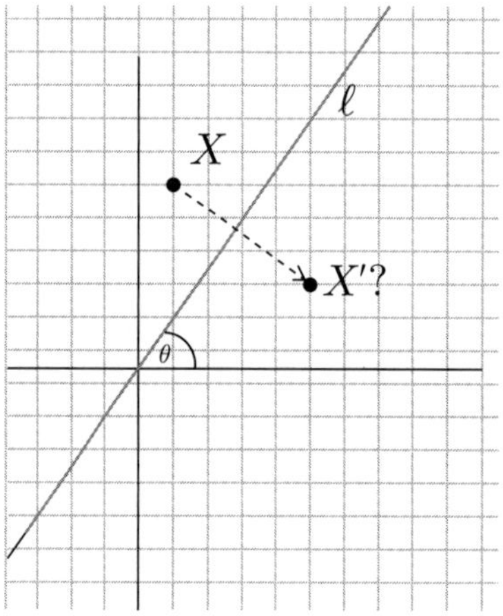


Now, follow the steps outlined above. In this case, $\ell$ is all multiples of $(3,4)$, so $\cos\theta = \frac{3}{5}$ and $\sin\theta = \frac{4}{5}$. Then the rotation matrix is

$$R_\theta = \begin{pmatrix} \frac{3}{5} & -\frac{4}{5} \\ \frac{4}{5} & \frac{3}{5} \end{pmatrix}$$

and

$$(R_\theta)^{-1} = (R_\theta)^\top = \begin{pmatrix} \frac{3}{5} & \frac{4}{5} \\ -\frac{4}{5} & \frac{3}{5} \end{pmatrix}$$

So, to reflect $(1,5)$ over $\ell$:

1. Rotate $(1,5)$ clockwise by $\theta$ by multiplying it by $R_\theta^{-1} = \begin{pmatrix} \frac{3}{5} & \frac{4}{5} \\ -\frac{4}{5} & \frac{3}{5} \end{pmatrix}$.

⟵ Rotating clockwise by $\theta$ ($R_\theta^{-1}$) is the same as rotating counterclockwise by $-\theta$ ($R_{-\theta}$).

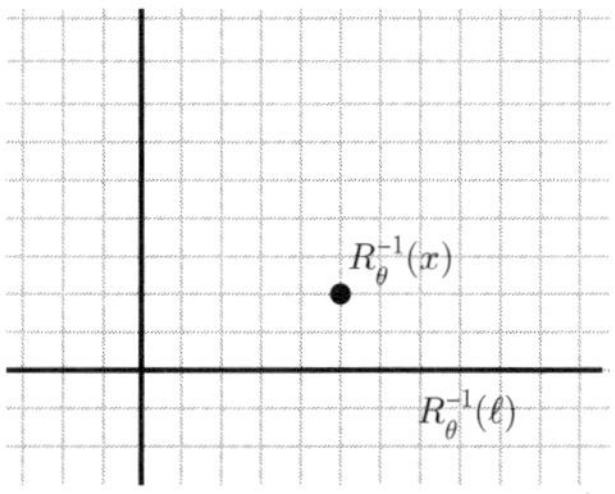


$$\begin{pmatrix} \frac{3}{5} & \frac{4}{5} \\ -\frac{4}{5} & \frac{3}{5} \end{pmatrix} \begin{pmatrix} 1 \\ 5 \end{pmatrix} = \begin{pmatrix} \frac{23}{5} \\ \frac{11}{5} \end{pmatrix}$$

2. Reflect that point over the $x$-axis by multiplying it by $F = \begin{pmatrix} 1 & 0 \\ 0 & -1 \end{pmatrix}$.

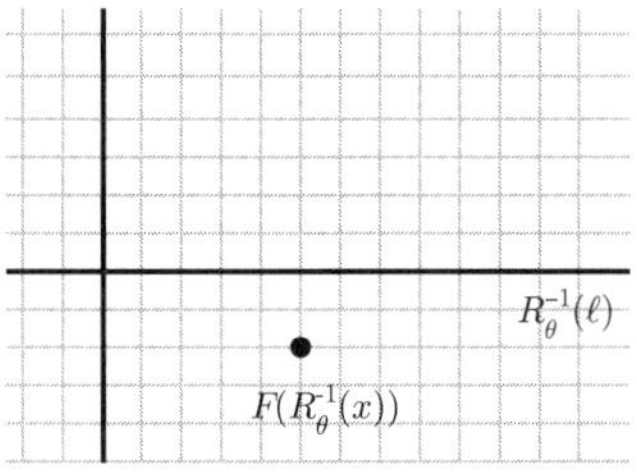


$$\begin{pmatrix} 1 & 0 \\ 0 & -1 \end{pmatrix} \begin{pmatrix} \frac{23}{5} \\ \frac{11}{5} \end{pmatrix} = \begin{pmatrix} \frac{23}{5} \\ -\frac{11}{5} \end{pmatrix}$$

3. Rotate that point counterclockwise by $\theta$ by multiplying it by $R_\theta = \begin{pmatrix} \frac{3}{5} & -\frac{4}{5} \\ \frac{4}{5} & \frac{3}{5} \end{pmatrix}$.

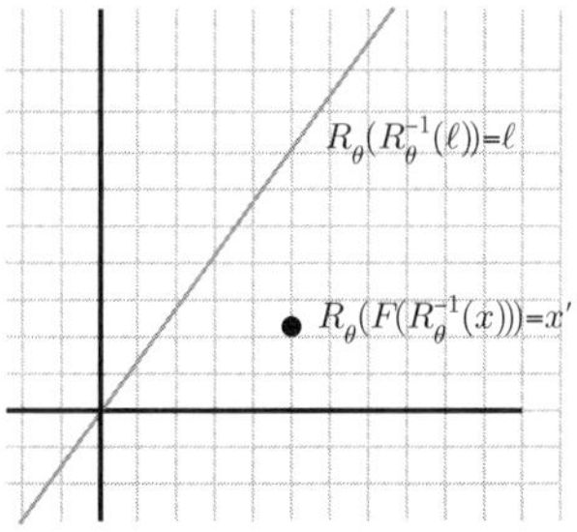


$$\begin{pmatrix} \frac{3}{5} & -\frac{4}{5} \\ \frac{4}{5} & \frac{3}{5} \end{pmatrix} \begin{pmatrix} \frac{23}{5} \\ -\frac{11}{5} \end{pmatrix} = \begin{pmatrix} \frac{115}{25} \\ \frac{59}{25} \end{pmatrix} = \begin{pmatrix} 4.52 \\ 2.36 \end{pmatrix}$$

So the process you followed was this:

$$R_\theta \left( F \left( R_\theta^{-1} \begin{pmatrix} 1 \\ 5 \end{pmatrix} \right) \right)$$

By associativity, this must also equal

$$\left(R_\theta F R_\theta^{-1}\right) \begin{pmatrix} 1 \\ 5 \end{pmatrix}$$

So $T = \left(R_\theta F R_\theta^{-1}\right)$. You can calculate the matrix product to get a single $2 \times 2$ matrix that shows the entire transformation $T$:

$$R_\theta F R_\theta^{-1} = \begin{pmatrix} \frac{3}{5} & -\frac{4}{5} \\ \frac{4}{5} & \frac{3}{5} \end{pmatrix} \begin{pmatrix} 1 & 0 \\ 0 & -1 \end{pmatrix} \begin{pmatrix} \frac{3}{5} & \frac{4}{5} \\ -\frac{4}{5} & \frac{3}{5} \end{pmatrix} = \begin{pmatrix} \frac{-7}{25} & \frac{24}{25} \\ \frac{24}{25} & \frac{7}{25} \end{pmatrix}$$

This matrix can be applied to any point $X$. As a check, compute $TX$ for $X = \begin{pmatrix} 1 \\ 5 \end{pmatrix}$: $\begin{pmatrix} \frac{-7}{25} & \frac{24}{25} \\ \frac{24}{25} & \frac{7}{25} \end{pmatrix} \begin{pmatrix} 1 \\ 5 \end{pmatrix} = \begin{pmatrix} \frac{115}{25} \\ \frac{59}{25} \end{pmatrix}$.

As expected, this is the same point found by following the steps earlier.

## Exercises

**1.** Find a matrix that defines each rotation:

**a.** $R_{60^\circ}$ **b.** $R_{45^\circ}$ **c.** $R_{90^\circ}$ **d.** $R_{120^\circ}$
**e.** $R_{240^\circ}$ **f.** $R_{180^\circ}$ **g.** $R_{270^\circ}$ **h.** $R_{225^\circ}$
**i.** $R_{330^\circ}$ **j.** $R_{24^\circ}$

**2.** Determine the image of $P$ under a rotation about $(0, 0)$ through an angle of $\theta$.

← Draw a picture.

**a.** $P = (\sqrt{3}, 1)$, $\theta = 30^\circ$ **b.** $P = (0, 2)$, $\theta = 45^\circ$
**c.** $P = (\sqrt{3}, 1)$, $\theta = -60^\circ$ **d.** $P = (-5, -2)$, $\theta = 180^\circ$

**3.** Suppose $A = (1, 2)$, $B = (3, -1)$, and $C = (5, 1)$. Find the image of $\triangle ABC$ under a rotation about $(0, 0)$ through an angle of $\theta$ if

**a.** $\theta = 90^\circ$ **b.** $\theta = 60^\circ$ **c.** $\theta = 135^\circ$

**4.** The image of $P$ under a rotation about $(0, 0)$ through $\theta$ is $P'$. Find $P$.

**a.** $\theta = 60^\circ$, $P' = (\sqrt{3}, 1)$ **b.** $\theta = 30^\circ$, $P' = (\frac{\sqrt{3}}{2}, \frac{3}{2})$

**5.** The image of $(2, -3)$ about the origin under a rotation of $\theta$ is $(3, 2)$. Find $\theta$.

**6.** Show that any $2 \times 2$ rotation matrix is an orthogonal matrix.

← See Exercise 24 from Lesson 5.2.

**7.** Find an exact value for each matrix.

**a.** $R_{75^\circ}$ **b.** $R_{105^\circ}$ **c.** $R_{195^\circ}$ **d.** $R_{165^\circ}$

← Your answer should be a matrix of numbers, with no trigonometric expressions.

**8.** Without a calculator, determine a single matrix $R_\theta$, $0° \le \theta < 360°$, that is equivalent to each expression:

**a.** $\begin{pmatrix} \frac{\sqrt{2}}{2} & -\frac{\sqrt{2}}{2} \\ \frac{\sqrt{2}}{2} & \frac{\sqrt{2}}{2} \end{pmatrix}^{47}$ **b.** $\begin{pmatrix} \frac{1}{2} & -\frac{\sqrt{3}}{2} \\ \frac{\sqrt{3}}{2} & \frac{1}{2} \end{pmatrix}^{500}$

**c.** $\begin{pmatrix} -\frac{1}{2} & \frac{\sqrt{3}}{2} \\ -\frac{\sqrt{3}}{2} & -\frac{1}{2} \end{pmatrix}^{23}$ **d.** $\begin{pmatrix} \frac{\sqrt{2}}{2} & \frac{\sqrt{2}}{2} \\ -\frac{\sqrt{2}}{2} & \frac{\sqrt{2}}{2} \end{pmatrix}^{10}$

**e.** $\begin{pmatrix} -1 & 0 \\ 0 & -1 \end{pmatrix}^{10}$ **f.** $\begin{pmatrix} 0 & -1 \\ 1 & 0 \end{pmatrix}^{13}$

**g.** $\begin{pmatrix} \frac{\sqrt{2}}{2} & -\frac{\sqrt{2}}{2} \\ \frac{\sqrt{2}}{2} & \frac{\sqrt{2}}{2} \end{pmatrix} \begin{pmatrix} \frac{\sqrt{3}}{2} & -\frac{1}{2} \\ \frac{1}{2} & \frac{\sqrt{3}}{2} \end{pmatrix} \begin{pmatrix} \frac{\sqrt{2}}{2} & \frac{\sqrt{2}}{2} \\ -\frac{\sqrt{2}}{2} & \frac{\sqrt{2}}{2} \end{pmatrix} \begin{pmatrix} 0 & -1 \\ 1 & 0 \end{pmatrix}$

**9.** For each of the following, first draw a picture, and then calculate the answer:

**a.** The line whose equation is $2x + 3y = 6$ is rotated 270° about the origin. What is the equation of the image?

**b.** The line whose equation is $5x + 2y = 6$ is rotated about the origin through an angle of $\tan^{-1} \frac{4}{3}$. What is the equation of the image?

**c.** The line whose equation is $X = (1, 2) + t(3, 1)$ is rotated about the origin through an angle of 45°. What is the equation of the image?

**10.** For each of the following, first calculate the answer, and *then* draw the picture *after* you do the calculations:

**a.** The graph of $x^2 - xy + y^2 = 16$ is rotated about the origin through an angle of 45°. What is the equation of the image?

**b.** The graph of

$$25x^2 - 120xy + 144y^2 - 416x + 559y + 52 = 0$$

is rotated about the origin through an angle of $\cos^{-1} \frac{5}{13}$. What is the equation of the image?

**11.** Find the image of $P = (-3, 5)$ under a rotation about point $C = (4, 1)$ through an angle of $\theta = \sin^{-1}\left(\frac{4}{5}\right)$.

⟵ Stuck because the center isn't the origin? Try conjugation!

**12.** Suppose $P'$ is the image of $P$ rotated 120° about the origin. If $P' = (-\sqrt{3}, 1)$, find $P$.

**13.** Point $P$ is rotated about center $C = (3, 2)$ through an angle of $\sin^{-1}\left(\frac{3}{5}\right)$ to obtain point $P'$. If $P' = (15, -10)$, find $P$.

**14.** Find the equation of the image of the set of points that satisfy

$$x^2 - 2xy + y^2 - \sqrt{2}x - \sqrt{2}y = 0$$

under a rotation of the plane about $(0, 0)$ through an angle of 45°.

**15.** Suppose $A = (1, 2)$, $B = (5, -3)$, and $C = (3, 4)$. Find the image of $\triangle ABC$ if it is rotated $90^\circ$ about $A$ and then reflected over the $y$-axis.

**16.** Suppose that $A = (1, 4)$. Find the image of $A$ if $A$ is reflected over the graph of

**a.** $12x = 5y$
**b.** $12x = -5y$
**c.** **Take It Further.** $12x + 5y = 60$

**17.** Let $A = (3, 1)$, $B = (5, 2)$, and $C = (3, 5)$. $\triangle ABC$ is reflected over $\overleftrightarrow{AC}$; find the new vertices.

**18.** Let $A = (3, 1)$, $B = (5, 2)$, and $C = (3, 5)$. $\triangle ABC$ is reflected over $\overleftrightarrow{AB}$; find the new vertices.

**19.** Let $A = (3, 1)$, $B = (5, 2)$, and $C = (3, 5)$. $\triangle ABC$ is reflected over $\overleftrightarrow{BC}$; find the new vertices.

**20.** Find a matrix that transforms $\mathbb{R}^2$ by reflecting a point over the graph of $y = x$, multiplies the image by 2, reflects this image over the $y$-axis, and then rotates this image $90^\circ$ about $(0, 0)$.

**21.** Let $A = (6, -8)$, $B = (18, 1)$, and $C = (10, 5)$. $\triangle ABC$ is reflected over $\overleftrightarrow{AB}$; find the new vertices.

**22.** Use Theorem 5.5 to prove the **angle-sum identities**.

$$\cos(\theta + \alpha) = \cos\theta\cos\alpha - \sin\theta\sin\alpha \quad \text{and}$$
$$\sin(\theta + \alpha) = \sin\theta\cos\alpha + \cos\theta\sin\alpha$$

**23.** Suppose you have a transformation $T$ on $\mathbb{R}^2$ that satisfies all the properties of Theorem 5.1, and suppose you know that $T(1, 0) = (8, -1)$ and $T(0, 1) = (4, 5)$.

Show that $T$ can be defined by a matrix, and give the specific matrix that defines $T$.

**24.** Suppose you have a transformation $T$ on $\mathbb{R}^2$ that satisfies all the properties of Theorem 5.1, and suppose you know that $T(1, 1) = (8, -1)$ and $T(0, 1) = (4, 5)$.

**a.** Find $T(9, 4)$.
**b.** Can $T$ be defined by a matrix?

**25.** **Take It Further.** Rotation matrices exist in $\mathbb{R}^3$ as well, but they have an *axis of rotation* instead of a center point. Find the $3 \times 3$ rotation matrices that rotate space through an angle $\theta$ about each of these possible axes of rotation:

**a.** the $x$-axis
**b.** the $y$-axis
**c.** the $z$-axis
**d.** the line defined by $X = t(1, 1, 1)$

**26.** Let $T$ be the transformation that rotates $\mathbb{R}^3$ $90^\circ$ about the $x$-axis, then $90^\circ$ about the $y$-axis, and then $90^\circ$ about the $z$-axis.

**a.** Find a matrix representation for $T$.
**b.** Find a fixed vector for $T$.
**c.** Is $T$ a rotation? Explain.

**27.** **a.** Are rotations commutative in $\mathbb{R}^2$?
**b.** Are rotations commutative in $\mathbb{R}^3$?

Explain with examples or drawings.

## 5.4 Determinants, Area, and Linear Maps

You are already familiar with the geometric concept of area. Vector algebra and matrices provide an easy way to calculate it in two and three dimensions. The ideas in this section will later be extended to higher dimensions.

**In this lesson, you will learn how to**

- find the area of a triangle or a parallelogram, given the coordinates of the vertices, using matrices
- find the area for the image of a triangle or a rectangle after a transformation
- understand the relation between area and determinant of $2 \times 2$ matrices

In Chapter 2, you used vector methods to find area. For instance, Theorem 2.9 states that, in $\mathbb{R}^3$, the area of the parallelogram spanned by the nonzero vectors $A$ and $B$ is $\|A \times B\|$. But how would you find the area spanned by vectors in $\mathbb{R}^2$? If you solved Exercise 5**i** from Lesson 2.5, you know how to do it.

⟵ Theorem 2.9 was introduced in Lesson 2.5.

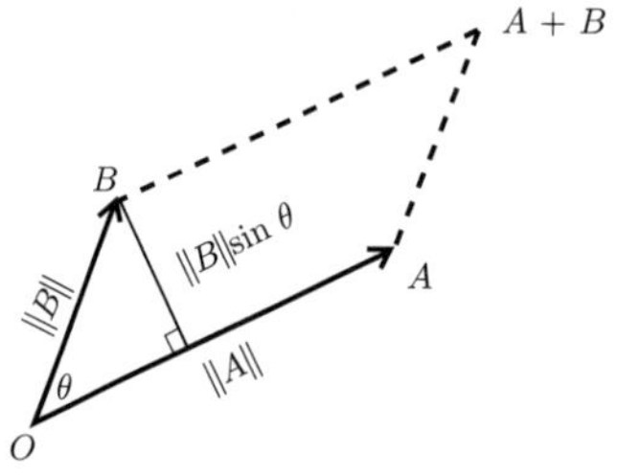


### Example

**Problem.** Find the area of the parallelogram spanned by $A(2,3)$ and $B(5,-1)$.

⟵ In exercises from Chapter 1 (such as Exercise 6 from Lesson 1.2), you may have found the area using the "surround it with a rectangle" method:

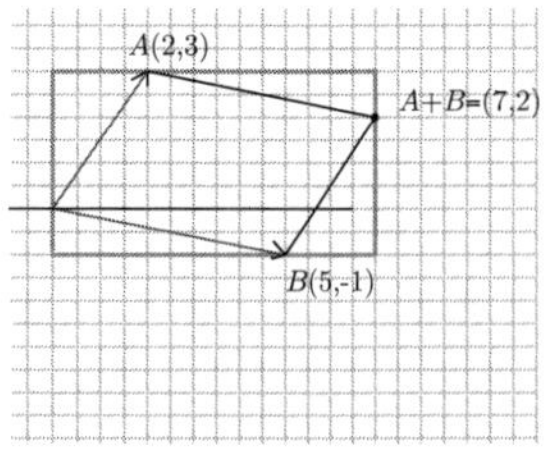


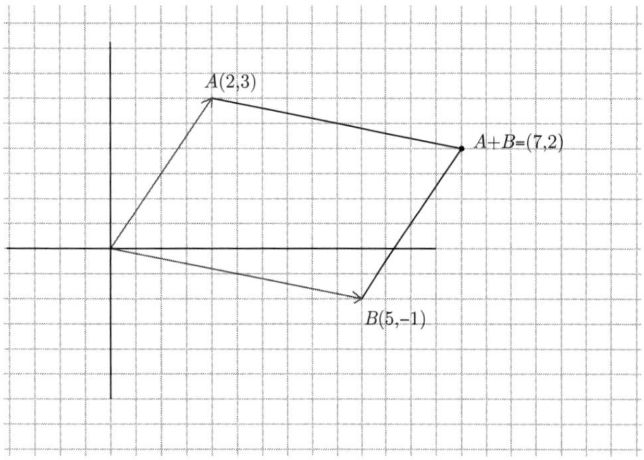

**Solution.** The idea is to embed the points into $\mathbb{R}^3$ so that you can take a cross product. Think of $\mathbb{R}^2$ as the $x$–$y$ plane in $\mathbb{R}^3$. So instead of $A(2,3)$ and $B(5,-1)$, think of $A(2,3,0)$ and $B(5,-1,0)$. The area of the parallelogram doesn't change, but in $\mathbb{R}^3$ you can take the cross product.

$$\begin{aligned} A \times B &= \left( \begin{vmatrix} 3 & 0 \\ -1 & 0 \end{vmatrix}, -\begin{vmatrix} 2 & 0 \\ 5 & 0 \end{vmatrix}, \begin{vmatrix} 2 & 3 \\ 5 & -1 \end{vmatrix} \right) \\ &= (0,0,-17) \end{aligned}$$

By Theorem 2.9, the area is the length of the cross product, and

$$\|A \times B\| = \|(0,0,-17)\| = 17$$

**Habits of Mind**

**Reason about calculations.** How could you have predicted that $A \times B$ would lie along the $z$-axis?

---

This method is perfectly general. Given any two points in $\mathbb{R}^2$, say $P = (a,b)$ and $Q = (c,d)$, the area of the parallelogram spanned by $P$ and $Q$ is

$$\begin{aligned} \|(a,b,0) \times (c,d,0)\| &= \left\| \left( \begin{vmatrix} b & 0 \\ d & 0 \end{vmatrix}, -\begin{vmatrix} a & 0 \\ c & 0 \end{vmatrix}, \begin{vmatrix} a & b \\ c & d \end{vmatrix} \right) \right\| \\ &= \|(0,0,ad-bc)\| = |ad-bc| \end{aligned}$$

This reasoning proves a theorem.

**Remember**

The determinant of $\begin{pmatrix} r & p \\ t & q \end{pmatrix}$ is the number $rq - tp$. The notation is either $\begin{vmatrix} r & p \\ t & q \end{vmatrix}$ or $\det \begin{pmatrix} r & p \\ t & q \end{pmatrix}$.

**Theorem 5.6**

*The area of the parallelogram spanned by $P = (a,b)$ and $Q = (c,d)$ is the absolute value of the determinant of the matrix whose columns are $P$ and $Q$:*

$$\left| \det \begin{pmatrix} a & c \\ b & d \end{pmatrix} \right| = |ad - bc|$$

$\longleftarrow$ Notice that the coordinates of the points are in the columns of the matrix. It would work just as well if they were in the rows (why?). Using columns makes things easier in the next part of the lesson.

**Minds in Action** Episode 18

*Derman, Tony, and Sasha are talking after school.*

TONY: So, absolute value of determinant is area. Does it act like area?

DERMAN: How does area act?

SASHA: Well, let's make a list. Here are some things that area should do:

*Sasha writes on the board.*

1. The area of the unit square should be 1.

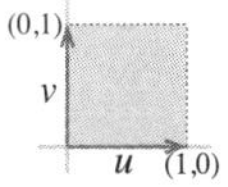


2. If you scale one side of a parallelogram by $c$, the area should scale by $c$.

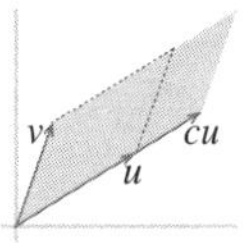

3. If you switch the order of the vectors that span the parallelogram, the area should stay the same.

4. If the vectors that span the parallelogram are scalar multiples of each other, the area should be 0.

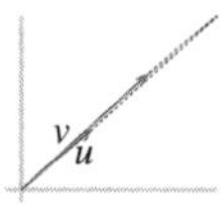


TONY: So, if absolute value of determinant is area, we better have properties like these:

*Tony writes beside each of Sasha's items*

1. The area of the unit square should be 1, so $\left|\det \begin{pmatrix} 1 & 0 \\ 0 & 1 \end{pmatrix}\right| = 1$.
2. If you scale one side of a parallelogram by $c$, the area should scale by $c$, so

$$\left|\det \begin{pmatrix} c \cdot r & p \\ c \cdot s & q \end{pmatrix}\right| = c \left|\det \begin{pmatrix} r & p \\ s & q \end{pmatrix}\right|$$

3. If you switch the order of the vectors that span the parallelogram, the area should stay the same.

$$\left|\det \begin{pmatrix} a & c \\ b & d \end{pmatrix}\right| = \left|\det \begin{pmatrix} c & a \\ d & b \end{pmatrix}\right|$$

4. If the vectors that span the parallelogram are scalar multiples of each other, the area should be 0.

$$\left|\det \begin{pmatrix} c \cdot r & r \\ c \cdot s & s \end{pmatrix}\right| = 0$$

DERMAN: I feel a theorem coming on . . .

---

As Derman predicted, here are the basic rules for determinants.

**Theorem 5.7 (Basic Rules of $2 \times 2$ Determinants)**
*Determinants have the following properties:*

(1) $\det \begin{pmatrix} 1 & 0 \\ 0 & 1 \end{pmatrix} = 1$

(2) $\det \begin{pmatrix} c \cdot r & p \\ c \cdot s & q \end{pmatrix} = c \det \begin{pmatrix} r & p \\ s & q \end{pmatrix}$ *"scalars come out"*

(3) $\det \begin{pmatrix} a & c \\ b & d \end{pmatrix} = -\det \begin{pmatrix} c & a \\ d & b \end{pmatrix}$ "det *is* ***alternating***"

(4) $\det \begin{pmatrix} c \cdot r & r \\ c \cdot s & s \end{pmatrix} = 0$

← Notice that the statements in the theorem do not involve absolute value. Do they imply Tony's claims?

### For You to Do

**1.** Prove Theorem 5.7.

### Developing Habits of Mind

**Use algebra to extend geometric ideas.** Theorem 5.7 is all about the algebra of determinants, but it was motivated by Sasha's list of geometric properties of area. Once again, geometry inspires algebra. It's possible to define the determinant of any square matrix, and you'll do exactly that in Chapter 9. These general determinants will have very similar properties to the ones in Theorem 5.7, all motivated by corresponding properties of (generalized) area. You haven't encountered it lately, but this will be another application of the "extension program" that you met in Chapters 1 and 2.

**Remember**

The extension program: Take a familiar geometric idea in two and three dimensions, find a way to describe it with vectors, and then use the algebra as the definition of the idea in higher dimensions.

Sometimes, it's the algebra that leads the way, and an algebraic result is motivated by algebra itself. For example, notice that part ((2)) of Theorem 5.7 is similar to part ((3)) of Theorem 5.1 from Lesson 5.2. Both say that "scalars come out." But for determinants, what's scaled is not the matrix—it's one *column*.

### For You to Do

**2.** **a.** Show that $\det\begin{pmatrix} 7+5 & 9 \\ 6+2 & 3 \end{pmatrix} = \det\begin{pmatrix} 7 & 9 \\ 6 & 3 \end{pmatrix} + \det\begin{pmatrix} 5 & 9 \\ 2 & 3 \end{pmatrix}$.

**b.** If $A$ and $B$ are $2 \times 2$ matrices, is $\det(A+B)$ equal to $\det(A) + \det(B)$?

If you analyze the calculations above, you see that it works in general.

**Theorem 5.8**

*Given two matrices with a common corresponding column, the determinant of the matrix formed by adding the two other corresponding columns is the sum of the two given matrices' determinants. To be precise:*

$$\det\begin{pmatrix} a+a' & c \\ b+b' & d \end{pmatrix} = \det\begin{pmatrix} a & c \\ b & d \end{pmatrix} + \det\begin{pmatrix} a' & c \\ b' & d \end{pmatrix}$$

### For Discussion

**3.** Find another place in this course, besides area and cross product, where determinants showed up.

### For You to Do

**4.** Find the area of the triangle whose vertices are $(2,3)$, $(7,-1)$, and $(4,6)$.

### Determinants and Linear Maps

There are many connections between determinants and linear maps. One has to do with area.

**Minds in Action** Episode 19

*Sasha, Tony, and Derman are still talking.*

TONY: Remember the example from Lesson 5.2? We looked at the effect of $M = \begin{pmatrix} 8 & -1 \\ 1 & 3 \end{pmatrix}$ when it is applied to the unit square.

DERMAN: We got the parallelogram whose vertices are $O$, $(8, 1)$, $(7, 4)$, and $(-1, 3)$.

*Derman draws on the board*

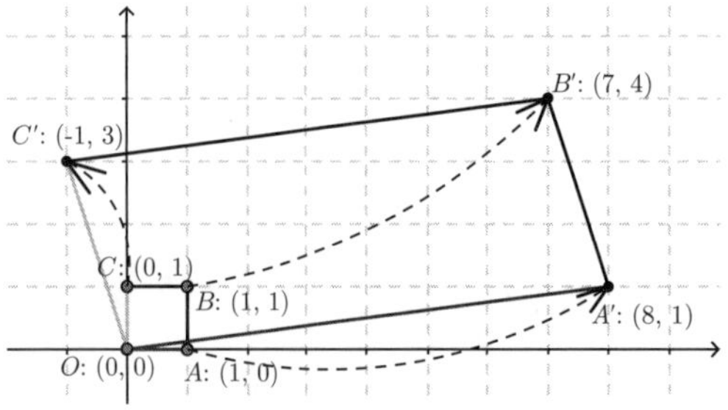


TONY: And, right after that, we were asked to find the area of the parallelogram.

SASHA: Oh . . . Now we know that it's just the absolute value of the determinant of the matrix whose columns are the points that generate the parallelogram. It's

$$\left|\det \begin{pmatrix} 8 & -1 \\ 1 & 3 \end{pmatrix}\right| = 25$$

Hey! That's just the absolute value of the determinant of $M$, our original matrix.

DERMAN: Maybe it's a coincidence.

TONY: I don't think so. Look, the generators of our new parallelogram are the points we get when we multiply $\begin{pmatrix} 1 \\ 0 \end{pmatrix}$ and $\begin{pmatrix} 0 \\ 1 \end{pmatrix}$ by $M$.

SASHA: And by the Pick-Apart Theorem, what you get are just the first and second columns of $M$.

⟵ The Pick-Apart Theorem is Theorem 4.8 from Lesson 4.5.

DERMAN: So the vertices that generate the new thing are the columns of $M$. That means that the area of the new thing is the absolute value of the determinant of $M$. So it's not a coincidence.

TONY: There are seldom coincidences in mathematics.

SASHA: I wonder . . . Applying $M$ to the vertices of the unit square multiplies its area by 25, the determinant of $M$. I wonder if multiplying by $M$ always multiplies area by 25.

DERMAN: Where do you get these ideas?

TONY: Well, before we jump into this, let's look at another example. Suppose we start with $A = (4, 3)$ and $B = (2, -5)$. We know that $O$, $A$, $A + B$, and $B$ form a parallelogram $\mathfrak{B}$ whose area is

$$\left|\det \begin{pmatrix} 4 & 2 \\ 3 & -5 \end{pmatrix}\right| = 26$$

Let's find the area of the image of $\mathfrak{B}$ under $M$ and see if it's $25 \cdot 26$.

---

**For You to Do**

**5.** Find the area of the image of $\mathfrak{B}$ under $M$ and see if it's $25 \cdot 26$.

---

Well, it's true, and it's true for any $2 \times 2$ matrix $M$: if $A$ and $B$ are points in $\mathbb{R}^2$, the area of the parallelogram spanned by $MA$ and $MB$ is $|\det(M)|$ times the area of the parallelogram spanned by $A$ and $B$. This happens thanks to the properties described in Theorems 5.7 and 5.8.

$\longleftarrow$ And there's more. Later you'll see how this generalizes to $\mathbb{R}^n$.

Suppose you have two points $A = (a, b)$ and $B = (c, d)$. The area of the parallelogram spanned by $A$ and $B$ is the absolute value of the determinant of the matrix $Q$ whose columns are $A$ and $B$.

$$\det Q = \left|\det \begin{pmatrix} a & c \\ b & d \end{pmatrix}\right| = |ad - bc|$$

Now suppose you multiply $A$ and $B$ by $M = \begin{pmatrix} 8 & -1 \\ 1 & 3 \end{pmatrix}$. The area of the parallelogram spanned by $MA$ and $MB$ is the absolute value of the determinant of the matrix $N$ whose columns are $MA$ and $MB$. The idea is to decompose $N$ using the Pick-Apart Theorem, and then to use what you know about determinants (Theorem 5.7).

$\longleftarrow$ $M$ could be any $2 \times 2$ matrix here.

You know that

$$\begin{aligned} MA &= \begin{pmatrix} 8 & -1 \\ 1 & 3 \end{pmatrix} \begin{pmatrix} a \\ b \end{pmatrix} \\ &= a \begin{pmatrix} 8 \\ 1 \end{pmatrix} + b \begin{pmatrix} -1 \\ 3 \end{pmatrix} \end{aligned}$$

Similarly,

$$\begin{aligned} MB &= \begin{pmatrix} 8 & -1 \\ 1 & 3 \end{pmatrix} \begin{pmatrix} c \\ d \end{pmatrix} \\ &= c \begin{pmatrix} 8 \\ 1 \end{pmatrix} + d \begin{pmatrix} -1 \\ 3 \end{pmatrix} \end{aligned}$$

So,

$$N = \left(a\begin{pmatrix}8\\1\end{pmatrix} + b\begin{pmatrix}-1\\3\end{pmatrix}, c\begin{pmatrix}8\\1\end{pmatrix} + d\begin{pmatrix}-1\\3\end{pmatrix}\right).$$

⟵ The notation here is a little clumsy. It means that the first column of $N$ is $a\begin{pmatrix}8\\1\end{pmatrix} + b\begin{pmatrix}-1\\3\end{pmatrix}$, and the second column is $c\begin{pmatrix}8\\1\end{pmatrix} + d\begin{pmatrix}-1\\3\end{pmatrix}$

Now calculate $\det N$ using the properties. First use Theorem 5.8 on the first column, breaking the determinant into the sum of the determinants of two matrices $R$ and $T$:

$$\begin{aligned}\det N &= \det\left(a\begin{pmatrix}8\\1\end{pmatrix} + b\begin{pmatrix}-1\\3\end{pmatrix}, c\begin{pmatrix}8\\1\end{pmatrix} + d\begin{pmatrix}-1\\3\end{pmatrix}\right)\\ &= \underbrace{\det\left(a\begin{pmatrix}8\\1\end{pmatrix}, c\begin{pmatrix}8\\1\end{pmatrix} + d\begin{pmatrix}-1\\3\end{pmatrix}\right)}_{R} + \underbrace{\det\left(b\begin{pmatrix}-1\\3\end{pmatrix}, c\begin{pmatrix}8\\1\end{pmatrix} + d\begin{pmatrix}-1\\3\end{pmatrix}\right)}_{T}\end{aligned}$$

Then use the same theorem on each of the second columns, breaking $\det R$ into the sum of two determinants:

$$\begin{aligned}\det R &= \det\left(a\begin{pmatrix}8\\1\end{pmatrix}, c\begin{pmatrix}8\\1\end{pmatrix} + d\begin{pmatrix}-1\\3\end{pmatrix}\right)\\ &= \det\left(a\begin{pmatrix}8\\1\end{pmatrix}, c\begin{pmatrix}8\\1\end{pmatrix}\right) + \det\left(a\begin{pmatrix}8\\1\end{pmatrix}, d\begin{pmatrix}-1\\3\end{pmatrix}\right)\end{aligned}$$

Next, use parts ((2)) and ((3)) of Theorem 5.7 to take out scalars and reduce this further:

$$\begin{aligned}\det R &= \det\left(a\begin{pmatrix}8\\1\end{pmatrix}, c\begin{pmatrix}8\\1\end{pmatrix}\right) + \det\left(a\begin{pmatrix}8\\1\end{pmatrix}, d\begin{pmatrix}-1\\3\end{pmatrix}\right)\\ &= ac\det\begin{pmatrix}8 & 8\\1 & 1\end{pmatrix} + ad\det\begin{pmatrix}8 & -1\\1 & 3\end{pmatrix}\\ &= ad\,\det M\end{aligned}$$

⟵ $\det\begin{pmatrix}8 & 8\\1 & 1\end{pmatrix} = 0$, right?

### For You to Do

**6.** Do the same thing for $\det T$: show that

$$\det T = -bc\,\det M$$

Putting it all together, you get

$$\begin{aligned}\det N &= ad\,\det M - bc\,\det M\\ &= (\det M)(ad - bc)\\ &= (\det M)\cdot(\det Q)\end{aligned}$$

**Remember**

$$Q = \begin{pmatrix}a & c\\b & d\end{pmatrix}$$

Even though $M$ was a specific matrix in this calculation, there's nothing special about it. The same argument works in general.

**Theorem 5.9 (Effect of a Linear Map on Area)**

*If $M$ is any $2 \times 2$ matrix and $A$ and $B$ are points, then the area of the parallelogram spanned by $MA$ and $MB$ is $|\det M|$ times the area of the parallelogram spanned by $A$ and $B$.*

***Proof.*** Using the above notation, the area of the parallelogram spanned by $MA$ and $MB$ is $|\det N|$ and the area of the parallelogram spanned by $A$ and $B$ is $|\det Q|$. But

$$|\det N| = |(\det M) \cdot (\det Q)| = |(\det M)| \cdot |(\det Q)|$$

■

### Facts and Notation

Theorem 5.9 holds more generally: if a matrix $M$ is applied to a figure $\mathfrak{F}$ that can be approximated to any degree of accuracy by squares (a circle, say), then the image of $\mathfrak{F}$ under $M$ is scaled by $|\det M|$.

### Minds in Action Episode 20

*Derman, Sasha, and Tony are thinking about this proof*

DERMAN: Well, that was enough of a workout for today.

SASHA: Wait, I see something.

DERMAN: Of course you see something.

SASHA: I think there's a bonus to all this. Remember that $N$ was defined as the matrix whose columns are $MA$ and $MB$.

TONY: Yes, the columns of $N$ are $M\begin{pmatrix} a \\ b \end{pmatrix}$ and $M\begin{pmatrix} c \\ d \end{pmatrix}$. Ah . . . and $Q$ is $\begin{pmatrix} a & c \\ b & d \end{pmatrix}$.

SASHA: So, by Pick-Apart, $N$ is $MQ$. And $\det N$ is $(\det M) \cdot (\det Q)$. See what I mean?

TONY: . . . and it will always work.

DERMAN: Here comes a theorem . . .

**Theorem 5.10 (Product Rule for Determinants)**

*If $M$ and $Q$ are $2 \times 2$ matrices,*

$$\det(MQ) = (\det M) \cdot (\det Q)$$

### For Discussion

**7.** Prove the Product Rule for Determinants.

### Developing Habits of Mind

**Look for proofs that extend.** Theorem 5.10 could be proved by direct calculation—just set up generic matrices $M$ and $Q$, multiply them, and show that the determinant of the product is the product of the determinants. This works (try it), but it has a couple of drawbacks:

1. It doesn't extend. There's a product rule for determinants of $n \times n$ matrices, and you'll see it in Chapter 9. But direct calculation would get way out of hand for any matrices larger than $2 \times 2$.
2. Direct calculation shows you *that* it's true but it doesn't show *why* it's true.

An argument based on the calculation right before Theorem 5.9 uses $M$ as a representation of a linear map and $Q$ as a storehouse for the vertices of a parallelogram. It then shows that the area of the parallelogram spanned by the columns of $Q$ gets scaled by the determinant of $M$ when $M$ is applied to that parallelogram.

Mathematicians accept any solid proof as valid, but proofs that reveal hidden meaning are more prized than those that don't.

## Exercises

**1.** Suppose $A = \begin{pmatrix} 1 \\ 5 \end{pmatrix}$ and $B = \begin{pmatrix} 7 \\ -2 \end{pmatrix}$. Find the area of

**a.** the parallelogram spanned by $A$ and $B$
**b.** the parallelogram spanned by $2A$ and $B$
**c.** the parallelogram spanned by $2A$ and $2B$
**d.** the parallelogram spanned by $2A$ and $-3B$
**e.** the parallelogram spanned by $MA$ and $MB$, where $M = \begin{pmatrix} 2 & 4 \\ 3 & 5 \end{pmatrix}$
**f.** the parallelogram spanned by $M^2A$ and $M^2B$, where $M = \begin{pmatrix} 2 & 4 \\ 3 & 5 \end{pmatrix}$
**g.** the parallelogram spanned by $M^\top A$ and $M^\top B$, where $M = \begin{pmatrix} 2 & 4 \\ 3 & 5 \end{pmatrix}$
**h.** the parallelogram spanned by $M^{-1}A$ and $M^{-1}B$, where $M = \begin{pmatrix} 2 & 4 \\ 3 & 5 \end{pmatrix}$
**i.** the parallelogram spanned by $A$ and $A + B$
**j.** the parallelogram spanned by $A$ and $A + 3B$
**k.** $\triangle OAB$

← Exercise 1 is more than a collection of fun calculations. There are hidden theorems in some of the parts. See if you can find them.

**2.** **a.** If you multiply a $2 \times 2$ matrix by a scalar $k$, what happens to its determinant?
**b.** Give a geometric interpretation of your answer to part **a** above. What does it say about area?

**3.** Suppose $A = (4, -3)$, $B = (1, 6)$, and $C = (3, 4)$. Compare the areas of the parallelograms spanned by $A$ and $B$, $A$ and $C$, and $A$ and $B + C$.

**4.** Suppose $A$, $B$, and $C$ are points in $\mathbb{R}^2$. Show that the area of the parallelogram spanned by $A$ and $B + C$ is equal to the area of the parallelogram spanned by $A$ and $B$ plus the area of the parallelogram spanned by $A$ and $C$ by

**a.** using geometry
**b.** using the algebra of determinants

**5.** Suppose $M = \begin{pmatrix} 3 & 0 \\ 0 & 2 \end{pmatrix}$ and $\mathfrak{C}$ is the unit circle: the graph of $x^2 + y^2 = 1$.

**a.** What is the equation for and the graph of the image of $\mathfrak{C}$ under $M$?
**b.** What is the area of the image of $\mathfrak{C}$ under $M$?

**6.** Develop a formula for the area of an ellipse in terms of the lengths of its semi-axes.

**7.** If $M$ and $N$ are $2 \times 2$ matrices, show that

$$\det(MN) = \det(NM)$$

even if $MN \neq NM$.

**8.** If $A$ is a $2 \times 2$ matrix that has an inverse, how is $\det A$ related to $\det(A^{-1})$? Prove what you say.

**9.** Suppose $M = \begin{pmatrix} 8 & -1 \\ 8 & -1 \end{pmatrix}$. What is the area spanned by $MA$ and $MB$ for any points $A$ and $B$? Illustrate geometrically.

**10.** If $M$ is a $2 \times 2$ matrix with determinant 0, show that $M$ maps all of $\mathbb{R}^2$ onto a line through the origin.

**11.** If $A$ is a $2 \times 2$ matrix, show that the area of the parallelogram spanned by the columns of $A$ is the same as the area of the parallelogram spanned by the rows of $A$.

**12.** Suppose that $A = \begin{pmatrix} 5 \\ 1 \end{pmatrix}$ and $B = \begin{pmatrix} 2 \\ 4 \end{pmatrix}$. Suppose further that $C = xA + yB = \begin{pmatrix} 19 \\ 11 \end{pmatrix}$ for numbers $x$ and $y$.

**a.** Find $x$ and $y$.
**b.** Find the area of the parallelograms determined by
(i) $A$ and $B$
(ii) $A$ and $C$
(iii) $B$ and $C$

**13.** Suppose that $A$ and $B$ are linearly independent vectors in $\mathbb{R}^2$ and that $C = xA + yB$ for numbers $x$ and $y$.

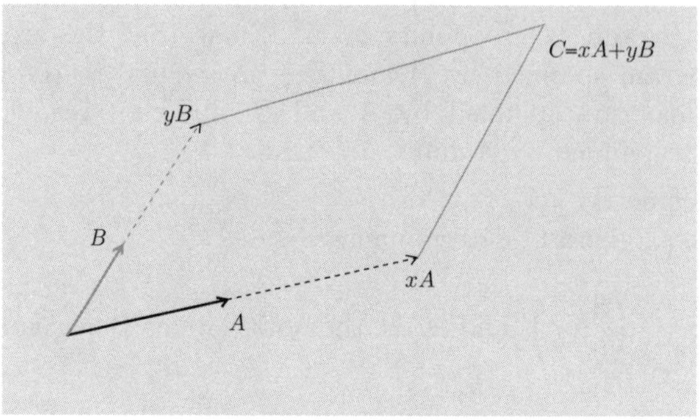


Show that

$$x = \frac{\det(C, B)}{\det(A, B)} \quad \text{and} \quad y = \frac{\det(A, C)}{\det(A, B)}$$

←
Here, $\det(A, B)$ means the determinant of the matrix whose columns are $A$ and $B$.

**a.** using algebra
**b.** using geometry

**14.** In Exercise 18 from Lesson 2.5, you saw that if $A, B$, and $C$ are vectors in $\mathbb{R}^3$, no two of which are collinear, then the volume of the parallelepiped determined by $A, B$, and $C$ is $|C \cdot (A \times B)|$.

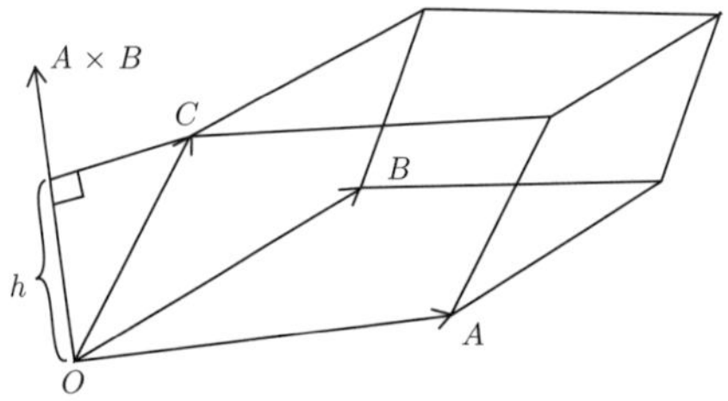


It would make sense, then, to extend the definition of determinant to $3 \times 3$ matrices by defining the determinant of a $3 \times 3$ matrix whose columns are $A$, $B$, and $C$ to be $C \cdot (A \times B)$. Find the determinant of each matrix:

←
You'll treat general determinants in depth in Chapter 9.

**a.** $\det \begin{pmatrix} 1 & 2 & 3 \\ 4 & 5 & 6 \\ 7 & 8 & 10 \end{pmatrix}$ **b.** $\det \begin{pmatrix} 2 & 2 & 3 \\ 8 & 5 & 6 \\ 14 & 8 & 10 \end{pmatrix}$

**c.** $\det \begin{pmatrix} 2 & 1 & 3 \\ 5 & 4 & 6 \\ 8 & 7 & 10 \end{pmatrix}$ **d.** $\det \begin{pmatrix} 6 & 2 & 3 \\ 7 & 5 & 6 \\ 8 & 8 & 10 \end{pmatrix}$

**e.** $\det \begin{pmatrix} 1+6 & 2 & 3 \\ 4+7 & 5 & 6 \\ 7+8 & 8 & 10 \end{pmatrix}$ **f.** $\det \begin{pmatrix} 1 & 2 & 3 \\ 4 & 5 & 6 \\ 7 & 8 & 9 \end{pmatrix}$

**15.** **a.** Which of the properties in Theorems 5.7 and 5.8 extend to $3 \times 3$ matrices?
**b.** **Take It Further.** Proofs?

## 5.5 Image, Pullback, and Kernel

Most of the matrix functions you have been working with thus far have been represented by square matrices. However, nonsquare matrices can also act as functions. With nonsquare matrices, the outputs are in a different dimension from the inputs, so lines and shapes acted upon by those matrices may also change dimensions.

**In this lesson, you will learn how to**

- find the image of a linear map using matrix multiplication
- find and interpret geometrically the set of pullbacks of a vector under a given matrix
- understand the relationship between the kernel of a matrix and the set of pullbacks of a vector under that matrix

### For You to Do

**1.** Consider the matrix $A = \begin{pmatrix} 1 & 2 & 3 \\ 4 & 5 & 6 \end{pmatrix}$. For what kinds of vectors $X$ is the product $AX$ defined, and what are the outputs?

### Definition

Let $A$ be an $m \times n$ matrix and let $X$ be a vector in $\mathbb{R}^n$. The vector $AX$ is the **image** under $A$ of $X$.

The **image of a matrix** $A$, denoted by $\text{Im}(A)$, is the set of vectors $B$ such that $AX = B$ for at least one vector $X$ in $\mathbb{R}^n$.

### Example 1

**Problem.** Let $X = \begin{pmatrix} -1 \\ 3 \\ 1 \end{pmatrix}$. Calculate its image under $A = \begin{pmatrix} 1 & 2 & 3 \\ 4 & 5 & 6 \end{pmatrix}$.

**Solution.** The image under $A$ of $X$ is the matrix product $AX$.

$$AX = \begin{pmatrix} 1 & 2 & 3 \\ 4 & 5 & 6 \end{pmatrix} \begin{pmatrix} -1 \\ 3 \\ 1 \end{pmatrix} = \begin{pmatrix} 8 \\ 17 \end{pmatrix}$$

$\longleftarrow$ Alternatively, you can say that $A$ *sends* $X$ to $\begin{pmatrix} 8 \\ 17 \end{pmatrix}$.

### For You to Do

**2.** Find the image under $A = \begin{pmatrix} 1 & 2 & 3 \\ 4 & 5 & 6 \end{pmatrix}$ of each vector:

**a.** $\begin{pmatrix} 2 \\ 3 \\ 4 \end{pmatrix}$ **b.** $\begin{pmatrix} -3 \\ 1 \\ 2 \end{pmatrix}$ **c.** $\begin{pmatrix} -1 \\ 4 \\ 6 \end{pmatrix}$ **d.** $\begin{pmatrix} 0 \\ 0 \\ 0 \end{pmatrix}$

### Developing Habits of Mind

**The extension program.** Most of the functions you have worked with so far have likely operated on numbers. But functions can operate on any mathematical objects: this chapter focuses on functions that take *vectors* as inputs and produce vectors as outputs.

← In Chapters 6 and 7, you'll extend this idea of function even further when you generalize the concept of vectors to things other than $n$-tuples of numbers.

Matrix multiplication defines such a function from $\mathbb{R}^n$ to $\mathbb{R}^m$—an $m \times n$ matrix times a column vector in $\mathbb{R}^n$ produces a column vector in $\mathbb{R}^m$. The function is defined by matrix multiplication. For example, the function

$$F(x, y, z) = (3y - 2x, z)$$

is a function from $\mathbb{R}^3$ to $\mathbb{R}^2$ and can be described completely by the matrix $A = \begin{pmatrix} -2 & 3 & 0 \\ 0 & 0 & 1 \end{pmatrix}$, since if $X = \begin{pmatrix} x \\ y \\ z \end{pmatrix}$, then $F(X) = AX = \begin{pmatrix} 3y - 2x \\ z \end{pmatrix}$.

← For this type of function, $F(X)$ is the same as the matrix product $AX$, and both will be used in this chapter.

A natural question emerges: can any function from $\mathbb{R}^n$ to $\mathbb{R}^m$ be defined using matrix multiplication? No. For example, the function defined by

$$F(x, y, z) = (2xz + 2yz, 2xy)$$

cannot be represented by a matrix.

← Why?

Many functions from $\mathbb{R}^n$ to $\mathbb{R}^m$ *do* have matrix representations, though. Such functions are called **linear maps**. Linear maps turn out to have many useful properties—properties that will be developed in depth in Chapter 8. For example, if $F$ is represented by matrix $A$, $F(O)$ must be $O$ (why?).

← Note that, depending on the dimensions of $A$, the input $O$ and the output $O$ could be different.

### For Discussion

**3.** Let $F$ be a function represented by the $3 \times 2$ matrix $A = \begin{pmatrix} 1 & 4 \\ 2 & 5 \\ 3 & 6 \end{pmatrix}$, and suppose $F(X) = \begin{pmatrix} 5 \\ 4 \\ 3 \end{pmatrix}$ and $F(Y) = \begin{pmatrix} -14 \\ -10 \\ -6 \end{pmatrix}$.

Determine each of the following.

**a.** $F(2X - 3Y)$ **b.** $F(-X)$ **c.** $F(O)$

### Facts and Notation

"$F$ is a function from $\mathbb{R}^3$ to $\mathbb{R}^2$" can be written as $F : \mathbb{R}^3 \to \mathbb{R}^2$.

This statement only names the function and its domain and range sets, not how it behaves. The following statement both names and defines a linear mapping from $\mathbb{R}^3$ to $\mathbb{R}^2$:

$$F : \mathbb{R}^3 \to \mathbb{R}^2 \text{ such that } (x, y, z) \mapsto (x + 2y + 3z, 4x + 5y + 6z)$$

Often the domain and range are not given, since they are implied by the variables in use.

$$F : (x, y, z) \mapsto (x + 2y + 3z, 4x + 5y + 6z)$$

$$(x, y, z) \overset{F}{\mapsto} (x + 2y + 3z, 4x + 5y + 6z)$$

This particular function is a linear mapping, so it can also be described using the matrix $A = \begin{pmatrix} 1 & 2 & 3 \\ 4 & 5 & 6 \end{pmatrix}$.

$$F : (x, y, z) \mapsto \begin{pmatrix} 1 & 2 & 3 \\ 4 & 5 & 6 \end{pmatrix} \begin{pmatrix} x \\ y \\ z \end{pmatrix}$$

$$F(x, y, z) = \begin{pmatrix} 1 & 2 & 3 \\ 4 & 5 & 6 \end{pmatrix} \cdot \begin{pmatrix} x \\ y \\ z \end{pmatrix}$$

$$F(X) = AX$$

It's said that $A$ **represents** the function $F$. So matrices aren't functions themselves, but they can represent functions, and the behavior of the function is entirely controlled by the matrix.

### For You to Do

**4.** If $X = \begin{pmatrix} -1 \\ 3 \\ 1 \end{pmatrix}$, the image under $A = \begin{pmatrix} 1 & 2 & 3 \\ 4 & 5 & 6 \end{pmatrix}$ of $X$ is $AX = \begin{pmatrix} 8 \\ 17 \end{pmatrix}$.

Find *all* vectors $Y$ for which the image under $A$ of $Y$ is $\begin{pmatrix} 8 \\ 17 \end{pmatrix}$.

### Developing Habits of Mind

**Look for structure.** In Chapter 4, you learned about various operations on matrices. You can add, scale, transpose, and (sometimes) multiply matrices. For square matrices, you can (sometimes) find inverses. As you will see (later in this course and in other courses), these operations carry over to operations on functions. But one operation is especially important for this chapter: *matrix multiplication corresponds to function composition.*

If $f : D \to D'$ and $g : D' \to D''$, then the function from $D$ to $D''$ defined by "first apply $f$, then apply $g$" is denoted by $g \circ f$:

$$g \circ f : D \to D'' \quad \text{by the rule} \quad g \circ f(x) = g(f(x))$$

---

The next theorem shows the corresponding statement using matrix multiplication.

### Theorem 5.11

*Suppose $F : \mathbb{R}^n \to \mathbb{R}^m$ is defined by a matrix $A$, and $G : \mathbb{R}^m \to \mathbb{R}^p$ is defined by matrix $B$. Then $G \circ F$ is represented by the matrix $BA$.*

⟵ See Exercise 12 from Lesson 5.1. By the way, what sizes are $A$ and $B$?

### For Discussion

**5.** Prove Theorem 5.11.

---

By definition, a function must have a unique output for any given input. However, the outputs may or may not come from unique inputs. This is still the case when dealing with functions whose inputs and outputs are vectors.

Suppose $AX = B$ for vectors $X$ and $B$ and matrix $A$. The vector $B$ is the unique **image** under $A$ of $X$. As problem 4 shows, though, $X$ may not be the only vector whose image is $B$.

### Definition

Let $A$ be an $m \times n$ matrix, let $X$ be a vector in $\mathbb{R}^n$, and let $B$ be a vector in $\mathbb{R}^m$. If $AX = B$, then $X$ is a **pullback** of $B$ under $A$.

⟵ Since $B$ is the image of $X$, sometimes $X$ is instead referred to as the **preimage** of $B$. Pullback is the more common term.

Note that the definition says "*a* pullback" and not "*the* pullback." A vector can have more than one pullback. This idea is not new: if $f : \mathbb{R} \to \mathbb{R}$ such that $f(x) = x^2$, then there are two pullbacks for 49. (What are they?)

### Example 2

**Problem.** Find the set of pullbacks of $B$ under $A$, where $A = \begin{pmatrix} 1 & 3 & 1 \\ 2 & 0 & 8 \end{pmatrix}$ and $B = \begin{pmatrix} 4 \\ 2 \end{pmatrix}$.

**Solution.** A pullback of $B$ under $A$ is any vector $X = \begin{pmatrix} x \\ y \\ z \end{pmatrix}$ so that $AX = B$. This is a system of equations:

$$\begin{pmatrix} 1 & 3 & 1 \\ 2 & 0 & 8 \end{pmatrix} \begin{pmatrix} x \\ y \\ z \end{pmatrix} = \begin{pmatrix} 4 \\ 2 \end{pmatrix}$$

There are multiple methods for solving it, but one way is to find the echelon form of an augmented matrix.

$$\begin{pmatrix} 1 & 3 & 1 & 4 \\ 2 & 0 & 8 & 2 \end{pmatrix} \rightarrow \begin{pmatrix} 1 & 0 & 4 & 1 \\ 0 & 1 & -1 & 1 \end{pmatrix}$$

The solution set is given by

$$x = 1 - 4z$$
$$y = 1 + z$$

The entire set of pullbacks of $B$ is

$$X = (1, 1, 0) + t(-4, 1, 1)$$

⟵ Sometimes the phrase *pullback* refers to this set, rather than any specific pullback.

### Example 3

**Problem.** Find the set of pullbacks of $B$ under $A$, where $A = \begin{pmatrix} 1 & 2 \\ 3 & 0 \\ 1 & 8 \end{pmatrix}$ and $B = \begin{pmatrix} 3 \\ 3 \\ 10 \end{pmatrix}$.

**Solution.** A pullback of $B$ under $A$ is any vector $X = \begin{pmatrix} x \\ y \end{pmatrix}$ so that $AX = B$. This is a system of equations:

$$\begin{pmatrix} 1 & 2 \\ 3 & 0 \\ 1 & 8 \end{pmatrix} \begin{pmatrix} x \\ y \end{pmatrix} = \begin{pmatrix} 3 \\ 3 \\ 10 \end{pmatrix}$$

Again, find the echelon form of the augmented matrix.

$$\begin{pmatrix} 1 & 2 & 3 \\ 3 & 0 & 3 \\ 1 & 8 & 10 \end{pmatrix} \rightarrow \begin{pmatrix} 1 & 0 & 0 \\ 0 & 1 & 0 \\ 0 & 0 & 1 \end{pmatrix}$$

The third equation listed, $0x + 0y = 1$, means there is no solution to the original system of equations. So vector $B$ has *no pullback*.

In both cases above, finding the set of pullbacks of $B$ under $A$ is equivalent to finding all solutions to the matrix equation $AX = B$. One key question for this chapter is under what conditions $AX = B$ will have no solutions, exactly one solution, or more than one solution.

### Developing Habits of Mind

**Find alternative methods that you can generalize.** The method of solution in the examples is effective, but a second method may prove more powerful and generalizable. It uses Theorem 4.16 from Lesson 4.6:

> *If* $AS = B$, *then* every *solution to* $AX = B$ *is found by letting* $X = S + K$, *where* $K$ *is in* $\ker(A)$.

So, if you can find one pullback, you can find them all! In the first example, you started with this system:

$$\begin{pmatrix}1 & 3 & 1\\ 2 & 0 & 8\end{pmatrix}\begin{pmatrix}x\\ y\\ z\end{pmatrix} = \begin{pmatrix}4\\ 2\end{pmatrix}$$

The solution $S = \begin{pmatrix}1\\ 1\\ 0\end{pmatrix}$ can be found by inspection, by writing $\begin{pmatrix}4\\ 2\end{pmatrix}$ as a linear combination of the columns of $A$.

The kernel of $A$ can be found using its echelon form:

$$\begin{pmatrix}1 & 3 & 1\\ 2 & 0 & 8\end{pmatrix} \to \begin{pmatrix}1 & 0 & 4\\ 0 & 1 & -1\end{pmatrix}$$

Based on the echelon form, the kernel is all vectors in the form $t(-4, 1, 1)$. Then, Theorem 4.16 states that the full set of solutions to $AX = B$ is

$$X = (1, 1, 0) + t(-4, 1, 1)$$

This method doesn't help much when there are no solutions. And when you want to solve a specific system, the direct approach using the augmented matrix is just as good as finding a solution and then the kernel. But Theorem 4.16 is theoretically important: given a matrix $A$, the pullback of any vector has the same *structure* as the pullback of any other vector.

⟵ It has the same structure algebraically and, when you can draw pictures, geometrically. If, for some $B$, the solution set to $AX = B$ is a line in $\mathbb{R}^3$, then the solution set to $AX = C$ is a line (in fact a parallel line) for any choice of $C$.

---

### Example 4

**Problem.** Let $A = \begin{pmatrix}1 & 2 & 3\\ 4 & 5 & 6\\ 7 & 8 & 9\end{pmatrix}$. For each of the following vectors, find the set of pullbacks under $A$.

**a.** $\begin{pmatrix}6\\ 15\\ 24\end{pmatrix}$ **b.** $\begin{pmatrix}5\\ 11\\ 17\end{pmatrix}$ **c.** $\begin{pmatrix}1\\ 1\\ 1\end{pmatrix}$

**Solution.** Find one solution to each system by inspection, and use the echelon form to find the kernel of $A$. $\begin{pmatrix}1 & 2 & 3\\ 4 & 5 & 6\\ 7 & 8 & 9\end{pmatrix} \to \begin{pmatrix}1 & 0 & -1\\ 0 & 1 & 2\\ 0 & 0 & 0\end{pmatrix}$. The kernel is given by the set of equations $x = z$, $y = -2z$, or $\ker(A) = t(1, -2, 1)$.

Using inspection to find a solution to each of the given systems (use linear combinations of the columns) gives the following sets of pullbacks:

**a.** $X = (1, 1, 1) + t(1, -2, 1)$

**b.** $X = (-1, 1, 0) + t(1, -2, 1)$

**c.** $X = (0, 1, 1) + t(1, -2, 1)$

⟵ What do these pullbacks look like geometrically?

Other answers are possible, but they will determine the same sets.

### For You to Do

**6.** Suppose $A = \begin{pmatrix} 1 & 2 & 3 \\ 4 & 5 & 6 \\ 7 & 8 & 9 \end{pmatrix}$. For each of the following vectors, find the set of pullbacks under $A$.

a. $\begin{pmatrix} 5 \\ 7 \\ 9 \end{pmatrix}$ b. $\begin{pmatrix} 3 \\ 3 \\ 3 \end{pmatrix}$ c. $\begin{pmatrix} 3 \\ 3 \\ 4 \end{pmatrix}$

**7.** Repeat problem 6 above using $A = \begin{pmatrix} 1 & 2 & 3 \\ 4 & 5 & 6 \\ 7 & 8 & 0 \end{pmatrix}$.

## Exercises

**1.** Let $A = \begin{pmatrix} 1 & 3 & 7 \\ 2 & 7 & 1 \end{pmatrix}$. Find the image under $A$ for each given vector.

a. $\begin{pmatrix} 1 \\ 4 \\ 1 \end{pmatrix}$ b. $\begin{pmatrix} 3 \\ 0 \\ 0 \end{pmatrix}$ c. $\begin{pmatrix} 1 \\ 0 \\ 1 \end{pmatrix}$

⟵ The matrix $A$ represents a function that maps $\mathbb{R}^3$ to $\mathbb{R}^2$.

**2.** Use the same $A$ from Exercise 1. Find the set of pullbacks under $A$ for each given vector.

a. $\begin{pmatrix} 4 \\ 9 \end{pmatrix}$ b. $\begin{pmatrix} 4 \\ -6 \end{pmatrix}$ c. $\begin{pmatrix} 11 \\ 0 \end{pmatrix}$

**3.** If $X$ is in the pullback of $\begin{pmatrix} 3 \\ 7 \end{pmatrix}$ under $A$ and $Y$ is in the pullback of $\begin{pmatrix} 8 \\ 1 \end{pmatrix}$ under $A$, find the image of the following under $A$.

a. $X + Y$ b. $2X$ c. $2X + 3Y$ d. $0X$

⟵ Again, use the same $A$ from Exercise 1.

**4.** Suppose $M = \begin{pmatrix} 1 & 2 \\ 3 & 0 \\ 1 & 8 \end{pmatrix}$. Show that the image of $M$ is a plane in $\mathbb{R}^3$ that contains the origin, and find an equation for that plane.

**Remember**

The **image of a matrix** $M$ is the set of all vectors $N$ that can be solutions to $MX = N$.

**5.** Let matrix multiplication by $B = \begin{pmatrix} 1 & 3 & 2 & 1 \\ 1 & 4 & 4 & 6 \\ 3 & 10 & 8 & 8 \end{pmatrix}$ represent a function from $\mathbb{R}^4$ to $\mathbb{R}^3$. Find the image under $B$ for each given vector.

a. $\begin{pmatrix} 1 \\ 4 \\ 2 \\ 1 \end{pmatrix}$ b. $\begin{pmatrix} 1 \\ 0 \\ 2 \\ 0 \end{pmatrix}$ c. $\begin{pmatrix} 1 \\ 1 \\ 1 \\ 1 \end{pmatrix}$

**6.** For each of the following vectors, find the set of pullbacks under $B = \begin{pmatrix} 1 & 3 & 2 & 1 \\ 1 & 4 & 4 & 6 \\ 3 & 10 & 8 & 8 \end{pmatrix}$, the same $B$ as in Exercise 5.

**a.** $\begin{pmatrix} 4 \\ 5 \\ 13 \end{pmatrix}$ **b.** $\begin{pmatrix} 4 \\ 2 \\ 0 \end{pmatrix}$ **c.** $\begin{pmatrix} 5 \\ 4 \\ 14 \end{pmatrix}$

**7.** Consider these four vectors:

$$A = \begin{pmatrix} 0 \\ 0 \end{pmatrix}, B = \begin{pmatrix} 1 \\ 0 \end{pmatrix}, C = \begin{pmatrix} 1 \\ 1 \end{pmatrix}, \text{ and } D = \begin{pmatrix} 0 \\ 1 \end{pmatrix}.$$

For each of the following matrices,

- compute the images $A'$ through $D'$ of these four vectors under the matrix
- draw figures $ABCD$ and $A'B'C'D'$
- describe how figure $A'B'C'D'$ is obtained from $ABCD$

**a.** $\begin{pmatrix} 2 & 0 \\ 0 & 2 \end{pmatrix}$ **b.** $\begin{pmatrix} 0 & 1 \\ 1 & 0 \end{pmatrix}$ **c.** $\begin{pmatrix} 1 & 0 \\ 0 & 1 \end{pmatrix}$ **d.** $\begin{pmatrix} 0 & 0 \\ 0 & 1 \end{pmatrix}$

**e.** $\begin{pmatrix} 2 & 0 \\ 0 & 3 \end{pmatrix}$ **f.** $\begin{pmatrix} -1 & 0 \\ 0 & 1 \end{pmatrix}$ **g.** $\begin{pmatrix} \frac{\sqrt{3}}{2} & -\frac{1}{2} \\ \frac{1}{2} & \frac{\sqrt{3}}{2} \end{pmatrix}$ **h.** $\begin{pmatrix} \frac{\sqrt{2}}{2} & -\frac{\sqrt{2}}{2} \\ \frac{\sqrt{2}}{2} & \frac{\sqrt{2}}{2} \end{pmatrix}$

**8.** For each $A$ and $X$, find the image of $X$ under $A$.

**a.** $A = \begin{pmatrix} 1 & 3 & -1 \\ 2 & 1 & 4 \end{pmatrix}$, $X = \begin{pmatrix} 1 \\ 0 \\ 3 \end{pmatrix}$

**b.** $A = \begin{pmatrix} -1 & 2 & 4 & 2 \\ 0 & 0 & 1 & 0 \\ 4 & 1 & 3 & 1 \end{pmatrix}$, $X = \begin{pmatrix} 1 \\ -1 \\ 1 \\ -1 \end{pmatrix}$

**c.** $A = \begin{pmatrix} 1 & 4 \\ 2 & 1 \\ -1 & 3 \end{pmatrix}$, $X = \begin{pmatrix} 3 \\ 7 \end{pmatrix}$

**d.** $A = \begin{pmatrix} 0 & 1 & 0 \\ 1 & 0 & 0 \\ 0 & 0 & 0 \end{pmatrix}$, $X = \begin{pmatrix} 2 \\ 3 \\ 4 \end{pmatrix}$

**9.** Find $\ker(A)$ for the given matrix $A$.

**a.** $A = \begin{pmatrix} 1 & 2 & 4 \\ 2 & 5 & 5 \end{pmatrix}$ **b.** $A = \begin{pmatrix} 1 & -1 & 2 & 4 \\ 3 & -2 & 5 & 6 \\ 1 & 2 & 4 & 5 \end{pmatrix}$

**c.** $A = \begin{pmatrix} 1 & -1 & 2 & 4 \\ 3 & -2 & 5 & 6 \\ 4 & -3 & 7 & 10 \end{pmatrix}$ **d.** $A = \begin{pmatrix} 4 & 1 & 8 \\ 1 & 4 & 6 \\ 3 & -3 & 2 \end{pmatrix}$

**10.** Let $A = \begin{pmatrix} 1 & 3 & 2 \\ 2 & 5 & 3 \\ 1 & 2 & 1 \end{pmatrix}$. For each of the given vectors, find the pullback.

$\longleftarrow$ In this context, "pullback" means the same as "the set of pullbacks."

**a.** $\begin{pmatrix} 3 \\ 5 \\ 2 \end{pmatrix}$ **b.** $\begin{pmatrix} 2 \\ 4 \\ 2 \end{pmatrix}$ **c.** $\begin{pmatrix} 5 \\ 8 \\ 3 \end{pmatrix}$

**d.** $\begin{pmatrix} 5 \\ 8 \\ 2 \end{pmatrix}$ **e.** $\begin{pmatrix} 0 \\ 0 \\ 0 \end{pmatrix}$ **f.** $\begin{pmatrix} 6 \\ 10 \\ 3 \end{pmatrix}$

**11.** Suppose that $A = \begin{pmatrix} 1 & 2 & 3 \\ 4 & 5 & 6 \\ 7 & 8 & 0 \end{pmatrix}$, that $X$ is in the pullback of $\begin{pmatrix} 1 \\ 5 \\ 2 \end{pmatrix}$, and that $Y$ is in the pullback of $\begin{pmatrix} -1 \\ 2 \\ 7 \end{pmatrix}$. Find the image of $4X - 5Y$ under $A$.

## 5.6 The Solution Set for $AX = B$

You started using matrices as tools to solve systems of equations. Now you can think of a matrix as representing a function that operates on vectors. The representations are a little different, but understanding both is helpful to determine the types of solutions each can generate.

**In this lesson, you will learn how to**

- determine whether a linear function is one-to-one
- find a matrix that represents a projection along a vector or a line
- describe the set of pullbacks of a linear function using formal function language

**Minds in Action** Episode 21

DERMAN: Let's review what we know about solving $AX = B$. That's a system of equations.

TONY: I remember that $AX = B$ has a solution whenever $B$ is a linear combination of the columns of $A$. Here, I'll make one using prime numbers for $A$ and something easy for $B$.

$$\begin{pmatrix} 2 & 3 & 5 \\ 7 & 11 & 13 \end{pmatrix} \begin{pmatrix} x \\ y \\ z \end{pmatrix} = \begin{pmatrix} 8 \\ 29 \end{pmatrix}$$

All right, find me a solution to this one.

*Derman thinks for a good, long time.*

DERMAN: The solution is $x = 1, y = 2, z = 0$.

TONY: Nice, but careful when you say *the* solution.

DERMAN: Right, there might be more than one. We have to look at the kernel of $A$.

TONY: And I think two things can happen. If the kernel is the zero vector, then that's the only solution. Zero's always in the kernel.

DERMAN: And if the kernel has other vectors in it, then there are more solutions, and you get them by adding anything in the kernel to the solution you already have.

TONY: Very smooth.

*Sasha walks over . . .*

SASHA: Let's review what we know about solving $AX = B$.

DERMAN: We just did that! It's a system of equations.

SASHA: But what about all this function stuff? What if we think of $A$ as a function?

DERMAN: I guess the language will be different, but it should be the same stuff. $AX = B$ still has a solution whenever $B$ is a linear combination of the columns of $A$.

SASHA: So, you're saying $B$ is in the image of $A$, $\text{Im}(A)$, if and when $B$ is a linear combination of $A$'s columns.

DERMAN: That's not what I said, but it sounds good. Here, look at this.

*Sasha looks at the equation that Tony wrote.*

SASHA: Ah. The solutions are the pullback of $B$ under $A$, if they exist.

TONY: Hey, she's right! By the way, there are solutions: $x = 1$, $y = 2$, $z = 0$ is one.

SASHA: Okay, great. Then we have to look at the kernel of $A$.

DERMAN: Oh, we've been here. Two things can happen. If the kernel is the zero vector, then that's the only ... uh, pullback. Zero's always in the kernel.

TONY: And if the kernel has other vectors in it, then the pullback set is larger, and you get anything in the pullback by adding anything in the kernel to the pullback you already have!

SASHA: Very smooth.

---

**For You to Do**

**1.** Find the complete set of pullbacks of $\begin{pmatrix} 8 \\ 29 \end{pmatrix}$ under $\begin{pmatrix} 2 & 3 & 5 \\ 7 & 11 & 13 \end{pmatrix}$.

← Alternatively, this could also have said "Find the pullback of ..." with the same intended meaning.

---

**For Discussion**

**2.** There are definitely systems with no solutions and vectors with no pullbacks. There are definitely systems with unique solutions, which occur when $\ker(A)$ is just the zero vector. Can there be systems with exactly four solutions? Your work in Exercise 6 from Lesson 3.5 may be helpful.

---

Derman, Tony, and Sasha have noticed that there are two ways to describe the solution set to $AX = B$—either in terms of linear equations or in terms of linear mappings. The next two theorems say essentially the same thing from these two points of view, and they summarize much of what has happened so far.

**Theorem 5.12 ($AX = B$, Function Version)**
*Suppose $A$ is $m \times n$. A vector $B$ in $\mathbb{R}^m$ is in the image of $A$ (so that $B$ has a nonempty pullback under $A$) if and only if $B$ is a linear combination of the columns of $A$. If $B$ is in the image of $A$, the kind of pullback that $B$ has is determined by $A$'s kernel. If $\ker(A)$ contains only the zero vector, $B$ has only one vector in its pullback, while if $\ker(A)$ contains nonzero vectors (so that it contains infinitely many vectors), then the pullback of $B$ is infinite.*

← $\ker(A)$ is either just the zero vector or contains infinitely many vectors.

**Theorem 5.13 ($AX = B$, System Version)**

*Suppose $A$ is $m \times n$, and suppose $B$ is a vector in $\mathbb{R}^m$. The system $AX = B$ has a solution if and only if $B$ is a linear combination of the columns of $A$. If $AX = B$ has a solution, the solution set is determined by $A$'s kernel. If $\ker(A)$ contains only the zero vector, $AX = B$ has only one solution, while if $\ker(A)$ contains nonzero vectors (so that it contains infinitely many vectors), then $AX = B$ has infinitely many solutions.*

⟵ Theorem 4.16 from Lesson 4.6 shows you how to construct all the solutions from one solution: you add your fixed solution to things in the kernel.

You can tell whether $\ker(A) = O$ or $\ker(A)$ is infinite by going through the process of computing it. The following theorem shows a useful way to tell just by looking at $A$.

**Theorem 5.14 ("Fatter Than Tall" or FTT)**

*If matrix $A$ is $m \times n$ with $m < n$, then $\ker(A)$ contains more than the zero vector.*

⟵ Another way to say this: a homogeneous system with more variables than equations has a nonzero solution.

Combining this theorem with Theorem 3.4 from Lesson 3.5, you get a useful corollary.

**Corollary 5.15**

*Any set of more than $n$ vectors in $\mathbb{R}^n$ is linearly dependent.*

---

**For Discussion**

**3.** Write out a careful explanation of why Theorem 5.14 is true in the case that $A$ is $3 \times 5$. Why is it true in general?

---

**Habits of Mind**

**Generalize from examples.** Find the kernel of a few FTT matrices and you'll see why the theorem is true.

Many functions on real numbers have different inputs with the same output: it is possible to have $x \neq y$ while still having $f(x) = f(y)$. But for some functions, this never happens: if the inputs are different, the outputs must be different. For functions that take vector inputs, this means the images of all input vectors are distinct, and all pullback sets contain a unique element. Functions that behave this way, regardless of the type of input, are called one-to-one.

⟵ What function matches the figure below? Is it one-to-one?

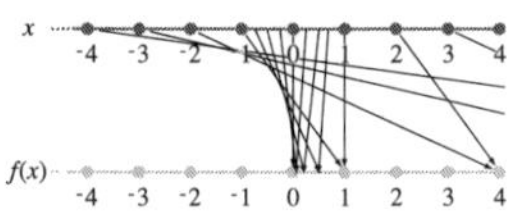


**Definition**

A function $F$ is **one-to-one** if, whenever $X$ and $Y$ are in the domain of $F$ and $X \neq Y$, then $F(X) \neq F(Y)$. Alternatively, if $F(X) = F(Y)$, then $X = Y$.

---

**Example**

**Problem.** Suppose $F : \mathbb{R}^2 \to \mathbb{R}^2$ is defined by $F(X) = \text{Proj}_{(1,1)} X$.

1. What is $\text{Im}(F)$?
2. Is $F$ one-to-one?

3. What is the pullback of $(2, 2)$?
4. Can $F$ be represented by a matrix? If so, find it. If not, explain why.

**Solution.** The first three questions can be answered from a picture of what $F$ does to points; it projects them onto the line along $B = (1, 1)$.

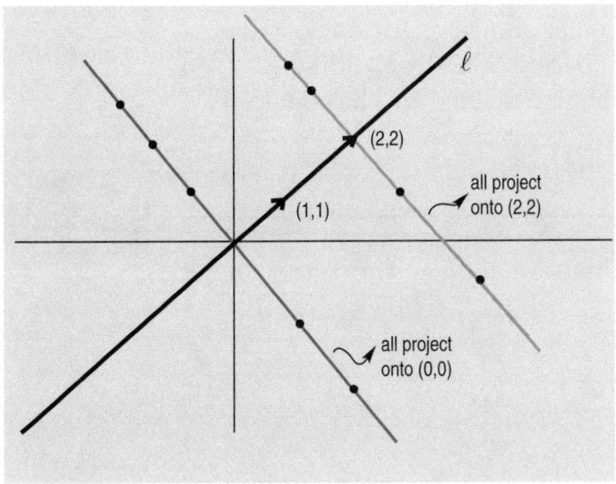


1. The image is the line $\ell : y = x$. The line passes through the origin in the direction of $B$, and its vector equation is $X = t(1, 1)$.
2. No, $F$ is not one-to-one, because all the points on a line perpendicular to $\ell$ get mapped to the same point on $\ell$. For example, $F(3, 3) = F(2, 4) = F(1, 5)$, so $F$ is not one-to-one.
3. The pullback of $(2, 2)$ is all points $X$ for which $F(X) = (2, 2)$. This is the line through $(2, 2)$ perpendicular to $\ell$. Its direction can be found by using dot product, and its vector equation is $X = (2, 2) + t(-1, 1)$.

The first three questions were answered geometrically, but all four questions can be answered algebraically using the definition of projection from Chapter 2. This builds a direct formula for $F(x, y)$ in terms of $x$ and $y$.

$$\begin{aligned} F(x, y) &= \text{Proj}_{(1,1)}(x, y) \\ &= \frac{(x, y) \cdot (1, 1)}{(1, 1) \cdot (1, 1)}(1, 1) \\ &= \frac{x + y}{2}(1, 1) \\ &= \left(\frac{x + y}{2}, \frac{x + y}{2}\right) \end{aligned}$$

Since both coordinates of $F(x, y)$ are linear in terms of $x$ and $y$, $F$ *can* be represented by a $2 \times 2$ matrix: $A = \begin{pmatrix} \frac{1}{2} & \frac{1}{2} \\ \frac{1}{2} & \frac{1}{2} \end{pmatrix}$. Knowing matrix $A$ allows you to gain additional insight on the first three questions.

1. $\text{Im}(F)$ is the set of all linear combinations of the columns of $A$: the multiples of $(\frac{1}{2}, \frac{1}{2})$. This is the same as the line $\ell$ defined above.
2. One way to detect whether $F$ is one-to-one is to see if $\ker(A)$ contains only the zero vector. In this case, $\ker(A)$ is the set of multiples of $(-1, 1)$, because $\text{rref}(A) = \begin{pmatrix} 1 & 1 \\ 0 & 0 \end{pmatrix}$. The pullback of anything in $\text{Im}(F)$ must be infinite, so $F$ is not one-to-one.
3. To find the pullback of $(2, 2)$, first find a particular solution. An obvious one is $(2, 2)$, but there are others. Then, the entire pullback is found by adding $(2, 2)$ to everything in $\ker(A)$. The pullback of $(2, 2)$ is the line defined by the equation

$$X = (2, 2) + t(-1, 1)$$

---

As seen in the example, if $F$ is a linear mapping, the kernel of its representative matrix offers the following quick test.

**Theorem 5.16**

*Let $F$ be a linear mapping represented by matrix $A$. $F$ is one-to-one if and only if the kernel of $A$ contains just $O$.*

---

### For You to Do

**4.** Prove Theorem 5.16.

---

### For Discussion

**5.** Theorems 5.14 and 5.16 can be combined to make an interesting corollary. What is it, and what does it imply?

---

### Developing Habits of Mind

**Use functions to prove facts about matrices.** There's a theorem in Chapter 4 that still hasn't been proved: Theorem 4.10 from Lesson 4.5 says that matrix multiplication is associative.

If $A$ is $m \times n$, $B$ is $n \times p$, and $C$ is $p \times q$, then $(AB)C = A(BC)$.

This can be proved right from the definition of matrix multiplication, but the proof is messy and not very enlightening. But if you look at matrices as functions, the proof comes from the following facts.

**Function composition is associative:** If you have three functions

$$f : D \to D', \ g : D' \to D'', \text{ and } h : D'' \to D'''$$

then

$$h \circ (g \circ f) = (h \circ g) \circ f$$

← Why is function composition associative? Look what each side of the equation does to any element of $D$.

**Matrix multiplication corresponds to function composition:** This is the content of Theorem 5.11 from Lesson 5.5.

**For Discussion**

**6.** Use this line of reasoning to prove that matrix multiplication is associative.

## Exercises

**1.** How many vectors are in the solution set to $AX = B$ for each system below? Justify your answer.

**a.** $A = \begin{pmatrix} 1 & 2 & 4 \\ 2 & 5 & 5 \end{pmatrix}$, $B = \begin{pmatrix} 2 \\ 0 \end{pmatrix}$

**b.** $A = \begin{pmatrix} 1 & 2 \\ 2 & 5 \end{pmatrix}$, $B = \begin{pmatrix} 2 \\ 0 \end{pmatrix}$

**c.** $A = \begin{pmatrix} 1 & 3 & 2 & -1 \\ 3 & 8 & 5 & -2 \\ 1 & 5 & 4 & -3 \end{pmatrix}$, $B = \begin{pmatrix} 6 \\ 12 \\ 3 \end{pmatrix}$

**d.** $A = \begin{pmatrix} 1 & 3 & 2 & -1 \\ 3 & 8 & 5 & -2 \\ 1 & 5 & 4 & -3 \end{pmatrix}$, $B = \begin{pmatrix} 4 \\ 11 \\ 6 \end{pmatrix}$

**e.** $A = \begin{pmatrix} 1 & 3 & 2 \\ 3 & 8 & 5 \\ 1 & 5 & 4 \end{pmatrix}$, $B = \begin{pmatrix} 4 \\ 11 \\ 6 \end{pmatrix}$

**f.** $A = \begin{pmatrix} 1 & 3 \\ 3 & 8 \\ 1 & 5 \end{pmatrix}$, $B = \begin{pmatrix} 4 \\ 11 \\ 6 \end{pmatrix}$

**g.** $A = \begin{pmatrix} 1 & 3 \\ 3 & 8 \\ 1 & 6 \end{pmatrix}$, $B = \begin{pmatrix} 4 \\ 11 \\ 6 \end{pmatrix}$

**2.** Explain why it is impossible for a system of four linear equations in four unknowns to have exactly two solutions.

**3.** Show that it is impossible for a linear map (a function represented by a matrix) $F : \mathbb{R}^3 \to \mathbb{R}^2$ to be one-to-one.

**4.** Suppose $F : \mathbb{R}^2 \to \mathbb{R}^3$ is defined by $X \overset{F}{\mapsto} \begin{pmatrix} 1 & 2 \\ 3 & 4 \\ 5 & 6 \end{pmatrix} X$.

**a.** Find $\ker(F)$.

**b.** Find $\operatorname{Im}(F)$.

**5.** The following are partial or complete definitions of functions. For each, decide whether it is one-to-one.

**a.** Defined by a matrix $A$ where $\ker(A) \neq O$

**b.** Defined by a matrix $A$ where $\ker(A) = O$

**c.** Defined by a $2 \times 3$ matrix $A$

**d.** Defined by $A = \begin{pmatrix} 0 & 1 & 0 \\ 1 & 0 & 0 \\ 0 & 0 & 0 \end{pmatrix}$

**e.** Defined by $A = \begin{pmatrix} 0 & -3 \\ 3 & 0 \end{pmatrix}$

**f.** Defined by a $2 \times 2$ matrix that projects $(x, y)$ onto $(1, 2)$

**6.** Suppose $F : \mathbb{R}^3 \to \mathbb{R}^3$ is defined by $X \overset{F}{\mapsto} \begin{pmatrix} 1 & 2 & 3 \\ 4 & 5 & 6 \\ 7 & 8 & 9 \end{pmatrix} X$.

**a.** Is $F$ one-to-one? Explain.
**b.** Find $\text{Im}(F)$.

**7.** Show that a function $F : \mathbb{R}^2 \to \mathbb{R}^3$ defined by a matrix cannot have all of $\mathbb{R}^3$ as its image.

**8.** Is there a $3 \times 3$ matrix $A$ and two vectors $B$ and $C$ so that $AX = B$ has exactly one solution and $AX = C$ has infinitely many? Justify your answer.

**9.** If $A$ is a $3 \times 3$ matrix whose columns are linearly independent, show that $AX = B$ has exactly one solution for every $B$ in $\mathbb{R}^3$.

**10.** Consider the two functions $S, T : \mathbb{R}^3 \to \mathbb{R}^3$ defined by

$$S(X) = \text{Proj}_{(1,2,-1)}(X)$$
$$T(X) = \text{Proj}_{(2,4,-2)}(X)$$

Show that $S$ and $T$ are represented by the same matrix. Explain how this can happen.

**11.** Suppose that $A$ is a matrix, $X$ is in the pullback of $\begin{pmatrix} 5 \\ 1 \\ 3 \end{pmatrix}$, $Y$ is in the pullback of $\begin{pmatrix} 3 \\ 2 \\ 0 \end{pmatrix}$, and $Z$ is in $\ker(A)$. Find the image of

**a.** $2X + 3Y$ **b.** $5Y$ **c.** $X - 2Y$ **d.** $X - 2Y + 3Z$

**12.** Suppose that $A$ is a matrix. If $X$ and $Y$ are both in the pullback of $B$, must $X + Y$ also be in the pullback of $B$? Must $3X$ be in the pullback of $B$? If so, prove it. If not, determine the conditions on $B$ that make it possible.

**13.** Suppose $R = \begin{pmatrix} 3 & 0 \\ 0 & 4 \end{pmatrix}$, $P = \begin{pmatrix} 3 & 4 \\ 2 & 3 \end{pmatrix}$, and $S = P^{-1}RP$.

**a.** Find all vectors $X$ so that $RX = kX$ for some number $k$.
**b.** Find all vectors $X$ so that $SX = kX$ for some number $k$.

**14.** Suppose $R = \begin{pmatrix} 1 & 0 \\ 0 & .5 \end{pmatrix}$, $P = \begin{pmatrix} 3 & 4 \\ 2 & 3 \end{pmatrix}$, and $S = P^{-1}RP$.

**a.** Find all vectors $X$ so that $RX = X$.
**b.** Find all vectors $X$ so that $SX = X$.
**c.** Find all vectors $X$ so that $RX = kX$ for some number $k$.
**d.** Find all vectors $X$ so that $SX = kX$ for some number $k$.
**e.** Calculate $R^{50}$ and $S^{50}$.

$\longleftarrow$
A vector $X$ so that $RX = X$ is a **fixed vector** for $R$.

# Chapter 5 Mathematical Reflections

These problems will help you summarize what you have learned in this chapter.

**1.** For each linear transformation matrix $M$,

(i) show what $M$ does to the unit square

(ii) describe the effect $M$ has in geometric language

**a.** $M = \begin{pmatrix} 4 & 0 \\ 0 & 3 \end{pmatrix}$ **b.** $M = \begin{pmatrix} 0 & -1 \\ 2 & 0 \end{pmatrix}$

**c.** $M = \begin{pmatrix} 1 & 2 \\ 0 & 1 \end{pmatrix}$ **d.** $M = \begin{pmatrix} -3 & 0 \\ 0 & 0 \end{pmatrix}$

**2.** **a.** Find the image of $P = (2, -4)$ under a rotation about $(0, 0)$ through an angle of $30°$.

**b.** The image of $P$ under a rotation about $(0, 0)$ through an angle of $120°$ is $P'$. If $P' = (-4, -2\sqrt{3})$, find $P$.

**3.** Let $A = (2, -4)$, $B = (4, 5)$, and $C = (1, -2)$. Find the area of the parallelogram spanned by

**a.** $A$ and $B$ **b.** $A$ and $C$ **c.** $A$ and $3B$ **d.** $2A$ and $3B$

**e.** $MA$ and $MB$, where $M = \begin{pmatrix} 1 & -1 \\ 1 & 1 \end{pmatrix}$

**4.** Suppose $F : \mathbb{R}^3 \to \mathbb{R}^3$ is defined by $X \overset{F}{\mapsto} \begin{pmatrix} 2 & 1 & -1 \\ 3 & 4 & 2 \\ 5 & 5 & 1 \end{pmatrix} X$.

**a.** Find $\ker(F)$.

**b.** Is $F$ one-to-one? Explain.

**c.** Find $\text{Im}(F)$.

**5.** How can you find a matrix that rotates points $\theta°$ about the origin?

**6.** What is the area of the parallelogram spanned by two given vectors in $\mathbb{R}^2$?

**7.** Let $A = \begin{pmatrix} 1 & 3 & 5 \\ 4 & 6 & 2 \end{pmatrix}$.

**a.** Find the image under $A$ for $\begin{pmatrix} 2 \\ 3 \\ -1 \end{pmatrix}$.

**b.** Find the set of pullbacks under $A$ for $\begin{pmatrix} 7 \\ 16 \end{pmatrix}$.

**Vocabulary**

In this chapter, you saw these terms and symbols for the first time. Make sure you understand what each one means, and how it is used.

- angle of rotation
- center of rotation
- conjugation
- fixed point
- fixed vector
- image of a matrix
- linear map
- linear transformation of $\mathbb{R}^n$
- one-to-one
- preimage
- pullback
- rotation

# Chapter 5 Review

In Lesson 5.2, you learned to

- describe what a $2 \times 2$ and a $3 \times 3$ linear transformation does to a triangle, the unit square, and the unit circle
- use the properties of linear transformations to show that not all transformations are linear
- find the matrices that produce scalings, reflections, and $90^\circ$ rotations

The following exercises will help you check your understanding.

**1.** For each linear transformation matrix, describe the effect it has in geometric language.

**a.** $\begin{pmatrix} 1 & 0 \\ 0 & 0 \end{pmatrix}$ **b.** $\begin{pmatrix} 0 & -1 \\ -1 & 0 \end{pmatrix}$ **c.** $\begin{pmatrix} 1 & 0 \\ 0 & -2 \end{pmatrix}$ **d.** $\begin{pmatrix} 0 & -3 \\ 3 & 0 \end{pmatrix}$

**e.** $\begin{pmatrix} 1 & 0 & 0 \\ 0 & 0 & 0 \\ 0 & 0 & 0 \end{pmatrix}$ **f.** $\begin{pmatrix} 2 & 0 & 0 \\ 0 & 2 & 0 \\ 0 & 0 & 0 \end{pmatrix}$ **g.** $\begin{pmatrix} -1 & 0 & 0 \\ 0 & -1 & 0 \\ 0 & 0 & 0 \end{pmatrix}$

**2.** For each given transformation of $\mathbb{R}^2$, find a corresponding $2 \times 2$ matrix.

**a.** Scale all points by a factor of 10.
**b.** Reflect all points over the graph of $y = -x$.
**c.** Reflect all points over the graph of $y = x$, and then reflect this image over the graph of $y = -x$.
**d.** Rotate all points $180^\circ$ counterclockwise.
**e.** Reflect all points over the $y$-axis, then scale this image by a factor of 2, and then reflect *this* image over the $x$-axis.

**3.** Let $M = \begin{pmatrix} 4 & -2 \\ 2 & 3 \end{pmatrix}$.

**a.** Show what $M$ does to the unit square.
**b.** Theorem 5.1 says that $M$ fixes the origin. Does it fix any other vectors?
**c.** Show what $M$ does to the unit circle.

In Lesson 5.3, you learned to

- find the matrix that defines a given rotation
- find one matrix for a set of successive rotations
- given a point, find its image after rotation
- find a point given its image and angle of rotation
- use conjugation to find the matrix that represents a reflection over a given line or a rotation about a given point

The following exercises will help you check your understanding.

**4.** Find a matrix that defines each rotation.

**a.** $R_{30^\circ}$ **b.** $R_{90^\circ}$ **c.** $R_{135^\circ}$ **d.** $R_{150^\circ}$
**e.** $R_{180^\circ}$ **f.** $R_{225^\circ}$ **g.** $R_{255^\circ}$ **h.** $R_{-60^\circ}$

**5.** Suppose $A = (0, 0)$, $B = (2, -3)$, and $C = (4, 1)$. Find the image of $\triangle ABC$ under a rotation about $(0, 0)$ through an angle of $\theta$ if

**a.** $\theta = 180^\circ$ **b.** $\theta = 45^\circ$ **c.** $\theta = 120^\circ$

**6.** The image of $P$ under a rotation of $45^\circ$ about $(0, 0)$ is $(2, 6)$. Find $P$.

In Lesson 5.4, you learned to

- find the area of a triangle or a parallelogram, given the coordinates of the vertices, using matrices
- find the area for the image of a triangle or a rectangle after a transformation
- understand the relation between area and determinant of $2 \times 2$ matrices

The following exercises will help you check your understanding.

**7.** Suppose $A = (-3, -5)$, $B = (1, -4)$, and $C = (3, -12)$. Find the area of the parallelogram spanned by

**a.** $A$ and $B$ **b.** $A$ and $C$ **c.** $A$ and $-B$ **d.** $3A$ and $3C$

**e.** $MA$ and $MB$, where $M = \begin{pmatrix} 2 & -1 \\ 1 & 2 \end{pmatrix}$

**8.** Suppose $A = (2, 5)$, $B = (6, 3)$, and $C = (4, -1)$. Find the area of $\triangle ABC$.

**9.** Let $A$ and $B$ be vectors in $\mathbb{R}^2$. Suppose that the area of the parallelogram spanned by $A$ and $B$ is 5. For each transformation matrix $M$, find the area of the parallelogram spanned by $MA$ and $MB$.

**a.** $M = \begin{pmatrix} 1 & 0 \\ 0 & -1 \end{pmatrix}$ **b.** $M = \begin{pmatrix} 2 & 0 \\ 0 & 2 \end{pmatrix}$

**c.** $M = \begin{pmatrix} 0 & 3 \\ 2 & 0 \end{pmatrix}$ **d.** $M = \begin{pmatrix} 2 & 4 \\ 3 & -1 \end{pmatrix}$

In Lesson 5.5, you learned to

- identify different notations for a function from $\mathbb{R}^n$ to $\mathbb{R}^m$, including its matrix representation
- find the image of a linear map using matrix multiplication
- find and interpret geometrically the set of pullbacks of a vector under a given matrix

The following exercises will help you check your understanding.

**10.** Let $A = \begin{pmatrix} 4 & 2 & -2 \\ 3 & -1 & 5 \end{pmatrix}$. Find the image under $A$ for each given vector.

**a.** $\begin{pmatrix} 2 \\ 0 \\ -3 \end{pmatrix}$ **b.** $\begin{pmatrix} 4 \\ -1 \\ 7 \end{pmatrix}$ **c.** $\begin{pmatrix} 0 \\ 0 \\ 0 \end{pmatrix}$

**11.** Find the set of pullbacks of $B$ under $A$ if

**a.** $A = \begin{pmatrix} 1 & -2 & -1 \\ -4 & 5 & -2 \end{pmatrix}$ and $B = \begin{pmatrix} -8 \\ -1 \end{pmatrix}$

**b.** $A = \begin{pmatrix} 3 & 1 & 5 \\ 4 & 1 & 4 \\ 2 & 1 & 6 \end{pmatrix}$ and $B = \begin{pmatrix} 4 \\ 2 \\ 6 \end{pmatrix}$

**c.** $A = \begin{pmatrix} 3 & 1 & 5 \\ 4 & 1 & 4 \\ 2 & 1 & 6 \end{pmatrix}$ and $B = \begin{pmatrix} 4 \\ 2 \\ 1 \end{pmatrix}$

**12.** Let $A = \begin{pmatrix} 2 & -1 & -5 \\ 1 & 3 & 1 \end{pmatrix}$, let $X$ be in the pullback of $\begin{pmatrix} 6 \\ -4 \end{pmatrix}$ under $A$, and let $Y$ be in the pullback of $\begin{pmatrix} -1 \\ 3 \end{pmatrix}$ under $A$.

**a.** Find $B$, the image of $2X + 3Y$ under $A$.
**b.** Find the set of pullbacks of $B$ under $A$.

In Lesson 5.6, you learned to

- determine whether a linear function is one-to-one
- find a matrix that represents a projection along a vector or a line
- describe the set of pullbacks of a linear function using formal function language

The following exercises will help you check your understanding.

**13.** How many vectors are in the solution set to $AX = B$ for each system below? Justify your answer.

**a.** $A = \begin{pmatrix} 1 & 1 & 5 \\ 2 & 3 & -2 \end{pmatrix}$, $B = \begin{pmatrix} 7 \\ 3 \end{pmatrix}$

**b.** $A = \begin{pmatrix} 1 & 1 & 5 \\ 2 & 3 & -2 \\ 3 & 4 & -4 \end{pmatrix}$, $B = \begin{pmatrix} 7 \\ 3 \\ 3 \end{pmatrix}$

**c.** $A = \begin{pmatrix} 1 & 1 & 5 \\ 2 & 3 & -2 \\ 3 & 4 & 3 \end{pmatrix}$, $B = \begin{pmatrix} 7 \\ 3 \\ 10 \end{pmatrix}$

**d.** $A = \begin{pmatrix} 1 & 1 & 5 \\ 2 & 3 & -2 \\ 3 & 4 & 3 \end{pmatrix}$, $B = \begin{pmatrix} 7 \\ 3 \\ -4 \end{pmatrix}$

**14.** Suppose $F : \mathbb{R}^3 \to \mathbb{R}^3$ is defined by $X \overset{F}{\mapsto} AX$. For each given $A$ and $B$,

(i) What is $\text{Im}(F)$?
(ii) Is $F$ one-to-one?
(iii) What is the pullback of vector $B$?

**a.** $A = \begin{pmatrix} 1 & 1 & -1 \\ 2 & 3 & -5 \\ -3 & -5 & 8 \end{pmatrix}$, $B = \begin{pmatrix} 2 \\ 2 \\ 7 \end{pmatrix}$

**b.** $A = \begin{pmatrix} 1 & 1 & -1 \\ 2 & 3 & -5 \\ -3 & -5 & 9 \end{pmatrix}$, $B = \begin{pmatrix} 2 \\ 2 \\ -2 \end{pmatrix}$

**15.** Suppose $F : \mathbb{R}^2 \to \mathbb{R}^2$ is defined by $F(X) = \text{Proj}_{(1,2)} X$.

**a.** Find a $2 \times 2$ matrix to represent $F$.
**b.** What is $\text{Im}F$?
**c.** Is $F$ one-to-one?
**d.** What is the pullback of $(2, 4)$?

# Chapter 5 Test

## Multiple Choice

**1.** Which linear transformation matrix corresponds to a reflection of all points over the graph of $y = x$?

**A.** $\begin{pmatrix} -1 & 0 \\ 0 & -1 \end{pmatrix}$ **B.** $\begin{pmatrix} 0 & -1 \\ -1 & 0 \end{pmatrix}$ **C.** $\begin{pmatrix} 0 & 1 \\ 1 & 0 \end{pmatrix}$ **D.** $\begin{pmatrix} 1 & 0 \\ 0 & -1 \end{pmatrix}$

**2.** Which point is the result when the matrix $M = \begin{pmatrix} 0 & -1 \\ 1 & 0 \end{pmatrix}$ is applied to the point $X = \begin{pmatrix} x \\ y \end{pmatrix}$?

**A.** $\begin{pmatrix} -x \\ y \end{pmatrix}$ **B.** $\begin{pmatrix} x \\ -y \end{pmatrix}$ **C.** $\begin{pmatrix} -y \\ x \end{pmatrix}$ **D.** $\begin{pmatrix} y \\ -x \end{pmatrix}$

**3.** Let $M = \begin{pmatrix} 0 & 1 \\ -1 & 0 \end{pmatrix}$. Which rotation about $(0, 0)$ is defined by matrix $M$?

**A.** $R_{45^\circ}$ **B.** $R_{90^\circ}$ **C.** $R_{180^\circ}$ **D.** $R_{270^\circ}$

**4.** Let $A = (-3, 4)$ and $B = (2, 7)$. What is the area of $\triangle AOB$?

**A.** 6.5 **B.** 13 **C.** 14.5 **D.** 29

**5.** If $X$ is in the pullback of $\begin{pmatrix} -5 \\ 11 \end{pmatrix}$ under $A$ and $Y$ is in the pullback of $\begin{pmatrix} -9 \\ 12 \end{pmatrix}$ under $A$, what is the image of $2X + Y$ under $A$?

**A.** $\begin{pmatrix} -28 \\ 46 \end{pmatrix}$ **B.** $\begin{pmatrix} -19 \\ 34 \end{pmatrix}$ **C.** $\begin{pmatrix} 1 \\ -6 \end{pmatrix}$ **D.** $\begin{pmatrix} 12 \\ 6 \end{pmatrix}$

**6.** Let $A = \begin{pmatrix} 2 & 3 & 1 \\ 3 & 5 & 3 \\ 5 & 7 & 1 \end{pmatrix}$ and $B = \begin{pmatrix} 2 \\ 2 \\ 6 \end{pmatrix}$. What is the solution to $AX = B$?

**A.** there is no solution
**B.** $X = O$
**C.** $X = (4, -3, 1)$
**D.** $X = (0, 1, -1) + t(4, -3, 1)$

## Open Response

**7.** Let $M = \begin{pmatrix} 16 & 3 \\ 5 & 2 \end{pmatrix}$.

**a.** Show what $M$ does to the unit square.
**b.** Theorem 5.1 says that $M$ fixes the origin. Does it fix any other vectors?
**c.** Show what $M$ does to the unit circle.

**8.** Let $P = (-1, 2)$.

**a.** Find the image of $P$ under a rotation about the origin through an angle of $120°$.

**b.** The image of $Q = (x, y)$ under a rotation about the origin through an angle of $30°$ is $P$. Find $Q$.

**9.** Suppose $A = (4, 1)$ and $B = (3, -2)$. Find the area of

**a.** the parallelogram spanned by $A$ and $B$

**b.** the parallelogram spanned by $5A$ and $5B$

**c.** the parallelogram spanned by $MA$ and $MB$, where $M = \begin{pmatrix} 2 & 1 \\ 4 & -2 \end{pmatrix}$

**10.** Let $A = \begin{pmatrix} 1 & 3 & 2 \\ 5 & 4 & -1 \end{pmatrix}$.

**a.** If $X = \begin{pmatrix} 1 \\ 2 \\ -5 \end{pmatrix}$, find the image under $A$ of $X$.

**b.** If $B = \begin{pmatrix} 7 \\ 2 \end{pmatrix}$, find the set of pullbacks of $B$ under $A$.

**11.** Suppose $F : \mathbb{R}^2 \to \mathbb{R}^3$ is defined by $X \overset{F}{\mapsto} \begin{pmatrix} 2 & 1 \\ 3 & -4 \\ 4 & 1 \end{pmatrix} X$.

**a.** Find $\ker(F)$.

**b.** Is $F$ one-to-one? Explain.

**c.** Find $\text{Im}(F)$.

**12.** Suppose $F : \mathbb{R}^2 \to \mathbb{R}^2$ is defined by $F(X) = \text{Proj}_{(3,1)} X$.

**a.** What is $\text{Im}F$?

**b.** Is $F$ one-to-one?

**c.** What is the pullback of $(6, 2)$?

# Cumulative Review

For problems 1–8, suppose $A = \begin{pmatrix} 2 & 1 & 1 \\ 1 & 1 & 0 \\ 2 & 0 & 3 \end{pmatrix}$ and $B = \begin{pmatrix} 2 & 3 & 1 \\ 2 & 3 & 0 \\ 1 & 2 & 1 \end{pmatrix}$. Find:

**Remember**

$B^2$ means $BB$.

**1.** $A^{-1}$ **2.** $B^{-1}$ **3.** $AB$ **4.** $BA$

**5.** $A^\top B^\top$ **6.** $(AB)^{-1}$ **7.** $2A - B$ **8.** $AB^2A$

For problems 9–11, consider the function from $\mathbb{R}^3$ to $\mathbb{R}^2$ defined by the matrix $A = \begin{pmatrix} 1 & 2 & -1 \\ 3 & 5 & 1 \end{pmatrix}$.

**9.** Find the image of $\begin{pmatrix} 2 \\ -1 \\ 3 \end{pmatrix}$.

**10.** Find the pullback of $\begin{pmatrix} -1 \\ 2 \end{pmatrix}$.

**11.** If $X$ is in the pullback of $\begin{pmatrix} 3 \\ 1 \end{pmatrix}$ and if $X'$ is in the pullback of $\begin{pmatrix} 2 \\ 4 \end{pmatrix}$, what is the image of $X + 3X'$.

**12.** Let $A = \begin{pmatrix} 1 & 2 & 3 \\ 4 & 5 & 6 \\ 7 & 8 & 0 \end{pmatrix}$. Find the solution to $AX = \begin{pmatrix} 6 \\ 15 \\ 15 \end{pmatrix}$.

$\longleftarrow$
This solution is unique.

For problems 13–15, classify the solution set of each system as one of the following: a) one solution, b) no solutions, and c) infinitely many solutions. Write a long explanation for each of your answers.

**13.** $\begin{pmatrix} 1 & 2 & 3 \\ 4 & 5 & 6 \\ 7 & 8 & 9 \end{pmatrix} \begin{pmatrix} x \\ y \\ z \end{pmatrix} = \begin{pmatrix} 5 \\ 11 \\ 17 \end{pmatrix}$

**14.** $\begin{pmatrix} 1 & 3 & 1 & 2 \\ 4 & 1 & -7 & -3 \\ 2 & 4 & 0 & 2 \end{pmatrix} \begin{pmatrix} x \\ y \\ z \\ w \end{pmatrix} = \begin{pmatrix} 7 \\ 6 \\ 12 \end{pmatrix}$

**15.** $\begin{pmatrix} 2 & 1 \\ 3 & 4 \\ 7 & 2 \end{pmatrix} \begin{pmatrix} x \\ y \end{pmatrix} = \begin{pmatrix} 4 \\ 11 \\ 11 \end{pmatrix}$

For problems 16–25, let $A = (4,1,2)$, $B = (3,4,-12)$, $C = (1,2,0)$, let $M$ be the midpoint of $\overrightarrow{AC}$, let $t = 2$, and let $\theta$ be the angle between $A$ and $2B$. Find

**16.** $A + 2B$ **17.** $(A - B) \cdot C$ **18.** $(tA - tB) \cdot C$

**19.** $d(A, C)$ **20.** $d(A, M)$ **21.** $\|\text{Proj}_B\, A\|$

**22.** $\text{Proj}_B\, 2A$ **23.** $A \times C$ **24.** $A \times tC$ **25.** $\cos\theta$

**26.** Find a nonzero vector orthogonal to $(1,4,7)$ and $(2,9,3)$.

**27.** Find a nonzero vector orthogonal to $(1,1,-1)$, $(-1,-2,4)$, and $(1,0,2)$.

**28.** Write $(0,4,-4)$ as a linear combination of $(1,0,2)$ and $(4,1,7)$.

**29.** If $A = (3,-8,5)$, $B = (8,4,5)$, and $C = (6,-4,-7)$, find the area of $\triangle ABC$.

**30.** If $A = (3,1,2)$, $B = (4,1,6)$, $C = (2,0,7)$, and $D = (4,y,z)$, find $y$ and $z$ if $\overrightarrow{AB} \parallel \overrightarrow{CD}$.

**31.** If $A = (5,8,10)$, $B = (3,4,7)$, and $C = (4,5,6)$, show that $\triangle ABC$ is a right triangle. Is $\triangle ABC$ isosceles?

**32.** $A$ and $B$ are vectors in $\mathbb{R}^n$. If $A$ is orthogonal to $B$, show that $A$ is orthogonal to any scalar multiple of $B$.

**33.** $A$ and $B$ are vectors in $\mathbb{R}^3$. Show that $A \times 2B = 2(A \times B)$.

# Cumulative Test

## Multiple Choice

**1.** Suppose $A = (2, -1)$ and $B = (-3, 4)$. If $\overrightarrow{AB}$ is equivalent to $\overrightarrow{OQ}$, which are the coordinates of $Q$?

A. $(-5, 5)$ B. $(-1, 3)$ C. $(1, -3)$ D. $(5, -5)$

**2.** A vector equation for line $\ell$ is $X = (3, -1) + k(2, -5)$. What is its coordinate equation?

A. $2x + 5y = -1$
B. $2x + 5y = 1$
C. $5x + 2y = -13$
D. $5x + 2y = 13$

**3.** Let $A = (0, -4, 5)$ and $B = (2, -1, 3)$. Which is $\|B - A\|$?

A. $\sqrt{3}$ B. $3$ C. $\sqrt{17}$ D. $17$

**4.** In $\mathbb{R}^3$, $A = (1, -2, 3)$ and $B = (1, 2, -1)$. Which is $\text{Proj}_B\, A$?

A. $\left(-\frac{3}{7}, -\frac{6}{7}, \frac{3}{7}\right)$
B. $\left(-\frac{3}{7}, \frac{6}{7}, -\frac{9}{7}\right)$
C. $(-1, -2, 1)$
D. $(-1, 2, -3)$

**5.** If $\theta$ is the angle between $A = (1, -1)$ and $B = (1, 3)$, what is $\cos\theta$?

A. $-\frac{\sqrt{10}}{10}$ B. $-\frac{2\sqrt{5}}{5}$ C. $-\frac{\sqrt{5}}{5}$ D. $-\frac{\sqrt{2}}{2}$

**6.** Line $\ell$ contains $(-2, 5)$ and is parallel to $(3, 4)$. Which point is on $\ell$?

A. $(-8, -3)$ B. $(-1, 14)$ C. $(5, -1)$ D. $(9, 1)$

**7.** Which equation describes all vectors that are orthogonal to $(1, -2, -3)$, $(2, 1, -1)$, and $(3, 4, 1)$?

A. $X = (0, 0, 0)$
B. $X = (1, -1, 1)$
C. $X = t(-1, 1, 0)$
D. $X = t(1, -1, 1)$

**8.** In $\mathbb{R}^3$, $A = (3, -1, 1)$, $B = (1, -2, 0)$, and $C = (2, 1, 1)$. Which of the following describes all linear combinations of $A$, $B$, and $C$?

A. the point $O$
B. the line $X = t(3, -1, 1)$
C. the plane $X = s(3, -1, 1) + t(1, -2, 0)$
D. all of $\mathbb{R}^3$

**9.** Let $A = \begin{pmatrix} 1 & -1 & 2 \\ 4 & -7 & 5 \\ 2 & -3 & 3 \end{pmatrix}$ and $B = \begin{pmatrix} 3 & -2 & 0 \\ 1 & 4 & -1 \\ 5 & -3 & 2 \end{pmatrix}$. What is the value of $A_{2*} \cdot B_{*1}$?

**A.** $-12$ **B.** 12 **C.** 26 **D.** 30

**10.** Suppose $A = \begin{pmatrix} 3 & -2 \\ -1 & 0 \end{pmatrix}$ and $B = \begin{pmatrix} 4 & -2 \\ 3 & -3 \end{pmatrix}$. Which of the following is equal to $(AB)^{\top}$?

**A.** $\begin{pmatrix} 6 & 0 \\ -4 & 2 \end{pmatrix}$

**B.** $\begin{pmatrix} 6 & -4 \\ 0 & 2 \end{pmatrix}$

**C.** $\begin{pmatrix} 14 & -8 \\ 12 & -6 \end{pmatrix}$

**D.** $\begin{pmatrix} 14 & 12 \\ -8 & -6 \end{pmatrix}$

**11.** $M = \begin{pmatrix} 0 & 1 \\ -1 & 0 \end{pmatrix}$ is a transformation matrix. What effect does $M$ have on point $P = \begin{pmatrix} x \\ y \end{pmatrix}$?

**A.** $M$ reflects $P$ over the $x$-axis.
**B.** $M$ reflects $P$ over the $y$-axis.
**C.** $M$ rotates $P$ through an angle of $90°$ counterclockwise.
**D.** $M$ rotates $P$ through an angle of $270°$ counterclockwise.

**12.** Let $A = \begin{pmatrix} -3 & 1 & 2 \\ -2 & 4 & -2 \\ 1 & 0 & -5 \end{pmatrix}$. What is the image under $A$ of $\begin{pmatrix} -2 \\ -8 \\ -8 \end{pmatrix}$?

**A.** $\begin{pmatrix} -2 \\ 0 \\ -2 \end{pmatrix}$ **B.** $\begin{pmatrix} 2 \\ 0 \\ 2 \end{pmatrix}$ **C.** $\begin{pmatrix} -18 \\ -12 \\ 38 \end{pmatrix}$ **D.** $\begin{pmatrix} 18 \\ 12 \\ 38 \end{pmatrix}$

## Open Response

**13.** In $\mathbb{R}^3$, let $A = (2, -3, 4)$ and $B = (-5, 5, 0)$.

**a.** Show that $(23, -27, 16)$ is a linear combination of $A$ and $B$.
**b.** Find a vector equation and a coordinate equation for the plane spanned by $A$ and $B$.

**14.** In $\mathbb{R}^3$, let $A = (0, 2, 1)$, $B = (6, 5, 3)$, and $C = (3, -1, 1)$. Show that $\triangle ABC$ is isosceles.

**15.** Characterize all vectors $X$ in $\mathbb{R}^3$ orthogonal to $A = (2, -1, 1)$ and $B = (4, 1, -3)$.

**16.** Suppose $A = (2, -3, 5)$ and $B = (1, -2, 4)$.

**a.** Find a nonzero vector orthogonal to both $A$ and $B$.

**b.** Find the area of the parallelogram with vertices $O$, $A$, $B$, and $A + B$.

**17.** Consider the system of equations below:

$$\begin{cases} x + 2y - 3z &= 4 \\ 3x + 4y - z &= -2 \end{cases}$$

**a.** Represent the system as an augmented matrix.

**b.** Reduce the augmented matrix to echelon form.

**c.** Find the solution to the system.

**d.** Does the graph of the solution set represent a point, a line, or a plane?

**18.** Given

$$A = \begin{pmatrix} 2 & -3 & 1 \\ 0 & -4 & 4 \\ 1 & -5 & 4 \end{pmatrix}$$

**a.** Find $\ker(A)$.

**b.** Are the columns of $A$ linearly dependent or linearly independent? Explain.

**19.** Let $A = \begin{pmatrix} 1 & 0 & 2 \\ 3 & 2 & -3 \\ -1 & -1 & 3 \end{pmatrix}$ and $B = \begin{pmatrix} -2 \\ 10 \\ -7 \end{pmatrix}$

**a.** Determine $A^{-1}$.

**b.** Solve the equation $AX = B$ by multiplying both sides of the equation by $A^{-1}$.

**20.** Consider the system

$$\begin{cases} x - 3y + z &= 0 \\ -2x + 5y - 5z &= -4 \\ x - 2y + 4z &= 4 \end{cases}$$

**a.** Rewrite the system in the form $AX = B$.

**b.** Find a solution to the system by writing $B$ as a linear combination of the columns of $A$.

**c.** Find all solutions to the system by finding the kernel.

**21.** Find the image of $P = (-4, 6)$ under a rotation of $30°$ about $(0, 0)$.

**22.** Find the set of pullbacks of $B = \begin{pmatrix} -12 \\ -5 \end{pmatrix}$ under $A = \begin{pmatrix} 1 & 2 & -5 \\ 2 & 3 & -2 \end{pmatrix}$.

CHAPTER

# 6 Markov Chains

In previous classes, you likely calculated the probability of a coin flip landing on heads or tails, or the results of rolling one or more number cubes. You can model some systems by identifying the various states the system might take. For instance, weather has states such as sunny, partly cloudy, mostly cloudy, overcast, rainy, snowy, and many more. And a number of factors go into determining what the weather is on any given day.

You can model such a system by determining the probability of what the weather might be tomorrow given what the weather is today. If you do that for all the various states, you can start to build your model to predict what the weather might be tomorrow, or the next day, or next week, based on these probabilities. (Of course, in this example, the further out you go, the less accurate your prediction is likely to be, but it's a start!)

In this chapter, you will look at how you can use matrices and their algebra that you learned in previous chapters to calculate probabilities for large systems.

**By the end of this chapter, you will be able to answer questions like these:**

1. How can I represent a system using a transition matrix?
2. What is the difference between an absorbing state and a transient state?
3. What is the average number of turns it would take to win the game *HiHo! Cherry-O*?

**You will build good habits and skills for ways to**

- use a model
- reason about calculations
- use precise language
- use linearity
- use properties
- seek structural similarity
- use forms with purpose

**Vocabulary and Notation**

- absorbing Markov chain
- absorbing state
- attractor
- Markov chain
- node
- probability vector
- random process
- steady state
- submatrix
- transient state
- transition graph
- transition matrix
- transition probability

## 6.1 Getting Started

Here are the rules for a short game called *Count to Five.* You start with zero points, and the goal is to score at least five points in the fewest turns. On each turn, roll a number cube and record the result.

| If you roll . . . | Then you . . . |
|---|---|
| 1 | add one point |
| 2 | add two points |
| 3 | add three points |
| 4 | subtract one point (but never drop below zero) |
| 5 | reset to zero |
| 6 | roll again immediately—this does not count as a turn |

The game ends when you reach at least five points.

← If you roll 3, 4, 2, 6, 2, you complete the game in four turns, not five.

**Remember**

A roll of 6 does not count as a turn.

**1.** **a.** Play the game three times, and for each time, write down the number of turns it took you to get at least five points.
**b.** Compute the average number of turns taken by your entire class.

**2.** Suppose you have two points while playing *Count to Five.*
**a.** Explain why the probability of having exactly four points next turn is $\frac{1}{5}$.
**b.** What is the probability of having zero points next turn? one? two? three? four? five?
**c.** What is the probability of having exactly three points *two* turns from now?
**d.** What is the probability of having zero points *two* turns from now? one? two? three? four? five or more?

← What is the sum of these six values?

**3.** Complete the following 6-by-6 table. Each column heading is the number of points you start a turn with, and each row heading is the number of points you end the same turn with. The entry in row $i$ and column $j$ is the probability of ending a turn with $i$ points when starting that turn with $j$ points.

| | **0** | **1** | **2** | **3** | **4** | **5** |
|---|---|---|---|---|---|---|
| **0** | .4 | | .2 | | | 0 |
| **1** | | | .2 | | | 0 |
| **2** | | | 0 | | | 0 |
| **3** | | | .2 | | | 0 |
| **4** | | | .2 | | | 0 |
| **5** | | | .2 | | | 1 |

**4.** Let $R$ be the 6-by-6 matrix of numbers found in Problem 3 above. Compute $R^2$ and compare the data to the numbers found in Problem **2d** above.

**5.** Suppose that if you're healthy today, there is a 90% chance you will be healthy again tomorrow (and a 10% chance of being sick). But if you're sick today, there is only a 30% chance you will be healthy tomorrow (and a 70% chance of being sick again).

**a.** You're healthy today. Find the probability that you'll be healthy two days from now.

**b.** Jeff is sick today. Assuming the same information, find the probability that Jeff will be healthy two days from now.

**c.** Find the probability that you'll be healthy three days from now.

**d.** **Take It Further.** Find the probability that you'll be healthy 10 days from now, and the probability that Jeff will be healthy 10 days from now.

**6.** Calculate each of the following.

**a.** $\begin{pmatrix} 0.9 & 0.3 \\ 0.1 & 0.7 \end{pmatrix}^2$ **b.** $\begin{pmatrix} 0.9 & 0.3 \\ 0.1 & 0.7 \end{pmatrix}^3$ **c.** $\begin{pmatrix} 0.9 & 0.3 \\ 0.1 & 0.7 \end{pmatrix}^{10}$

Roger and Serena are playing a game of tennis. The first player to win four points wins the game, but a player must win by two. Roger and Serena have each won three points, so neither of them can win on the next point.

⟵ In tennis, this situation is called *deuce*.

⟵ Serena is serving. In general, the server has an advantage in tennis.

**7.** Suppose Serena is 60% likely to win any point, and the score is tied.

**a.** Find the probability that Serena wins the game on the next two points.

**b.** Find the probability that Roger wins the game on the next two points.

**c.** Find the probability that after the next two points, the game is tied again.

There are five possible situations that can come up during Roger and Serena's game, once they are tied with three points each. Each situation can be labeled:

- (**T**): the game is tied
- (**S1**): Serena, the server, is ahead by one
- (**R1**): Roger, the receiver, is ahead by one
- (**SW**): Serena wins by going ahead by two
- (**RW**): Roger wins by going ahead by two

**8.** Draw a graph depicting what can happen in the game of problem 7. Include the five labels T, S1, R1, SW, and RW (in any order you prefer).

**9.** **Take It Further.** Suppose the probability that Serena wins a point is $p$, and the players are tied with three points each. Find the probability that Serena wins the game, in terms of $p$.

**10.** **a.** Suppose

$$M = \begin{pmatrix} 0.5 & 0.4 & 0.5 \\ 0.2 & 0.3 & 0.5 \\ 0.3 & 0.3 & 0 \end{pmatrix}$$

Show that $M - I$ has a nonzero kernel, where $I$ is the $3 \times 3$ identity matrix.

**b.** **Take It Further.** Suppose $M$ is any $n \times n$ matrix with the property that the sum of the entries in any column is 1. Show that $M - I$ has a nonzero kernel, where $I$ is the $n \times n$ identity matrix.

## 6.2 Random Processes

In the Getting Started lesson, you used matrices to keep track of various probabilities. In this lesson, you will formalize the properties of these matrices and see how they are useful for making predictions of outcomes over longer periods of time.

**In this lesson, you will learn how to**

- identify a random process and its various states
- model a random process using probability
- follow a Markov chain model across a number of states

Suppose that when you wake up every day, you decide whether you are healthy or sick. You make a list day by day of your status. After four weeks you might have a list like this one:

> healthy, healthy, sick, healthy, healthy, healthy, healthy, healthy, sick, sick, sick, healthy, healthy, healthy, healthy, healthy, healthy, healthy, sick, sick, sick, sick, healthy, healthy, healthy, healthy, healthy, healthy . . .

This is an example of a **random process**, which consists of a set of *states* and a sequence of *observations* of those states. In this case, the states are *healthy* and *sick* (H and S for short) and the observations are your day-to-day status.

**Minds in Action** Episode 22

DERMAN: So, being healthy or sick is random, like flipping a coin?

SASHA: A little, I guess. If you're healthy one day, you might be sick the next day.

DERMAN: Maybe it's like flipping an unfair coin. I wouldn't want to be sick half the time.

SASHA: There may be more to it, also. It seems like if you're already sick one day, you're more likely to *stay* sick the next day. And if you're healthy, you'll probably stay healthy.

DERMAN: I see what you mean. You could be healthy for weeks, then sick for five days in a row. Maybe you flip one coin if you're healthy, and a different coin if you're sick.

SASHA: Why does it always have to be about coins?

### Example

Jennie is an excellent fast-pitch softball pitcher. She has three pitches: a fastball, a changeup, and a drop curve. Jennie mixes up her pitches to try to fool hitters. Here's a sample of 24 pitches she threw during a game, in the order she threw them:

F, F, C, D, F, F, F, D, C, F, F, C,
D, F, C, F, D, F, F, D, C, F, D, F

← Don't worry if you can't tell a fastball and a drop curve apart! In the list of pitches, F is a fastball, C is a changeup, and D is a drop curve.

This can be thought of as a random process with three states and the sequence of observations is the sequence of pitches.

**Problem.** Jennie just threw a fastball. About how likely do you think it is that the next pitch will be a fastball? a changeup? a drop curve?

**Solution.** There is no way to know the exact probabilities, but an approximation can be found by looking at the 24 pitches thrown. Thirteen of the 24 pitches are fastballs. Of these, 12 are followed by other pitches:

$$\begin{array}{ll} F \to F: & 5 \\ F \to C: & 3 \\ F \to D: & 4 \end{array}$$

According to the data, the probability that the next pitch will be a fastball is *about* $\frac{5}{12}$, while the probability that it is a changeup is *about* $\frac{3}{12}$ and the probability that it is a drop curve is *about* $\frac{4}{12}$.

A larger sample of 1000 pitches would give more confident approximate probabilities, but even those can't be considered exact.

### For You to Do

1. After Jennie throws a drop curve, about how likely is it that the next pitch will be a fastball? a changeup? another drop curve?

Random processes can be classified as *continuous* or *discrete*. The temperature at a specific location is an example of a continuous random process, since it changes as a function of a continuously changing variable—time. But, if you measured the temperature every hour on the hour, those observations would form a discrete random process.

← Think of a continuous random process as a continuous function whose domain is $\mathbb{R}$. A discrete random process is like a function whose domain is the nonnegative integers.

In some random processes, the probability of moving between states can also depend on external factors, such as the time of day or time of year when measuring temperature. But in certain random processes, the probability of going from one state to the next depends only on their current state: the probability of having four points at the end of a turn of *Count to Five* depends only on the current number of points. With a finite number of states, such a random process is called a **Markov chain**.

**Definition**

A **Markov chain** is a discrete random process with a finite number of states having the property that the probability of entering a state for the next observation depends only on the current state.

**Developing Habits of Mind**

**Use a model.** Markov chains are often used to model processes. Jennie doesn't throw pitches at random, but a Markov chain can analyze her overall pitching patterns. In a Markov chain, you would assume knowledge of the probabilities described in the example above. Then, you could use the Markov chain to generate new sequences of pitches, or to predict the overall percentage of fastballs.

Suppose that each day for two years, you kept track of whether you were healthy or sick each day. You might observe that if you were healthy on a given day, you were healthy the next day 90% of the time; but if you were sick that day, you were healthy the next day only 30% of the time. These observations could be used as the basis for a Markov chain that could then predict what percentage of days you will be healthy in the future.

**For Discussion**

**2.** Suppose you are healthy today.

**a.** Using the percentages given above, follow the random process to determine whether you are sick or healthy for the next 28 days.

**b.** With the data from your class, determine the overall approximate percentage of days that you will be healthy under this random process.

## Exercises

**1.** Calculate each of the following:

**a.** $\begin{pmatrix} 0.9 & 0.3 \\ 0.1 & 0.7 \end{pmatrix}^{15}$ **b.** $\begin{pmatrix} 0.9 & 0.3 \\ 0.1 & 0.7 \end{pmatrix}^{20}$ **c.** $\begin{pmatrix} 0.9 & 0.3 \\ 0.1 & 0.7 \end{pmatrix}^{28}$

← What does Exercise 1 have to do with this lesson?

**2.** Consider the matrix $D = \begin{pmatrix} 1 & 0 \\ 0 & 0.6 \end{pmatrix}$.

**a.** Compute $D^2$ without a calculator.

**b.** As $n$ grows larger, describe what happens to $D^n$.

**3.** Suppose $p_1$ is the probability you are healthy three days from now, and $p_2$ is the probability you are sick three days from now. Given the information from Getting Started, calculate, in terms of $p_1$ and $p_2$,

**a.** the probability of being healthy four days from now

**b.** the probability of being sick four days from now

**4.** Compute this matrix product:

$$\begin{pmatrix} 0.9 & 0.3 \\ 0.1 & 0.7 \end{pmatrix} \begin{pmatrix} p_1 \\ p_2 \end{pmatrix}$$

**5.** Let $A = \begin{pmatrix} 0.9 & 0.3 \\ 0.1 & 0.7 \end{pmatrix}$ and $P = \begin{pmatrix} 3 & 3 \\ -1 & 3 \end{pmatrix}$.

**a.** Calculate $P^{-1}$.
**b.** Calculate $PAP^{-1}$.
**c.** Calculate $(PAP^{-1})(PAP^{-1})$.
**d.** Calculate $PA^2P^{-1}$.
**e.** As $n$ grows larger, describe what happens to $\left(PAP^{-1}\right)^n$.

**6.** Give some examples of processes that are Markov chains, and others that are not.

**7.** Derman has a game he plays with a penny and a nickel. He flips a coin and then uses the result to determine which coin to flip next. If it's heads, the next flip is the penny; if it's tails, the next flip is the nickel. But there's a secret: the penny is unfairly weighted to flip heads 75% of the time.

← You can model Derman's unfair penny by flipping two coins at once. If they don't *both come up tails*, then that counts as a "head" for the unfair penny.

Derman flips the penny first . . .

**a.** If you play this game for 25 flips, would you expect more heads or more tails? Is it guaranteed?
**b.** Play the game through 25 flips and count the number of heads.
**c.** Collect the results from a large group to estimate the probability that any one flip comes heads in this game.
**d.** **Take It Further.** As the game continues, determine the long-term probability that any one flip comes heads.

← The answer should be between 0.5 and 0.75 . . . why?

**8.** Calculate to four decimal places each of the following:

**a.** $\begin{pmatrix} 0.75 & 0.5 \\ 0.25 & 0.5 \end{pmatrix}^2$ **b.** $\begin{pmatrix} 0.75 & 0.5 \\ 0.25 & 0.5 \end{pmatrix}^3$

**c.** $\begin{pmatrix} 0.75 & 0.5 \\ 0.25 & 0.5 \end{pmatrix}^4$ **d.** $\begin{pmatrix} 0.75 & 0.5 \\ 0.25 & 0.5 \end{pmatrix}^{10}$

**9.** A local restaurant awards free lunches once a month in a drawing. Each month, 10% of the people who are eligible win a free lunch. Once you win the lunch, you can't win it again. Andrea enters the drawing every month, hoping for a free lunch.

**a.** To four decimal places, find the probability that Andrea has *not* won a free lunch in the first two months.
**b.** . . . in the first three months.
**c.** . . . in the first five months.
**d.** Andrea says that since there is a 10% chance of winning each month, she's sure to win sometime within the next 10 months. Is this accurate? Why or why not?

**10.** Calculate to four decimal places each of the following:

**a.** $\begin{pmatrix} 0.9 & 0 \\ 0.1 & 1 \end{pmatrix}^2$ **b.** $\begin{pmatrix} 0.9 & 0 \\ 0.1 & 1 \end{pmatrix}^3$

**c.** $\begin{pmatrix} 0.9 & 0 \\ 0.1 & 1 \end{pmatrix}^4$ **d.** $\begin{pmatrix} 0.9 & 0 \\ 0.1 & 1 \end{pmatrix}^{10}$

**11.** A website has a home page with links to two other pages. From each page, you can click through to the next page, or click to return to the home page.

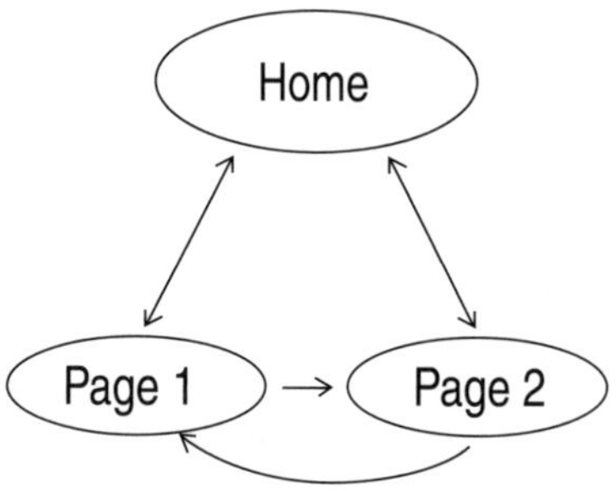


Suppose you are at the home page now, and begin randomly clicking on links.

**a.** Complete this table with the probability that you are at the home page after $n$ clicks.

| # Clicks | $P$(home page) |
|---|---|
| 0 | 1 |
| 1 | |
| 2 | |
| 3 | |
| 4 | |
| 5 | |

**b.** Explain why, in the long run, the probability of being at the home page tends toward $\frac{1}{3}$.

**12.** Here is a $3 \times 3$ matrix:

$$A = \begin{pmatrix} 0 & \frac{1}{2} & \frac{1}{2} \\ \frac{1}{2} & 0 & \frac{1}{2} \\ \frac{1}{2} & \frac{1}{2} & 0 \end{pmatrix}$$

Find a formula for $A^n$ in terms of $n$. Calculating the exact value of $A^n$ for small powers of $n$ should help.

← Take advantage of the massive symmetry of $A$ . . .

**13.** Here is a $3 \times 3$ matrix:

$$A = \begin{pmatrix} 0 & x & x \\ x & 0 & x \\ x & x & 0 \end{pmatrix}$$

Find a formula for $A^n$ in terms of $x$ and $n$.

**14.** Repeat Exercise 11 with a website that has a home page with links to three other pages. From each page, you can click through to the next page, or click to return to the home page.

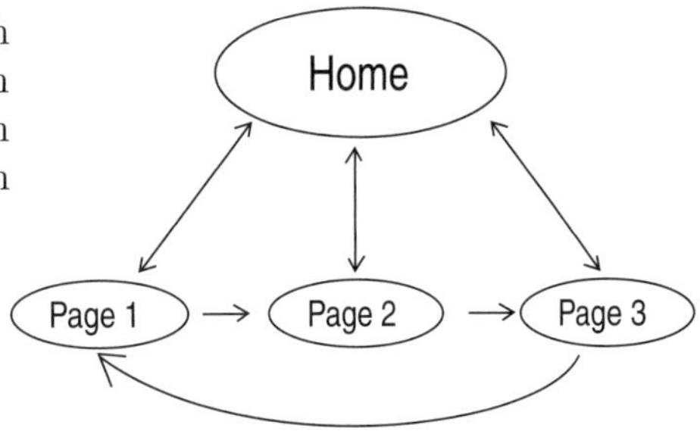


**15.** Calculate to four decimal places each of the following:

**a.** $\begin{pmatrix} 0 & \frac{1}{2} & \frac{1}{2} & \frac{1}{2} \\ \frac{1}{3} & 0 & 0 & \frac{1}{2} \\ \frac{1}{3} & \frac{1}{2} & 0 & 0 \\ \frac{1}{3} & 0 & \frac{1}{2} & 0 \end{pmatrix}^2$ **b.** $\begin{pmatrix} 0 & \frac{1}{2} & \frac{1}{2} & \frac{1}{2} \\ \frac{1}{3} & 0 & 0 & \frac{1}{2} \\ \frac{1}{3} & \frac{1}{2} & 0 & 0 \\ \frac{1}{3} & 0 & \frac{1}{2} & 0 \end{pmatrix}^3$

**c.** $\begin{pmatrix} 0 & \frac{1}{2} & \frac{1}{2} & \frac{1}{2} \\ \frac{1}{3} & 0 & 0 & \frac{1}{2} \\ \frac{1}{3} & \frac{1}{2} & 0 & 0 \\ \frac{1}{3} & 0 & \frac{1}{2} & 0 \end{pmatrix}^4$ **d.** $\begin{pmatrix} 0 & \frac{1}{2} & \frac{1}{2} & \frac{1}{2} \\ \frac{1}{3} & 0 & 0 & \frac{1}{2} \\ \frac{1}{3} & \frac{1}{2} & 0 & 0 \\ \frac{1}{3} & 0 & \frac{1}{2} & 0 \end{pmatrix}^{10}$

⟵ What about a formula for $\begin{pmatrix} 0 & x & x & x \\ y & 0 & 0 & x \\ y & x & 0 & 0 \\ y & 0 & x & 0 \end{pmatrix}^n$ ?

**16.** Repeat Exercises 11 and 15 with a website that has a home page with links to four other pages. From each page, you can click through to the next page, or click to return to the home page. You will need to construct appropriate matrices for the second exercise.

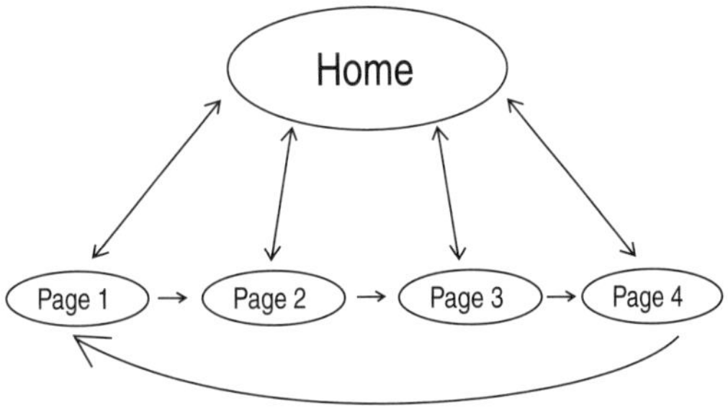


**17.** Give a set of specific assumptions that could be used to form a Markov chain that models Jennie's pitch selection in the Example on page 283.

⟵ One possible assumption (not a complete set) is that every time Jennie throws a fastball, her next pitch is a drop curve 30% of the time.

**18.** Some Markov chains have **absorbing states**—states that, when entered, are never exited.

**a.** Of the examples you've seen so far (in the Getting Started lesson and in this lesson, including exercises), which have absorbing states?

**b.** Give another example of a Markov chain with at least one absorbing state.

⟵ One major distinction among Markov chains is whether or not they have absorbing states. This will become a focus later in the module.

**19.** Many board games (such as *Chutes and Ladders*) can be modeled completely by very large Markov chains. Pick five examples of board games, and determine whether they can be modeled *completely* by

Markov chains, can be modeled *partially* by Markov chains, or are unrelated to Markov chains.

**20.** Model the tennis situation from Getting Started by using a spinner, a die, or a calculator's random number generator.

**a.** Play the game 10 times and record the winner and the total number of points played.
**b.** Using data from others, determine the approximate probability that Serena wins the game. Is it higher or lower than 60%?
**c.** How often did a game last exactly six points?
**d.** How often did a game last exactly seven points?
**e.** How often did a game last eight or more points?

⟵
If using a calculator that gives random numbers between 0 and 1, numbers below 0.6 award points to Serena, and numbers above 0.6 award points to Roger. If using a die, a roll of 1-3 is a point for Serena, a roll of 4-5 is a point for Roger, and, like *Count to Five*, a roll of 6 is ignored.

**21.** Let $T$ be the 5-by-5 matrix

$$T = \begin{pmatrix} 1 & 0.6 & 0 & 0 & 0 \\ 0 & 0 & 0.6 & 0 & 0 \\ 0 & 0.4 & 0 & 0.6 & 0 \\ 0 & 0 & 0.4 & 0 & 0 \\ 0 & 0 & 0 & 0.4 & 1 \end{pmatrix}$$

**a.** Calculate $T^2$ without the aid of a calculator.
**b.** Use a calculator to compute $T^{20}$, and write each entry to four decimal places.

**22.** A simple board game has five spaces with each player starting on A. On a turn, flip a coin; tails is a one-space move, and heads is a two-space move, counterclockwise. If you land on "flip again," do so immediately; this is part of the same turn. After D, players continue to A.

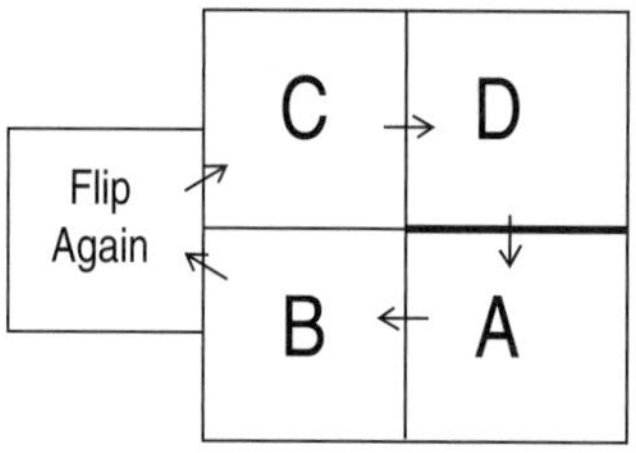


**a.** After one turn, give the probabilities that a player could be on A, B, C, and D.
**b.** After two turns, give the probabilities.
**c.** After three turns, give the probabilities.
**d.** **Take It Further.** After 100 turns, give the approximate probabilities.

⟵
One of these probabilities is zero.

**23.** **Take It Further.** Consider the tennis example from Getting Started. Suppose Serena is behind by one point. Either Roger wins the game with the next point, or the game returns to a tie.

**a.** If Serena is 60% likely to win each point, find the probability that she wins the game to four decimal places.

**b.** There is a probability $p$ that makes it exactly 50% likely for Serena to win the game. Find $p$ to four decimal places.

## 6.3 Representations of Markov Chains

The changes in the different states of a Markov procedure can be modeled in very useful ways using both graphs and matrices. In this lesson, you will represent Markov chains using both models, and you will see some ways to simulate a Markov chain using random numbers.

**In this lesson, you will learn how to**

- make a transition graph to visualize a Markov chain
- create a probability matrix from a transition graph
- use random numbers to simulate the different stages of a Markov chain

### Transition Graphs

The simplest way to visualize a small Markov chain is with a **transition graph**. This graph models the situation from Exercise 5 from Lesson 6.1.

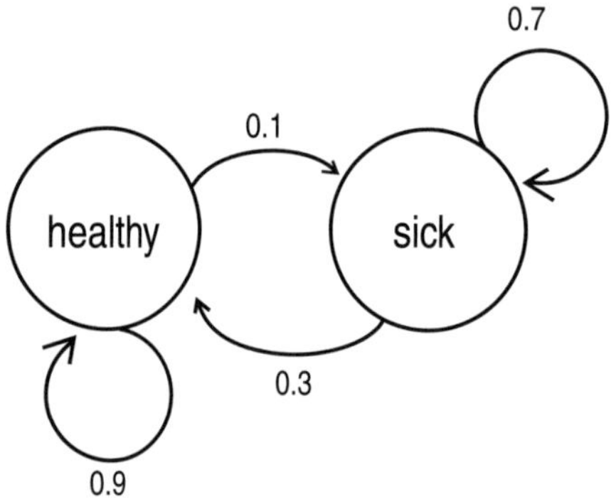


Transition graph for Exercise 5 from Lesson 6.1

Each state of the Markov chain is identified with a circle, usually called a **node**. Here, the nodes are labeled *healthy* and *sick*. The graph also includes several transition arrows, each marked with its **transition probability**, which must be between 0 and 1 (inclusive). The sum of the transition probabilities leaving each node must be exactly 1.

⟵ Usually, arrows are only drawn for nonzero probabilities.

**For You to Do**

1. Draw a transition graph for the tennis game from problem 7 from Lesson 6.1. Use five nodes labeled T, S1, R1, SW, RW. Assume that Serena is 60% likely to win any point.

⟵ You may want to order the nodes differently. Note that there must be at least one transition arrow leaving SW and RW.

### For Discussion

**2.** **a.** Use the transition graph for the tennis game to determine the probability that Serena wins the game in two points.

**b.** Use the transition graph to determine the probability that the game remains a tie after two points.

### Transition Matrices

The **transition matrix** of a Markov chain contains the same information as the graph. It is a square matrix with as many columns as the number of states in the Markov chain. Each column contains all the transition probabilities for a specific state.

### Example 1

This 2-by-2 matrix includes the information for the transition graph at the start of this lesson.

$$M = \begin{array}{c} \\ \text{healthy} \\ \text{sick} \end{array} \begin{array}{c} \textbf{healthy} \quad \text{sick} \\ \begin{pmatrix} \mathbf{0.9} & 0.3 \\ \mathbf{0.1} & 0.7 \end{pmatrix} \end{array}$$

⟵ The first column gives the transition probabilities from the *healthy* state.

The matrix above is labeled with its states. This won't be necessary once you decide on a standard ordering of the states. The first column $M_{*1}$ of transition matrix $M$ gives the transition probabilities from state 1 to each state in order. The second column $M_{*2}$ gives transition probabilities from state 2 to each state, and so on.

⟵ Suppose there is no way to go from state $a$ to state $b$. Then either $M_{ba}$ or $M_{ab}$ must be 0. Which one?

A transition matrix is more convenient than a graph in many respects, especially when it comes to calculation. Often, a transition graph is built to aid in the construction of the corresponding transition matrix, and is then discarded.

### For You to Do

**3.** For each transition graph, construct the transition matrix. The order of the states is given by the labels on the states.

**a.**

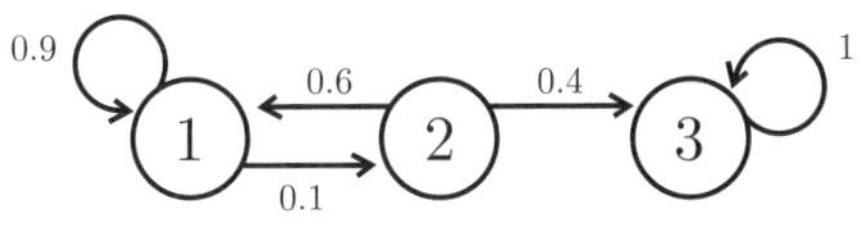


**b.**

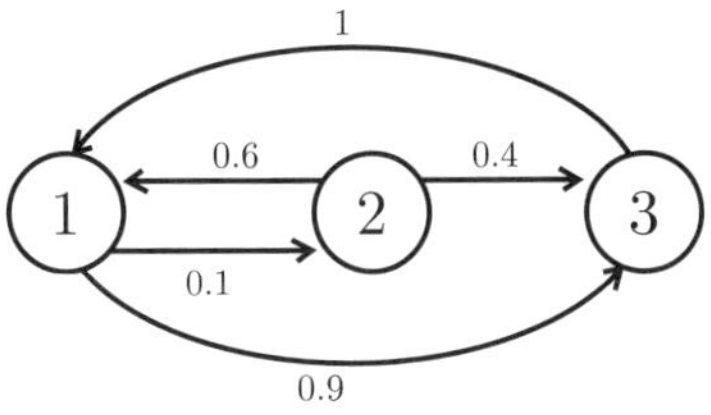

**c.**

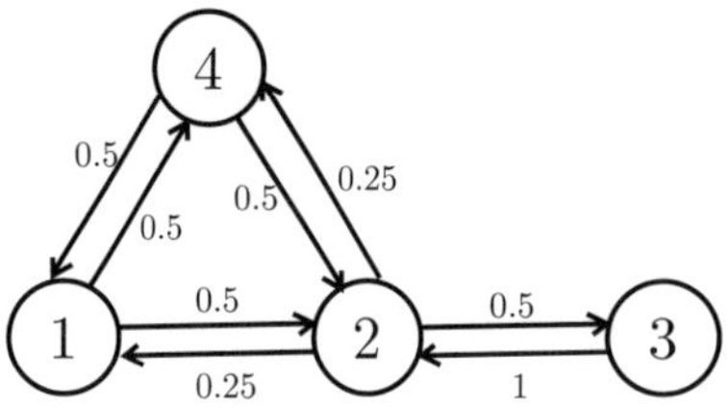


**d.**

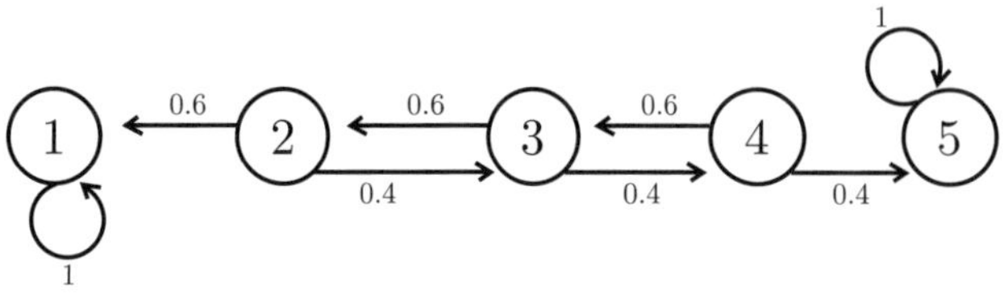


### Example 2

In the softball example from the last lesson, Jennie (or her catcher) decides what sequence of pitches to throw. A Markov chain can model these decisions. After deciding the state order {fastball, changeup, drop curve}, here is a possible 3-by-3 transition matrix.

$$M = \begin{pmatrix} 0.5 & 0.4 & 0.5 \\ 0.2 & 0.3 & 0.5 \\ 0.3 & 0.3 & 0 \end{pmatrix}$$

← All entries in $M$ are nonnegative and the sum of the entries in each column is exactly 1. Why must the entries in a column sum to 1?

**Problem.** What does $M_{31}$ tell you?

**Solution.** The value of $M_{31}$ is 0.3, and it tells you that if Jennie thows a fastball (state 1), her next pitch will be a drop curve (state 3) 30% of the time. This information is found in the first column, third row of the transition matrix.

### For You to Do

**4.** Based on the transition matrix above, what is the probability of a changeup if a fastball has just been thrown? What is the probability of two consecutive drop curves?

**5.** If Jennie throws a fastball, what is the probability that she will throw a fastball *two* pitches later?

The entries in a transition matrix can also be described using the notation of conditional probability:

$$M_{31} = P(\text{state } 3|\text{state } 1)$$

← The notation $P(B|A)$ stands for the probability that event $B$ occurs, given that event $A$ has already occurred. Say it like this: "The probability of $B$, given $A$."

The value of $M_{31}$ is the probability of a transition into state 3, given that the process is currently in state 1. In a Markov chain, each transition probability never changes, so a fixed matrix can be written to model a Markov chain. In general,

$$M_{ij} = P(\text{state } i|\text{state } j)$$

## Exercises

Exercises 1 through 5 refer to this graph of five states. For each of the given rules, determine a $5 \times 5$ transition matrix using the state order $\{s_1, s_2, s_3, s_4, s_5\}$. It may be helpful to first draw a transition graph for some situations.

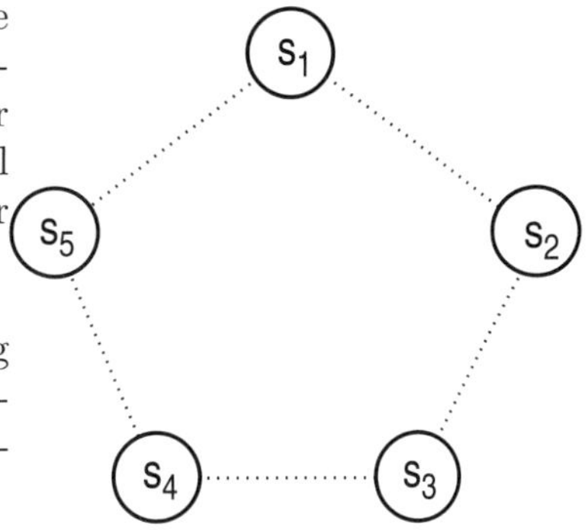


**1.** The probability of moving clockwise is $\frac{1}{2}$, and the probability of moving counterclockwise is $\frac{1}{2}$.

**2.** The probability of moving clockwise is $\frac{3}{4}$, and the probability of moving counterclockwise is $\frac{1}{4}$.

**3.** The probability of moving clockwise is $\frac{1}{2}$, and the probability of moving counterclockwise is $\frac{1}{2}$, except for $s_3$, which cannot be escaped from.

**4.** The probability of moving clockwise is $\frac{1}{2}$, and the probability of moving counterclockwise is $\frac{1}{2}$, except that $s_1$ feeds $s_3$ and no other states, and $s_3$ feeds $s_1$ and no other states.

**5.** The probability of moving clockwise is $\frac{2}{5}$, and the probability of moving counterclockwise is $\frac{3}{5}$, except for $s_1$ and $s_5$, which cannot be escaped from.

**6.** Build a transition graph for the small board game from Exercise 22 from the previous lesson.

**7.** Build a $4 \times 4$ transition matrix for the game from Exercise 22 from the previous lesson.

⟵ Why is the transition matrix $4 \times 4$ and not $5 \times 5$?

**8.** In a transition graph, the sum of the probabilities leaving a node must be 1. Does this also apply to the probabilities entering a node? Why or why not?

**9.** Make a transition graph that corresponds to the $3 \times 3$ transition matrix given in Example 2 from this lesson.

**10.** Draw a transition graph for *Count to Five* from Getting Started. There should be six states.

⟵ What about a transition matrix?

**11.** Consider the transition graph given in the For You to Do problem **3d** from this lesson.

   **a.** If you are in state $s_2$ now, what is the probability you will be in $s_4$ after two transitions?

   **b.** If you are in state $s_2$ now, what states could you be in after two transitions, and with what probabilities?

**c.** Repeat this calculation for all states $s_i$ and build a $5 \times 5$ matrix containing the results. Each column should give the five probabilities for each starting state; your answers to part **b** should form the second column of the matrix.

$$\begin{pmatrix} & 0.6 & & 0 & 0 \\ & & & & \\ 0 & & & & \\ 0 & & & & \\ & & 0.16 & & 1 \end{pmatrix}$$

⟵ Some entries have been given to help you get started. **Note:** This is *not* the transition matrix for the graph, but it may be related to it...

**12.** For each of the transition matrices $T$ you found in the For You to Do problem 3 from this lesson, compute $T^2$.

**13.** Write the transition matrix for Example 1 from this lesson using the state order {*sick, healthy*}.

**14.** Write the $5 \times 5$ transition matrix for the tennis example from Getting Started using the state order {SW, S1, T, R1, RW}.

**15.** Write the transition matrix for the tennis example using the state order {S1, T, R1, SW, RW}.

**16.** Construct a $3 \times 3$ matrix $T$ based on Example 2 from this lesson giving the probabilities for what Jennie will throw two pitches from now. For example, $T_{11} = 0.48$, and this is the probability that if Jennie throws a fastball now, she will throw another fastball two pitches later.

**17.** Given matrix $M$, calculate $M^2$ and $M^{10}$.

$$M = \begin{pmatrix} 0.5 & 0.4 & 0.5 \\ 0.2 & 0.3 & 0.5 \\ 0.3 & 0.3 & 0 \end{pmatrix}$$

**18.** In a high school, 90% of the students advance from one grade to the next: freshmen, sophomores, juniors, and seniors. 90% of seniors graduate. Of the 10% who don't advance, all freshmen and sophomores repeat, while 5% of juniors and seniors repeat. The other 5% drop out.

**a.** Build a transition graph for this situation. How many states should there be? Remember that all states must have transition arrows leaving them with probabilities summing to 1.
**b.** Build a transition matrix for this situation. Label the states.
**c.** Does this Markov chain have absorbing states? Explain.

**19.** **a.** If a student is a freshman today, what can happen after two years, and with what probabilities?
**b.** If a student is a junior today, what can happen after two years, and with what probabilities?

**20.** **Write About It.** Why are transition matrices always square?

## Perspective: Simulating Markov Chains

Markov chains can be run as experiments, conducted by hand or with the aid of a computer. Simulation uses a list of random numbers between zero and one. Random numbers can be generated by computer, or even using a phone book! Take a random "white" page (of residential listings) and look at only the last four digits. If the phone number is 978-555-5691, the last four digits are 5691. Then, put a decimal point in front of the four digits to get 0.5691, a random number between zero and one.

$\longleftarrow$
Do you think this method actually produces good random numbers? What if you used a "yellow" page with business listings?

Here's a sample set of 10 numbers from a phone book:

$$0.6224,\ 0.5711,\ 0.1623,\ 0.9062,\ 0.3186,$$
$$0.1076,\ 0.2482,\ 0.5610,\ 0.4593,\ 0.7568$$

Here's how to use these numbers to simulate 10 pitches from Jennie's softball game, using the Markov chain seen earlier. First, decide what state to start in. In this case, assume her first pitch is a fastball.

As a reminder, the $3 \times 3$ transition matrix for this situation is

$$\begin{pmatrix} 0.5 & 0.4 & 0.5 \\ 0.2 & 0.3 & 0.5 \\ 0.3 & 0.3 & 0 \end{pmatrix}$$

These steps use the transition matrix to simulate 10 pitches:

1. Take the first random number from the list and call it $s$. Here, $s = 0.6224$.
2. Look at the column in the transition matrix for the Markov chain corresponding to the current state. Since we are starting in *fastball*, that column is $\begin{pmatrix} 0.5 \\ 0.2 \\ 0.3 \end{pmatrix}$.
3. From top to bottom, add as many entries in the column as it takes to get a number that is greater than or equal to $s$. In this case, that takes two numbers: $0.5 + 0.2 \geq s$; just 0.5 isn't enough. The number of entries you added is the destination state number. Therefore, you transition to *changeup*.
4. Repeat these steps as long as you want using the next random number in the list as $s$. The next number would use the data from the second column of the transition matrix, $\begin{pmatrix} 0.4 \\ 0.3 \\ 0.3 \end{pmatrix}$, corresponding to the *changeup* state.

$\longleftarrow$
This step is selecting the state to transition into. In this case, it gives a probability of 0.5 of remaining a *fastball*, a 0.2 probability of transitioning to a *changeup*, and a 0.3 probability of transitioning to a *drop curve*.

**For You to Do**

**6.** Determine when the first transition to the *drop curve* state will occur.

This is the result when using the 10 phone book numbers listed above.

← Why does 0.6224 give a C? Why does 0.5711 give a second C?

| start | 0.6224 | 0.5711 | 0.1623 | 0.9062 | 0.3186 |
|---|---|---|---|---|---|
| F | C | C | F | D | F |

| 0.1076 | 0.2482 | 0.5610 | 0.4593 | 0.7568 |
|---|---|---|---|---|
| F | F | C | C | D |

You can use the same random numbers and simulate the two-state sick-healthy model. As before, you must pick a starting state. If you start as healthy, you start the simulation process using the first column of the transition matrix

$$\begin{pmatrix} 0.9 & 0.3 \\ 0.1 & 0.7 \end{pmatrix}$$

This is the result when using the 10 phone book numbers listed above.

| start | 0.6224 | 0.5711 | 0.1623 | 0.9062 | 0.3186 |
|---|---|---|---|---|---|
| healthy | healthy | healthy | healthy | sick | sick |

| 0.1076 | 0.2482 | 0.5610 | 0.4593 | 0.7568 |
|---|---|---|---|---|
| healthy | healthy | healthy | healthy | healthy |

### For You to Do

**7.** **a.** Use as many of the random numbers as are needed to simulate the tennis game with the transition matrix you constructed in the For You to Do problem **3d** from this lesson. Assume the game starts tied.

**b.** Repeat the simulation of the tennis game five more times using different pages of a phone book.

**8.** Using a calculator or computer, simulate a 100-pitch softball game for Jennie.

## 6.4 Applying Matrix Algebra to Markov Chains

One reason that matrices are so useful is that you can calculate with them. In this lesson, you'll see how calculations with a transition matrix can give you information about the Markov chains that it represents.

**In this lesson, you will learn how to**

- use matrix algebra with transition matrices to calculate the probability of future states of a Markov chain
- understand the relationship between the power of a transition matrix and probabilities of various states

In the previous lesson, you learned that each column of a transition matrix gives the probabilities for any given starting state, so the entries in each column are nonnegative and sum to 1. Such vectors are called *probability vectors*.

**Definition**

A **probability vector** is a vector of nonnegative real numbers whose elements add to 1.

**For Discussion**

**1.** Is $\begin{pmatrix} 1 \\ 0 \end{pmatrix}$ a probability vector?

**2.** Write a general expression for a two-term probability vector, and a three-term probability vector. Use as few variables as possible.

Recall the two-state Markov chain that models being healthy or sick. Its $2 \times 2$ transition matrix is

$$M = \begin{pmatrix} 0.9 & 0.3 \\ 0.1 & 0.7 \end{pmatrix}$$

$M_{*1}$, the first column, is a probability vector that determines transitions from the healthy state.

**For You to Do**

**3.** Suppose the probability that Aaron is healthy today is $p_1$ and the probability he is sick is $p_2$. As a linear combination of $p_1$ and $p_2$, find the probability that Aaron is healthy tomorrow, and (separately) the probability that Aaron is sick tomorrow.

The calculations in the For You to Do problems should remind you of matrix multiplication. Given a probability vector, you can multiply it by the transition matrix to produce a new probability vector for the next observation:

$$\begin{pmatrix} 0.9 & 0.3 \\ 0.1 & 0.7 \end{pmatrix} \begin{pmatrix} p_1 \\ p_2 \end{pmatrix}$$

← If they don't remind you of matrix multiplication, try the same calculation on a Markov chain with more than two states.

### For You to Do

4. After one day, the probability of being healthy is 0.9 and the probability of being sick is 0.1. Use matrix multiplication to determine the probability of being healthy after two days, and the probability of being sick after two days.

The same process can be followed for any transition matrix and for any probability vector.

### Example

Local biologists are tracking the movement of deer between Big Creek, Tall Pines, and Hunters' Canyon. The graph below shows the probability of movement from one year to the next.

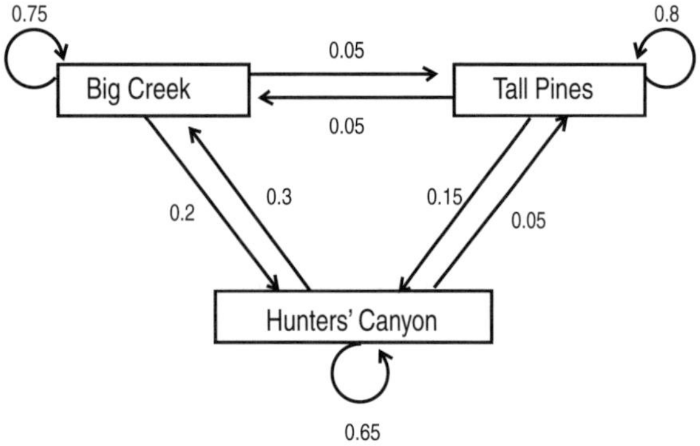


**Problem.** Last year, 60% of the deer were in Big Creek, 10% in Tall Pines, and 30% in Hunters' Canyon. Determine this year's distribution of the deer.

**Solution Method 1.** Use the transition graph provided. The proportion of deer in Big Creek is found by multiplying last year's proportions by the probabilities on the transition arrows leading to Big Creek:

$$\mathbf{0.75} \cdot 0.6 + \mathbf{0.05} \cdot 0.1 + \mathbf{0.3} \cdot 0.3 = 0.545$$

You can use similar calculations to find the other proportions. For Tall Pines:

$$\mathbf{0.05} \cdot 0.6 + \mathbf{0.8} \cdot 0.1 + \mathbf{0.05} \cdot 0.3 = 0.125$$

For Hunters' Canyon:

$$\mathbf{0.2} \cdot 0.6 + \mathbf{0.15} \cdot 0.1 + \mathbf{0.65} \cdot 0.3 = 0.33$$

The sum of these proportions is $0.545 + 0.125 + 0.33 = 1$, as expected.

**Solution Method 2.** Build a transition matrix. You will perform the same calculations as in Solution Method 1 above if you multiply the transition matrix by the probability vector. Using the state order {*Big Creek, Tall Pines, Hunters' Canyon*}, the transition matrix $T$ and the probability vector $P$, given as the starting distribution, are:

$$T = \begin{pmatrix} 0.75 & 0.05 & 0.3 \\ 0.05 & 0.8 & 0.05 \\ 0.2 & 0.15 & 0.65 \end{pmatrix}, \qquad P = \begin{pmatrix} 0.6 \\ 0.1 \\ 0.3 \end{pmatrix}$$

⟵ Note that the numbers in $T$ and $P$ are all found in the calculations from Solution Method 1. The numbers from $T$ are also in the transition graph, and the numbers from $P$ are taken from last year's distribution.

Multiply $T$ by $P$ to calculate this year's distribution:

$$TP = \begin{pmatrix} 0.75 & 0.05 & 0.3 \\ 0.05 & 0.8 & 0.05 \\ 0.2 & 0.15 & 0.65 \end{pmatrix} \begin{pmatrix} 0.6 \\ 0.1 \\ 0.3 \end{pmatrix} = \begin{pmatrix} 0.545 \\ 0.125 \\ 0.33 \end{pmatrix}$$

This gives the same distribution as Solution Method 1, and is more easily adapted to different situations or distributions.

---

**For You to Do**

5. **a.** What will *next* year's distribution of the deer be?
   **b.** The year after that?
   **c.** How might you determine the distribution 10 years from now?

---

Solution Method 2 suggests the following theorem for calculating the probabilities in a Markov chain.

**Theorem 6.1**

*Given a Markov chain with transition matrix $T$ and probability vector $P$, the probability vector for the next observation is $TP$.*

---

**For Discussion**

6. Prove Theorem 6.1.

---

Theorem 6.1 states that if you start with a transition matrix $T$ and a probability vector $P$, $TP$ is the probability vector for the next observation. Here's what happens if you apply this theorem to find the probability vector that follows $TP$:

$$T(TP) = (TT)P = T^2P$$

⟵ Make sure you can explain why these steps are valid.

Therefore, $T^2P$ is another probability vector, two observations ahead. This can continue for any number of observations, leading to a very powerful theorem.

**Theorem 6.2 (The $T^nP$ Theorem)**

*Given a Markov chain with transition matrix $T$ and probability vector $P$, the probability vector $n$ observations from now is $T^nP$.*

### For Discussion

7. a. Use the $T^n P$ Theorem to determine the probability that if you are healthy today, you will be healthy ten days from now.
   b. What happens to the healthy-sick probability vectors as the number of days increases?

⟵ If you are healthy today, what probability vector should you use for $P$?

⟵ Do the healthy-sick vectors approach a limit?

### Developing Habits of Mind

**Reason about calculations.** What if all the deer from the Example on page 298 started out in Big Creek? Then, the probability vector would be $\begin{pmatrix} 1 \\ 0 \\ 0 \end{pmatrix}$, a unit vector. Perform the matrix multiplication to get

$$\begin{pmatrix} 0.75 & 0.05 & 0.3 \\ 0.05 & 0.8 & 0.05 \\ 0.2 & 0.15 & 0.65 \end{pmatrix} \begin{pmatrix} 1 \\ 0 \\ 0 \end{pmatrix} = \begin{pmatrix} 0.75 \\ 0.05 \\ 0.2 \end{pmatrix}$$

It's the first column of transition matrix $T$. Thinking about linear combinations can help here: the unit vector says the result will be 1 times the first column of $T$, plus 0 times the second column, plus 0 times the third column. The Pick-Apart Theorem from Chapter 4 comes in handy here:

$$(AB)_{*i} = A \cdot B_{*i}$$

You can calculate new probability vectors for multiple starting vectors at once. Multiply $T$ by a matrix made of column vectors. Here's an example:

$$\begin{pmatrix} 0.75 & 0.05 & 0.3 \\ 0.05 & 0.8 & 0.05 \\ 0.2 & 0.15 & 0.65 \end{pmatrix} \begin{pmatrix} 0.6 & 1 \\ 0.1 & 0 \\ 0.3 & 0 \end{pmatrix} = \begin{pmatrix} 0.545 & 0.75 \\ 0.125 & 0.05 \\ 0.33 & 0.2 \end{pmatrix}$$

In particular, if you started with all the unit vectors, you'd get

$$\begin{aligned} TI &= \begin{pmatrix} 0.75 & 0.05 & 0.3 \\ 0.05 & 0.8 & 0.05 \\ 0.2 & 0.15 & 0.65 \end{pmatrix} \begin{pmatrix} 1 & 0 & 0 \\ 0 & 1 & 0 \\ 0 & 0 & 1 \end{pmatrix} \\ &= \begin{pmatrix} 0.75 & 0.05 & 0.3 \\ 0.05 & 0.8 & 0.05 \\ 0.2 & 0.15 & 0.65 \end{pmatrix} \end{aligned}$$

Of course, $T$ comes back! But what if you used these new vectors to calculate the next observation? You'd use $T$ as input:

$$\begin{aligned} TT &= \begin{pmatrix} 0.75 & 0.05 & 0.3 \\ 0.05 & 0.8 & 0.05 \\ 0.2 & 0.15 & 0.65 \end{pmatrix} \begin{pmatrix} 0.75 & 0.05 & 0.3 \\ 0.05 & 0.8 & 0.05 \\ 0.2 & 0.15 & 0.65 \end{pmatrix} \\ &= \begin{pmatrix} 0.625 & 0.1225 & 0.4225 \\ 0.0875 & 0.65 & 0.0875 \\ 0.2875 & 0.2275 & 0.49 \end{pmatrix} \end{aligned}$$

Squaring the original transition matrix gives a new set of probability vectors, and these vectors give the correct probabilities after two transitions. Higher powers give even more long-term behavior and results. This leads to a useful theorem, the *Matrix Power Formula*.

⟵ You may have noticed this behavior in some exercises from the previous lesson.

**Theorem 6.3 (The Matrix Power Formula)**

*If $T$ is a transition matrix for a Markov chain, the columns of $T^n$ are the probability vectors for the $n^{th}$ observation. Put another way, $(T^n)_{ij}$ is the probability of being in state $i$, starting from state $j$, after $n$ observations.*

⟵ $T^n$ is the transition matrix from observation $n$ to observation $n+1$.

**For You to Do**

**8.** **a.** Square the $3 \times 3$ transition matrix from the softball example given in Example 2 from Lesson 6.3:

$$\begin{pmatrix} 0.5 & 0.2 & 0.3 \\ 0.4 & 0.3 & 0.3 \\ 0.5 & 0.5 & 0 \end{pmatrix}$$

**b.** What is the probability that if Jessie throws a drop curve, she will throw another drop curve exactly two pitches later?

## Exercises

**1.** For the healthy-sick example, assume you are healthy today. Compute the probability that you will be healthy in

**a.** 3 days **b.** 4 days **c.** 5 days
**d.** 20 days **e.** 50 days

**2.** Repeat Exercise 1, but assume you are sick today.

⟵ The last column of $T$ should be the unit vector $\begin{pmatrix} 0 \\ 0 \\ 0 \\ 0 \\ 0 \\ 1 \end{pmatrix}$. Once you have five points, you never go back . . .

**3.** In Exercise 10 from Lesson 6.3, you drew a transition graph for the *Count to Five* game from Getting Started.

Construct a 6-by-6 transition matrix $T$ for the game, with the state order from zero points to five points.

**4.** **a.** Revisit Exercise 2d from Getting Started. If you have two points now, write down the probabilities in two turns: the probability of having zero points in two turns, one point in two turns, and so on.

**b.** Let $T$ be the matrix from Exercise 3. Compute $T^2$. What are the elements of its third column, and what do they mean?

**5.** A **steady state** is a probability vector $X$ such that for the transition matrix $T$, $TX = X$. Find a steady state for each of the following matrices $T$.

**a.** $T = \begin{pmatrix} 0.9 & 0.3 \\ 0.1 & 0.7 \end{pmatrix}$

**b.** $T = \begin{pmatrix} 0.75 & 0.5 \\ 0.25 & 0.5 \end{pmatrix}$

**c.** $T = \begin{pmatrix} 0.5 & 0.4 & 0.5 \\ 0.2 & 0.3 & 0.5 \\ 0.3 & 0.3 & 0 \end{pmatrix}$

**d.** $T = \begin{pmatrix} 1 & 0.6 & 0 & 0 & 0 \\ 0 & 0 & 0.6 & 0 & 0 \\ 0 & 0.4 & 0 & 0.6 & 0 \\ 0 & 0 & 0.4 & 0 & 0 \\ 0 & 0 & 0 & 0.4 & 1 \end{pmatrix}$

**6.** For each transition matrix $T$ in Exercise 5, compute $T^{50}$ with the help of a calculator. Do you see anything interesting?

Exercises 7 through 9 use the Example on page 298.

**7.** **a.** In 1999, all the deer were in Hunters' Canyon. Determine the distribution of deer in 2000, 2001, and 2002.

**b.** Compute $T^3$ for the transition matrix given in the Example.

**c.** If all the deer were in Big Creek three years ago, what proportion of deer are now in Big Creek?

**d.** If all the deer started in Tall Pines three years ago, what proportion of deer are now in Big Creek?

**8.** How could you determine what will happen to the deer population in the long run? Does your answer depend on the starting distribution of deer?

**9.** **a.** Suppose the deer population this year is given by the vector $\begin{pmatrix} 0.7 \\ 0.2 \\ 0.1 \end{pmatrix}$. Find next year's deer population.

**b.** Show that next year's deer population can be expressed as a linear combination of the columns of the transition matrix $T$.

**c.** If this year's population is given by the vector $\begin{pmatrix} a \\ b \\ 1-a-b \end{pmatrix}$, find next year's deer population.

**10.** Prove the following theorem.

**Theorem 6.4**

*If $X$ is a steady state for the Markov chain with transition matrix $T$, then $X$ is in the kernel of the matrix $T - I$ (where $I$ is the corresponding identity matrix).*

**11.** Let $T$ be the $4 \times 4$ transition matrix from Exercise 6 in the last lesson.

**a.** Compute $T^2 \cdot \begin{pmatrix} 1 \\ 0 \\ 0 \\ 0 \end{pmatrix}$ and interpret the results.

**b.** Compute $T^3 \cdot \begin{pmatrix} 1 \\ 0 \\ 0 \\ 0 \end{pmatrix}$ and interpret the results.

**c.** Compute $T^2 \cdot \begin{pmatrix} 0 \\ 1 \\ 0 \\ 0 \end{pmatrix}$ and interpret the results.

**d.** Compute $T^{10}$.

**e.** Find a steady state for $T$.

**12.** Serena and Roger are playing tennis. They are tied with three points each, and Serena is 60% likely to win any given point.

**a.** Use a transition matrix to determine the probability that Serena wins the game sometime within four points.

**b.** Find the probability that Roger wins the game sometime within six points.

**c.** Find the probability that Serena wins the game within 100 points.

⟵ You'll need to calculate $T^4$ for some transition matrix . . . then what?

**13.** Find the probability that Serena will win the game if she is given a one-point lead.

**14.** Find the probability that Roger will win the game if he is given a one-point lead.

**15.** **a.** Compute the probability that you finish the game in three turns or less when playing *Count to Five.*

**b.** Compute the probability that you finish the game in *exactly* four turns when playing *Count to Five.*

**16.** For each transition matrix $T$ in Exercise 5, find *all* possible steady states.

**17.** Let $T = \begin{pmatrix} 0.9 & 0.3 \\ 0.1 & 0.7 \end{pmatrix}$.

**a.** Find $\text{Tr}(T)$ and $\det(T)$.

**b.** Find the two values of $k$ such that the equation $TP = kP$ has a nonzero solution.

$$\begin{pmatrix} 0.9 & 0.3 \\ 0.1 & 0.7 \end{pmatrix} \begin{pmatrix} x \\ y \end{pmatrix} = \begin{pmatrix} kx \\ ky \end{pmatrix}$$

**18.** Let $T = \begin{pmatrix} a & 1-b \\ 1-a & b \end{pmatrix}$, a generic $2 \times 2$ transition matrix.

**a.** Find $\text{Tr}(T)$ and $\det(T)$.

**b.** Find the two values of $k$, in terms of $a$ and $b$, such that the equation $TP = kP$ has a nonzero solution.

$$\begin{pmatrix} a & 1-b \\ 1-a & b \end{pmatrix} \begin{pmatrix} x \\ y \end{pmatrix} = \begin{pmatrix} kx \\ ky \end{pmatrix}$$

**19.** Let $T = \begin{pmatrix} 0.75 & 0.05 & 0.3 \\ 0.05 & 0.8 & 0.05 \\ 0.2 & 0.15 & 0.65 \end{pmatrix}$.

Find all three values of $k$ such that $TP = kP$ has a nonzero solution. (Hint: Can you rewrite the equation so the right side is the zero vector?)

**20.** Consider the situation of Exercise 18 from the previous lesson. Use Theorem 6.2 to determine each of the following:

**a.** The probability that a junior will graduate in two more years.
**b.** The probability that a freshman will graduate in four years.
**c.** The probability that a freshman will graduate in five years or less.
**d.** The probability that a freshman will graduate in six years or less.
**e.** The probability that a freshman will graduate at all.

$\longleftarrow$
What power of the transition matrix could give this answer?

**21.** A town's population is 10,000. Today, 9000 people are healthy and 1000 people are sick.

**a.** Using the $2 \times 2$ transition matrix for this example, determine the number of healthy and sick people for each of the next ten days.
**b.** What is happening in the long run?

**22.** Let $T$ be a matrix with nonnegative real entries, and suppose there is an $n$ so that all the entries of $T^n$ are nonzero.

**a.** Explain why all the entries of $T^n$ must be positive.
**b.** **Take It Further.** Prove that for any $k > n$, all the entries of $T^k$ must be positive.

**23.** Let $P = \begin{pmatrix} 0.5 \\ 0.2 \\ 0.3 \end{pmatrix}$ and $T = \begin{pmatrix} 0.75 & 0.05 & 0.3 \\ 0.05 & 0.8 & 0.05 \\ 0.2 & 0.15 & 0.65 \end{pmatrix}$.

**a.** Calculate $T^2P$.
**b.** Find three probability vectors $V_1, V_2, V_3$ and three nonzero constants $c_1, c_2, c_3$ so that $P = c_1V_1 + c_2V_2 + c_3V_3$.
**c.** For the vectors you found, calculate $c_1T^2V_1 + c_2T^2V_2 + c_3T^2V_3$.

**24.** Show that, in general, if $P = c_1V_1 + c_2V_2 + c_3V_3$, then

$$T^nP = c_1T^nV_1 + c_2T^nV_2 + c_3T^nV_3$$

## 6.5 Regular Markov Chains

In a previous lesson, you learned that whether you are sick or healthy today has very little effect on whether you are sick or healthy a long time from now (say, 100 days). This lesson focuses on Markov chains with that property.

**In this lesson, you will learn how to**

- recognize when a Markov chain is *regular*
- determine the *attractor* for a regular Markov chain

**Definition**

A probability vector $X$ is a **steady state** for a transition matrix $T$ if and only if $TX = X$.

**In-Class Experiment**

Here are three transition graphs. For each,

- build the $3 \times 3$ transition matrix $T$
- determine all possible steady states
- calculate $T^{50}$

**a.**

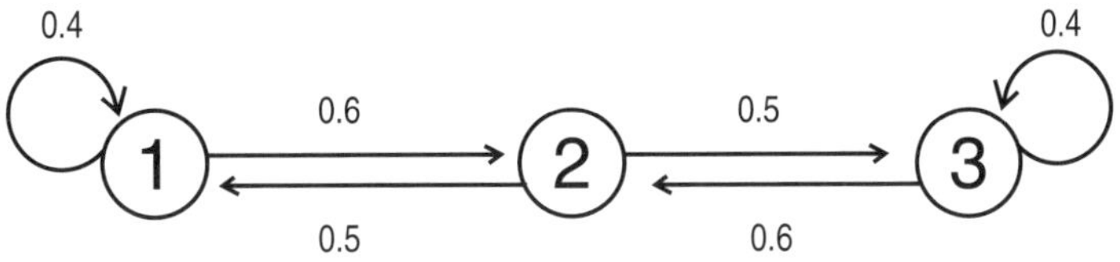


**b.**

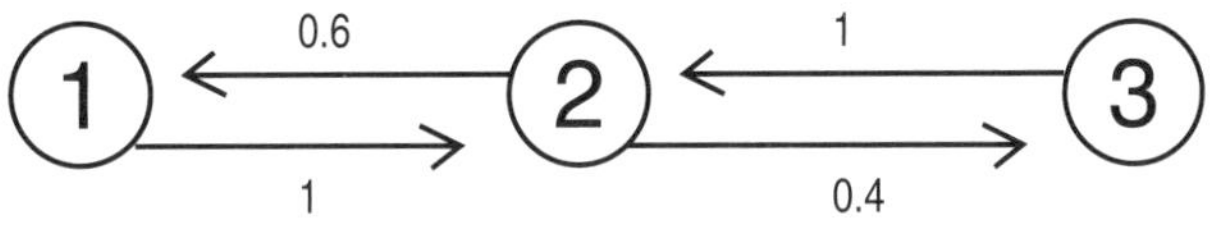


**c.**

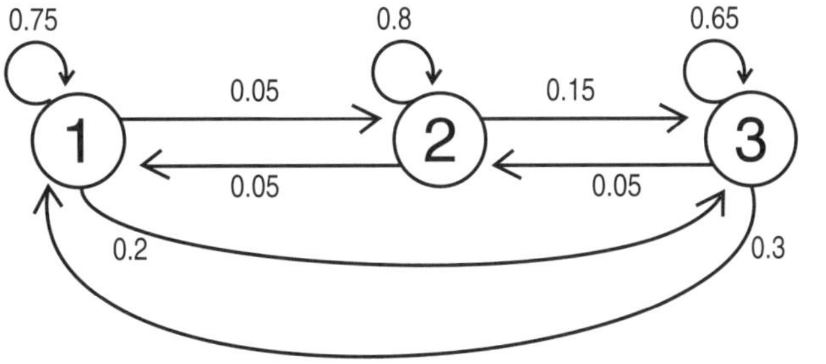

A Markov chain with the property that, no matter what the starting distribution is, it always converges to the same state is called a *regular Markov chain* due to its "regular" long-term behavior.

### Developing Habits of Mind

**Use precise language.** What does it mean to say that "it always converges to the same state"? One way to think about it is through transition matrices. Suppose you have a Markov chain, regular in the above sense, and you start with some probability vector $P$ that describes the current state. If the transition matrix is $T$, then the probabilities after $n$ observations is $T^n P$. To say that "$T^n P$ converges on a state $V$" means that

*You can make $\|T^n P - V\|$ as small as you want by making $n$ big enough.*

This is often written as a limit:

$$\lim_{n \to \infty} T^n P = V.$$

⟵ Two real numbers are close if the absolute value of their difference is small. Thinking of $\mathbb{R}$ as $\mathbb{R}^1$ and extending this idea, it makes sense to call two vectors $A$ and $B$ in $\mathbb{R}^n$ **close** if the distance between them $\|B - A\|$ is small.

#### Definition

A Markov chain with transition matrix $T$ is **regular** if and only if there is a vector $V$ with the property that, for any probability vector $P$,

$$\lim_{n \to \infty} T^n P = V$$

The results from calculating $T^{50}$ in the In-Class Experiment show the same "regular" behavior that some, but not all, Markov chains exhibit.

⟵ If the columns of $T^n$ all converge to some vector $V$, will $T^n P$ also converge for $V$ for any vector $P$?

There is another characterization of regular chains that only involves matrices.

#### Theorem 6.5

*A Markov chain with transition matrix $T$ is regular if and only if there exists an $n$ such that $T^n$ contains no zeros.*

⟵ In some linear algebra books, a matrix $T$ is called regular if, for some integer $n$, $T^n$ contains only positive entries.

Theorem 6.5 does not seem to match the intuitive idea of "regularity" described in the definition of *regular* earlier in this lesson, but it is possible to prove that the definition is equivalent to the intuitive idea. The proof is beyond the scope of this book, but here are some ideas that might make it plausible.

Suppose $T$ is a transition matrix where $T^n$ has no zeros. Think about how a column of $T^{n+1}$ is generated: it is a linear combination of the columns of $T^n$ whose coefficients are all positive and sum to 1.

What are the consequences of this? It means the columns of $T^{n+1}$ are more of a "mix" than the columns of $T^n$, and particularly large or small columns will be leveled off by the linear combination. As $n$ grows, the columns of $T^n$ get very close to one another, inducing "regularity."

On the flip side, you might like to think about nonregular Markov chains, and why their transition matrices don't induce this behavior.

**For You to Do**

1. Use Theorem 6.5 to find some examples of regular Markov chains and *non*regular Markov chains that you have seen in this module. For each, take the transition matrix $T$ and calculate $T^{50}$.

**Example 1**

The first transition graph in the In-Class Experiment corresponds to this 3-by-3 transition matrix:

$$T = \begin{pmatrix} 0.4 & 0.5 & 0 \\ 0.6 & 0 & 0.6 \\ 0 & 0.5 & 0.4 \end{pmatrix}$$

**Problem.** If you start in the first state, what are the long-term probabilities of being in each state? What if you start in the second or third?

**Solution.** Use the $T^n P$ Theorem. Calculate $T^n P$ using probability vector $P = \begin{pmatrix} 1 \\ 0 \\ 0 \end{pmatrix}$ and a high power of $n$. Here are the results for $n = 20$ and 21.

← Theorem 6.2 from Lesson 6.4.

$$T^{20}P = \begin{pmatrix} 0.4 & 0.5 & 0 \\ 0.6 & 0 & 0.6 \\ 0 & 0.5 & 0.4 \end{pmatrix}^{20} \begin{pmatrix} 1 \\ 0 \\ 0 \end{pmatrix} \approx \begin{pmatrix} 0.3125 \\ 0.3750 \\ 0.3125 \end{pmatrix}$$

← The "$\approx$" here means that the result is close to the right-hand sides.

$$T^{21}P = \begin{pmatrix} 0.4 & 0.5 & 0 \\ 0.6 & 0 & 0.6 \\ 0 & 0.5 & 0.4 \end{pmatrix}^{21} \begin{pmatrix} 1 \\ 0 \\ 0 \end{pmatrix} \approx \begin{pmatrix} 0.3125 \\ 0.3750 \\ 0.3125 \end{pmatrix}$$

If you start in the first state, the long-term probability of being in the first state is roughly 0.3125, the second state is roughly 0.375; and the third state is roughly 0.3125.

If starting in the second state, use initial probability vector $P = \begin{pmatrix} 0 \\ 1 \\ 0 \end{pmatrix}$ instead. This time $T^{20}P$ is

← Note that $T^{20}\begin{pmatrix} 0 \\ 1 \\ 0 \end{pmatrix}$ is the second column of $T^{20}$ (why?).

$$\begin{pmatrix} 0.4 & 0.5 & 0 \\ 0.6 & 0 & 0.6 \\ 0 & 0.5 & 0.4 \end{pmatrix}^{20} \begin{pmatrix} 0 \\ 1 \\ 0 \end{pmatrix} \approx \begin{pmatrix} 0.3125 \\ 0.3750 \\ 0.3125 \end{pmatrix}$$

To four decimal places, this is the same result as starting in the first state.

Starting with probability vector $\begin{pmatrix} 0 \\ 0 \\ 1 \end{pmatrix}$ to represent the third state, the same result occurs. The starting state does not seem to influence the long-term behavior of this Markov chain.

### For You to Do

**2.** **a.** Calculate $T^{20}$.

**b.** Calculate $T^{20} \cdot \begin{pmatrix} .4 \\ .4 \\ .2 \end{pmatrix}$.

**3.** Algebraically, find a state, a probability vector $V = \begin{pmatrix} x \\ y \\ z \end{pmatrix}$ such that $TV = V$.

⟵ How does the above example and Chapter 4's "Pick-Apart Theorem" (Theorem 4.8) make the calculation in problem 2 simpler?

Recall that the standard basis vectors in $\mathbb{R}^3$ (written as columns) are

$$E_1 = \begin{pmatrix} 1 \\ 0 \\ 0 \end{pmatrix}, \quad E_2 = \begin{pmatrix} 0 \\ 1 \\ 0 \end{pmatrix}, \quad \text{and} \quad E_3 = \begin{pmatrix} 0 \\ 0 \\ 1 \end{pmatrix}$$

In Example 1, you saw that, for large values of $n$, if you apply $T^n$ to any of these vectors, you get something very close to

$$V = \begin{pmatrix} .3125 \\ .3750 \\ .3125 \end{pmatrix} = \begin{pmatrix} \frac{5}{16} \\ \frac{3}{8} \\ \frac{5}{16} \end{pmatrix}$$

This vector $V$ is a steady state for $T$.

But it also seems that $T^n$ times *any* probability vector is close to this steady state. Linearity can help to explain this.

### Developing Habits of Mind

**Use linearity.** Any vector $\begin{pmatrix} a \\ b \\ c \end{pmatrix}$ can be written in terms of $E_1, E_2$, and $E_3$:

$$\begin{pmatrix} a \\ b \\ c \end{pmatrix} = a \begin{pmatrix} 1 \\ 0 \\ 0 \end{pmatrix} + b \begin{pmatrix} 0 \\ 1 \\ 0 \end{pmatrix} + c \begin{pmatrix} 0 \\ 0 \\ 1 \end{pmatrix}$$

⟵ This example uses a probability vector in $\mathbb{R}^3$, but the argument is general. See Exercise 24 from Lesson 6.4.

Consider a probability vector $P$. It can be written as

$$P = aE_1 + bE_2 + cE_3$$

where $a + b + c = 1$ (why)? Now use matrix algebra to calculate $T^n P$ in terms of the unit coordinate vectors.

$$\begin{aligned} T^n P &= T^n(aE_1 + bE_2 + cE_3) \\ &= T^n(aE_1) + T^n(bE_2) + T^n(cE_3) \\ &= a(T^n E_1) + b(T^n E_2) + c(T^n E_3) \end{aligned}$$

⟵ This is true regardless of the size of $T$. So, for any $P$, $T^n P$ is a weighted average of the columns of $T^n$.

This might not seem like an improvement, but you saw that, for large $n$, each of the vectors $T^n E_1$, $T^n E_2$, and $T^n E_3$ is very close to the steady state $V$. It follows that

$$T^n P = a(T^n E_1) + b(T^n E_2) + c(T^n E_3)$$

is very close to

$$T^n P = a(V) + b(V) + c(V) = (a + b + c)V = V$$

since $a + b + c = 1$.

The example used so far in this lesson is, in fact, general: for regular Markov chains, it doesn't matter what probability vector you start with. The long-term behavior always approaches a single steady state.

**Theorem 6.6**

*Given a regular Markov chain with transition matrix $T$, there is an* ***attractor*** *$V$, a probability vector such that for any given probability vector $P$,*

$$\lim_{n \to \infty} T^n P = V$$

*This vector $V$ is the only probability vector that is a steady state for such a Markov chain.*

← This is difficult to prove. But you can prove parts of it. See Exercises 16, 17, and 18 for some ideas related to the proof. The full proof is outside the scope of this book.

**For Discussion**

In the In-Class Experiment, you noticed that for some Markov chains, the columns of $T^n$ end up being nearly identical. For the Markov chain from Example 1, $T^{20}$ is

$$T^{20} \approx \begin{pmatrix} 0.3125 & 0.3125 & 0.3125 \\ 0.3750 & 0.3750 & 0.3750 \\ 0.3125 & 0.3125 & 0.3125 \end{pmatrix}$$

← The columns of $T^n$ don't always behave this way! This only works for regular Markov chains.

4. Explain why for *any* regular Markov chain, the columns of high powers of its transition matrix must be nearly identical. (Hint: What would happen otherwise?)

**For You to Do**

5. Find the one steady state $V$ for the healthy-sick example from previous lessons. Pick five different starting probability vectors $P$, and, for each, find the smallest value of $n$ so that $\|T^n P - V\| < .001$.

**Example 2**

Consider this 3-by-3 transition matrix seen earlier to model Jessie's pitch selection:

$$T = \begin{pmatrix} 0.5 & 0.4 & 0.5 \\ 0.2 & 0.3 & 0.5 \\ 0.3 & 0.3 & 0 \end{pmatrix}$$

**Problem.** Show that this Markov chain is regular, and then find the attractor $V$.

**Solution.** To show that the Markov chain is regular, find a power of $T$ such that $T^n$ contains no zeros. For this Markov chain, $T^2$ is sufficient. You previously determined

$$T^2 = \begin{pmatrix} 0.48 & 0.47 & 0.45 \\ 0.31 & 0.32 & 0.25 \\ 0.21 & 0.21 & 0.3 \end{pmatrix}$$

Since $T^2$ has no zeros, the Markov chain with transition matrix $T$ is regular.

By Theorem 6.6, the attractor $V$ is the only steady state (the only probability vector that solves $TV = V$). Here are three ways to find $V$.

**Solution Method 1.** Let $V = \begin{pmatrix} a \\ b \\ c \end{pmatrix}$ and set up a system of equations to solve.

$$\begin{aligned} 0.5a + 0.4b + 0.5c &= a \\ 0.2a + 0.3b + 0.5c &= b \\ 0.3a + 0.3b \qquad\quad &= c \end{aligned}$$

This system doesn't have a single solution, but you can add one more, $a + b + c = 1$, to get a unique solution:

$$a = \frac{55}{117},\ b = \frac{35}{117},\ c = \frac{27}{117}$$

⟵ Alternatively, write $V = \begin{pmatrix} a \\ b \\ 1-a-b \end{pmatrix}$ from the start to produce a 2-by-2 system of equations with one solution.

**Solution Method 2.** Approximate $V$ by determining $T^n$ for a high power of $n$. Since the Markov chain is regular, each column of $T^n$ should be approximately equal to $V$.

$$T^{20} \approx \begin{pmatrix} 0.47009 & 0.47009 & 0.47009 \\ 0.29915 & 0.29915 & 0.29915 \\ 0.23077 & 0.23077 & 0.23077 \end{pmatrix}$$

All the columns give the same answer to five decimal places, so any column vector is a good approximation to $V$.

**Remember**

This won't happen unless the Markov chain is regular! This method only approximates $V$.

**Solution Method 3.** Use linear algebra to simplify the equation $TV = V$.

$$\begin{aligned} TV &= V \\ TV &= IV && \text{(for some identity matrix)} \\ TV - IV &= O && \text{(a zero vector)} \\ (T - I)V &= O \end{aligned}$$

Find the kernel of $T - I$.

$$T - I = \begin{pmatrix} -0.5 & 0.4 & 0.5 \\ 0.2 & -0.7 & 0.5 \\ 0.3 & 0.3 & -1 \end{pmatrix}$$

One method is to take the augmented matrix

$$A = \begin{pmatrix} -0.5 & 0.4 & 0.5 & \mathbf{0} \\ 0.2 & -0.7 & 0.5 & \mathbf{0} \\ 0.3 & 0.3 & -1 & \mathbf{0} \end{pmatrix}$$

**Remember**

$(T - I)V = O$ is of the form $AV = O$, so you can apply any of the methods from Chapters 3 and 4.

⟵ Note how similar the matrix $T - I$ is to the equations of Solution Method 1!

and calculate its echelon form

$$\text{rref}(A) = \begin{pmatrix} 1 & 0 & \frac{-55}{27} & 0 \\ 0 & 1 & \frac{-35}{27} & 0 \\ 0 & 0 & 0 & 0 \end{pmatrix}$$

The third column of this matrix, along with the fact that the sum of the elements of $V$ must add to 1, allow it to be found uniquely.

### For You to Do

**6.** Using more than one method, find the attractor $V$ for the deer example from the previous lesson. The 3-by-3 transition matrix is

$$T = \begin{pmatrix} 0.75 & 0.05 & 0.3 \\ 0.05 & 0.8 & 0.05 \\ 0.2 & 0.15 & 0.65 \end{pmatrix}$$

## Exercises

In Exercises 1 through 7, use the transition graph to determine whether each Markov chain is regular. If it is regular, find the unique steady state. If it is not regular, change the matrix (as little as possible) to make it regular.

← Hmm, thanks to Theorem 6.5, regularity is determined by the transition matrix . . .

**1.**

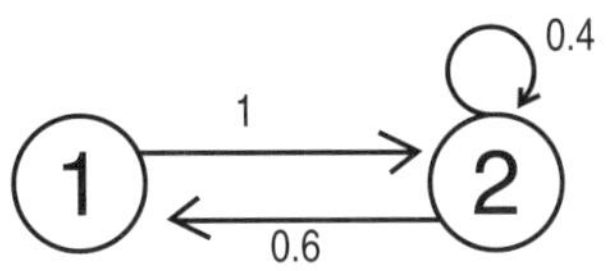


**2.**

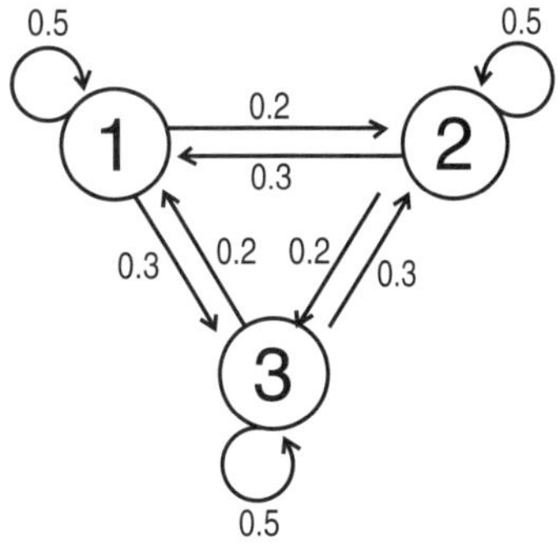


**3.**

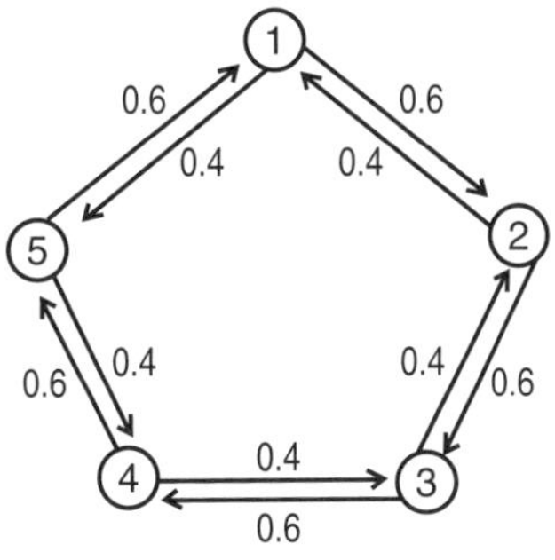


**4.**

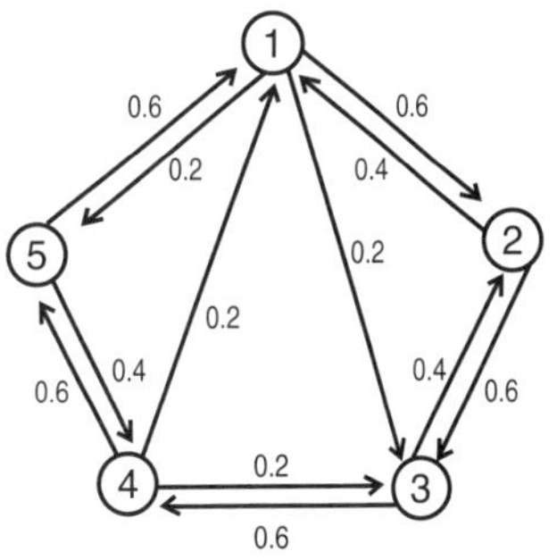

**5.**

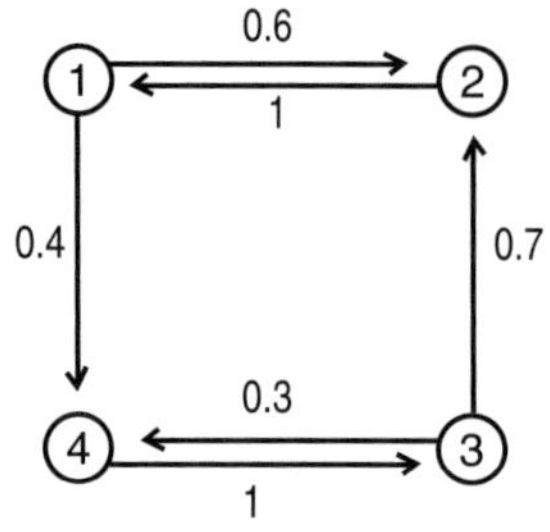


**6.**

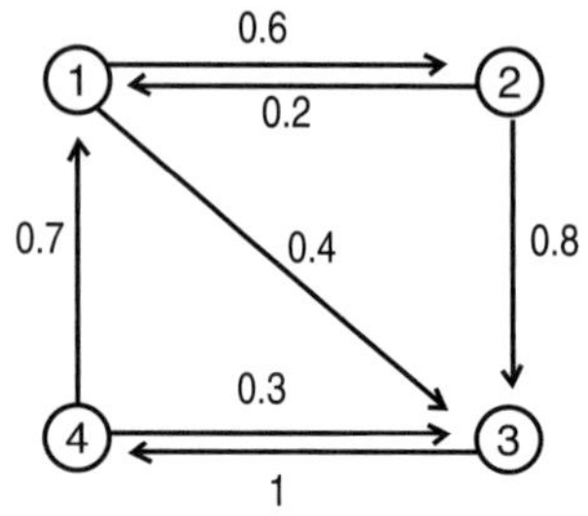


**7.**

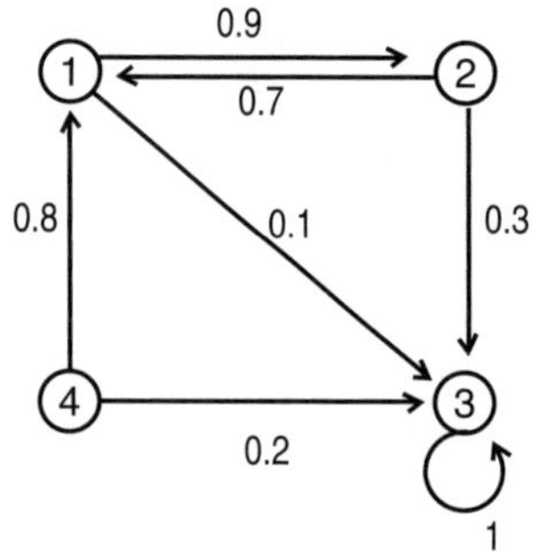


**8.** Suppose $V$ is an attractor (see Theorem 6.6 for the definition). Prove that $V$ is a steady state for this Markov chain.

**9.** Suppose $V$ and $W$ are two distinct steady states for a Markov chain. Prove that this Markov chain cannot be regular by showing that the required statement

$$\lim_{n\to\infty} T^n P = V$$

is not true for all probability vectors $P$.

**10.** Give justifications for each algebra step in the Developing Habits of Mind titled "Use linearity" on page 308. Be careful about whether each variable and calculation talks about a scalar, a vector, or a matrix.

**11.** Consider the five-square game first seen in Exercise 22 from Lesson 6.2.

- **a.** Show that the Markov chain for this situation is regular.
- **b.** Find the steady state.
- **c.** In this game, each time the player lands on A they are given \$200. If the game lasts 100 turns, about how much money would you expect a player to earn during the game?

**12.** The first player to land on any square in the game from Exercise 11 may purchase a contract that pays them each time a player lands on that space. Here are the contracts and their payouts:

| | |
|---|---|
| A: | costs \$500, pays \$100 |
| B: | costs \$700, pays \$150 |
| C: | costs \$1500, pays \$200 |
| D: | costs \$2000, pays \$300 |

Which square is the best investment? Use the steady state to help you decide.

⟵ How many turns will it take for you to recoup the investment you make when buying each space?

**13.** Explain why *Count to Five* and the tennis example from Getting Started do *not* correspond to regular Markov chains.

**14.** **a.** Given the transition graph below, determine whether or not the Markov chain is regular.

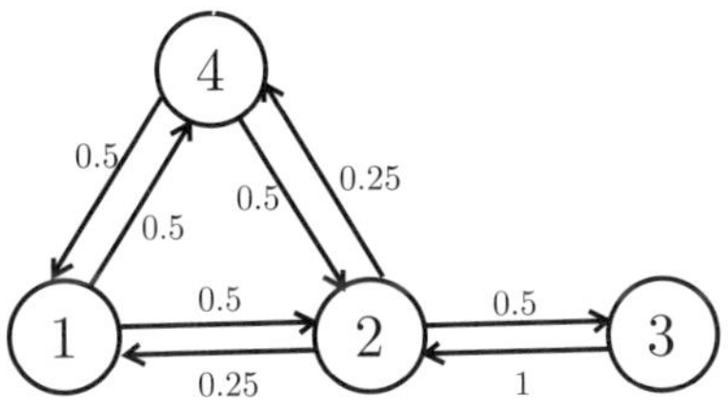


**b.** In the long run, what percentage of the time will you be in state 4?

**15.** In Exercise 13, you showed that the transition matrix for *Count to Five* does not correspond to a regular Markov chain.

**a.** Prove or disprove the following:

If the transition matrix $T$ for a Markov chain has $T_{ii} = 1$ for some $i$, then $T$ is *not* the transition matrix for a regular Markov chain.

**b.** If the transition matrix $T$ has any nondiagonal $T_{ij} = 1$, is it still possible for $T$ to correspond to a regular Markov chain?

⟵ In other words, $T$ has a 1 on its diagonal. What would the column that includes that 1 look like?

**16.** Suppose $T$ is a transition matrix, and all the elements of $T$ are nonzero.

**a.** Explain why $T$ must correspond to a regular Markov chain.

**b.** Prove that every element in $T$ satisfies $0 < T_{ij} < 1$.

**17.** Suppose $T$ is a transition matrix, and all elements in $T$ are strictly between 0 and 1. Additionally, $T$ has a maximal element, some $T_{ij}$ larger than all other elements of $T$.

**a.** Prove that the largest entry in $T^2$ is less than the largest entry in $T$.

**b.** Prove that if $T^k$ has a maximal element, then the largest entry in $T^{k+1}$ is less than the largest element in $T^k$.

⟵ **Hint:** Use a linear combination of the rows or columns of $T$.

**18.** **Take It Further.** Suppose that $T$ is the transition matrix for a regular Markov chain with attractor $V$. Without using Theorem 6.6, show that $T(V) = V$.

**19.** Suppose that $T$ is the transition matrix for a regular Markov chain with attractor $V$. Show that every fixed vector for $T$ is of the form $kV$ for some number $k$.

**20.** Let $T = \begin{pmatrix} 0.9 & 0.3 \\ 0.1 & 0.7 \end{pmatrix}$.

**a.** Find a matrix $A$ such that $TA = A \begin{pmatrix} 1 & 0 \\ 0 & 0.6 \end{pmatrix}$.

**b.** Find the two values of $k$ such that $TX = kX$ has a nonzero solution.

**21.** Let $T = \begin{pmatrix} 0.9 & 0.3 \\ 0.1 & 0.7 \end{pmatrix}$, $D = \begin{pmatrix} 1 & 0 \\ 0 & 0.6 \end{pmatrix}$. Let $A$ be an invertible matrix such that $TA = AD$.

**a.** Calculate $A^{-1}$.

**b.** Show that $T = ADA^{-1}$.

**c.** Show that $T^2 = AD^2A^{-1}$.

**d.** Show that $T^n = AD^nA^{-1}$.

**e.** Suppose $n$ is large. What, approximately, is $D^n$?

**f.** If $n$ is large, evaluate the right side without a calculator to give an estimate for $T^n$.

**Remember**

An invertible matrix $A$ is one with an inverse, $A^{-1}$. Use the matrix you found in Exercise 20.

**22.** **Take It Further.** The game of *Monopoly* is closely related to the five-square game described earlier.

**a.** Build a large transition matrix for *Monopoly*.

**b.** Use technology to provide evidence that this Markov chain is regular, and to approximate its steady state.

**c.** What property on the board will recoup its hotel cost most quickly?

## 6.6 Absorbing Markov Chains

Some random processes include states that only transition to themselves. For instance, Markov chains describing games that declare a winner typically include such states, since once a player is declared a winner, they stay the winner (and the game usually ends). In this lesson, you will explore the nature of such Markov chains and the states they contain.

**In this lesson, you will learn how to**

- understand and identify absorbing and transient states of a random process
- use transition matrices to identify absorbing states

### For You to Do

**1.** If you haven't already, build the 6-by-6 transition matrix for the *Count to Five* game from Lesson 6.1, Getting Started.

**2.** Using the transition matrix, determine the probability of going from zero points to five points in the following number of turns.

**a.** two or fewer **b.** three or fewer **c.** four or fewer
**d.** 10 or fewer **e.** 100 or fewer

Here's the transition graph for *Count to Five*.

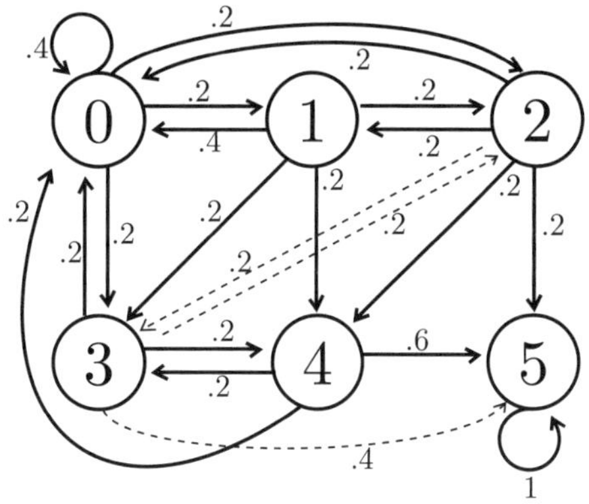

Once you have five points . . . that's it!

It is impossible to escape from the five-point state. Every other state has at least one transition away from itself.

### For Discussion

**3.** Give some other examples of situations where an inescapable state might occur.

An inescapable state is also called an **absorbing state**.

### Definition

An **absorbing state** is a state in a random process whose only transition is to itself.

A **transient state** is a state that has transitions to other states (that is, a state that is nonabsorbing).

An **absorbing Markov chain** is a Markov chain where all states have a (possibly indirect) path to an absorbing state.

The tennis example from Getting Started is an absorbing Markov chain with two absorbing states (SW and RW) and three transient states.

---

**Minds in Action** Episode 23

DERMAN: Hey, I think I can identify the absorbing states by looking at the transition matrix.

SASHA: Okay, how?

DERMAN: It's any column with a 1. If its probability is one, you have no choice!

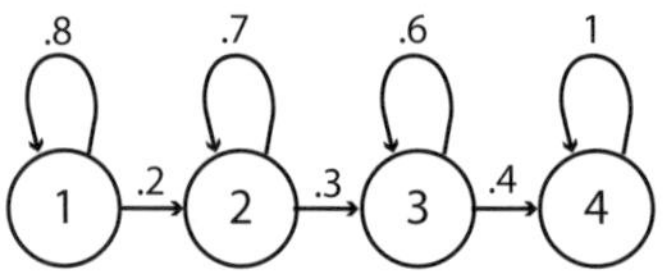


$$T = \begin{pmatrix} .8 & 0 & 0 & 0 \\ .2 & .7 & 0 & 0 \\ 0 & .3 & .6 & 0 \\ 0 & 0 & .4 & 1 \end{pmatrix}$$

SASHA: I believe you . . . mostly.

DERMAN: Come on, you know I'm right. The 1 forces you to go wherever it says.

SASHA: Now *that* I believe. But "wherever it says" might cause trouble. Here, let me change yours a little:

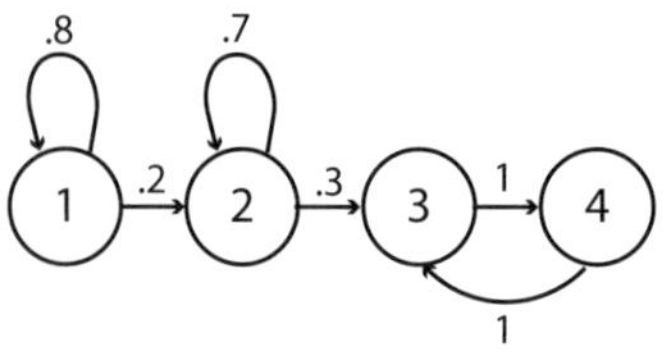


$$T = \begin{pmatrix} .8 & 0 & 0 & 0 \\ .2 & .7 & 0 & 0 \\ 0 & .3 & 0 & 1 \\ 0 & 0 & 1 & 0 \end{pmatrix}$$

Now the third and fourth columns would each have a 1 in them, but they're not absorbing states: each 1 forces you to go *somewhere else*.

DERMAN: Oh. I see what you're getting at. I have a new idea: if the one is on the diagonal, then it's an absorbing state, because it's forcing you to stay there. Otherwise, the 1 forces you to keep moving somewhere else.

SASHA: Like a webpage with only one link.

DERMAN: Or a boring board game.

---

Derman and Sasha's ideas lead to this theorem.

**Theorem 6.7**

*For a Markov chain with transition matrix $T$, state $i$ is an absorbing state if and only if $T_{ii} = 1$.*

### For Discussion

**4.** What are the transient and absorbing states in Derman and Sasha's examples?

**5.** What are the transient and absorbing states in the tennis example from Getting Started?

**6.** **a.** Can a Markov chain have no absorbing states? Explain.
**b.** Can a Markov chain have no transient states? Explain.

### Developing Habits of Mind

**Use properties.** One interesting question is whether or not an absorbing Markov chain can be regular. Both absorbing and regular chains can be classified by their transition matrices.

- A Markov chain is *regular* if for some $n$, the power $T^n$ of its transition matrix has no zeros.
- A Markov chain is *absorbing* if its transition matrix $T$ has a 1 on its diagonal.

For any absorbing state, the corresponding column of the transition matrix is a unit vector. Here's one example of a four-state transition matrix with state 3 absorbing:

$$T = \begin{pmatrix} 0.6 & 0.2 & 0 & 0 \\ 0.3 & 0.4 & 0 & 0 \\ 0.1 & 0.4 & 1 & 0.5 \\ 0 & 0 & 0 & 0.5 \end{pmatrix}$$

Now calculate $T^2$:

$$T^2 = \begin{pmatrix} 0.42 & 0.2 & 0 & 0 \\ 0.3 & 0.22 & 0 & 0 \\ 0.28 & 0.58 & 1 & 0.75 \\ 0 & 0 & 0 & 0.25 \end{pmatrix}$$

The third column of $T^2$ is the same unit vector! Recall the Pick-Apart Theorem from Chapter 4: $(AB)_{*j} = A \cdot B_{*j}$. For the $T$ given above, $T^2_{*3}$ is given by

$$T^2_{*3} = T \cdot T_{*3} = \begin{pmatrix} 0.6 & 0.2 & 0 & 0 \\ 0.3 & 0.4 & 0 & 0 \\ 0.1 & 0.4 & 1 & 0.5 \\ 0 & 0 & 0 & 0.5 \end{pmatrix} \begin{pmatrix} 0 \\ 0 \\ 1 \\ 0 \end{pmatrix}$$

Thinking about linear combinations, any product $TV$ is a linear combination of the columns of $T$, with the values in $V$ giving the combination. This multiplication says to use 1 times the third column, and 0 times the others. In other words, $T^2_{*3}$ is exactly the third column of $T$.

The same logic applies to $T^3 = (T^2) \cdot T$ or any power of $T$. The third column of $T^n$ will *always* be $\begin{pmatrix} 0 \\ 0 \\ 1 \\ 0 \end{pmatrix}$. This proves that $T$ cannot be the transition matrix of a regular Markov chain! Why? Theorem 6.5 from Lesson 6.5 says that it could only be regular if some $T^n$ had no zeros. Here, $T^n$ always has at least one zero entry.

The argument made for $T$ applies to any transition matrix with a 1 on its diagonal, so *an absorbing Markov chain can never be regular* and vice versa.

$\longleftarrow$ In the exercises, you'll be asked to show that this is *not* always true when $T$ has a 1 that isn't on the diagonal.

Markov chains can be regular or absorbing, never both. It is important to recognize that this isn't the entire story: many Markov chains are neither regular nor absorbing. Here is a simple example from earlier in this module:

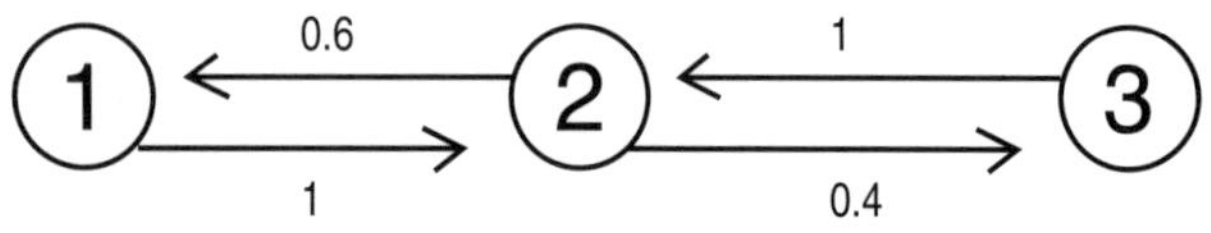


This Markov chain has no absorbing states, but each power $T^n$ of its transition matrix will have several zeros. Therefore, it is neither regular nor absorbing.

$\longleftarrow$ Suppose you started in state 1 of this diagram. Where could you be after $n$ transitions?

In the last lesson, you explored high powers of transition matrices for regular Markov chains. For regular chains, the probability vectors in each column tend toward the same steady state. Because an absorbing Markov chain can't be regular, this won't happen for an absorbing Markov chain. But what does happen?

**Remember**

Regular Markov chains require some $n$ for which $T^n$ has no zeros.

### Example

Roger and Serena are playing a game of tennis. The first player to win four points wins the game, but a player must win by two. Roger and Serena have each won three points, so neither of them can win on the next point. Serena is 60% likely to win any point.

**Problem.** Find the probability that Serena wins in 20 points or fewer.

**Solution.** Here is the transition graph and matrix for this game. The situation described corresponds to starting in state 3.

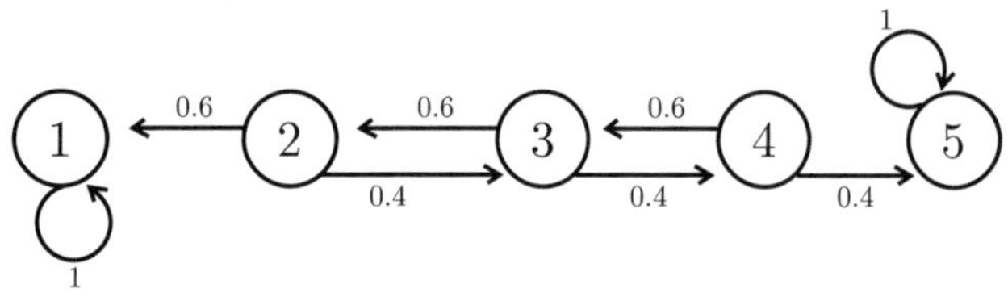

$$T = \begin{pmatrix} 1 & 0.6 & \mathbf{0} & 0 & 0 \\ 0 & 0 & \mathbf{0.6} & 0 & 0 \\ 0 & 0.4 & \mathbf{0} & 0.6 & 0 \\ 0 & 0 & \mathbf{0.4} & 0 & 0 \\ 0 & 0 & \mathbf{0} & 0.4 & 1 \end{pmatrix}$$

The absorbing states are states 1 and 5, with state 1 being a win for Serena. The question is asking for the probability of starting in state 3 and ending in state 1 after 20 transitions.

This is precisely the question answered by the $T^nP$ Theorem (Theorem 6.2). Since they are starting at state 3, the probability vector $P$ is $P = \begin{pmatrix} 0 \\ 0 \\ 1 \\ 0 \\ 0 \end{pmatrix}$. So the probability vector 20 transitions from now ($n = 20$) would be $T^{20}P$, which is just the third column of $T^{20}$. Here is $T^{20}$ to five decimal places:

$$T^{20} \approx \begin{pmatrix} 1 & 0.87655 & \mathbf{0.69186} & 0.41482 & 0 \\ 0 & 0.00032 & \mathbf{0} & 0.00049 & 0 \\ 0 & 0 & \mathbf{0.00065} & 0 & 0 \\ 0 & 0.00022 & \mathbf{0} & 0.00032 & 0 \\ 0 & 0.12291 & \mathbf{0.30749} & 0.58437 & 1 \end{pmatrix}$$

There is about a 69.2% chance that Serena has won (state 1), a 30.7% chance that Roger has won (state 5), and less than a 0.1% chance that the game is tied (state 3). The way the game is played makes it impossible for either player to be one point ahead after exactly 20 points.

$\longleftarrow$ With a finite number of states and transitions, a probability of 0 means the given transition is impossible. A higher $n$ could be used to further reduce the chances the game continues. Try $T^{100}$ on a calculator . . .

Column 3 of $T^{20}$ gives the probability vector after 20 transitions, when starting in state 3. The same can be said for any column of $T^{20}$ for any desired starting state. Note especially columns 1 and 5, which are locked in place as unit vectors. This is the effect of an absorbing state—if you start there, you must remain there.

**For You to Do**

**7.** If Roger starts with a one-point lead, find the probability that he wins within 25 points, the probability that he loses within 25 points, and the probability that the game is still going 25 points later.

### Transient States

So far the focus has been more on the absorbing states. Now you will take a closer look at transient states.

### For You to Do

**8.** In *Count to Five*, determine the probability that you have exactly zero points

**a.** ... when the game begins **b.** ... after one turn

**c.** ... after two turns **d.** ... after three turns

**e.** ... after four turns

**9.** Kevin claims that on average, players spend two turns during *Count to Five* with no points. Use the results from problem 8 to explain why the actual average must be larger than two.

### Developing Habits of Mind

**Reason about calculations.** Absorbing states act like vacuums. In an absorbing Markov chain, the probability of being in a transient state instead of an absorbing state continues to drop as the number of transitions increases, and the only steady states for absorbing Markov chains are those with zeros for all transient states.

The long-term probability of being in a transient state of an absorbing Markov chain always approaches zero. But an interesting question is the expected number of times each transient state will be visited before reaching an absorbing state. This could tell you the expected length of *Count to Five*, or how many times, on average, the tennis game will be tied.

You can compute the expected number of times an event occurs by adding the probability it occurs at each opportunity. For example, if there is a 60% chance of rain each day for four days, the expected number of days of rain is $0.6 \cdot 4 = 2.4$.

Given transition matrix $T$, $T_{ij}^n$ tells you the probability that if you start in state $j$, after $n$ turns you'll be in state $i$. For example, here are $T$ and $T^2$ for *Count to Five*:

$$T = \begin{pmatrix} .4 & .4 & .2 & .2 & .2 & 0 \\ .2 & 0 & .2 & 0 & 0 & 0 \\ .2 & .2 & 0 & .2 & 0 & 0 \\ .2 & .2 & .2 & 0 & .2 & 0 \\ 0 & .2 & .2 & .2 & 0 & 0 \\ 0 & 0 & .2 & .4 & .6 & 1 \end{pmatrix} \quad T^2 = \begin{pmatrix} \mathbf{.32} & .28 & .24 & .16 & .12 & 0 \\ .12 & .12 & .04 & .08 & .04 & 0 \\ \mathbf{.16} & .12 & .12 & .04 & .08 & 0 \\ .16 & .16 & .12 & .12 & .04 & 0 \\ .12 & .08 & .08 & .04 & .04 & 0 \\ .12 & .24 & .4 & .56 & .68 & 1 \end{pmatrix}$$

So $T_{11}^2$ says there is a 32% chance that if you started with zero points (the first state), you will have zero points after two turns. $T_{31}^2$ says there is a 16% chance that if you started with zero points, you will have two points after two turns.

Suppose you start in state 1, and you want to determine the expected number of times state 1 occurs. You can compute this by adding the probability it occurs in the first transition, the second transition, the third ... with no limit. Using a capital $E$ for the expected number of

$\longleftarrow$ The value is automatically at least 1, since state 1 occurs immediately at the start.

times, you get

$$E_{11} = 1 + T_{11} + T_{11}^2 + T_{11}^3 + \cdots \quad (1)$$

$$E_{11} = 1 + \sum_{k=1}^{\infty} T_{11}^k \quad (2)$$

$$E_{11} = \sum_{k=0}^{\infty} T_{11}^k \quad (3)$$

**For Discussion**

**10.** Explain each of the three equations above. What changes between equations (2) and (3)?

In the same manner, a matrix $E$ can be computed where $E_{ij}$ gives the expected number of times state $i$ occurs, given starting state $j$:

$$E = \sum_{k=0}^{\infty} T^k$$

$$E = I + T + T^2 + T^3 + T^4 + \cdots$$

⟵ What is $T^0$? For any real number $x$, $x^0 = 1$, but $T$ is a matrix . . .

It is possible to approximate $E$ by taking a finite number of terms from the infinite sum. For example, here is $I+T+T^2+T^3+T^4$ for the transition matrix $T$ given above:

$$I + T + T^2 + T^3 + T^4 = \begin{pmatrix} 2.198 & 1.099 & 0.762 & 0.603 & 0.496 & 0 \\ 0.493 & 1.269 & 0.363 & 0.162 & 0.107 & 0 \\ \mathbf{0.581} & 0.518 & 1.269 & 0.360 & 0.158 & 0 \\ 0.617 & 0.581 & 0.493 & 1.243 & 0.338 & 0 \\ 0.280 & 0.422 & 0.386 & 0.323 & 1.099 & 0 \\ 0.830 & 1.110 & 1.728 & 2.309 & 2.802 & 5 \end{pmatrix}$$

For example, a player starting *Count to Five* with zero points will spend an average of 0.581 turns with two points, up to and including the fourth turn.

A better approximation can be found by repeating this process up to a high power of $T$:

$$\sum_{k=0}^{100} T^k = \begin{pmatrix} 3.121 & 1.907 & 1.387 & 1.069 & 0.838 & 0 \\ 0.827 & 1.561 & 0.590 & 0.329 & 0.231 & 0 \\ 1.012 & 0.896 & 1.561 & 0.578 & 0.318 & 0 \\ 1.110 & 1.012 & 0.827 & 1.491 & 0.520 & 0 \\ 0.590 & 0.694 & 0.595 & 0.480 & 1.214 & 0 \\ 94.341 & 94.931 & 96.040 & 97.052 & 97.879 & 101 \end{pmatrix}$$

### For Discussion

**11.** Why are the totals for the last row so large?

---

The transition matrix for *Count to Five* has its transient states listed first and its absorbing state last. It is helpful to organize the transition matrix for an absorbing Markov chain in this way. Look back at the tennis example: the transient states are **S1, T, R1**, and the absorbing states are **SW** and **RW**. Here is the transition matrix rewritten in this state order:

$$T = \begin{array}{c} \\ \text{S1} \\ \text{T} \\ \text{R1} \\ \text{SW} \\ \text{RW} \end{array} \begin{array}{c} \begin{array}{ccccc} \text{S1} & \text{T} & \text{R1} & \text{SW} & \text{RW} \end{array} \\ \begin{pmatrix} \mathbf{0} & \mathbf{0.6} & \mathbf{0} & 0 & 0 \\ \mathbf{0.4} & \mathbf{0} & \mathbf{0.6} & 0 & 0 \\ \mathbf{0} & \mathbf{0.4} & \mathbf{0} & 0 & 0 \\ 0.6 & 0 & 0 & 1 & 0 \\ 0 & 0 & 0.4 & 0 & 1 \end{pmatrix} \end{array}$$

This matrix carries all the information of the original, but the order makes the distinction between transient and absorbing states clearer. A 3-by-3 submatrix $Q$ containing only the information about transient states is highlighted above.

← A **submatrix** is just a matrix made up of rows and columns picked from a larger matrix.

### For You to Do

**12.** Calculate $T^{20}$ for the above matrix. Compare to the result found in the Example from this lesson.

---

The transition matrix for any absorbing Markov chain can be written in the form

$$T = \begin{pmatrix} Q & \mathbf{0} \\ R & I \end{pmatrix}$$

with four submatrices. Each of these gives a specific piece of information:

- $Q$ is a square matrix of transition probabilities from *transient states to transient states*
- $R$ is a matrix of transition probabilities from *transient states to absorbing states*
- $\mathbf{0}$ is matrix of zeros reflecting the impossibility of transitioning from *absorbing states to transient states*
- $I$ is an identity matrix, reflecting the fact that absorbing states feed only themselves.

This breakdown allows you to focus on one piece of the matrix. If you're interested in the behavior of transient states, focus on matrix $Q$. If you're interested in the behavior of absorbing states, focus on matrix $R$.

### For You to Do

**13.** Calculate each of the following:

**a.** $(1 + r + r^2 + \cdots + r^n)(1 - r)$, where $r$ is a real number.

**b.** $(I + T + T^2 + \cdots + T^n)(I - T)$, where $T$ is a matrix and $I$ is the corresponding identity matrix.

### Developing Habits of Mind

**Seek structural similarity.** In an earlier course, you likely used this formula for the sum of a geometric series with common ratio $r$:

$$1 + r + r^2 + \cdots + r^n = \frac{1 - r^{n+1}}{1 - r}$$

Two important questions arise here: is there a similar formula for matrices, and what happens with the infinite sum?

For matrices, the result from the For You to Do problem 13 above suggests this rule:

$$I + T + T^2 + \cdots + T^n = \frac{I - T^{n+1}}{I - T}$$

But this makes no sense. There's no such thing as dividing by a matrix. Like solving $AX = B$, this is done by multiplying by an inverse matrix:

$$I + T + T^2 + \cdots + T^n = (I - T^{n+1}) \cdot (I - T)^{-1}$$

There's still trouble: $I - T$ might not have an inverse. For the matrix of *Count to Five*, here is $I - T$:

$$I - T = \begin{pmatrix} .6 & -.4 & -.2 & -.2 & -.2 & 0 \\ -.2 & 1 & -.2 & 0 & 0 & 0 \\ -.2 & -.2 & 1 & -.2 & 0 & 0 \\ -.2 & -.2 & -.2 & 1 & -.2 & 0 \\ 0 & -.2 & -.2 & -.2 & 1 & 0 \\ 0 & 0 & -.2 & -.4 & -.6 & 0 \end{pmatrix}$$

Not good. The last column is all zeros. Therefore $T$ has a nonzero kernel and cannot have an inverse. The same thing happens for any absorbing state of any absorbing Markov chain!

The formula seems doomed, but not so: use matrix $Q$ instead! This is the matrix of only transient state behavior, so it doesn't have this problem. It's not easy, but you can prove that $(I - Q)$ must have an inverse. So the formula is valid, but for $Q$ instead of $T$:

⟵ See the exercises, especially 22.

$$I + Q + Q^2 + Q^3 + \cdots + Q^n = (I - Q^{n+1}) \cdot (I - Q)^{-1}$$

The rule for infinite geometric series can apply as long as the entries in $Q$ tend toward zero, which they must for transient states. And that formula is much cleaner.

$$I + Q + Q^2 + Q^3 + \cdots = (I) \cdot (I - Q)^{-1} = (I - Q)^{-1}$$

Astoundingly, this formula *works* and gives the expected number of times any transient state is visited before an absorbing state is entered.

**For You to Do**

**14.** For the 6-by-6 matrix from *Count to Five*, determine the 5-by-5 submatrix $Q$, and then compute $(I - Q)^{-1}$. Compare to the summation given earlier.

**15.** What is the average total length of a game of *Count to Five*?

## Exercises

**1.** **a.** Draw and label a transition graph for a Markov chain with at least two absorbing states.
**b.** Build the transition matrix that corresponds to your transition graph.

**2.** For each of the transition graphs in Exercises 1–7 in Lesson 6.5, determine whether the corresponding Markov chain is absorbing.

**3.** Of the three transition graphs pictured below, one corresponds to a regular Markov chain, one to an absorbing chain, and one to a chain that is neither regular nor absorbing. Decide which is which.

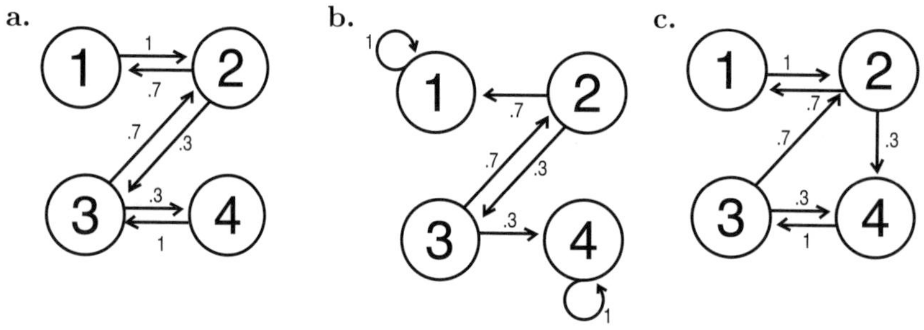


**4.** Draw a 4-state transition graph that corresponds to
**a.** a regular Markov chain
**b.** an absorbing Markov chain
**c.** a chain that is neither regular nor absorbing

⟵ Look back for other examples of chains that are neither regular nor absorbing.

**5.** Find all steady states for the 4-state Markov chain Derman drew in Minds in Action—Episode 23.

**6.** Find all steady states for the 4-state Markov chain Sasha drew in Minds in Action—Episode 23.

**7.** Show, with a transition matrix or a transition graph, that it *is* possible for a Markov chain to have an absorbing state while a subset of the Markov chain is regular.

⟵ The overall Markov chain is *not* regular in this case.

**8.** A simpler version of *Count to Five* is *Flip to Five*. Start with zero points. Flip a coin; if it's heads, add one point; if it's tails, add two. When you reach five points or more, stop.

**a.** Build a 6-by-6 transition matrix for this game.
**b.** Determine the probability that you finish the game in three or fewer turns.
**c.** Determine the probability that you finish the game in five or fewer turns.
**d.** Determine the average length of a game of *Flip to Five*.

**9.** **a.** Use an 11-by-11 transition matrix to determine the average length of a game of *Flip to Ten*, which plays like *Flip to Five* but doesn't end until you reach 10 points.
**b.** Suppose you have five points when playing *Flip to Ten*. Determine the average number of turns remaining in the game.

**10.** In For You to Do problem 15, you determined the average total length of a game of *Count to Five*. The method of solution only uses one column of that matrix. Use the entire matrix to determine the average number of turns remaining in *Count to Five* if you have

**a.** zero points **b.** one point **c.** two points
**d.** three points **e.** four points

**11.** You saw this reordered transition matrix for the tennis example earlier in the lesson with the 3-by-3 square matrix $Q$ in bold.

$$T = \begin{array}{c} \\ \text{S1} \\ \text{T} \\ \text{R1} \\ \text{SW} \\ \text{RW} \end{array} \begin{array}{c} \begin{array}{ccccc} \text{S1} & \text{T} & \text{R1} & \text{SW} & \text{RW} \end{array} \\ \begin{pmatrix} \mathbf{0} & \mathbf{0.6} & \mathbf{0} & 0 & 0 \\ \mathbf{0.4} & \mathbf{0} & \mathbf{0.6} & 0 & 0 \\ \mathbf{0} & \mathbf{0.4} & \mathbf{0} & 0 & 0 \\ 0.6 & 0 & 0 & 1 & 0 \\ 0 & 0 & 0.4 & 0 & 1 \end{pmatrix} \end{array}$$

**a.** Determine $I - Q$ and $(I - Q)^{-1}$.
**b.** If the players start the game tied, determine the average number of ties that will occur. (Hint: This answer is found in one of the entries of $(I - Q)^{-1}$.)
**c.** If the players start the game tied, determine the average number of points played when Serena is leading by 1.
**d.** If the players start the game tied, determine the average number of points played when Roger is leading by 1.
**e.** If the players start the game tied, determine the average number of points played.

**12.** Suppose Serena and Roger are equally likely to win points, instead of Serena winning 60% of the points. The players start the game tied.

**a.** Build a transition matrix $T$ for this situation, ordered the same way as the one above.
**b.** Determine the average number of points played.
**c.** What is the probability that Serena wins the game?

**13.** The game of *HiHo! Cherry-O* plays a lot like *Count to Five*. On each turn, you may do one of seven things.

- Add one point.
- Add two points.
- Add three points.
- Add four points.
- Subtract one point (but never drop below zero).
- Subtract two points (but never drop below zero).
- Reset score to zero.

The game ends when the score reaches 10 or more.

**a.** Build an 11-by-11 transition matrix for this game. Like *Count to Five*, the states are the number of possible points in the game (0 through 10).

**b.** Determine the probability of completing the game in four or fewer turns.

**c.** Determine the probability that the game will *not* be complete within 20 turns.

**d.** Use the methods of this lesson to determine the average number of turns taken in the game.

**14.** **Take It Further.** Suppose that the probability that Serena wins a point is $p$ and the probability that Roger wins is $(1-p)$. The players start the game tied.

**a.** Build a transition matrix $T$ for this situation, ordered the same way as the one from Exercise 11.

**b.** Determine the average number of points played, in terms of $p$.

**c.** What is the probability that Serena wins the game, in terms of $p$?

**15.** Consider an absorbing Markov chain with $n$ states whose only absorbing state is state $i$. Consider some other state $j$.

**a.** Explain why there must be a path from $j$ to $i$, possibly through several other states. (Hint: What does the definition of *absorbing Markov chain* say?)

**b.** Let $t$ be the length of the *shortest* path from $j$ to $i$. What is the largest possible value of $t$? (Hint: Could the shortest such path double back on itself?)

**16.** Let $T$ be the transition matrix for a Markov chain with one absorbing state $i$. Use the result of Exercise 15 to prove the following.

**Theorem 6.8**

*An $n$-state Markov chain with transition matrix $T$ and single absorbing state $i$ is an absorbing Markov chain if and only if all entries in $T_{i*}^{n-1}$ are nonzero.*

**Remember**

The notation $T_{i*}$ means the $i^{th}$ row of $T$.

**17.** Let $S$ be the set of absorbing states for a Markov chain.

**a.** Explain why the Markov chain is absorbing if and only if there is a path from every state $j$ to at least one of the states in $S$.

**b.** Must there be a path from every state $j$ to *all* of the absorbing states in $S$? Explain or give an example.

**c.** Let $T_S^k$ be the sum of the rows of $T^k$ that correspond to the absorbing states in $S$. Prove the following.

**Theorem 6.9**
*An $n$-state Markov chain with transition matrix $T$ and set of absorbing states $S$ is an absorbing Markov chain if and only if all entries in $T_S^{n-1}$ are nonzero.*

$\longleftarrow$
Why is $(n-1)$ specifically picked here as the power of $T$?

**18.** The tennis example from this module has 5 states. Here is $T^4$ with the transient states first, then the absorbing states:

$$T^4 = \begin{array}{c} \\ \text{S1} \\ \text{T} \\ \text{R1} \\ \text{SW} \\ \text{RW} \end{array} \begin{array}{c} \begin{array}{ccccc} \text{S1} & \text{T} & \text{R1} & \text{SW} & \text{RW} \end{array} \\ \begin{pmatrix} .1152 & 0 & .1728 & 0 & 0 \\ 0 & .2304 & 0 & 0 & 0 \\ .0768 & 0 & .1152 & 0 & 0 \\ .744 & .5328 & .216 & 1 & 0 \\ .064 & .2368 & .496 & 0 & 1 \end{pmatrix} \end{array}$$

**a.** Jesch claims that no matter what the situation in the tennis game, there is less than a 1 in 4 chance that the game will last through the next four points. Is he right? Explain using $T^4$ above.

**b.** Jesch also claims that no matter what the situation in the tennis game, there is less than a 1 in 12 chance that the game will last through the next eight points. Is he right?

$\longleftarrow$
It is possible to answer this question using only $T^4$. Look for that before calculating $T^8$.

**19.** Consider an absorbing Markov chain with $n$ states and transition matrix $T$.

**a.** Prove that there is a number $p < 1$ such that the probability of remaining in transient states after $n-1$ transitions *must* be less than or equal to $p$.

**b.** Prove that, given this $p < 1$, the probability of remaining in transient states after $2(n-1)$ transitions must be less than or equal to $p^2$.

**c.** The probability of remaining in transient states after $k(n-1)$ transitions must be less than or equal to what expression?

**d.** Prove that the probability of remaining in transient states forever is zero.

For an absorbing Markov chain, when states are ordered according to whether they are transient or absorbing, the transition matrix can be written in the form

$$T = \begin{pmatrix} Q & \mathbf{0} \\ R & I \end{pmatrix}$$

Here, $Q$ is a square matrix of transition probabilities from transient states to transient states.

**20.** **a.** Write the 4-by-4 transition matrix for the Markov chain whose transition graph is pictured in Exercise 3**b** in the above form.

**b.** For this transition matrix, calculate $Q^{20}$.

**21.** Use the result of Exercise 19 to prove this theorem.

**Theorem 6.10**

*Given an absorbing Markov chain with transition matrix $T$, if $Q$ is the square submatrix of transition probabilities from transient states to transient states, then*

$$\lim_{n\to\infty} Q^n = \mathbf{0}$$

**22.** Let $Q$ be a matrix such that

$$\lim_{n\to\infty} Q^n = \mathbf{0}$$

**a.** Show that

$$(I - Q) \cdot (I + Q + Q^2 + Q^3 + \cdots + Q^n + \cdots) = I$$

**b.** Explain how the above can be used to show that

$$(I + Q + Q^2 + Q^3 + \cdots + Q^n + \cdots) = (I - Q)^{-1}$$

**23. Take It Further.** Prove that if $Q$ is the matrix of transition probabilities from transient states to transient states, then $|\det(Q)| < 1$.

⟵ **Hint:** How does $\det(Q^2)$ compare to $\det(Q)$?

Consider the situation in Exercise 18 from Lesson 6.5.

**24.** Construct a 6-by-6 transition matrix $T$ for this Markov chain, using the state order {freshman, sophomore, junior, senior, graduated, dropped out}.

**25.** Determine the expected number of years a freshman will spend in each of the four grade levels.

**26.** Calculate $T^{30}_{51}$ and interpret its meaning.

**27.** The game of *Chutes and Ladders* can be modeled completely as an absorbing Markov chain.

**a.** Why can this game be modeled as a Markov chain at all?
**b.** Why can this game be modeled as an absorbing Markov chain?

**28. Take It Further.**

**a.** Using technology, build a large transition matrix for *Chutes and Ladders*. Even though there are 100 squares on the board, there will not be 100 states in the matrix, because some squares cannot be landed on. For example, square 4 cannot be landed on, since it immediately sends the player to square 14.
**b.** Determine the probability that the game ends within 10 turns.
**c.** Determine the probability that the game does *not* end within 100 turns.
**d.** Using the methods of this lesson, determine the average number of turns taken in the game before it ends by reaching square 100.

## 6.7 The World Wide Web: Markov and PageRank

When you search for particular words on the internet, search engines find millions of pages that match your terms. One way of organizing results uses regular Markov chains.

**In this lesson, you will learn how to**

- see an application of Markov chains to a difficult problem
- understand how the PageRank algorithm uses Markov chains to model user behavior

With the growth of the internet in the mid-1990s, sites to search the World Wide Web quickly became popular. The earliest such sites were created by hand: one popular site was titled "Yet Another Hierarchically Organized Oracle," or Yahoo for short. Further growth led to the creation of *search engines*, which might find all the webpages with keywords such as "tennis." But as the Web kept growing, the quality of searches decreased, because it was difficult to determine, algorithmically, what webpages were the most popular or trustworthy.

In 1998, two graduate students presented a solution to the problem based on regular Markov chains. Their solution treats each webpage as a state in a gigantic regular Markov chain, and then determines its unique steady state. Each page is ranked according to its steady-state probability, from high to low, and higher-ranked pages are displayed earlier in search results. Their search engine was called Google, and the quality of its search results made it the most frequently used website in the 2000s.

### For You to Do

This graph shows five webpages and the one-way links between them.

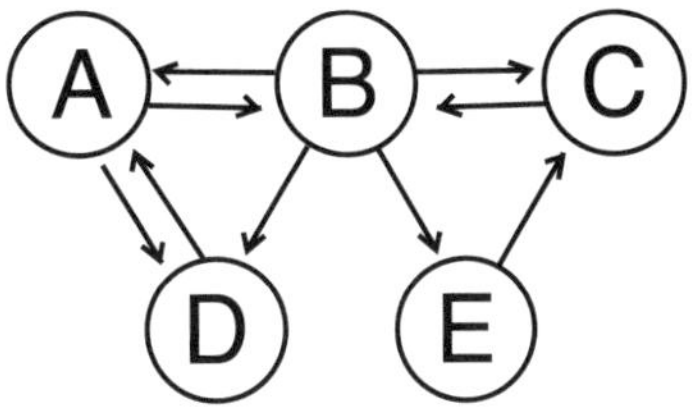


1. Based on the links, rank the pages from 1 to 5 in order of how frequently you think they would be visited.

2. Build $H$, a 5-by-5 transition matrix for these webpages. Assume that each link is clicked with equal probability. For example, the first column of $H$ corresponds to page A. Its values include a 0.5 probability of transitioning to B, a 0.5 probability of transitioning to D, and zeros otherwise.

**3.** Let $H$ be the transition matrix you calculated in problem 2. Calculate $H^{20}$. What does this matrix represent and how can it be used to answer problem 1?

---

This description comes from the 1998 paper by Sergey Brin and Lawrence Page (for whom PageRank might be named):

> PageRank can be thought of as a model of user behavior. We assume there is a "random surfer" who is given a webpage at random and keeps clicking on links, never hitting "back" but eventually gets bored and starts on another random page. The probability that the random surfer visits a page is its PageRank. ... A page can have a high PageRank if there are many pages that point to it, or if there are some pages that point to it and have a high PageRank. Intuitively, pages that are well cited from many places around the web are worth looking at. Also, pages that have perhaps only one citation from something like the Yahoo! homepage are also generally worth looking at.

← Lawrence Page, Sergey Brin, Rajeev Motwani, and Terry Winograd (1999) "The PageRank Citation Ranking: Bringing Order to the Web." Technical Report. Stanford InfoLab.

Problems 1–3 above suggest that a random Web surfer can be modeled using a transition matrix with columns containing zeros and $\frac{1}{n}$, where $n$ is the number of links away from each page. (For the above graph, $n = 2$ for page A.) However, there are two complications.

← The $\frac{1}{n}$ method doesn't account for what is described above when the surfer "eventually gets bored."

- **Some pages don't have links.** Over half the sites on the World Wide Web are dead ends with no links. PageRank models this by assuming the random Web surfer will pick a page at random, assigning an equal probability $\frac{1}{p}$ to each transition, where $p$ is the total number of pages.
- **Behavior does not produce a regular Markov chain.** Picture a webpage whose only link is a "Reload" button. According to the linking model, this page would be an absorbing state! Since a Markov chain cannot be both regular and absorbing, it would not be possible to compute PageRank without dealing with this situation. PageRank deals with this by giving a "damping factor" $d = 0.2$, a probability that instead of using a page's links, the next page is determined randomly.

← Google actually used $d = 0.15$, but $d = 0.2$ is a little easier to work with.

The probability of reaching *any* page from any other page is, therefore, at least $\frac{d}{p}$, where $p$ is the total number of pages. If there are $n$ links leaving a page, linked pages will be reached with probability $\frac{d}{p} + \frac{1-d}{n}$. (Why?)

---

**For Discussion**

**4.** Why do the changes listed above guarantee a transition matrix that corresponds to a regular Markov chain?

---

Once the transition matrix $T$ is constructed, the *PageRank vector* $v$ is the unique vector such that $Tv = v$. Without both of the changes listed above, this vector wouldn't be unique, and it would not be possible to correctly decide for all pairs of pages whether one was more or less likely to be visited than another. Computing the PageRank vector was at the heart of the execution and success of Google. Because of this, the end result of the PageRank calculation has been called the "$25 billion eigenvector."

← Google's own website states that they "became successful precisely because we were better and faster at finding the right answer than other search engines at the time." PageRank has evolved somewhat, but the Markov concept is still at its heart.

**Example**

**Problem.** Find the PageRank vector for the following Web graph, assuming damping factor $d = 0.2$.

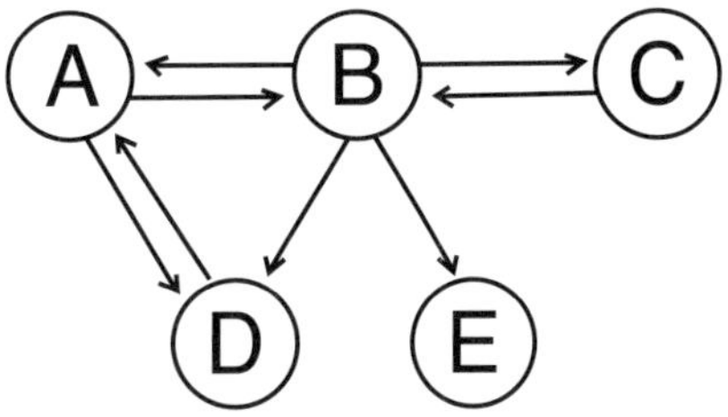


**Solution.** With five pages, $\frac{d}{p} = \frac{0.2}{5} = 0.04$. The value $1 - d = 0.8$ gives the probability that a normal link is followed instead, and if a page has $n$ links, each will have probability $0.04 + \frac{0.8}{n}$. If a page has no links, all pages will have equal probability $\frac{1}{5}$.

- Page A links to B and D, so each linked page has transition probability 0.44, while each unlinked page has transition probability 0.04.
- Page B links to A, C, D, and E, so each linked page has transition probability 0.24, while each unlinked page has transition probability 0.04.
- Page C links only to B, so B has transition probability 0.84, while other pages have transition probability 0.04.
- Page D links only to A, so A has transition probability 0.84, while other pages have transition probability 0.04.
- Page E links nowhere, so all pages have transition probability 0.2.

Here is the 5-by-5 transition matrix $T$:

$$T = \begin{pmatrix} 0.04 & 0.24 & 0.04 & 0.84 & 0.2 \\ 0.44 & 0.04 & 0.84 & 0.04 & 0.2 \\ 0.04 & 0.24 & 0.04 & 0.04 & 0.2 \\ 0.44 & 0.24 & 0.04 & 0.04 & 0.2 \\ 0.04 & 0.24 & 0.04 & 0.04 & 0.2 \end{pmatrix}$$

There are several methods for finding the vector $v$ that solves $Tv = v$. If $Tv = v$, then $Tv = Iv$ and $(T - I)v = \mathbf{0}$. The PageRank vector can be found by determining the kernel of $(T - I)$.

As seen in Lesson 6.5, an approximation to $v$ can be found by computing high powers of $T$. For example, here is $T^{20}$:

$$T^{20} \approx \begin{pmatrix} 0.291 & 0.291 & 0.291 & 0.291 & 0.291 \\ 0.262 & 0.262 & 0.262 & 0.262 & 0.262 \\ 0.110 & 0.110 & 0.110 & 0.110 & 0.110 \\ 0.227 & 0.227 & 0.227 & 0.227 & 0.227 \\ 0.110 & 0.110 & 0.110 & 0.110 & 0.110 \end{pmatrix}$$

The approximation shows that page A has the highest PageRank, followed by B and D, with C and E tied much lower.

### Developing Habits of Mind

**Use forms with purpose.** The example above uses a 5-by-5 transition matrix, but the Web has billions of pages! It is prohibitively difficult to compute $T^n$ for such a matrix, and even more difficult to find the kernel of $(T - I)$. Another method is used instead: a starting vector $v_0$ is chosen, and then $v_1 = Tv_0$ is computed. Then $v_2 = Tv_1$, and several more, until the results have "settled." The last of these vectors should be very close to the correct steady state. Why does this work? As seen earlier, if $T$ is the transition matrix of a regular Markov chain, then the columns of $T^n$ are almost identical for large enough $n$, and $T^n v$ will be almost identical for any probability vector $v$.

This method is more efficient since the entire matrix $T$ need not be recomputed time and again.

There is one significant shortcut used in PageRank calculation. Since more than half the pages on the Web are dead ends like page E from the example above, all such pages can be joined as a "chunk" (since they all have the same link behavior) with their probabilities computed after the fact. See Exercises 10 and 11 for more about this.

## Exercises

1. For each given Web graph, determine the PageRank vector using damping factor $d = 0.2$.

   **a.**

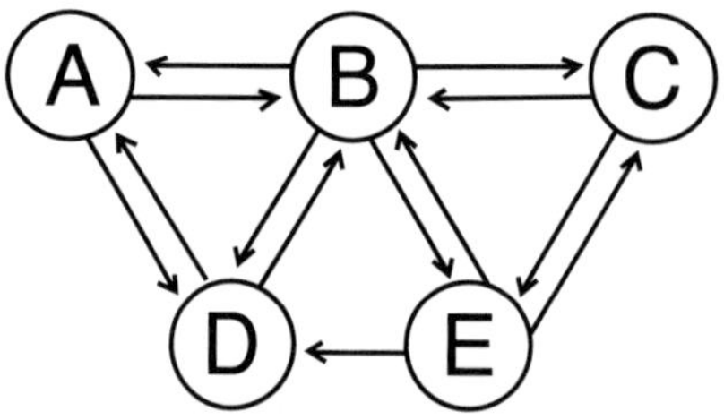

**b.**

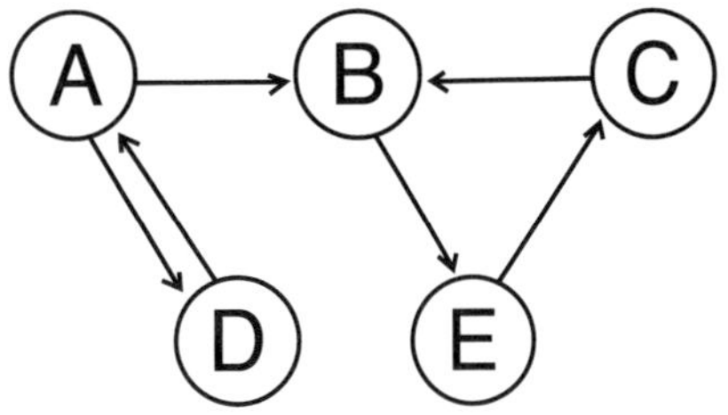


**c.**

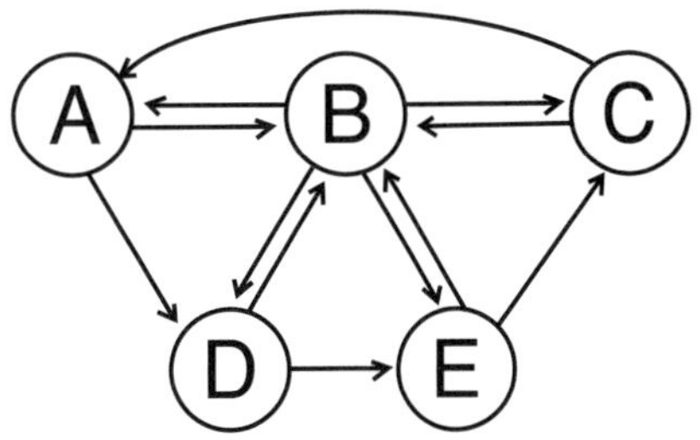


**d.**

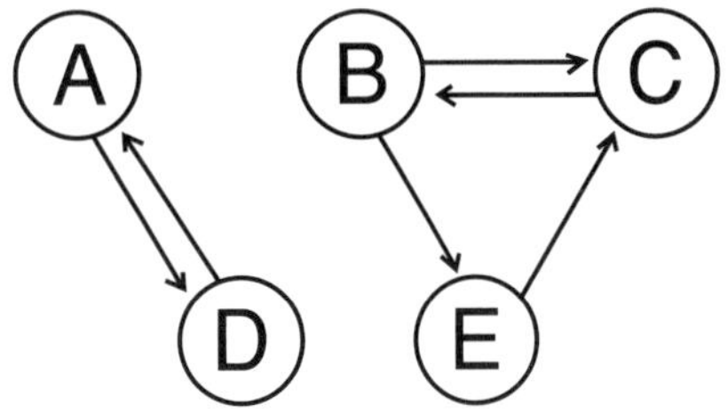


**e.**

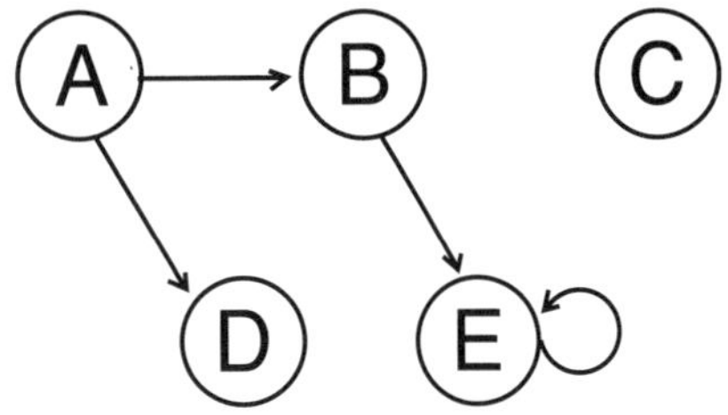


**2.** Use the map from Exercise **1d** for the following:

**a.** Recompute the PageRank using damping factor $d = 0.15$.

**b.** Recompute the PageRank using damping factor $d = 0.8$. What happens?

**c.** Recompute the PageRank using damping factor $d = 0$ (following links only). What happens?

**3.** Sally says that the nature of PageRank makes it impossible for a page to have greater than 0.5 probability in the PageRank vector. Is she right? Explain.

4. **Take It Further.** If $d = 0.2$ is the damping factor, determine the *maximum* possible PageRank a page can have when there are $n$ total pages.

5. Let $T$ be the transition matrix from the example in this lesson.

$$T = \begin{pmatrix} 0.04 & 0.24 & 0.04 & 0.84 & 0.2 \\ 0.44 & 0.04 & 0.84 & 0.04 & 0.2 \\ 0.04 & 0.24 & 0.04 & 0.04 & 0.2 \\ 0.44 & 0.24 & 0.04 & 0.04 & 0.2 \\ 0.04 & 0.24 & 0.04 & 0.04 & 0.2 \end{pmatrix}$$

Also consider these probability vectors:

$$V_1 = \begin{pmatrix} 1 \\ 0 \\ 0 \\ 0 \\ 0 \end{pmatrix}, V_2 = \begin{pmatrix} 0 \\ 1 \\ 0 \\ 0 \\ 0 \end{pmatrix}, V_3 = \begin{pmatrix} 0.2 \\ 0.2 \\ 0.2 \\ 0.2 \\ 0.2 \end{pmatrix}, V_4 = \begin{pmatrix} 0 \\ 0 \\ .5 \\ 0 \\ .5 \end{pmatrix}$$

← With no information about the behavior of a network, $V_3$ is a good starting "guess" for PageRank. When "updating" PageRank for new links and changes to the Web, the previous PageRank was used as the starting guess.

**a.** Calculate $T^n \cdot V_1$ for $n = 1, 2, 3, 4, 5, 20$. Describe what happens.
**b.** Repeat for $V_2$, $V_3$, and $V_4$.
**c.** For each of $V_1$ through $V_4$, determine the smallest value of $n$ for which all the elements of $T^n \cdot V_i$ are within 0.01 of the PageRank vector

$$V = \begin{pmatrix} 0.291 \\ 0.262 \\ 0.110 \\ 0.227 \\ 0.110 \end{pmatrix}$$

6. Build a small Web graph using about 10 pages from your school's website.

**a.** Build the transition matrix for your Web graph with damping factor $d = 0.2$. Which page has the highest PageRank? the lowest?
**b.** Repeat with damping factor $d = 0$. Are there any changes?

7. Build a small network for "following" in a social media service such as Twitter. The network should include between 10 and 20 people and all their interconnections. Use the PageRank algorithm with $d = 0.2$ to determine a popularity ranking for the people you chose.

← **Note:** "Following" isn't always both ways. A can follow B even if B doesn't follow A. Treat this like you would a "link" from page A to page B.

8. Look back at the airport network from Lesson 4.1. Consider each flight from one city to another to be a "link" like a link between webpages.

← Webpages can also have multiple links to the same page . . .

**a.** Determine the PageRank order of the six cities using $d = 0.2$. According to these flights, which airport will be the most crowded for this airline?
**b.** Repeat using $d = 0$. Is it reasonable to have a damping factor used in this situation?
**c.** What other assumptions are made that could affect the true ranking of these six cities?

**9.** Build another small network of links, and then compute the PageRank of the network. Here are some suggestions:

- Facebook friends or other two-way social media
- Text messages (count the number of messages sent from A to B in one day as the number of links)
- Cell phone calls
- Blog references
- YouTube "watch this next" lists
- News sites' "see also" listings
- Thesaurus lists of synonyms

**10.** Consider the Example from this lesson. A more efficient way of computing the PageRank for pages without links is available, and the method works by looking at the three possible ways to reach page E.

**a.** The first way is a link from "anywhere". Explain why the probability of reaching E this way is $\frac{0.2}{5} = 0.04$, using the damping factor given in the example.

**b.** The second way is a link from page B. The long-term probability of reaching B is 0.262 (as computed in the example). Show that the probability of reaching E this way is 0.0524.

**c.** The third way is a "link" from page E; E can be thought of as having links to *every* page, including itself since, when E is reached, all pages are equally likely to be next. If the long-term probability of reaching E is $p$, show that the probability of reaching E from itself is $0.16p$.

**d.** Solve the equation $p = 0.04 + 0.0524 + 0.16p$ to determine the long-term probability of reaching E.

**11.** The link network below includes three pages with no links.

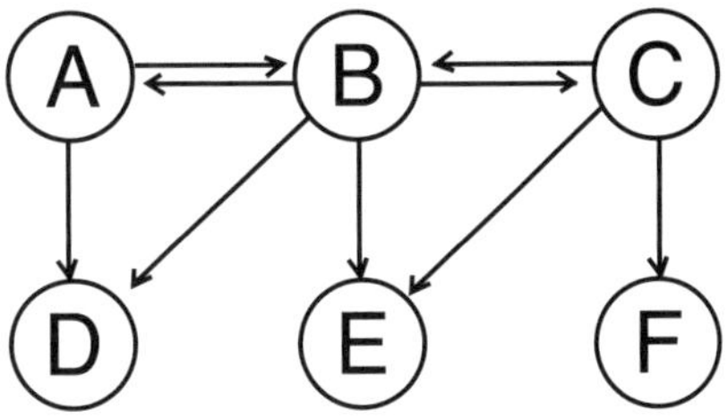


**a.** Build a new network with three "pages," the two pages with real links, and one "page" representing the set of pages with no links. Include the probabilities of moving from the no-link "page" to all other pages, including itself.

**b.** Using the new network, compute the long-term probability of reaching each of the three "pages."

**c.** Use the method of Exercise 10 to determine the long-term probability of reaching each of the three actual no-link pages.

← **Note:** The probability of going from the no-link page to itself is *not* $\frac{1}{3}$.

# Chapter 6 Mathematical Reflections

These problems will help you summarize what you have learned in this chapter:

**1.** Suppose that if you're healthy today, there is a 90% chance you will be healthy again tomorrow (and a 10% chance of being sick). But if you're sick today, there is only a 30% chance you will be healthy tomorrow (and a 70% chance of being sick again).

**a.** You're healthy today. Find the probability that you'll be healthy two days from now.
**b.** Jeff is sick today. Assuming the same information, find the probability that Jeff will be healthy two days from now.
**c.** Find the probability that you'll be healthy three days from now.

**2.** Roger and Serena are playing a game of tennis. The first player to win four points wins the game, but a player must win by two. Roger and Serena have each won three points, so neither of them can win on the next point. Suppose Serena and Roger are equally likely to win points. The players start tied.

**a.** Build a transition matrix $T$ for this situation.
**b.** Determine the average number of points played.
**c.** What is the probability that Serena wins the game?

**3.** After one day, the probability of being healthy is 0.9 and the probability of being sick is 0.1. Use matrix multiplication to determine the probability of being healthy after two days, and the probability of being sick after two days.

**4.** A **steady state** is a probability vector $X$ such that for the transition matrix $T$, $TX = X$. Find a steady state for each of the following matrices $T$.

**a.** $T = \begin{pmatrix} 0.9 & 0.3 \\ 0.1 & 0.7 \end{pmatrix}$

**b.** $T = \begin{pmatrix} 0.75 & 0.5 \\ 0.25 & 0.5 \end{pmatrix}$

**c.** $T = \begin{pmatrix} 0.5 & 0.4 & 0.5 \\ 0.2 & 0.3 & 0.5 \\ 0.3 & 0.3 & 0 \end{pmatrix}$

**d.** $T = \begin{pmatrix} 1 & 0.6 & 0 & 0 & 0 \\ 0 & 0 & 0.6 & 0 & 0 \\ 0 & 0.4 & 0 & 0.6 & 0 \\ 0 & 0 & 0.4 & 0 & 0 \\ 0 & 0 & 0 & 0.4 & 1 \end{pmatrix}$

**5.** How can I represent a system using a transition matrix?

### Vocabulary

In this chapter, you saw these terms and symbols for the first time. Make sure you understand what each one means, and how it is used.

- absorbing Markov chain
- absorbing state
- attractor
- Markov chain
- node
- probability vector
- random process
- steady state
- submatrix
- transient state
- transition graph
- transition matrix
- transition probability

6. What is the difference between an absorbing state and a transient state?

7. What is the average number of turns it would take to win the game *HiHo! Cherry-O*?

CHAPTER

# 7 Vector Spaces

There are some significant structural similarities between vectors and matrices. For example, you can add two vectors and you can add two matrices. You can scale a vector and you can scale a matrix.

This structure can also be found in other mathematical systems: you can add and scale polynomials; you can add and scale complex numbers; you can add and scale trigonometric functions. The properties described in Theorems 1.2 and 4.1 carry over to adding and scaling in these other systems.

Many systems behave this same way with respect to addition and scaling. These structural similarities were noticed by mathematicians for some time, and formalized in the late $19^{\text{th}}$ century by Giuseppe Peano when he described a *vector space*.

**By the end of this chapter, you will be able to answer questions like these:**

1. What properties are necessary for a system to be a vector space?
2. How can you determine whether a set is a generating system for a given vector space?
3. What is the coordinate vector for $v = \begin{pmatrix} 5 & 0 \\ 2 & 3 \end{pmatrix}$ with respect to the basis $B = \left\{ \begin{pmatrix} 0 & 0 \\ 1 & 1 \end{pmatrix}, \begin{pmatrix} 1 & 0 \\ 0 & 1 \end{pmatrix}, \begin{pmatrix} 1 & 1 \\ 0 & 0 \end{pmatrix}, \begin{pmatrix} 1 & 0 \\ 0 & 0 \end{pmatrix} \right\}$?

**You will build good habits and skills for ways to**

- look for similarity in structure
- find general results
- reason about calculations
- note general results

### Vocabulary and Notation

- basis for a vector space
- column space
- coordinate vector
- dimension of a vector space
- finite dimensional
- generating system
- linear span
- $\mathrm{Mat}_{m\times n}$
- $\mathbb{R}_n[x]$
- row space
- subspace
- vector space
- zero vector

# 7.1 Getting Started

## Exercises

1. Find numbers $a$, $b$, and $c$ that make each statement true or show that there are no such numbers:

   **a.** $a(1,2,-1,3)+b(1,0,1,0)+c(2,1,0,0)=(3,5,-1,9)$
   **b.** $a(1,2,-1,3)+b(1,0,1,0)+c(2,1,0,0)=(13,8,1,6)$
   **c.** $a(1,2,-1,3)+b(1,0,1,0)+c(2,1,0,0)=(16,13,0,15)$
   **d.** $a(1,2,-1,3)+b(1,0,1,0)+c(2,1,0,0)=(16,13,0,16)$
   **e.** $a(1,2,-1,3)+b(1,0,1,0)+c(2,1,0,0)=(0,0,0,0)$

2. Find numbers $a$, $b$, and $c$ that make each statement true or show that there are no such numbers:

   **a.** $a\begin{pmatrix}1 & 2\\ -1 & 3\end{pmatrix}+b\begin{pmatrix}1 & 0\\ 1 & 0\end{pmatrix}+c\begin{pmatrix}2 & 1\\ 0 & 0\end{pmatrix}=\begin{pmatrix}3 & 5\\ -1 & 9\end{pmatrix}$

   **b.** $a\begin{pmatrix}1 & 2\\ -1 & 3\end{pmatrix}+b\begin{pmatrix}1 & 0\\ 1 & 0\end{pmatrix}+c\begin{pmatrix}2 & 1\\ 0 & 0\end{pmatrix}=\begin{pmatrix}13 & 8\\ 1 & 6\end{pmatrix}$

   **c.** $a\begin{pmatrix}1 & 2\\ -1 & 3\end{pmatrix}+b\begin{pmatrix}1 & 0\\ 1 & 0\end{pmatrix}+c\begin{pmatrix}2 & 1\\ 0 & 0\end{pmatrix}=\begin{pmatrix}16 & 13\\ 0 & 15\end{pmatrix}$

   **d.** $a\begin{pmatrix}1 & 2\\ -1 & 3\end{pmatrix}+b\begin{pmatrix}1 & 0\\ 1 & 0\end{pmatrix}+c\begin{pmatrix}2 & 1\\ 0 & 0\end{pmatrix}=\begin{pmatrix}16 & 13\\ 0 & 16\end{pmatrix}$

   **e.** $a\begin{pmatrix}1 & 2\\ -1 & 3\end{pmatrix}+b\begin{pmatrix}1 & 0\\ 1 & 0\end{pmatrix}+c\begin{pmatrix}2 & 1\\ 0 & 0\end{pmatrix}=\begin{pmatrix}0 & 0\\ 0 & 0\end{pmatrix}$

3. Find numbers $a$, $b$, and $c$ that make each statement true or show that there are no such numbers:

   **a.** $a(x^3+2x^2-x+3)+b(x^3+x)+c(2x^3+x^2)=3x^3+5x^2-x+9$
   **b.** $a(x^3+2x^2-x+3)+b(x^3+x)+c(2x^3+x^2)=13x^3+8x^2+x+6$
   **c.** $a(x^3+2x^2-x+3)+b(x^3+x)+c(2x^3+x^2)=16x^3+13x^2+15$
   **d.** $a(x^3+2x^2-x+3)+b(x^3+x)+c(2x^3+x^2)=16x^3+13x^2+16$
   **e.** $a(x^3+2x^2-x+3)+b(x^3+x)+c(2x^3+x^2)=0$

4. Show that every vector in $\mathbb{R}^4$ is a linear combination of these four vectors:

$$(1,1,0,0),\ (0,1,1,0),\ (0,0,1,1),\ (0,0,0,1)$$

5. Show that every $2\times 2$ matrix is a linear combination of these four matrices:

$$\begin{pmatrix}1 & 1\\ 0 & 0\end{pmatrix},\ \begin{pmatrix}0 & 1\\ 1 & 0\end{pmatrix},\ \begin{pmatrix}0 & 0\\ 1 & 1\end{pmatrix},\ \begin{pmatrix}0 & 0\\ 0 & 1\end{pmatrix}$$

   ⟵ What should a "linear combination of matrices" mean?

6. Show that every polynomial of degree 3 or less is a linear combination of these four polynomials:

$$x^3+x^2,\ x^2+x,\ x+1,\ 1$$

   ⟵ What should a "linear combination of polynomials" mean?

$\mathbb{R}^3$ is **closed** under addition: the sum of any two vectors in $\mathbb{R}^3$ is always in $\mathbb{R}^3$. $\mathbb{R}^3$ is also closed under scalar multiplication: if $X \in \mathbb{R}^3$, $cX \in \mathbb{R}^3$ for any number $c$. Not all sets are closed under these operations: for example, odd numbers aren't closed under addition.

For each set given in Exercises 7–18, determine if it is closed under addition, and determine if it is closed under scalar multiplication.

**7.** The $x$–$y$ plane in $\mathbb{R}^3$

**8.** The set of vectors of the form $(a, 2a)$ in $\mathbb{R}^2$

**9.** The plane with equation $x + y + z = 1$ in $\mathbb{R}^3$

← The set here is all points in the plane with equation $x + y + z = 1$. Pick any two points in this plane $X$ and $Y$ . . . is $X + Y$ also in this plane?

**10.** The set of $2 \times 2$ matrices of the form $\begin{pmatrix} a & b \\ b & a \end{pmatrix}$

**11.** The set of polynomials with degree less than or equal to 3

**12.** The set of polynomials with degree 3 or greater

**13.** $\ker \begin{pmatrix} 1 & 2 & 3 \\ 4 & 5 & 6 \\ 7 & 8 & 9 \end{pmatrix}$

**14.** The set of $2 \times 2$ matrices that have determinant 1

**15.** The set of $3 \times 3$ symmetric matrices

← Is the sum of two symmetric matrices also symmetric? Is 3 times a skew-symmetric matrix also skew-symmetric?

**16.** The set of $3 \times 3$ skew-symmetric matrices

**17.** The line in $\mathbb{R}^3$ through $(5, 1, 2)$ in the direction of $(4, -1, 5)$

**18.** The line in $\mathbb{R}^3$ through $O$ in the direction of $(4, -1, 5)$

## 7.2 Introduction to Vector Spaces

You have seen that some of the basic operations and properties of vectors can be extended to matrices, since you can think of a matrix as a set of vectors (row or column vectors) lined up together. In this lesson, you will see that the basic rules of vectors can also be extended to many other systems.

**In this lesson, you will learn how to**

- tell what the structural similarities for different vector spaces are
- determine whether or not a given set is a vector space

In 1888, Giuseppe Peano defined a *vector space* to be any system that had the basic rules for arithmetic of points (Theorem 1.2).

### Definition

Let $V$ be a nonempty set of objects on which the operations of addition and scalar multiplication are defined. $V$ is a **vector space**, and the objects in $V$ are **vectors**, if these properties are true for all $v, w, u \in V$ and for all numbers $c$ and $d$:

(1) $V$ is *closed* under addition and scalar multiplication. This is, if $v, w \in V$, then $v + w \in V$ and $cv \in V$ for all numbers $c$.

(2) $v + w = w + v$ for all $v, w \in V$.

(3) $v + (w + u) = (v + w) + u$ for all $v, w, u \in V$.

(4) There is an element $\mathbf{0}$, called the **zero vector**, in $V$ so that for all $v \in V$, $v + \mathbf{0} = v$.

(5) For every $v \in V$, there is an element denoted by $-v$ with the property that $v + -v = \mathbf{0}$.

(6) $c(v + w) = cv + cw$ for all numbers $c$ and elements $v$, $w$ of $V$.

(7) $(c + d)v = cv + dv$ for all numbers $c, d$ and elements $v$ of $V$.

(8) $c(dv) = cd(v)$ for all numbers $c, d$ and elements $v$ of $V$.

(9) $1v = v$ for all elements $v$ in $V$.

⟵ If you think of $V$ as $\mathbb{R}^n$ (so that the elements of $V$ are $n$-tuples), this definition is almost identical to the basic rules of arithmetic for points (Theorem 1.2). Those rules are also essentially the same as the basic rules for matrix algebra (Theorem 4.1), since you can think of an $n$-dimensional vector as a $1 \times n$ matrix.

⟵ The zero vector for $\mathbb{R}^n$ is just the origin, $O$.

⟵ The vector $-v$ is called the "negative of $v$." How many negatives can a vector have? See Exercise 36.

⟵ As usual, "numbers $c$ and $d$" imply that $c$ and $d$ are real numbers. Such a vector space is called a **real vector space**. There are more general types.

### Developing Habits of Mind

**Seek structural similarity.** Complex numbers, polynomials, vectors, and matrices all have surface features that look quite different, and each has operations that are not shared by the others. But if you calculate enough with polynomials and complex numbers, you begin to feel that there are some underlying similarities in how "the operations go." For example, you may think of a complex number as working like a polynomial, with the extra layer that $i^2 = -1$.

Developing the habit of abstracting what's common in seemingly different situations has led to some major breakthroughs in mathematics. Still, it took quite a bit of effort to describe the common ground precisely. Developing a list of properties like those in Theorems 1.2 and 4.1 didn't come in a flash of insight. The list evolved and was polished gradually over time in an effort to precisely describe what mathematicians wanted to say.

### Example 1

Consider the set of $2 \times 3$ matrices, and define the operations of addition and scalar multiplication as usual.

← In this book, the set of all $m \times n$ matrices is denoted by $\text{Mat}_{m\times n}$.

**Problem.** Verify property (1) of the definition of a vector space to help show that the set of $2 \times 3$ matrices is a vector space under these operations.

**Solution.** Re-read property (1), letting $V$ be the set of $2 \times 3$ matrices.

> The set of $2 \times 3$ matrices is *closed* under addition and scalar multiplication. This means that if $M, N$ are $2 \times 3$ matrices, then $M + N$ is a $2 \times 3$ matrix, and $cM$ is a $2 \times 3$ matrix for all numbers $c$.

So here are some generic $2 \times 3$ matrices $M$ and $N$:

$$M = \begin{pmatrix} d & e & f \\ g & h & i \end{pmatrix}, \qquad N = \begin{pmatrix} n & o & p \\ q & r & s \end{pmatrix}$$

So if $M + N$ and $cM$ are also $2 \times 3$ matrices, the property is verified.

$$M + N = \begin{pmatrix} d+n & e+o & f+p \\ g+q & h+r & i+s \end{pmatrix}, \qquad cM = \begin{pmatrix} cd & ce & cf \\ cg & ch & ci \end{pmatrix}$$

Both $M + N$ and $cM$ are $2 \times 3$ matrices of real numbers, so property (1) is true for this (possible) vector space. Note that this relies on knowledge of closure of the real numbers under addition and multiplication.

### For Discussion

1. What $2 \times 3$ matrix serves as the zero vector?
2. Verify property (5) for the set of $2 \times 3$ matrices with the usual operations.

In this vector space, the vectors are $2 \times 3$ matrices. In general, the elements of a vector space won't look anything like what you think of as a vector—an $n$-tuple of numbers in $\mathbb{R}^n$. *A vector is just an element of a vector space,* regardless of what that space looks like.

← Later in this chapter, you'll see that, in an important way, the vectors in any vector space can be "viewed as" $n$-tuples.

### Example 2

**Problem.** Consider $\mathbb{R}_3[x]$, the set of polynomials in $x$ with degree less than or equal to 3. Define the operations of addition and scalar multiplication as usual. Verify properties (2), (4), and (5) of the definition of a vector space to help show that $\mathbb{R}_3[x]$ is a vector space under these operations.

← A *polynomial in x* is a polynomial with real coefficients using $x$ as its variable. $x^3 + 13x - \frac{1}{2}$ is in $\mathbb{R}_3[x]$, but $x^2 - x^4$ isn't, and neither is $x^2 - xy$. It's a useful convention to define the degree of 0, the zero polynomial, to be $-\infty$. Among other things, this guarantees that the zero polynomial is in $\mathbb{R}_n[x]$ for every positive integer $n$.

**Solution.** The vectors of $\mathbb{R}_3[x]$ are polynomials in the form $ax^3 + bx^2 + cx + d$, where $a$ through $d$ are real numbers.

**Property (2):** Let $v = ax^3 + bx^2 + cx + d$ and $w = ex^3 + fx^2 + gx + h$. Then

$$\begin{aligned} v + w &= (ax^3 + bx^2 + cx + d) + (ex^3 + fx^2 + gx + h) \\ &= (ax^3 + ex^3) + (bx^2 + fx^2) + (cx + gx) + (d + h) \\ &= (a + e)x^3 + (b + f)x^2 + (c + g)x + (d + h) \\ w + v &= (ex^3 + fx^2 + gx + h) + (ax^3 + bx^2 + cx + d) \\ &= (ex^3 + ax^3) + (fx^2 + bx^2) + (gx + cx) + (h + d) \\ &= (e + a)x^3 + (f + b)x^2 + (g + c)x + (h + d) \end{aligned}$$

In simpler terms, $\mathbb{R}_3[x]$ is commutative under addition because it relies on real-number addition, term by term.

← Again, the property relies on an equivalent property of real numbers.

**Property (4):** Let $v = ax^3 + bx^2 + cx + d$. Find an element $z \in \mathbb{R}_3[x]$ with

$$(ax^3 + bx^2 + cx + d) + z = (ax^3 + bx^2 + cx + d)$$

So the question is, "What do I add to $ax^3 + bx^2 + cx + d$ to get $ax^3 + bx^2 + cx + d$?" The zero vector of $\mathbb{R}_3[x]$ is the zero polynomial: $\mathbf{0} = 0x^3 + 0x^2 + 0x + 0$.

← Because the degree of 0 is $-\infty$, 0 is in $\mathbb{R}_3[x]$.

**Property (5):** Let $v = ax^3 + bx^2 + cx + d$. You want to find a polynomial you can add to this to get $\mathbf{0}$. As you might expect, you can negate each coefficient (or multiply the whole thing by $-1$):

$$\begin{aligned} &(ax^3 + bx^2 + cx + d) + (-ax^3 - bx^2 - cx - d) \\ &\qquad = (a - a)x^3 + (b - b)x^2 + (c - c)x + d \\ &\qquad = 0x^3 + 0x^2 + 0x + 0 \\ &\qquad = \mathbf{0} \end{aligned}$$

So for any $v \in \mathbb{R}_3[x]$, $v + (-1)v = \mathbf{0}$.

← To show that $\mathbb{R}_3[x]$ is a vector space under these operations, the remaining properties still need to be verified.

### For You to Do

**3.** Is $\mathbb{R}_3[x]$ closed under scalar multiplication? Is it closed under polynomial multiplication?

**4.** Is the set of polynomials whose degree is *exactly* 3 a vector space? Explain.

In the previous example, it turned out that in $\mathbb{R}_3[x]$ the negative of a vector (a polynomial) is the same as $-1$ times that vector. You can check that this is also true in $\mathbb{R}^3$ and in $\text{Mat}_{2\times 3}$. In fact, it's true in every vector space.

**Theorem 7.1**

*If $V$ is a vector space and $v$ is a vector in $V$, then $-v = (-1)v$.*

***Proof.*** First, think about what the theorem is saying. You want to show that the negative of a vector is what you get when you scale the vector by $-1$. But "the negative of a vector" is defined by its behavior: a vector $w$ is the negative of $v$ if, when you add it to $v$, you get $\mathbf{0}$. So, to see if $(-1)v$ is the same as $-v$, add $(-1)v$ to $v$ and see if you get $\mathbf{0}$.

← By Exercise 36, there is only one negative for each vector.

$$\begin{aligned} (-1)v + v &= (-1)v + 1v && \text{why?} \\ &= ((-1)+1)v && \text{why?} \\ &= 0v && \text{why?} \\ &= \mathbf{0} && \text{Theorem 7.2 from Exercise 35} \end{aligned}$$ ■

### For You to Do

**5.** Complete the missing reasons above.

**Remember**

$V$ is a vector space so it satisfies all nine properties given earlier.

### Developing Habits of Mind

**Find general results.** Theorem 7.1 is one of the main reasons that the concept of a vector space is so useful. This theorem about vector spaces didn't use any particular vector space! This means it is true in *every* vector space. This one theorem gives a result about $\mathbb{R}^n$, $\text{Mat}_{m\times n}$, $\mathbb{R}_3[x]$, and every other vector space you'll ever meet. So, you don't have to prove the same thing multiple times—you prove it once and then are guaranteed that it holds in every system that satisfies the definition of vector space.

### Minds in Action Episode 24

*Sasha, Tony, and Derman are pondering this idea of vector space.*

DERMAN: So if anything can be a vector, isn't *everything* a vector space?

TONY: I don't know—it's a brand new idea. Try making something up!

DERMAN: All right, let's do ordered pairs $(x, y)$. What did we call that, $\mathbb{R}^2$?

SASHA: Well, you need operations for addition and scalar multiplication.

DERMAN: I'll make up something silly: you switch the coordinates each time you make a calculation! That means

$$(a, b) + (c, d) = (d + b, a + c)$$

$$c(a, b) = (cb, ca)$$

TONY: Okay, so you're saying that's a vector space?

DERMAN: Sure, why wouldn't it be?

SASHA: It has to satisfy those nine properties we just read.

DERMAN: It satisfies property (1) for sure! If $v$ and $w$ are ordered pairs, then $v + w$ and $cv$ are ordered pairs.

TONY: Fine, but what about property (2)?

DERMAN: Hmm. I'll try an example: $v = (2, 5)$ and $w = (10, 3)$. Then

$$v + w = (3 + 5, 2 + 10) = (8, 12)$$
$$w + v = (5 + 3, 10 + 2) = (8, 12)$$

DERMAN: Yes! This is definitely a vector space!

SASHA: Try property (9).

DERMAN: That should be easy. If $v = (2, 5)$, then $1 \cdot v = (1 \cdot 5, 1 \cdot 2) = (5, 2)$.

DERMAN: Shoot. That was supposed to equal $(2, 5)$, since it should be $1v = v$.

SASHA: That kills it. It's not a vector space.

DERMAN: Really? I guess it does. One counterexample to any of the properties voids the whole thing.

---

### For You to Do

6. For each of the nine properties, determine whether it is true or false using Derman's system. For each property that is false, provide a counterexample or explain why the property cannot be satisfied.

---

### Example 3

Let $\mathbb{R}[x]$ denote the set of polynomials in $x$ with no limit on degree, and define addition and scalar multiplication in the usual way.

**Problem.** Is $\mathbb{R}[x]$ a vector space under these operations?

**Solution.** You must show that properties (1)–(9) are satisfied. In this case, any polynomial is a vector. While formal proofs of each property are possible, a sketch is provided below.

**Property (1).** Is the sum of two polynomials another polynomial? If you scale a polynomial, do you get another polynomial? Yes, every time; so $\mathbb{R}[x]$ is closed under adding and scaling.

**Properties (2) and (3).** Addition of polynomials is commutative and associative.

**Property (4).** As seen in the last example, the zero vector for $\mathbb{R}[x]$ is the *zero polynomial*: the number 0.

$\longleftarrow$ Numbers are polynomials. Nonzero numbers are polynomials of degree 0. The zero polynomial has, by convention, degree $-\infty$.

**Property (5).** Given a polynomial $f$, the coefficients of $-1f$ are the opposites of the coefficients of $f$. Therefore, the coefficients of $f + -1f$ must all be zero.

**Properties (6) and (7).** Because the polynomials' coefficients are real numbers, you can use all the properties of $\mathbb{R}$. An example may help, such as

$$f = 2x^2 + 5x + 6, g = 4x^3 - 2x^2 - 5x - 7, c = 5, d = 3$$

The details for the specific example can help you construct a proof for general polynomials like $f = a_n x^n + a_{n-1}x^{n-1} + \cdots + a_1 x + a_0$.

**Property (8).** As with properties (6) and (7), a specific example can help generate a detailed proof.

**Property (9).** In $1f$, each coefficient of $f$ is multiplied by 1, leaving it identical. Then $1f = f$ since all its coefficients are equal.

Since $\mathbb{R}[x]$ satisfies all nine properties, it is a vector space, and the "vectors" are polynomials of any degree.

---

**Minds in Action** Episode 25

*Derman, Tony, and Sasha are still thinking.*

TONY: I dunno. This all seems to me like you're just changing the way things look. I mean, $2 \times 2$ matrices are just vectors in $\mathbb{R}^4$. I can write $\begin{pmatrix} 1 & 2 \\ 3 & 4 \end{pmatrix}$ or $(1, 2, 3, 4)$. I add them the same way no matter how I write them, and I scale them the same way, too.

DERMAN: You can multiply matrices.

TONY: Sure, but in this vector space world, you only look at adding and scaling. They're the same, I tell you.

DERMAN: Yeah, and $\mathbb{R}_3[x]$—it's just $\mathbb{R}^4$, too. If I call it $x^3 + 2x^2 + 3x + 4$ or $(1, 2, 3, 4)$, it doesn't matter. Adding and scaling work the same way.

TONY: It's all just $\mathbb{R}^n$. A rose by any other name . . . .

SASHA: What about $\mathbb{R}[x]$?

*Long pause . . .*

TONY: Well, that's different.

DERMAN: Let's go to lunch.

---

## Exercises

1. Verify that properties (2)–(9) of a vector space hold for the set of $2 \times 3$ matrices under addition and scalar multiplication.

2. Verify properties (1), (6), (7), (8), and (9) for $\mathbb{R}_3[x]$, the set of polynomials whose degree is at most 3, under the usual definition of addition and scalar multiplication.

In Exercises 3–34, determine if the description gives a vector space. Give a counterexample for each example that is not a vector space, and give a good explanation for each that is a vector space. Unless otherwise stated, addition and scalar multiplication are defined as expected.

3. $V = \mathbb{R}_4[x]$. This is the set of polynomials whose degree is at most 4.

4. $V =$ the set of polynomials of degree at *least* 4.

5. $V = \mathbb{C}$, the complex numbers.

6. $V$ is the set of ordered pairs $(x, y)$, but both $x$ and $y$ must be integers.

7. $V$ is the set of ordered pairs $(x, y)$ with $y = 0$.

8. $V$ is the set of ordered pairs $(x, y)$ with $2x + 3y = 12$.

9. $V$ is the set of ordered pairs $(x, y)$ with $2x + 3y = 0$.

10. $V$ is the set of ordered triples $(x, y, z)$ with $z = 0$.

11. $W$ is the set of ordered triples $(x, y, z)$ with $x + y = z$.

12. $U$ is the set of ordered triples $(x, y, z)$ with $x + y + z = 5$.

13. $S$ is the set of ordered triples $(x, y, z)$ with $x + y + z = 0$.

$\longleftarrow$ In Exercises 10–13, $V$, $W$, $U$, and $S$ are planes in $\mathbb{R}^3$.

14. $V$ is the set of $3 \times 3$ lower triangular matrices.

15. $V$ is the set of $3 \times 3$ diagonal matrices.

16. $V$ is the set of $3 \times 3$ symmetric matrices.

17. $V$ is the set of $3 \times 3$ skew-symmetric matrices.

18. $V$ is the set of all polynomials $f$ with $f(3) = 0$.

19. $V$ is the set of all polynomials $f$ with $f(3) = 1$.

20. $V$ is the set of $2 \times 2$ matrices in the form $\begin{pmatrix} a & 1 \\ 1 & d \end{pmatrix}$.

21. $W = \ker \begin{pmatrix} 1 & 2 & 3 \\ 4 & 5 & 6 \\ 7 & 8 & 9 \end{pmatrix}$.

22. $U = \ker \begin{pmatrix} 1 & 2 & 3 \\ 4 & 5 & 6 \\ 7 & 8 & 0 \end{pmatrix}$.

**23.** $S$ is the set of all linear combinations of the rows of

$$\begin{pmatrix} 1 & 2 & 3 \\ 4 & 5 & 6 \\ 7 & 8 & 9 \end{pmatrix}$$

**24.** $T$ is the set of all linear combinations of the columns of

$$\begin{pmatrix} 1 & 2 & 3 \\ 4 & 5 & 6 \\ 7 & 8 & 9 \end{pmatrix}$$

**25.** $V$ is the set of all vectors $B$ that can be written as $\begin{pmatrix} 1 & 2 & 3 \\ 4 & 5 & 6 \\ 7 & 8 & 9 \end{pmatrix} \begin{pmatrix} x \\ y \\ z \end{pmatrix}$ for some vector $\begin{pmatrix} x \\ y \\ z \end{pmatrix}$.

**26.** $V$ is the set of solutions $\begin{pmatrix} x \\ y \\ z \end{pmatrix}$ to the system

$$\begin{pmatrix} 1 & 4 & 2 \\ 3 & 0 & 1 \\ 4 & 4 & 3 \end{pmatrix} \begin{pmatrix} x \\ y \\ z \end{pmatrix} = \begin{pmatrix} 2 \\ -1 \\ 1 \end{pmatrix}$$

**27.** $W$ is the set of solutions $\begin{pmatrix} x \\ y \\ z \end{pmatrix}$ to the system

$$\begin{pmatrix} 1 & 4 & 2 \\ 3 & 0 & 1 \\ 4 & 4 & 3 \end{pmatrix} \begin{pmatrix} x \\ y \\ z \end{pmatrix} = \begin{pmatrix} 0 \\ 0 \\ 0 \end{pmatrix}$$

**28.** $U$ is the set of solutions $\begin{pmatrix} x \\ y \\ z \end{pmatrix}$ to the system

$$\begin{pmatrix} 1 & 4 & 2 \\ 3 & 0 & 1 \\ 4 & 4 & 3 \end{pmatrix} \begin{pmatrix} x \\ y \\ z \end{pmatrix} = \begin{pmatrix} 5 \\ 3 \\ 7 \end{pmatrix}$$

**29.** $V$ is the set of ordered pairs $(x, y)$ with addition as usual, but scaling is defined by the rule $c(x, y) = (2cx, 2cy)$.

**30.** $V$ is the set of ordered pairs $(x, y)$ with addition as usual, but scaling is defined by the rule $c(x, y) = (cx, 0)$.

**31.** **Take It Further.** $V$ is the set of $2 \times 2$ matrices $\begin{pmatrix} a & b \\ c & d \end{pmatrix}$ such that $ad - bc = 1$.

**32.** $V$ is the set of ordered pairs $(x, y)$ with addition according to the unusual rule

$$(a, b) + (c, d) = (ac - bd, ad + bc)$$

and scalar multiplication as usual.

**33. Take It Further.** $V$ is the set of all functions $f$ that map the interval from 0 to 1 into the real numbers

$$f : [0, 1] \rightarrow \mathbb{R}$$

where addition and scalar multiplication are defined as in precalculus:

$$f + g : [0, 1] \rightarrow \mathbb{R} \text{ is defined by } (f + g)(x) = f(x) + g(x)$$
$$cf : [0, 1] \rightarrow \mathbb{R} \text{ is defined by } (cf)(x) = c(f(x))$$

**34. Take It Further.** The Fibonacci sequence is

$$1, 1, 2, 3, 5, 8, 13, \ldots$$

Each term after the second is the sum of the two terms before it. The Fibonacci sequence starts with $1, 1, \ldots$ but "Fibonacci-like" sequences start with *any* two real numbers and use the same rule. Here is one:

$$\frac{1}{2}, \frac{3}{2}, 2, \frac{7}{2}, \frac{11}{2}, 9, \ldots$$

Let $F$ be the set of all Fibonacci-like sequences. Addition is defined by adding two sequences term by term, and scaling by $c$ means to multiply every term in the sequence by $c$.

**35.** The steps below lead you through a proof of the following theorem.

**Theorem 7.2**
*If $v$ is a vector in vector space $V$, then*

$$0v = \mathbf{0}$$

← Why is the second zero in bold, but not the first?

**a.** Use property (7) to show that $0v + 0v = 0v$.
**b.** Add $-0v$ on the left of both sides of the above equation to show that this implies $0v = \mathbf{0}$.

**36.** The following properties are true in any vector space $V$. Prove them using the nine properties of a vector space.

← Here, all variables stand for vectors in $V$.

**a.** If $v + w = v$ and $v + u = v$, then $w = u$. (In other words, there cannot be more than one distinct zero vector.)
**b.** If $v + w = b$ and $v + u = b$, then $w = u$. (Hint: Use part **a**.)
**c.** A vector can have only one negative: if $v+w = \mathbf{0}$ and $v+u = \mathbf{0}$, then $u = w$.

**37. Take It Further.** Suppose $S$ is the set of polynomials of degree at least 5, together with the zero polynomial, 0. Is $S$ a vector space? Explain your answer.

**38.** **Take It Further.** Patrick says that it shouldn't be necessary to check if a vector space is closed under scaling.

PATRICK: Look: $3v = v + v + v$ and $5v = v + v + v + v + v$. Scaling is repeated addition, so if it's closed under addition, it's closed under scaling.

Either show that Patrick is right, or prove him wrong.

## 7.3 Subspaces

The elements of a subset of a general vector space have the same arithmetic operations as the enclosing space. How can you tell if that subset is also a vector space? Do you have to check and make sure that all nine properties are satisfied?

**In this lesson, you will learn how to**

- determine if a subset of a vector space is also a vector space
- identify the key properties you need to test in order to prove that a subset of a vector space is a vector space in its own right

Many of the exercises in the last lesson asked you to determine whether or not a subset of $\mathbb{R}^n$ was a vector space under the usual operations of addition and scalar multiplication. Knowing that you are working within $\mathbb{R}^n$ makes proving some of the nine properties easier, since the known properties of $\mathbb{R}^n$ carry through to any of its subsets. Any subset of $\mathbb{R}^n$ must already have commutativity, associativity, distributivity, and multiplicative identity.

$\longleftarrow$ Six of the nine properties covered, just like that!

The biggest property to worry about is closure under the two operations. If a subset of $\mathbb{R}^n$ fails to be a vector space under the usual operations, it is due to lack of closure on one or both operations.

### Example

**Problem.** Let $S$ be the set of ordered triples $(x, y, z)$ with $x + y + z = 3$. Is $S$ a vector space under the usual operations of addition and scalar multiplication?

**Solution.** No. Find a counterexample by picking $v = (1, 1, 1)$ and $w = (3, 0, 0)$. By definition, $v, w \in S$. If $S$ is to be a vector space, then $v + w = (4, 1, 1)$ must be in $S$ as well. But for this vector, $x + y + z = 6$, and therefore $v + w \notin S$. (Why, exactly, do you know $v + w$ can't be in $S$?)

$\longleftarrow$ Either of these counterexamples is enough to prove that $S$ is *not* a vector space, since it fails property (1), closure under addition and scalar multiplication.

$S$ is not closed under scalar multiplication, either. If $v = (1, 1, 1)$, then $cv = (c, c, c)$, which is only in $S$ when $c = 1$.

Note that there was no need to check if $S$ is associative, commutative, or distributive; these properties are inherited from $\mathbb{R}^n$. More generally, any subset $S$ of a vector space $V$, using the same operations as the vector space, must inherit $V$'s commutativity, associativity, distributivity, and multiplicative identity. More about this shortly.

$\longleftarrow$ If there were a counterexample inside $S$, that counterexample would prove that $V$ wasn't a vector space!

As the example shows, though, not every subset of a vector space $V$ is itself a vector space. Subsets that *are* vector spaces are called subspaces.

**Definition**

A subset $S$ of a vector space $V$ is a **subspace** of $V$ if it is a vector space under the same operations defined on $V$.

**Remember**

A vector space is, by definition, nonempty.

**Minds in Action** Episode 26

DERMAN: Sasha, you've convinced me that not everything can be a subspace.

SASHA: And why is that?

DERMAN: Because not everything is a vector space. If the thing I pick isn't a vector space, it definitely can't be a subspace.

SASHA: Very smooth. Can you think of any subspaces?

DERMAN: I thought that set of polynomials . . . what did we call it . . .

*Derman flips back to Example 2 from Lesson 7.2 . . .*

DERMAN: Yeah, I think $\mathbb{R}_3[x]$ is a subspace of something. It's restrictive.

SASHA: You'd have to find some other vector space that contains $\mathbb{R}_3[x]$.

DERMAN: I've got it! It's $\mathbb{R}[x]$, the set of polynomials!

SASHA: Nice, I think that works. Is $\mathbb{R}_3[x]$ a subset of $\mathbb{R}[x]$?

DERMAN: Well sure! $\mathbb{R}[x]$ is all polynomials, and $\mathbb{R}_3[x]$ is all polynomials of degree less than or equal to 3. If it's in $\mathbb{R}_3[x]$, it *has to* be in $\mathbb{R}[x]$.

SASHA: For this to work, $\mathbb{R}_3[x]$ and $\mathbb{R}[x]$ both have to be vector spaces.

DERMAN: But we proved that in the last lesson. That means $\mathbb{R}_3[x]$ is a subspace of $\mathbb{R}[x]$! Sweet.

SASHA: Nicely done.

**For You to Do**

1. Find a different subset of $\mathbb{R}[x]$ that is also a subspace.
2. Find a different set, other than $\mathbb{R}[x]$, in which $\mathbb{R}_3[x]$ is a subspace.
3. Find a subset of $\mathbb{R}[x]$ that is *not* a subspace of $\mathbb{R}[x]$.

**Facts and Notation**

From now on, if $V$ is a vector space and $S$ is a subset of $V$ that is up for consideration as a subspace of $V$, assume, so that it's not necessary to say it every time, that the addition and scalar multiplication in $S$ are the same ones as those in $V$.

$\longleftarrow$ **Notation**: "$S$ is a subset of $V$" is often denoted by "$S \subset V$".

### For Discussion

4. Suppose $S$ is a nonempty subset of a vector space $V$. Why do properties ((2)), ((3)), ((6)), ((7)), ((8)), and ((9)) in the definition of vector space from Lesson 7.2 automatically hold in $S$, just because they hold in $V$?

← Another way to say it: $S$ *inherits* these properties from $V$.

So, if it's already known that $S$ is a subset of $V$, and $V$ is a vector space, then only three properties must be demonstrated about $S$ for it to be a subspace:

- $S$ must be closed under addition and scalar multiplication.
- $S$ must contain the zero vector $\mathbf{0}$, and since $S$ is part of $V$, it must be the same zero vector $V$ uses.
- $S$ must contain inverses: for any $s \in S, s + (-1)s = \mathbf{0}$.

The other properties are inherited from $V$. However, the list can be shortened even further.

### Lemma 7.3

*If $S$ is a nonempty subset of a vector space $V$, and $S$ is closed under scalar multiplication, then $S$ contains $\mathbf{0}$ and inverses.*

***Proof.*** Consider any element $s \in S$ (since $S$ is nonempty, such an element must exist). $S$ is closed under scalar multiplication, so $cs$ exists for any real number $c$. Particularly, consider $c = 0$ and $c = -1$:

- Let $c = 0$. $0s$ must be in $S$. As you might suspect, $0s = \mathbf{0}$, the zero vector, so $\mathbf{0} \in S$. Proving that $0s = \mathbf{0}$ requires some finesse of the other properties of vector space $V$, and a sketch of the proof is found in Exercise 35 in the last lesson. The zero vector is contained in $S$.
- Let $c = -1$. $(-1)s$ must be in $S$. Theorem 7.1 proved that for any $v \in V$, $-v = (-1)v$. Any $s \in S$ is also in $V$, so $-s = (-1)s$. Now consider $s + (-1)s$:

$$\begin{aligned} s + (-1)s &= s + (-s) && \text{Theorem 7.1} \\ &= \mathbf{0} && \text{by definition of negative} \end{aligned}$$

  $S$ has inverses: for any $s \in S, s + (-1)s = \mathbf{0}$.

■

So, rather than checking nine properties, you can whittle the list down to two.

### Theorem 7.4

*A nonempty subset $S$ of a vector space $V$ is a subspace of $V$ if and only if it is closed under addition and scalar multiplication.*

### Developing Habits of Mind

**Reason about calculations.** Lemma 7.3 above states that if $S$ is closed under scalar multiplication, it has an identity and inverses. What about addition? To show that $S$ is a subspace of $V$, you must still show that $S$ is closed under addition. If it isn't, $S$ violates property (1) and cannot be a vector space.

Still, Theorem 7.3 makes life a lot easier. Look back over the exercises from the last lesson. Most of them describe subspaces! Anything starting with "$V$ is the set of ordered pairs $(x, y)$ with . . . " describes a possible subspace of the known vector space $\mathbb{R}^2$. The same is true for ordered triples, polynomials, matrices, and many more. You can now answer these exercises by showing that $V$ is a subspace of the known vector space. You must show $V$ is closed under addition and scalar multiplication.

$\longleftarrow$
Suppose $V$ is the set of all polynomial functions $f$ so that either $f(3) = 0$ or $f(4) = 0$. Show that $V$ is closed under scalar multiplication but not under addition.

As always, *be careful*: it must already be clear that the enclosing set is a vector space. Typically, the enclosing set is $\mathbb{R}^n$ or a polynomial or matrix space, using standard operations. But if the enclosing space or the operations are unusual, watch out!

### For Discussion

**5.** Look back at the exercises from Lesson 7.2. Find as many examples of subspaces as you can.

## Exercises

$\longleftarrow$
As with the last lesson, if not otherwise stated, assume that the standard operations for addition and scalar multiplication are in use.

**1.** Let $V$ be a vector space.

**a.** Is $V$ a subspace of $V$? Explain.

**b.** Let $S$ contain only the zero vector of $V$. Show that $S$ is a subspace of $V$.

In Exercises 2–5, determine which are subspaces of $\mathbb{R}^3$.

**2.** All vectors of the form $(a, 0, 0)$

**3.** All vectors of the form $(a, 1, 1)$

**4.** All vectors of the form $(a, b, c)$, where $b = a + c$

**5.** All vectors of the form $(a, b, c)$, where $b = a + c + 1$

In Exercises 6–10, determine which are subspaces of $\text{Mat}_{2\times2}(\mathbb{R})$, the set of $2 \times 2$ matrices with real entries.

**6.** All matrices of the form $\begin{pmatrix} 0 & a \\ b & 0 \end{pmatrix}$

**7.** All matrices of the form $\begin{pmatrix} a & b \\ c & d \end{pmatrix}$, where $a, b, c$, and $d$ are integers

**8.** All matrices of the form $\begin{pmatrix} a & b \\ c & d \end{pmatrix}$, where $a + d = 0$

**9.** All matrices of the form $\begin{pmatrix} a & 3b \\ 2a - b & a + b \end{pmatrix}$, where $a, b \in \mathbb{R}$

**10.** All $2 \times 2$ matrices $A$ so that $A^\top = A$

$\longleftarrow$
What's another name for a matrix with $A^\top = A$?

**11.** Which of the following are subspaces of $\mathbb{R}_3[x]$, the set of polynomials with degree less than or equal to 3?

- **a.** All polynomials $a_3x^3 + a_2x^2 + a_1x + a_0$, where $a_0, a_1, a_2$, and $a_3$ are integers
- **b.** All polynomials $a_3x^3+a_2x^2+a_1x+a_0$, where $a_0+a_1+a_2+a_3=0$
- **c.** All polynomials $f(x) = a_3x^3+a_2x^2+a_1x+a_0$, where $f(3) = 0$
- **d.** All polynomials $f(x) = a_3x^3+a_2x^2+a_1x+a_0$ that are divisible by $2x - 3$

**12.** Which of the following are subspaces of $\mathbb{R}^\mathbb{R}$, the set of real-valued functions?

- **a.** All functions $f$ so that $f(x) < 0$ for all $x \in \mathbb{R}$
- **b.** All constant functions
- **c.** All even functions: functions $f$ such that $f(x) = f(-x)$ for all $x \in \mathbb{R}$

**13.** Consider $S$, the set of solutions to

$$\begin{pmatrix} 3 & 7 & -3 \\ 1 & 1 & 4 \\ 4 & 2 & 4 \end{pmatrix} \begin{pmatrix} x \\ y \\ z \end{pmatrix} = \begin{pmatrix} 4 \\ 5 \\ 6 \end{pmatrix}$$

Is $S$ a subspace of $\mathbb{R}^3$? Explain how you know.

**14.** Consider $V = \mathbb{R}_2[x]$ and $S$, the set of polynomials of the form $ax^2 + bx$ with $a, b \in \mathbb{R}$. Is $S$ a subspace of $V$? Explain how you know.

**15.** Show that the kernel of $\begin{pmatrix} 1 & 4 & 7 \\ 2 & 8 & 3 \\ 1 & 4 & -4 \end{pmatrix}$ is a subspace of $\mathbb{R}^3$.

**16.** **Take It Further.** Show that the kernel of *any* $3 \times 3$ matrix is a subspace of $\mathbb{R}^3$, or find a counterexample.

In Exercises 17–31, determine whether $S$ is a subspace of $V$.

**17.** $S$ is the set of ordered triples $(x, y, x + y)$, $V$ is $\mathbb{R}^3$.

**18.** $S$ is the set of ordered triples $(x, y, z)$ with $x - y + 2z = 0$, $V$ is $\mathbb{R}^3$.

**19.** $S$ is the set of ordered 4-tuples $(x, y, z, w)$ with $x - y + z - w = 4$, $V$ is $\mathbb{R}^4$.

**20.** $S$ is the set of ordered 4-tuples $(x, y, x - y, x + 2y)$, $V$ is $\mathbb{R}^4$.

**21.** $S$ is the set of ordered triples $(x, y, z)$ with $x+y = z$ *and* $2x+y = 2z$, $V$ is $\mathbb{R}^3$.

**22.** $S$ is the set of matrices of the form $\begin{pmatrix} a & b \\ a & a+b \end{pmatrix}$ with $a, b \in \mathbb{R}$, $V$ is $\mathrm{Mat}_{2\times 2}(\mathbb{R})$, the set of $2 \times 2$ matrices.

**23.** $S$ is the set of matrices of the form $\begin{pmatrix} x & x-y \\ 1 & 2x \end{pmatrix}$ with $x, y \in \mathbb{R}$, $V$ is $\mathrm{Mat}_{2\times 2}(\mathbb{R})$, the set of $2 \times 2$ matrices.

**24.** $S$ is the set of matrices of the form $\begin{pmatrix} x & x-y \\ 0 & 2x \end{pmatrix}$ with $x, y \in \mathbb{R}$, $V$ is $\mathrm{Mat}_{2\times 2}(\mathbb{R})$, the set of $2 \times 2$ matrices.

**25.** $S$ is the set of linear combinations of the columns of $\begin{pmatrix} 1 & 4 & 2 \\ 3 & 1 & 4 \\ 4 & 5 & 6 \end{pmatrix}$, $V$ is $\mathbb{R}^3$.

$\longleftarrow$ In the next lesson, this set will be called the **column space** of a given matrix.

**26.** $S$ is the kernel of $\begin{pmatrix} 1 & 4 \\ 7 & 2 \\ 1 & 3 \end{pmatrix}$, $V$ is $\mathbb{R}^3$.

**27.** $S$ is the set of vectors in $\mathbb{R}^3$ perpendicular to $(4, 1, 3)$, $V$ is $\mathbb{R}^3$.

**28.** **Take It Further.** $S$ is the set of vectors in $\mathbb{R}^3$ perpendicular to $(a, b, c)$, $V$ is $\mathbb{R}^3$.

**29.** $S$ is the set of polynomials in the form $a(7x^2+1)+b(2x^3-3x^2+5)$, where $a, b \in \mathbb{R}$, $V$ is $\mathbb{R}_3[x]$.

**30.** $S$ is the set of vectors of the form $a(0, 7, 0, 1) + b(2, -3, 0, 5)$ where $a, b \in \mathbb{R}$, $V$ is $\mathbb{R}^4$.

**31.** $S$ is the set of solutions to $\begin{pmatrix} 4 & 1 & 7 \\ 3 & 1 & 0 \\ 7 & -1 & 2 \end{pmatrix} \begin{pmatrix} x \\ y \\ z \end{pmatrix} = \begin{pmatrix} 3 \\ 1 \\ 2 \end{pmatrix}$, $V$ is $\mathbb{R}^3$.

**32.** Find three spaces, $A, B$, and $C$, so that $A$ is a subspace of $B$ and $B$ is a subspace of $C$. Is $A$ also a subspace of $C$?

**33.** Are subspaces transitive? In other words, if $A$ is a subspace of $B$ and $B$ is a subspace of $C$, does that force $A$ to be a subspace of $C$?

## 7.4 Linear Span and Generating Systems

You already know how to tell if a vector is a linear combination of a set of given vectors. This lesson will look at all vectors that are a linear combination of a given set of vectors.

**In this lesson, you will learn how to**

- find a generating system for a vector space
- determine if a given vector is in the linear span of a set of vectors
- determine if a set of vectors generates a vector space

### For You to Do

1. Show that the point $(11, -1, 16)$ lies in the plane spanned by the vectors $(1, -3, 4)$ and $(2, 2, 1)$.
2. Show that there exist real numbers $c_1$ and $c_2$ so that
$$11x^2 - x + 16 = c_1(x^2 - 3x + 4) + c_2(2x^2 + 2x + 1)$$
3. Show that there is a solution to
$$\begin{pmatrix} 1 & 2 \\ -3 & 2 \\ 4 & 1 \end{pmatrix} \begin{pmatrix} x \\ y \end{pmatrix} = \begin{pmatrix} 11 \\ -1 \\ 16 \end{pmatrix}$$

The three For You to Do problems above are very similar. The term **linear combination** applies to vectors, but something more general is happening in all three problems. If you think of the more general concept of **vector** used in this chapter, these three problems are essentially identical. The general concept of linearity leads to this definition of **linear span**.

$\longleftarrow$ You could say they are *isomorphic*. More on that later.

### Definition

Let $v_1, v_2, \ldots, v_n$ be elements of vector space $V$. The **linear span** $S$ of this set of vectors is the set of all combinations in the form

$$c_1 v_1 + c_2 v_2 + \cdots + c_n v_n$$

where $c_i \in \mathbb{R}$.

The notation $L\{v_1, v_2, \ldots, v_n\}$ represents the linear span of a set of vectors.

If $S$ is the linear span of a set of vectors $\{v_1, v_2, \ldots, v_n\}$, then $v \in S$ if it can be written as

$\longleftarrow$ *Vector* here means the more general concept, so it could be a polynomial, matrix, or something else.

$$v = c_1 v_1 + c_2 v_2 + \cdots + c_n v_n$$

for real numbers $c_1$ through $c_n$.

### Example 1

**Problem.** Consider polynomials $v_1 = x^2+1$, $v_2 = x^2-1$, and $v_3 = x^2+x+1$. Let $S$ be the linear span of $v_1, v_2$, and $v_3$. Which of the following expressions are in $S = L\{x^2+1, x^2-1, x^2+x+1\}$?

**a.** $2x^2$ **b.** $5x^2+6x+5$ **c.** $x^3$ **d.** 0

**Solution.**

**a.** To show that $2x^2$ is in $S$, find $a, b, c$ with

$$a(x^2+1)+b(x^2-1)+c(x^2+x+1)=2x^2$$

The left side can be expanded and then collected term-by-term to get

$$(a+b+c)x^2+(c)x+(a-b+c)=2x^2$$

$2x^2$ is in $S$ if there exist $a, b, c$ that solve this system:

$$\begin{aligned} a+b+c &= 2 \\ c &= 0 \\ a-b+c &= 0 \end{aligned}$$

The solution is $a=1, b=1, c=0$. Since a solution exists, $2x^2$ is in the linear span of these vectors. That is,

$$1(x^2+1)+1(x^2-1)+0(x^2+x+1)=2x^2$$

a fact that you can check.

**b.** Similarly, $5x^2+6x+5$ is in $S$ if there is a solution to this system:

$$\begin{aligned} a+b+c &= 5 \\ c &= 6 \\ a-b+c &= 5 \end{aligned}$$

The solution is $a=-1, b=0, c=6$, and, sure enough,

$$-1(x^2+1)+0(x^2-1)+6(x^2+x+1)=5x^2+6x+5$$

**c.** $x^3$ is *not* in $S$: since $a, b$, and $c$ can only be real numbers, there is no way for $a(x^2+1)+b(x^2-1)+c(x^2+x+1)$ to have degree higher than 2. If you wanted to include $x^3$ in the linear span of some set of vectors, at least one of the vectors would have to have degree 3 or higher. (Why?)

**d.** $\mathbf{0}$ is definitely in $S$:

$$\mathbf{0}=0(x^2+1)+0(x^2-1)+0(x^2+x+1)$$

Similarly, the zero vector for a vector space must be part of the linear span of any nonempty set of vectors.

← You could also use $c_1, c_2, c_3$ here, but with a low number of variables, subscripts are often unnecessary.

← Here, the term *vectors* is being used in the sense of elements of a vector space.

**Remember**

In $\mathbb{R}[x]$, the zero vector, $\mathbf{0}$, is the number 0.

← Can you write $\mathbf{0}$ as a linear combination of these three polynomials in another way?

### For Discussion

**4.** Suppose $p = kx^2 + mx + n$. Is $p \in S$ for any choice of real numbers $k, m, n$? Some of your work from previous chapters may be helpful.

When working with matrices, two specific vector spaces come up frequently. One vector space's elements are the rows of the matrix, and the other's elements are the columns.

### Definition

Given an $m \times n$ matrix $A$, the **row space** of $A$ is the linear span of $m$ vectors, the rows of $A$. The **column space** of $A$ is the linear span of $n$ vectors, the columns of $A$.

Put another way, the row space of $A$ is the set of linear combinations of the rows of $A$. The column space of $A$ is the set of linear combinations of the columns of $A$.

### For You to Do

**5.** Write down a $2 \times 3$ matrix, and then determine its row space and column space. Then, describe the row space geometrically within $\mathbb{R}^2$, and describe the column space geometrically within $\mathbb{R}^3$.

The vectors that are used to construct a linear span come from a vector space $V$. Since $V$ is closed under addition and scalar multiplication, the linear span $S$ must be contained in $V$. It makes sense, then, to ask whether a linear span is a subspace of $V$. As it turns out, it is.

### Theorem 7.5

*Let $v_1, v_2, \ldots, v_n$ be vectors in a vector space $V$. If $S = L\{v_1, v_2, \ldots, v_n\}$, then $S$ is a subspace of $V$.*

***Proof.*** As seen in the last lesson, it suffices to show that $S$ is closed under addition and scalar multiplication. If $s \in S$, then $s = c_1v_1 + c_2v_2 + \cdots + c_nv_n$ for real numbers $c_1$ through $c_n$.

- **Addition.** Let $s, t \in S$; prove that $(s + t) \in S$. To do this, write $(s + t)$ in terms of the initial vectors $v_1$ through $v_n$:

$$\begin{array}{rcccccc} s = & c_1v_1+ & c_2v_2+ & \cdots & + & c_nv_n \\ t = & d_1v_1+ & d_2v_2+ & \cdots & + & d_nv_n \\ (s+t) = & (c_1+d_1)v_1+ & (c_2+d_2)v_2+ & \cdots & + & (c_n+d_n)v_n \end{array}$$

  The properties needed to combine terms on the right come from the known properties of $V$, especially property (7): $cv + dv = (c + d)v$ for any $v \in V$.

Since $(s+t)$ can be written in terms of $v_1$ through $v_n$, it is an element of the linear span $S$.

- **Scalar multiplication.** Let $s \in S$; prove that $ks \in S$ for any real number $k$. This can be quickly done by calculating $ks$ directly:

$$\begin{array}{rcccccc} s = & c_1v_1+ & c_2v_2+ & \cdots & + & c_nv_n \\ ks = & (kc_1)v_1+ & (kc_2)v_2+ & \cdots & + & (kc_n)v_n \end{array}$$

Again, this relies on properties of $V$, particuarly property (8): $c(dv) = cd(v)$ for any real numbers $c, d$, and $v \in V$.

Since you have shown $S$ is closed under addition and scalar multiplication, $S$ is a subspace of $V$. ∎

### Developing Habits of Mind

**Seek structural similarity.** The definition of linear span is closely tied to the definition of linear combination. And many of the results are similar. Theorem 7.5 can be thought of as a parallel to these statements about linear combinations:

> If $v$ and $w$ are linear combinations of a set of vectors, then so are $v + w$ and $kv$ for any real number $k$.

If you add a new vector to a set, does it automatically increase the set of linear combinations? Try a numerical example. Is $(7, 8, 9)$ a linear combination of $(1, 2, 3)$ and $(3, 3, 3)$? It is (right?). What, then, does the set of linear combinations of $(1, 2, 3)$, $(3, 3, 3)$, and $(7, 8, 9)$ look like? Does adding $(7, 8, 9)$ increase the set of linear combinations?

Try again, adding $(5, 6, 9)$, which isn't a linear combination of $(1, 2, 3)$ and $(3, 3, 3)$. Now what happens?

If you keep thinking about linear combinations, you'll have a strong sense of what happens with the more general concept of linear span.

The experiment above helps motivate the following theorem and corollary.

**Theorem 7.6**

*Let $S = L\{v_1, v_2, \ldots, v_n\}$ and $v \in S$. Then $S$ is also the linear span of $\{v_1, v_2, \ldots, v_n, v\}$. That is, $S = L\{v_1, v_2, \ldots, v_n, v\}$*

***Proof.*** To prove that $S$ is the linear span of $\{v_1, v_2, \ldots, v_n, v\}$, there are two parts.

- If $s \in S$, then $s = c_1v_1 + c_2v_2 + \cdots + c_nv_n + c_{n+1}v$. There is already a way to write $s$ in terms of $v_1$ through $v_n$, so let $c_{n+1} = 0$. Therefore, $s$ can be written in terms of $c_1$ through $c_{n+1}$. Anything in $S$ is in the linear span of $\{v_1, v_2, \ldots, v_n, v\}$.

- If $s = c_1v_1 + c_2v_2 + \cdots + c_nv_n + c_{n+1}v$, then $s$ can also be written in terms of just $v_1$ through $v_n$. Since $v \in S$, it can be rewritten in terms of $v_1$ through $v_n$:

$$\begin{aligned} s &= c_1v_1 + c_2v_2 + \cdots + c_nv_n + c_{n+1}v \\ &= c_1v_1 + c_2v_2 + \cdots + c_nv_n + c_{n+1}(d_1v_1 + d_2v_2 + \cdots + d_nv_n) \\ &= (c_1 + c_{n+1}d_1)v_1 + (c_2 + c_{n+1}d_2)v_2 + \cdots + (c_n + c_{n+1}d_n)v_n \end{aligned}$$

  The actual coefficients on the last line aren't important, but the fact that $s$ can be written only in terms of $v_1$ through $v_n$ is. Since it can be written this way, $s \in S$.

← To show that something is in the linear span of $v_1$ through $v_n$, show that it can be written as $c_1v_1 + \cdots + c_nv_n$.

Any element in $S$ is in the linear span of $\{v_1, v_2, \ldots, v_n, v\}$, and any element in the linear span must also be in $S$. Therefore, the two sets are equal, and $S$ is the linear span. ■

This quick corollary follows, since the zero vector must always be in any nonempty vector space.

**Corollary 7.7**

*Let $S = L\{v_1, v_2, \ldots, v_n\}$. Then $S$ is also $L\{v_1, v_2, \ldots, v_n, \mathbf{0}\}$.*

Theorem 7.6 shows that there can be more than one set of vectors with the same linear span. Of particular interest are sets of vectors whose linear span covers an entire vector space $V$.

**Definition**

Let $v_1, v_2, \ldots, v_n$ be vectors in a vector space $V$. The set of vectors $\{v_1, v_2, \ldots, v_n\}$ is a **generating system** for $V$ if $L\{v_1, v_2, \ldots, v_n\} = V$.

**Example 2**

**Problem.** Find a generating system for $\mathbb{R}_2[x]$, the set of polynomials of degree less than or equal to 2.

**Solution.** There are many possible answers. One simple answer is

$$\begin{aligned} v_1 &= x^2 \\ v_2 &= x \\ v_3 &= 1 \end{aligned}$$

Consider an element in $\mathbb{R}_2[x]$, say $v = 3x^2 - 5x + 7$. Then $v = 3v_1 - 5v_2 + 7v_3$, so $v$ is in the linear span of $v_1, v_2$, and $v_3$.

In fact, any $v = ax^2 + bx + c$ can be written as $v = av_1 + bv_2 + cv_3$, proving that any element in $\mathbb{R}_2[x]$ is part of the linear span of these vectors. And since each $v_i \in \mathbb{R}_2[x]$, there is a guarantee that the linear span can't be anything larger. These vectors are a generating system for $\mathbb{R}_2[x]$.

**For You to Do**

6. Find two *other* generating systems for $\mathbb{R}_2[x]$.

**Minds in Action** Episode 27

SASHA: What'd you get for your generating system?

DERMAN: You first.

SASHA: I looked at Example 1 from Lesson 7.4. I am pretty sure this is a generating system:

$$v_1 = x^2 + 1, v_2 = x^2 - 1, v_3 = x^2 + x + 1$$

DERMAN: Hmm, yours only has three things in it, just like the simple one they gave with $x^2, x$, and 1.

SASHA: Oh. Doesn't yours also have three vectors?

DERMAN: I'm not even sure how many vectors mine has.

SASHA: Well, read me your list.

DERMAN: Okay, I got $x^2, x, 1, 2x^2, 2x, 2, 3x^2, 3x, 3, x+1, 2x+1, x^2+1, -x, -2x$. I could keep going like this for a long time.

SASHA: I'll bet. How can your giant list be a generating system?

DERMAN: It just has to be true that you can make anything $ax^2 + bx + c$ with the pieces. They never said how many pieces.

SASHA: Yeah, but some of your pieces seem unnecessary. You don't need $2x^2$ if you already have $x^2$. And you don't need $x + 1$ if you already have $x$ and 1.

DERMAN: So what? It's still a generating system, right?

SASHA: I guess it is. But isn't it a nuisance having all that extra stuff? I'd think you'd want it to be as small as possible.

DERMAN: Small like yours, you mean. Do you know yours is the smallest? Could it be done with just two vectors instead of three?

SASHA: I have no idea.

**For Discussion**

7. Is there a pair of polynomials $v_1$ and $v_2$ such that $L\{v_1, v_2\} = \mathbb{R}_2[x]$, or are three necessary for a generating system?

## Exercises

1. Prove Corollary 7.7 using Theorem 7.6.

2. Prove that Sasha's choices from the dialogue above are a generating system for $\mathbb{R}_2[x]$.

For Exercises 3–6, determine whether or not the given vector is in the given linear span.

**3.** $(3, 6, 9)$ in $L\{(1, 4, 6), (2, 5, 8)\}$

**4.** $(3, 6, 0)$ in $L\{(1, 4, 7), (2, 5, 8)\}$

**5.** $x^2 + 1$ in $L\{1 + x, 1 - x\}$

**6.** $x^2 + 1$ in $L\{1 + x, 1 - x, x^2\}$

For each vector space described in Exercises 7–28, find a generating system.

**7.** $\mathrm{Mat}_{3\times 3}(\mathbb{R})$

**8.** Ordered triples $(x, y, z)$ with $z = 0$

**9.** $\mathbb{R}^4$

**10.** Ordered triples in the form $(a, b, 2a + 3b)$

**11.** Matrices of the form $\begin{pmatrix} a & b \\ b & b-a \end{pmatrix}$

**12.** Matrices in the form $\begin{pmatrix} a & b+a & a-b \\ b & a-b & 2a-b \end{pmatrix}$

**13.** Ordered triples $(x, y, z)$ with $x + y - 2z = 0$

**14.** $\ker \begin{pmatrix} 1 & 2 & 3 \\ 4 & 5 & 6 \end{pmatrix}$

**15.** $\ker \begin{pmatrix} 1 & 2 & 3 \\ 4 & 5 & 6 \\ 7 & 8 & 0 \end{pmatrix}$

**16.** $\ker \begin{pmatrix} 1 & 2 & 3 \\ 4 & 5 & 6 \\ 7 & 8 & 9 \end{pmatrix}$

**17.** The column space of $\begin{pmatrix} 1 & 2 & 3 \\ 4 & 5 & 6 \\ 7 & 8 & 9 \end{pmatrix}$

**18.** The column space of $\begin{pmatrix} 1 & 2 & 3 & 2 \\ 4 & 5 & 6 & 0 \\ 7 & 8 & 9 & 3 \end{pmatrix}$

**19.** The column space of $\begin{pmatrix} 1 & 2 & 3 & 4 \\ 4 & 5 & 6 & 10 \\ 7 & 8 & 9 & 16 \end{pmatrix}$

**20.** The set of $3 \times 3$ diagonal matrices

**21.** The set of $3 \times 3$ symmetric matrices

**22.** The set of $3 \times 3$ upper triangular matrices

**23.** The set of $3 \times 3$ skew-symmetric matrices

**24.** The set of strictly upper triangular $4 \times 4$ matrices

**25.** $\mathbb{R}_2[x]$ **26.** $\mathbb{R}_3[x]$ **27.** $\mathbb{R}_4[x]$ **28.** $\mathbb{R}[x]$

For Exercises 29–38, determine whether the given set is a generating system for the corresponding vector space. Justify your answers.

**29.** $\{(1,0,0),(0,1,0),(0,0,1)\}$ for $\mathbb{R}^3$

**30.** $\{(1,10,0),(0,1,1),(0,0,1)\}$ for $\mathbb{R}^3$

**31.** $\{(1,2,3),(4,5,6),(7,8,9)\}$ for $\mathbb{R}^3$

**32.** $\{(1,2),(3,6)\}$ for $\mathbb{R}^2$

**33.** $\{(1,1,0,0),(0,1,1,0),(0,0,1,0),(0,0,0,1)\}$ for $\mathbb{R}^4$

**34.** $\left\{\begin{pmatrix}1 & 0\\0 & 0\end{pmatrix}, \begin{pmatrix}0 & 1\\0 & 0\end{pmatrix}, \begin{pmatrix}0 & 0\\1 & 0\end{pmatrix}, \begin{pmatrix}0 & 0\\0 & 1\end{pmatrix}\right\}$ for $\operatorname{Mat}_{2\times 2}\mathbb{R}$

**35.** $\left\{\begin{pmatrix}1 & 2\\0 & 0\end{pmatrix}, \begin{pmatrix}0 & 1\\2 & 0\end{pmatrix}, \begin{pmatrix}0 & 1\\1 & 0\end{pmatrix}\right\}$ for $\operatorname{Mat}_{2\times 2}\mathbb{R}$

**36.** $\left\{\begin{pmatrix}1 & 0\\0 & 0\end{pmatrix}, \begin{pmatrix}0 & 1\\0 & 0\end{pmatrix}, \begin{pmatrix}0 & 0\\1 & 2\end{pmatrix}, \begin{pmatrix}0 & 0\\0 & 1\end{pmatrix}\right\}$ for $\operatorname{Mat}_{2\times 2}\mathbb{R}$

**37.** $\{x^2, x-1, 1\}$ for $\mathbb{R}_2[x]$

**38.** $\{1+x, 1-x, x^2\}$ for $\mathbb{R}_2[x]$

**39.** Find a generating system for the column space of $\begin{pmatrix}1 & 2 & 3\\4 & 5 & 6\\7 & 8 & 9\end{pmatrix}$.

**40.** Find an *independent* generating system for the column space of $\begin{pmatrix}1 & 2 & 3\\4 & 5 & 6\\7 & 8 & 9\end{pmatrix}$.

**41.** Find an *independent* generating system for the row space of $\begin{pmatrix}1 & 2 & 3\\4 & 5 & 6\\7 & 8 & 9\end{pmatrix}$.

**Remember**

A set of vectors is *independent* if every nonzero linear combination of the vectors is nonzero . . . or if its kernel is zero . . . or . . .

For Exercises 42–44, prove that each set is a generating system for the corresponding vector space.

**42.** $\{(1, 0, -1), (0, 1, 2)\}$ for the row space of $\begin{pmatrix} 1 & 2 & 3 \\ 4 & 5 & 6 \\ 7 & 8 & 9 \end{pmatrix}$

**43.** $\{(1, 1, 2), (-3, -2, 2), (11, 8, 11)\}$ for $\mathbb{R}^3$

**44.** $\{1 + x, 1 - x, 1 + x^2\}$ for $\mathbb{R}_2[x]$

## 7.5 Bases and Coordinate Vectors

In past chapters, you have become very familiar with linear dependence and independence of $n$-tuples. This chapter has introduced more general vector spaces. It is natural to extend the definitions of linear dependence, independence, and generating system to sets of vectors in a general vector space $V$. Then each linearly independent generating system for $V$ will be used to build a structure-preserving relation between $V$ and the more familiar $n$-tuples.

**In this lesson, you will learn how to**

- determine whether a set of vectors is a basis for a vector space
- find the dimension of a vector space
- determine the dimension of the row and column spaces for a given matrix, as well as the dimension of its kernel
- find the coordinate vector of any vector given a basis for a vector space

### For You to Do

**1.** Which of these are generating systems for $\mathbb{R}_2[x]$ under the standard operations?

**a.** $\{x^2+1, 2x+1\}$
**b.** $\{x^2+1, x^2-1, x^2+x+1\}$
**c.** $\{x^2+1, x^2-1, x^2+x+1, 2x+1\}$
**d.** $\{x^2, x, 1\}$
**e.** $\{x^2, x, 1, 2x^2, 2x, 2\}$

The problem above shows that there are many generating systems for a given vector space, but some seem more useful than others. The key lies in an understanding of linear independence. Recall the definitions of linear dependence and independence from Chapter 3:

- Vectors $A_1, A_2, \ldots, A_k$ are **linearly dependent** if there are numbers $c_1, c_2, \ldots, c_k$ that are *not all zero* so that $c_1A_1+c_2A_2+\cdots+c_kA_k = O$, where $O = (0, 0, \ldots, 0)$.
- On the other hand, the vectors are **linearly independent** if the only solution to $c_1A_1 + c_2A_2 + \cdots + c_kA_k = O$ is $c_1 = c_2 = \cdots = c_k = 0$.

This definition carries over quite nicely into vector spaces.

**Definition**

Given a vector space $V$ with elements $v_1, v_2, \ldots, v_k$,

- Vectors $v_1, v_2, \ldots, v_k$ are **linearly dependent** if there are numbers $c_1, c_2, \ldots, c_k$ that are *not all zero* so that $c_1v_1 + c_2v_2 + \cdots + c_kv_k = \mathbf{0}$.
- On the other hand, the vectors are **linearly independent** if the only solution to $c_1v_1 + c_2v_2 + \cdots + c_kv_k = \mathbf{0}$ is $c_1 = c_2 = \cdots = c_k = 0$.

← This equation uses whatever addition and scalar multiplication operations are defined for the vector space.

For all practical purposes, this is the same definition. And the corollaries that follow are the same, too; of particular use is a theorem that if any of $v_1$ through $v_k$ can be written as a linear combination of the others, the set is linearly dependent.

See Theorem 3.2 from Lesson 3.4.

**For Discussion**

**2.** For each set of vectors in For You to Do problem 1, determine whether it is linearly independent over $\mathbb{R}_2[x]$ under the standard operations.

When a generating system of vectors is linearly independent, it is called a *basis*.

**Definition**

Let $v_1, v_2, \ldots, v_n$ be vectors in a vector space $V$. The set of vectors $\{v_1, v_2, \ldots, v_n\}$ is a **basis for vector space $V$** if they are linearly independent *and* form a generating system for $V$. In this book, a basis is often denoted by gothic letters, like $\mathfrak{B}$.

← A basis for $V$ is a linearly independent generating system.

**Facts and Notation**

Every vector space does indeed have a basis. Unfortunately, this fact is difficult to prove, so consider it an assumption for this course.

While all vector spaces have a basis, some vector spaces do not have *finite* bases—bases that contain a finite number of vectors. A good example is $\mathbb{R}[x]$, the vector space of polynomials of any degree. A vector space with a finite basis is called **finite dimensional**.

← . . . and a vector space that does not have a finite basis is called **infinite dimensional**.

The following lemma will be useful in the rest of this chapter. It is a direct consequence of the Fatter Than Tall Theorem (Theorem 5.14 from Lesson 5.6) and a restatement of a corollary (Corollary 5.15) to that theorem.

**Lemma 7.8 (Formerly Corollary 5.15)**
*Any set of more than $n$ vectors in $\mathbb{R}^n$ is linearly dependent.*

**Minds in Action** Episode 28

*Tony, Sasha, and Derman are contemplating bases.*

TONY: So, Lemma 7.8 implies that any set of more than $n$ vectors in $\mathbb{R}^n$ can't be a basis for $\mathbb{R}^n$. There are no six-vector bases of $\mathbb{R}^4$.

DERMAN: Can *three* vectors be a basis for $\mathbb{R}^4$?

TONY: I don't think so . . . I bet I could prove that if you gave me some time. Let me think about it tonight.

DERMAN: OK, how about this one: if $n$ vectors in $\mathbb{R}^n$ are linearly independent, *must* they form a basis?

SASHA: (*after they pause to think*) Yes. And I think we can prove it. Suppose, just for example, that you have four linearly independent vectors in $\mathbb{R}^4$, say $\{v_1, v_2, v_3, v_4\}$. To be a basis for $\mathbb{R}^4$, I need to show that it generates $\mathbb{R}^4$. So, pick a vector $v$ in $\mathbb{R}^4$. I want to show that it can be written as a combination of my alleged basis.

← Sasha is using lowercase letters, even though she imagines that the vectors are in $\mathbb{R}^4$.

DERMAN: Wow, "alleged." That's police talk.

SASHA: My idea is that, by the lemma, $\{v_1, v_2, v_3, v_4, v\}$, five vectors in $\mathbb{R}^4$, must be linearly dependent.

TONY: Bingo. Yes, that means that you get a combination like

$$c_1v_1 + c_2v_2 + c_3v_3 + c_4v_4 + cv = \mathbf{0}$$

with the $c$'s not all zero. So, solve this for $v$—subtract $cv$ from both sides and divide both sides by $-c$. You've got $v$ as a combo of the alleged basis.

SASHA: Blah. All that linearly dependent means is that at least *one* of the coefficients in

$$c_1v_1 + c_2v_2 + c_3v_3 + c_4v_4 + cv = \mathbf{0} \qquad (*)$$

is not 0. But what if $c$ is 0? Then you can't solve for $v$ because you can't divide by $-c$. I feel like we're on the right track, but this doesn't quite work.

*More thinking . . .*

DERMAN: Look, if $c = 0$, equation $(*)$ becomes

$$c_1v_1 + c_2v_2 + c_3v_3 + c_4v_4 = \mathbf{0}$$

TONY: Bingo again. But because $\{v_1, v_2, v_3, v_4\}$ is independent, we have $c_1 = c_2 = c_3 = c_4 = 0$. So if $c = 0$, then *all* the $c$'s would be zero, $c_1 = c_2 = c_3 = c_4 = c = 0$ and that can't be.

SASHA: Great. So, $c \neq 0$ and we can solve for $v$. Done.

DERMAN: What were we trying to prove again?

---

The reasoning in this episode leads to the following theorem.

**Theorem 7.9**

*Any set of $n$ linearly independent vectors in $\mathbb{R}^n$ is a basis for $\mathbb{R}^n$.*

### For You to Do

1. Use the reasoning in the dialogue to prove Theorem 7.9.

### Example 1

**Problem.** For each set of vectors, determine whether it is a basis for $\mathbb{R}^3$.

1. $\{(1,0,1),(0,2,1)\}$
2. $\{(1,0,1),(1,0,-1),(1,1,1)\}$
3. $\{(1,0,1),(1,0,-1),(1,1,1),(0,2,1)\}$

**Solution.** For the first two sets of vectors, write them as the columns of a matrix. To test for independence, reduce the matrix to echelon form.

1. These two vectors are the columns of the $3 \times 2$ matrix $\begin{pmatrix} 1 & 0 \\ 0 & 2 \\ 1 & 1 \end{pmatrix}$. It is possible to see that these are independent by inspection, or find the echelon form:

$$\begin{pmatrix} 1 & 0 \\ 0 & 2 \\ 1 & 1 \end{pmatrix} \rightarrow \begin{pmatrix} 1 & 0 \\ 0 & 1 \\ 0 & 0 \end{pmatrix}$$

However, these two vectors do not form a generating system for $\mathbb{R}^3$, since the set of linear combinations forms a plane. Any point not on that plane is not a linear combination of the columns of the original matrix, and hence these vectors *are not* a basis for $\mathbb{R}^3$ since they are not a generating system.

← What's an equation of the form $X \cdot N = d$ for this plane?

2. These three vectors are the columns of the $3 \times 3$ matrix $M = \begin{pmatrix} 1 & 1 & 1 \\ 0 & 0 & 1 \\ 1 & -1 & 1 \end{pmatrix}$. Reduce this matrix to echelon form:

$$\begin{pmatrix} 1 & 1 & 1 \\ 0 & 0 & 1 \\ 1 & -1 & 1 \end{pmatrix} \rightarrow \begin{pmatrix} 1 & 0 & 0 \\ 0 & 1 & 0 \\ 0 & 0 & 1 \end{pmatrix}$$

Since the echelon form is the identity matrix, you can invoke the TFAE Theorem (Theorem 4.19 from Lesson 4.6) to conclude that the vectors are independent. They also form a generating system for $\mathbb{R}^3$. To see this, let $B$ be any vector in $\mathbb{R}^3$, written as a column. By the same Theorem 4.19, $MX = B$ has a (unique) solution. But $MX$ is a linear combination of the columns of $M$. So, the columns of $M$ (the original three vectors) generate $\mathbb{R}^3$. Therefore, these three vectors *are* a basis for $\mathbb{R}^3$.

← $MX$ is a linear combination of the columns of $M$ by the Pick-Apart Theorem (Theorem 4.8 from Lesson 4.5).

3. These are four vectors in $\mathbb{R}^3$, so by Lemma 5.15, they are linearly dependent and hence can't be a basis.

### Developing Habits of Mind

**Note general results.** The logic in the last part of the example above does not consider the actual vectors! Lemma 5.15 implies that any set of more than $n$ vectors cannot be a basis for $\mathbb{R}^n$.

An important question remains:

*Is the same story true for other vector spaces? If so, how would you calculate the right $n$ for a particular vector space?*

This question will lead to the main result of this lesson. You will show that *any two bases of the same vector space contain the same number of vectors*. This is one of the most important results in linear algebra. The remainder of this lesson is technical, so be prepared. It will help, as you go through it, to keep several numerical examples at hand.

⟵ Attention is restricted to finite bases in this chapter.

The observations made about $\mathbb{R}^n$ are surprisingly applicable to other vector spaces, and the key is the **coordinate vector**. Given a basis $\mathfrak{B} = \{v_1, v_2, \ldots, v_n\}$ for a vector space $V$, consider any vector $v \in V$. Since $\mathfrak{B}$ is a generating system, then there must exist coefficients so that

$$v = c_1v_1 + c_2v_2 + \cdots + c_nv_n$$

### Definition

Let $\mathfrak{B} = \{v_1, v_2, \ldots, v_n\}$ be a basis for a vector space $V$. For any vector $v \in V$, its **coordinate vector with respect to $\mathfrak{B}$** is an element of $\mathbb{R}^n$, $(c_1, c_2, \ldots, c_n)$, such that

$$v = c_1v_1 + c_2v_2 + \cdots + c_nv_n$$

The coordinate vector for $v$ with respect to $\mathfrak{B}$ is denoted by $\underline{v}_{\mathfrak{B}}$.

### Facts and Notation

It takes a little time to get used to the "underline" notation $\underline{v}_{\mathfrak{B}}$. Think of it as "the coordinates of $v$ with respect to $\mathfrak{B}$."

### For You to Do

**3.** Consider $\mathbb{R}_2[x]$. One basis for this vector space is

$$\mathfrak{B} = \{x^2 + 1, x^2 - 1, x^2 + x + 1\}$$

⟵ You haven't proven that this is a basis, but you will prove it by the end of the lesson.

For each vector $v \in \mathbb{R}_2[x]$ below, find $\underline{v}_{\mathfrak{B}}$.

**a.** $2x^2 + 2$ **b.** $2$ **c.** $3x^2 - 5x + 7$

**d. Take It Further.** $ax^2 + bx + c$ in terms of $a, b$, and $c$

### Minds in Action — Episode 29

SASHA: Wait a minute. This definition says "its coordinate vector." I'm not convinced these are unique.

DERMAN: You should be more trusting.

SASHA: Yeah, yeah. Could we prove they are unique?

DERMAN: What if two coordinate vectors corresponded to the same vector?

SASHA: That's a great idea. If that happened with the coordinate vectors, we could look at . . . (*Sasha mumbles a bit while thinking.*)

### For Discussion

4. Construct a proof that for any basis, the corresponding coordinate vectors must be unique. That is, show that if $V$ is a vector space with basis $\mathfrak{B}$, every vector $w$ in $V$ has one and only one coordinate vector with respect to $\mathfrak{B}$.

← See Exercise 27.

### For You to Do

Suppose $V$ is a vector space with basis $\mathfrak{B} = \{v_1, \ldots, v_n\}$.

5. Show that $\underline{\mathbf{0}}_{\mathfrak{B}} = O = (0, 0, 0, \ldots, 0)$.
6. Show that every vector in $\mathbb{R}^n$ is the coordinate vector for some vector in $V$.

Assigning coordinate vectors with respect to a basis is a way to take the vectors in any vector space $V$ and make them look like the vectors from Chapter 1: $n$-tuples in $\mathbb{R}^n$. And the correspondence $v \mapsto \underline{v}_{\mathfrak{B}}$ is a one-to-one correspondence—every vector in $V$ corresponds to exactly one $n$-tuple, and vice versa. This correspondence does more than match up vectors in $V$ with $n$-tuples. It is also **structure preserving** in the sense of the following theorem.

← Tony had a premonition about this idea in Episode 25.

**Theorem 7.10**

*Let $V$ be a vector space with basis $\mathfrak{B}$. If $v, w \in V$ have coordinate vectors $c$ and $d$, then*

(1) *$v + w$ has coordinate vector $c + d$*

(2) *$kv$ has coordinate vector $kc$*

*In other words,*

(1) $\underline{v + w}_{\mathfrak{B}} = \underline{v}_{\mathfrak{B}} + \underline{w}_{\mathfrak{B}}$

(2) $\underline{kv}_{\mathfrak{B}} = k\underline{v}_{\mathfrak{B}}$

← Think of this theorem as the statement, "coordinate vectors respect structure."

← In other words, the map $v \mapsto \underline{v}_{\mathfrak{B}}$ enjoys the "linearity" properties of Theorem 5.1 from Lesson 5.2.

***Proof.*** Both proofs rely on the fact that $v, w \in V$, a known vector space. Let $v$ have coordinate vector $c = (c_1, c_2, \ldots, c_n)$ and $w$ have coordinate vector $d = (d_1, d_2, \ldots, d_n)$. This means

$$\begin{aligned} v &= c_1v_1 + c_2v_2 + \cdots + c_nv_n \\ w &= d_1v_1 + d_2v_2 + \cdots + d_nv_n \end{aligned}$$

where $v_1$ through $v_n$ are the vectors in basis $\mathfrak{B}$. To prove that $v + w$ has coordinate vector $c + d$, compute the vector directly.

$$\begin{aligned} v + w &= (c_1v_1 + \cdots + c_nv_n) + (d_1v_1 + \cdots + d_nv_n) \\ &= (c_1v_1 + d_1v_1) + (c_2v_2 + d_2v_2) + \cdots + (c_nv_n + d_nv_n) \\ &= (c_1 + d_1)v_1 + (c_2 + d_2)v_2 + \cdots + (c_n + d_n)v_n \end{aligned}$$

Therefore, the coordinate vector of $v + w$ is $(c_1 + d_1, c_2 + d_2, \ldots, c_n + d_n)$, which equals $c + d$. ∎

### For Discussion

**7.** In a similar fashion, prove the second half of Theorem 7.10.

So, the correspondence that assigns vectors in $V$ to coordinate vectors in $\mathbb{R}^n$ is very tight:

1. it associates each vector in $V$ with exactly one $n$-tuple, and vice versa, and
2. it associates addition in $V$ with addition in $\mathbb{R}^n$ and scalar multiplication in $V$ with scalar multiplication in $\mathbb{R}^n$.

Such a correspondence, one that maps objects to objects and operations to operations, is called a **representation**, and the correspondence $v \mapsto \underline{v}_{\mathfrak{B}}$ is said to be a "representation of $V$".

⟵ Because of item **1a**, the representation is said to be **faithful**.

Theorem 7.10 leads to the following corollary.

### Corollary 7.11

*Suppose $V$ is a vector space, $\mathfrak{B}$ is a basis, $w_1, \ldots, w_k$ is a collection of vectors, and $c_1, \ldots, c_k$ is a collection of scalars. Then*

$$\underline{c_1w_1 + \cdots + c_kw_k}_{\mathfrak{B}} = c_1\underline{w_1}_{\mathfrak{B}} + \cdots + c_1\underline{w_k}_{\mathfrak{B}}$$

### For You to Do

**8.** Use Theorem 7.10 to prove Corollary 7.11.

Corollary 7.11 leads to the following theorem.

**Theorem 7.12**

*Suppose $V$ is a vector space with a basis $\mathfrak{B}$ that contains $n$ vectors. A collection of vectors $w_1, \ldots, w_k$ is linearly independent in $V$ if and only if its coordinate vectors with respect to $\mathfrak{B}$ are linearly independent in $\mathbb{R}^n$.*

← In short, this theorem states that "coordinate vectors respect linear independence."

**For Discussion**

**9.** Write out a careful proof of Theorem 7.12.

**Minds in Action** Episode 30

*Next day, Tony comes into class, all excited.*

TONY: I was up all night thinking of what we were discussing in Episode 28. Remember? Derman said, "Can *three* vectors be a basis for $\mathbb{R}^4$?" and I said, "I don't think so . . . I bet I could prove that if you gave me some time. Let me think about it tonight." Well, I think I have it. The key is Theorem 7.12.

DERMAN: You remember episode numbers?

TONY: Yes. Anyway, my idea goes like this: we know that one basis for $\mathbb{R}^4$ contains four vectors:

$$\mathfrak{Z} = \{(1,0,0,0),(0,1,0,0),(0,0,1,0),(0,0,0,1)\}$$

Suppose you had another basis with only three vectors in it, say

← Tony is imagining that each $v_i$ is a vector in $\mathbb{R}^4$.

$$\mathfrak{B} = \{v_1, v_2, v_3\}$$

Since $\mathfrak{B}$ generates $\mathbb{R}^4$, I can write each element of $\mathfrak{Z}$ as a combination of vectors in $\mathfrak{B}$, like this:

$$\begin{aligned}(1,0,0,0) &= a_{11}v_1 + a_{12}v_2 + a_{13}v_3\\ (0,1,0,0) &= a_{21}v_1 + a_{22}v_2 + a_{23}v_3\\ (0,0,1,0) &= a_{31}v_1 + a_{32}v_2 + a_{33}v_3\\ (0,0,0,1) &= a_{41}v_1 + a_{42}v_2 + a_{43}v_3\end{aligned}$$

Derman, would you please write out the coordinate vectors for each element of $\mathfrak{Z}$ with respect to $\mathfrak{B}$?

*Derman writes on the board with a little flourish.*

$$\begin{aligned}\underline{(1,0,0,0)}_{\mathfrak{B}} &= (a_{11}, a_{12}, a_{13})\\ \underline{(0,1,0,0)}_{\mathfrak{B}} &= (a_{21}, a_{22}, a_{23})\\ \underline{(0,0,1,0)}_{\mathfrak{B}} &= (a_{31}, a_{32}, a_{33})\\ \underline{(0,0,0,1)}_{\mathfrak{B}} &= (a_{41}, a_{42}, a_{43})\end{aligned}$$

TONY: Thank you. Now Sasha, what do you see here?

SASHA: Oh, very smooth, Tony. On the right, I see four vectors in $\mathbb{R}^3$. By Lemma 7.8, these four vectors are linearly dependent. I get it . . .

TONY: So, by Theorem 7.12, the vectors on the left are linearly dependent. But these are our basis $\mathfrak{Z}$ vectors, and we know that they are independent.

DERMAN: That's crazy.

TONY: Well, it's a contradiction. It says that the alleged basis $\mathfrak{B}$ isn't a basis after all.

SASHA: So there are no three-vector bases of $\mathbb{R}^4$. This must work in general.

DERMAN: I feel a theorem coming on.

---

**Theorem 7.13 (Invariance of Dimension)**

*Let $\mathfrak{Z}$ and $\mathfrak{B}$ be two bases for a vector space $V$. $\mathfrak{Z}$ and $\mathfrak{B}$ must have* exactly *the same number of elements.*

### For You to Do

**10.** Write out a careful proof of Theorem 7.13.

---

**Definition**

The **dimension** of a vector space $V$, denoted by $\dim(V)$, is the number of elements in any basis of $V$.

**Remember**

This book considers only finite-dimensional vector spaces.

### Example 2

**Problem.** Determine the dimension of the vector space of $2 \times 2$ matrices.

**Solution.** The dimension of a vector space is the number of elements in any basis. Find a basis for $\text{Mat}_{2\times 2}(\mathbb{R})$, the set of $2 \times 2$ matrices. One obvious basis is

$$\left\{ \begin{pmatrix} 1 & 0 \\ 0 & 0 \end{pmatrix}, \begin{pmatrix} 0 & 1 \\ 0 & 0 \end{pmatrix}, \begin{pmatrix} 0 & 0 \\ 1 & 0 \end{pmatrix}, \begin{pmatrix} 0 & 0 \\ 0 & 1 \end{pmatrix} \right\}$$

These four matrices are independent, since there is no nonzero linear combination of these matrices that produces a matrix of all zeros. Additionally, these four matrices form a generating system; the linear span of these four matrices is the set of all possible $2 \times 2$ matrices.

Therefore, this is a basis for $\text{Mat}_{2\times 2}(\mathbb{R})$, and the dimension of this vector space is 4.

---

### Developing Habits of Mind

**Seek similar structures.** It's a good idea to stand back from all the details and look at the main results of this lesson and how they relate to the rest of the course.

1. In Chapter 1, you learned about the algebra of $\mathbb{R}^n$ and developed some basic rules for calculating with vectors.
2. In Chapter 4, you developed an algebra of matrices, and these same rules for calculating held.
3. In this chapter, you defined a vector space to be *any* structure that enjoyed these rules for calculation.
4. Now things come full circle. These arbitrary vector spaces (at least the finite-dimensional ones) are really $\mathbb{R}^n$ in disguise. The coordinate vector map allows you to represent any vector space as $\mathbb{R}^n$ for some $n$.
5. And with any finite-dimensional vector space $V$, you can associate a *dimension*—an integer $n$ that is the size of any basis for $V$.

$\longleftarrow$ These basic rules are listed in Theorem 1.2 (see Lesson 1.2).

As a final application of all this, you can add to the TFAE theorem of Chapter 4 (Theorem 4.19 from Lesson 4.6).

**Theorem 7.14 (The TFAE Theorem)**
*The following statements are all equivalent for an $n \times n$ matrix $A$:*

(1) *The columns of $A$ are linearly independent.*

(2) $\operatorname{rref}(A) = I$.

(3) $A^{-1}$ *exists.*

(4) $AX = B$ *has a unique solution for any* $B$.

(5) $\ker(A) = O$.

(6) *The dimension of the column space of $A$ is $n$.*

**For You to Do**

**11.** Suppose the dimension of the column space of $A$ is $n$. Show that at least one of the other conditions in Theorem 7.14 is true—and therefore, all are true.

## Exercises

For Exercises 1–16, determine whether the set is a basis for the given vector space.

**1.** $\{(1,1,0),(0,1,1),(0,0,1)\}$ for $\mathbb{R}^3$

**2.** $\{(1,4),(2,3)\}$ for $\mathbb{R}^2$

**3.** $\left\{\begin{pmatrix}1&0\\0&0\end{pmatrix},\begin{pmatrix}0&1\\1&0\end{pmatrix},\begin{pmatrix}0&0\\1&1\end{pmatrix},\begin{pmatrix}0&0\\0&1\end{pmatrix}\right\}$ for $\operatorname{Mat}_{2\times 2}(\mathbb{R})$, the set of $2\times 2$ matrices

**4.** $\{(1,2,3),(4,5,6),(7,8,9)\}$ for $\mathbb{R}^3$

**5.** $\{(1,2,3),(4,5,6),(7,8,0)\}$ for $\mathbb{R}^3$

**6.** $\{(1,2,3),(4,5,6),(7,8,0),(1,4,7)\}$ for $\mathbb{R}^3$

**7.** $\{1+x, 1-x, x^2\}$ for $\mathbb{R}_3[x]$

**8.** $\{1+x, 1-x, x^2\}$ for $\mathbb{R}_2[x]$

**9.** $\{(1,0,1),(3,1,2)\}$ for $L\{(1,0,1),(3,1,2),(4,1,3)\}$

**10.** $\{(1,2,3),(2,5,3),(3,7,8)\}$ for $\mathbb{R}^3$

**11.** $\{(1,3,2),(2,5,4),(7,2,-6),(3,-1,4)\}$ for $\mathbb{R}^3$

**12.** $\{x^2+1, 3-x^2, x^3\}$ for $\mathbb{R}_3[x]$

**13.** $\{x^2+x, 3x-x^2, x^3\}$ for $\mathbb{R}_3[x]$

**14.** $\{(1,3,1),(2,0,4),(7,2,6),(3,-1,4)\}$ for $\mathbb{R}^3$

**15.** $\left\{\begin{pmatrix}1 & 1\\ 0 & 0\end{pmatrix}, \begin{pmatrix}0 & 1\\ 0 & 0\end{pmatrix}, \begin{pmatrix}0 & 0\\ 1 & 1\end{pmatrix}, \begin{pmatrix}0 & 0\\ 0 & 1\end{pmatrix}\right\}$ for $\mathrm{Mat}_{2\times 2}(\mathbb{R})$

**16.** $\left\{\begin{pmatrix}2\\1\\3\end{pmatrix}, \begin{pmatrix}4\\0\\1\end{pmatrix}, \begin{pmatrix}3\\1\\2\end{pmatrix}\right\}$ for the column space of $\begin{pmatrix}2 & 6 & 3 & 4\\ 1 & 1 & 1 & 0\\ 3 & 4 & 2 & 1\end{pmatrix}$

**Remember**

The *column space* of a matrix is the linear span of its column vectors.

For Exercises 17–25, find a basis for the given vector space.

**17.** The set of $3 \times 3$ symmetric matrices

**18.** The set of $4 \times 4$ upper triangular matrices

**19.** $\ker \begin{pmatrix}1 & 3 & 1\\ 2 & 0 & 4\end{pmatrix}$

**20.** $L\{(1,2,3),(4,5,6),(7,8,9)\}$

**21.** The column space of $\begin{pmatrix}1 & 2 & 5 & 3\\ 3 & 0 & 3 & 3\\ 1 & 4 & 9 & 5\end{pmatrix}$

**22.** The column space of $\begin{pmatrix}1 & -1 & 3\\ 5 & 6 & 4\\ -1 & -8 & 6\end{pmatrix}$

**23.** The set of vectors in $\mathbb{R}^3$ perpendicular to both $(1,4,3)$ and $(2,9,1)$

**24.** The row space of $\begin{pmatrix}1 & 1 & 0 & 2\\ 2 & 3 & 4 & 1\\ 1 & 0 & 1 & 0\\ 4 & 4 & 5 & 3\end{pmatrix}$

**25.** $L\left\{\begin{pmatrix}1 & 0\\ 1 & 2\end{pmatrix}, \begin{pmatrix}3 & -1\\ 4 & 6\end{pmatrix}, \begin{pmatrix}1 & -1\\ 3 & 4\end{pmatrix}, \begin{pmatrix}2 & 0\\ 2 & 4\end{pmatrix}\right\}$

⟵ A linear span of matrices? Sure, these are *vectors* inside the vector space $\mathrm{Mat}_{2\times 2}(\mathbb{R})$.

**26.** Find a basis for $\mathbb{R}^3$ that contains $(1, 3, 4)$ and $(2, 5, 1)$.

**27.** Let $V$ be a vector space, and let $\mathfrak{B} = \{v_1, v_2, \ldots, v_n\}$ be a basis for $V$. Show that any $v$ in $V$ can be written as a linear combination of the $\{v_1, v_2, \ldots, v_n\}$ *in exactly one way.*

For Exercises 28–37, find the coordinate vector for $v$ with respect to the given basis.

**28.** $v = (1, 2, 3), \quad B = \{(1, 0, 0), (0, 1, 0), (0, 0, 1)\}$

**29.** $v = (1, 2, 3), \quad B = \{(1, 1, 0), (0, 1, 1), (0, 0, 1)\}$

**30.** $v = (1, 3, 1, 2), \quad B = \{(1, 1, 0, 0), (0, 1, 1, 0), (0, 0, 1, 1), (0, 0, 0, 1)\}$

**31.** $v = (1, 3, 1), \quad B = \{(1, 3, 1), (2, 0, 4), (1, 2, 3)\}$

**32.** $v = (1, 2, 3), \quad B = \{(1, 2, 0), (3, 1, 4), (-1, 4, 6)\}$

**33.** $v = 4x^3 + 3x^2 - 2x + 3, \quad B = \{1, 1 + x, x + x^2, x^2 + x^3\}$

**34.** $v = \begin{pmatrix}4 & 3\\ -2 & 3\end{pmatrix}, \quad B = \left\{\begin{pmatrix}1 & 1\\ 0 & 0\end{pmatrix}, \begin{pmatrix}0 & 1\\ 1 & 0\end{pmatrix}, \begin{pmatrix}0 & 0\\ 1 & 1\end{pmatrix}, \begin{pmatrix}0 & 0\\ 0 & 1\end{pmatrix}\right\}$

**35.** $v = \begin{pmatrix}2 & 1\\ -1 & 4\end{pmatrix}, \quad B = \left\{\begin{pmatrix}1 & 1\\ 0 & 0\end{pmatrix}, \begin{pmatrix}0 & 1\\ 1 & 0\end{pmatrix}, \begin{pmatrix}0 & 0\\ 1 & 1\end{pmatrix}, \begin{pmatrix}0 & 0\\ 0 & 1\end{pmatrix}\right\}$

**36.** $v = \begin{pmatrix}3 & 4\\ 0 & 4\end{pmatrix}, \quad B = \left\{\begin{pmatrix}1 & 1\\ 0 & 0\end{pmatrix}, \begin{pmatrix}0 & 1\\ 1 & 0\end{pmatrix}, \begin{pmatrix}0 & 0\\ 1 & 1\end{pmatrix}, \begin{pmatrix}0 & 0\\ 0 & 1\end{pmatrix}\right\}$

**37.** $v = x^2 - 3x + 4, \quad B = \{1 + x, 1 - x, x^2 + 1\}$

**38.** Let $V$ be a vector space with basis $\mathfrak{B} = \{v_1, v_2, v_3\}$. Suppose that $v$ and $w$ are vectors in $V$ such that $\underline{v_{\mathfrak{B}}} = (1, -1, 4)$ and $\underline{w_{\mathfrak{B}}} = (1, 4, 6)$. Calculate each of the following:

**a.** $\underline{(2v - w)_{\mathfrak{B}}}$ **b.** $\underline{(v - 2w)_{\mathfrak{B}}}$ **c.** $\underline{v_{1\mathfrak{B}}}$ **d.** $\underline{(v - v_2 + v_3)_{\mathfrak{B}}}$

⟵ Even though you don't know what $v_1$ is, what is $\underline{v_{1\mathfrak{B}}}$?

In Exercises 39–52, find the dimension of each vector space.

**39.** $\mathbb{R}^n$ **40.** $\mathbb{R}_5[x]$ **41.** $\mathbb{R}_n[x]$

**42.** $\mathrm{Mat}_{3\times 2}(\mathbb{R})$ **43.** $\mathrm{Mat}_{m\times n}(\mathbb{R})$

**44.** The set of $3 \times 3$ symmetric matrices

**45.** The set of $n \times n$ symmetric matrices

**46.** The set of $n \times n$ diagonal matrices

**47.** The set of $n \times n$ scalar matrices

**48.** The set of $3 \times 3$ upper triangular matrices

**49.** The set of $n \times n$ upper triangular matrices

**50.** The set of $n \times n$ skew-symmetric matrices

**51.** All matrices of the form $\begin{pmatrix} a & b & c \\ b & a & a+c \end{pmatrix}$

**52.** $L\{(1,3,1,0),(2,0,1,4),(3,3,2,4),(1,-3,0,4)\}$

**53.** Given $A = \begin{pmatrix} 1 & 3 & 2 & 4 & 3 \\ 3 & 10 & -1 & 6 & 1 \\ 2 & 7 & -3 & 2 & -2 \\ 5 & 16 & 3 & 14 & 7 \end{pmatrix}$

- **a.** Find $\dim(\ker A)$.
- **b.** Find the *row rank* of $A$, the dimension of its row space.
- **c.** Find the *column rank* of $A$, the dimension of its column space.

For Exercises 54–60, determine the dimensions of the kernel, the row space, and the column space of the given matrix $A$.

**54.** $A = \begin{pmatrix} 1 & 2 & 3 \\ 4 & 5 & 6 \\ 7 & 8 & 9 \end{pmatrix}$

**55.** $A = \begin{pmatrix} 1 & 2 & 3 & 4 \\ 4 & 5 & 6 & 7 \\ 7 & 8 & 9 & 10 \end{pmatrix}$

**56.** $A = \begin{pmatrix} 1 & 2 & 3 \\ 4 & 5 & 6 \\ 7 & 8 & 0 \end{pmatrix}$

**57.** $A = \begin{pmatrix} 1 & 3 & 0 & 1 \\ 2 & 1 & 4 & 2 \\ 3 & 4 & 4 & 3 \end{pmatrix}$

**58.** $A = \begin{pmatrix} 1 & 3 & 0 & 1 & 2 \\ 3 & 10 & 1 & 4 & 1 \\ 2 & 7 & 1 & 3 & -1 \end{pmatrix}$

**59.** $A = \begin{pmatrix} 1 & 3 & 0 & 1 & 2 \\ 3 & 10 & 1 & 4 & 1 \\ 2 & 7 & 1 & 3 & -1 \\ 3 & 13 & 1 & 5 & 3 \end{pmatrix}$

**60.** $A = \begin{pmatrix} 1 & 3 & 0 & 1 & 2 \\ 3 & 10 & 1 & 4 & 1 \\ 2 & 7 & 1 & 3 & -1 \\ 3 & 13 & 1 & 5 & 3 \\ 4 & 13 & 1 & 5 & 4 \end{pmatrix}$

**61.** Look back at the results from Exercises 54–60.

- **a.** Is there a relationship between the total number of columns in $A$, the dimension of the kernel of $A$, and the row rank of $A$?
- **b.** Is there a relationship between the row rank and column rank of $A$?
- **c.** **Take It Further.** Can you prove that your relationships are true?

**62.** Determine each of the following for $A = \begin{pmatrix} 1 & 3 & 4 & 1 & 2 \\ 3 & 2 & 7 & 1 & 3 \\ 2 & -1 & 3 & 0 & 1 \\ 4 & 5 & 11 & 2 & 5 \end{pmatrix}$.

**a.** $\dim(\ker A)$
**b.** The column rank of $A$
**c.** The row rank of $A$
**d.** A basis for $\ker A$
**e.** A basis for the column space of $A$
**f.** A basis for the row space of $A$

# Chapter 7 Mathematical Reflections

These problems will help you summarize what you have learned in this chapter.

1. Consider the set $V$ of $4 \times 1$ matrices and define the operations of addition and scalar multiplication as usual. Verify that properties (1), (2), (4), and (5) hold for the set $V$.

2. Let $S$ be the set of $4 \times 1$ matrices of the form $\begin{pmatrix} a \\ 0 \\ a \\ 0 \end{pmatrix}$ and let $V$ be the set of $4 \times 1$ matrices with real entries. Is $S$ a subspace of $V$? Explain.

3. Let $S = L\{(1,2,4),(-1,1,3),(-2,1,4)\}$.
   **a.** Is $(-1,4,-1)$ in $S$? Explain.
   **b.** Is $S$ a generating system for $\mathbb{R}^3$? Explain.

4. Show that $\{x^2, 3x^2+x, 2x^2-x+1\}$ is a basis for $\mathbb{R}_2[x]$.

5. Let $v = \begin{pmatrix} 2 & -1 \\ 3 & 6 \end{pmatrix}$ and $B = \left\{\begin{pmatrix} 1 & 0 \\ 1 & 0 \end{pmatrix}, \begin{pmatrix} 0 & 1 \\ 0 & 0 \end{pmatrix}, \begin{pmatrix} 1 & 1 \\ 0 & 1 \end{pmatrix}, \begin{pmatrix} 0 & 1 \\ 0 & 1 \end{pmatrix}\right\}$.
   Find the coordinate vector for $v$ with respect to basis $B$.

6. What properties are necessary for a system to be a vector space?

7. How can you determine whether a set is a generating system for a given vector space?

8. Find the coordinate vector for $v = \begin{pmatrix} 5 & 0 \\ 2 & 3 \end{pmatrix}$ with respect to the basis
   $B = \left\{\begin{pmatrix} 0 & 0 \\ 1 & 1 \end{pmatrix}, \begin{pmatrix} 1 & 0 \\ 0 & 1 \end{pmatrix}, \begin{pmatrix} 1 & 1 \\ 0 & 0 \end{pmatrix}, \begin{pmatrix} 1 & 0 \\ 0 & 0 \end{pmatrix}\right\}$.

**Vocabulary**

In this chapter, you saw these terms and symbols for the first time. Make sure you understand what each one means, and how it is used.

- basis for a vector space
- column space
- coordinate vector
- dimension of a vector space
- finite dimensional
- generating system
- linear span
- $\text{Mat}_{m \times n}$
- $\mathbb{R}_n[x]$
- row space
- subspace
- vector space
- zero vector

# Chapter 7 Review

In Lesson 7.2, you learned to

- understand the structural similarities for different vector spaces
- determine whether or not a given set is a vector space

The following problems will help you check your understanding.

**1.** Verify that properties (1), (2), (4), (5), and (6) of a vector space hold for the set of polynomials whose degree is at most 2, under the usual definition of addition and scalar multiplication.

**2.** Determine whether or not $V$ is a vector space. If it is not a vector space, give a counterexample. If it is a vector space, give a good explanation to justify your answer.

**a.** $V = \mathbb{Q}$, the set of rational numbers.
**b.** $V =$ the set of $1 \times 3$ matrices.
**c.** $V =$ the set of ordered pairs $(x, y)$ with $x + y = 0$.
**d.** $V =$ the set of ordered pairs $(x, y)$ with $x + y = 1$.

**3.** Let $V =$ the set of $2 \times 2$ matrices in the form $\begin{pmatrix} a & b \\ c & d \end{pmatrix}$ such that $d = 0$.

**a.** Verify that properties (1), (4), and (5) hold for $V$.
**b.** If $d = 1$, would properties (1), (4), and (5) still hold? Explain.

In Lesson 7.3, you learned to

- determine if a subset of a vector space is also a subspace
- identify the key properties you need to test in order to prove a subset of a vector space is a subspace

The following problems will help you check your understanding.

**4.** Determine whether each given set is a subspace of the vector space $V = \mathbb{R}[x]$. Explain your reasoning.

**a.** All polynomials of degree at most 2
**b.** All polynomials of degree 2
**c.** All polynomials of degree at least 2
**d.** $\mathbb{R}[x]$

**5.** Determine which of the following are subspaces of $\mathbb{R}^2$. Explain how you know.

**a.** All vectors of the form $(x, 0)$
**b.** All vectors of the form $(x, 1)$
**c.** All vectors of the form $(x, y)$, where $x + 2y = 3$
**d.** All vectors of the form $(x, y)$, where $x + 2y = 0$

**6.** Let $S$ be the set of matrices of the form $\begin{pmatrix} a & a \\ b & b \end{pmatrix}$ and $V = \text{Mat}_{2\times 2}(\mathbb{R})$, the set of $2 \times 2$ matrices. Show that $S$ is a subspace of $V$.

In Lesson 7.4, you learned to

- find a generating system for a vector space
- determine if a given vector is in the linear span of a set of vectors
- determine if a set of vectors generates a vector space

The following problems will help you check your understanding.

**7.** Find a generating system for each vector space.

**a.** Matrices of the form $\begin{pmatrix} a & a \\ b & a \end{pmatrix}$

**b.** Ordered triples $(x, y, z)$ with $x + y + z = 0$

**c.** $\ker \begin{pmatrix} -1 & 2 & -2 \\ 3 & -5 & 4 \\ 1 & -1 & 0 \end{pmatrix}$

**d.** The column space of $\begin{pmatrix} 2 & -1 & 0 \\ 3 & 2 & -2 \\ 0 & 1 & -4 \end{pmatrix}$

**8.** Determine whether the given vector is in the given linear span.

**a.** $(1, 3, -5)$ in $L\{(3, 2, -1), (2, -1, 4)\}$

**b.** $x - 2$ in $L\{2x, x - 1\}$

**9.** Determine whether the given set is a generating system for the corresponding vector space. Justify your answers.

**a.** $\{(2, 3), (3, 4)\}$ for $\mathbb{R}^2$

**b.** $\{(3, 3, 0), (0, 2, 1), (0, -3, -2)\}$ for $\mathbb{R}^3$

In Lesson 7.5, you learned to

- determine whether a set of vectors is a basis for a vector space
- find the dimension of a vector space
- determine the dimension of the row and column spaces for a given matrix, as well as the dimension of its kernel

The following problems will help you check your understanding.

**10.** Determine whether the set is a basis for the given vector space.

**a.** $\{(1, 2, 3), (2, 0, -2), (-1, 4, 3)\}$ for $\mathbb{R}^3$

**b.** $\{(1, 2), (3, -1), (4, 0)\}$ for $\mathbb{R}^2$

**c.** $\left\{ \begin{pmatrix} 2 & 0 \\ 0 & 0 \end{pmatrix}, \begin{pmatrix} 0 & 1 \\ 0 & 1 \end{pmatrix}, \begin{pmatrix} 0 & 0 \\ 1 & 0 \end{pmatrix}, \begin{pmatrix} 0 & 2 \\ 1 & 0 \end{pmatrix} \right\}$ for $\text{Mat}_{2\times 2}(\mathbb{R})$

**d.** $\left\{x^2 + x, x - 1\right\}$ for $\mathbb{R}_2[x]$

**11.** For each vector space, find a basis and state the dimension of the vector space.

**a.** $L\{(-1,2,4),(3,1,0),(1,5,8)\}$

**b.** The column space of $\begin{pmatrix} 2 & 1 & -3 \\ 1 & 3 & 1 \\ 3 & 4 & -2 \end{pmatrix}$

**c.** $\ker \begin{pmatrix} 1 & -3 & 4 \\ 2 & 3 & -1 \end{pmatrix}$

**12.** Find the coordinate vector for $v$ with respect to the given basis.

**a.** $v = (-2,-14,-2)$, $B = \{(2,-1,5),(0,1,3),(4,5,7)\}$

**b.** $v = 4x^2 - 2x - 5$, $B = \{x^2, x+1, x-2\}$

**c.** $v = \begin{pmatrix} 3 & 6 \\ 1 & 5 \end{pmatrix}$, $B = \left\{ \begin{pmatrix} 0 & 1 \\ 0 & 0 \end{pmatrix}, \begin{pmatrix} 0 & 0 \\ 1 & 1 \end{pmatrix}, \begin{pmatrix} 0 & 1 \\ 0 & 1 \end{pmatrix}, \begin{pmatrix} 1 & 1 \\ 0 & 0 \end{pmatrix} \right\}$

# Chapter 7 Test

## Multiple Choice

**1.** Which is *not* a property of a vector space $V$?

**A.** There is an element $\mathbf{0}$ in $V$ called the zero vector.
**B.** For every $v \in V$, there is an element $-v$ with the property $v + -v = 0$.
**C.** For all $v$ and $w \in V$, $v \cdot w = w \cdot v$.
**D.** $c(v + w) = cv + cw$ for all numbers $c$ and elements $v$, $w$ of $V$.

**2.** Which of the following is a subspace of $\mathbb{R}^2$?

**A.** The set of ordered pairs $(x, y)$ with $x > 0$ and $y > 0$
**B.** The set of ordered pairs $(x, y)$ with $y = 2x$
**C.** The set of ordered pairs $(x, y)$ with $x = 1$
**D.** The set of ordered pairs $(x, y)$ with $y > x$

**3.** Let $S = L\{x^2, x + 2\}$. Which of the following is in $S$?

**A.** $x - 2$
**B.** $x^2 + 4x - 8$
**C.** $2x^2 - 3x - 6$
**D.** $x^3 + 2x^2$

**4.** Three vectors form a basis for $\mathbb{R}^3$. If two of those vectors are $(1, 0, -1)$ and $(2, 1, 3)$, which could be the third vector?

**A.** $(2, 0, -2)$ **B.** $(2, -1, -1)$ **C.** $(3, 1, 2)$ **D.** $(3, 2, 7)$

**5.** What is the dimension of $\mathbb{R}_4[x]$?

**A.** 0 **B.** 3 **C.** 4 **D.** 5

**6.** Let $v = (2, 1, -5)$ and $B = \{(2, 0, 1), (1, 0, 3), (1, 1, 2)\}$. What is the coordinate vector for $v$ with respect to basis $B$?

**A.** $(0, 1, 1)$ **B.** $(1, 1, -1)$ **C.** $(2, -3, 1)$ **D.** $(3, -2, -2)$

## Open Response

**7.** Let $V = \mathbb{R}_1[x]$ be the set of polynomials whose degree is at most 1. Verify that each property of a vector space given below holds for $V$.

**a.** The set is closed under addition and scalar multiplication.
**b.** $v + w = w + v$ for all $v, w \in V$.
**c.** $c(dv) = cd(v)$ for all numbers $c, d$ and elements $v$ of $V$.

**8.** Let $V = \text{Mat}_{2\times2}(\mathbb{R})$ and let $S =$ the set of $2 \times 2$ matrices in the form $\begin{pmatrix} a & 0 \\ 0 & b \end{pmatrix}$. Show that $S$ is a subspace of $V$.

**9.** Find a generating system for each vector space.

**a.** Matrices in the form $\begin{pmatrix} a & b \\ 2a + b & 2a - b \end{pmatrix}$
**b.** Ordered triples $(x, y, z)$ with $x + y = 0$

**10.** Show that $\{x^3 + 1, x^2 - x, x + 1, x - 1\}$ is a basis for $\mathbb{R}_3[x]$.

**11.** Find a basis for the column space of $\begin{pmatrix} 2 & -3 & 3 \\ 0 & 2 & -2 \\ -1 & 2 & -2 \end{pmatrix}$.

**12.** Determine the dimension of the kernel of $A = \begin{pmatrix} 1 & 2 & 5 \\ 5 & 4 & 7 \\ 3 & 1 & 0 \end{pmatrix}$.

CHAPTER

# 8 Bases, Linear Mappings, and Matrices

In the last chapter, you learned how abstract vector spaces can be generated by a linearly independent system of vectors called a *basis*. But a vector space can have many different bases. In this chapter, you will see how these different bases have a geometric interpretation.

The concept of linear maps and linear transformations will also be extended to general vector spaces. These maps are defined by their action on a basis, so they will have different representations as matrices according to which basis is picked for the domain and the range. You will see that these different representations are related in a very nice way. And matrices will also be used to switch among the different representations.

**By the end of this chapter, you will be able to answer questions like these:**

1. How is the dimension of the row space of a matrix related to the dimension of its kernel?
2. How are different matrix representations for the same linear map related?
3. Let $M = \begin{pmatrix} 9 & 4 \\ -12 & -5 \end{pmatrix}$ and $D = \begin{pmatrix} 3 & 0 \\ 0 & 1 \end{pmatrix}$. Is $M$ is similar to $D$?

**You will build good habits and skills for ways to**

- reason about calculations
- ensure statements are consistent
- use different strategies to establish a proof
- find associations between different concepts

### Vocabulary and Notation

- bijective linear map
- blow up to a basis
- change of basis matrix
- eigenvalue
- eigenvector, characteristic vector
- identity mapping
- invariant line
- linear map
- matrix for a transformation with respect to a basis
- maximal linearly independent set
- row and column rank
- similar matrices
- structure-preserving representation

# 8.1 Getting Started

## Exercises

**1.** Suppose $A = \begin{pmatrix} 1 & 2 & -2 & 4 & 7 \\ 2 & 3 & -1 & 5 & 8 \\ 3 & 5 & -3 & 9 & 15 \end{pmatrix}$. Find the dimension of

**a.** the row space of $A$
**b.** the column space of $A$
**c.** the kernel of $A$

**2.** Find a basis for $\mathbb{R}^3$ that contains $(1,2,3)$ and $(2,3,4)$. Explain your method.

**3.** Let $\mathfrak{R} = \{(1,0,2,0), (0,1,0,3), (2,2,4,3), (1,1,0,-3), (0,0,1,4), (1,1,1,1)\}$.

**a.** Show that $\mathfrak{R}$ is not a basis for $\mathbb{R}^4$.
**b.** Show that $\mathfrak{R}$ is a generating system for $\mathbb{R}^4$.
**c.** Find a subset of $\mathfrak{R}$ that is a basis for $\mathbb{R}^4$. Explain your method.

**4.** Let $\mathfrak{S} = \{x^2+1, x^2-1, x^3+1\}$.

**a.** Show that $\mathfrak{S}$ is not a basis for $\mathbb{R}_3[x]$.
**b.** Show that $\mathfrak{S}$ is linearly independent.
**c.** Find a basis for $\mathbb{R}_3[x]$ that contains $\mathfrak{S}$. Explain your method.

**5.** Let $\mathfrak{T} = \left\{ \begin{pmatrix} 1 & 1 \\ 1 & 1 \end{pmatrix}, \begin{pmatrix} 1 & 1 \\ 1 & 2 \end{pmatrix}, \begin{pmatrix} 0 & 2 \\ 2 & 4 \end{pmatrix}, \begin{pmatrix} 2 & 1 \\ 1 & 3 \end{pmatrix}, \begin{pmatrix} 2 & 5 \\ 5 & 0 \end{pmatrix} \right\}$.

**a.** Show that $\mathfrak{T}$ is not a basis for the vector space of all $2 \times 2$ symmetric matrices.
**b.** Show that $\mathfrak{T}$ is a generating system for the vector space of all $2 \times 2$ symmetric matrices.
**c.** Find a subset of $\mathfrak{T}$ that is a basis for the vector space of $2 \times 2$ symmetric matrices. Explain your method.

**6.** Suppose that $V$ is a vector space, $v$, $w$ are vectors in $V$, and $c$ is a scalar. Show that $L\{v, w\} = L\{v, w + cv\}$.

**7.** Show that any set consisting of one nonzero vector is linearly independent.

**8.** Suppose that $V$ is a vector space and $\{v_1, v_2, \ldots, v_m\}$ is a generating system for $V$. Suppose further that $v_m$ is a combination of $\{v_1, v_2, \ldots, v_{m-1}\}$. Show that if you eliminate $v_m$ from the set, you still have a generating system for $V$.

**9.** Let $V$ be a vector space of dimension $n$ and let

$$\mathfrak{B} = \{v_1, v_2, \ldots, v_n\}$$

be a basis for $V$. Show that a set of vectors

$$\mathfrak{B}' = \{w_1, w_2, \ldots, w_n\}$$

in $V$ is a basis for $V$ if and only if the set of coordinate vectors

$$\{\underline{w_1}_{\mathfrak{B}}, \underline{w_2}_{\mathfrak{B}}, \dots, \underline{w_n}_{\mathfrak{B}}\}$$

is a basis for $\mathbb{R}^n$.

**10.** Suppose $T : \mathbb{R}^4 \to \mathbb{R}^3$ is defined by

$$T(x, y, z, w) = (2x + y + w, y + 2z + w, x + y + z + w)$$

and let

$$\mathfrak{B} = \{(1,1,0,0), (0,1,1,0), (0,0,1,1), (0,0,0,1)\} \quad \text{and}$$
$$\mathfrak{B}' = \{(1,1,0), (0,1,1), (0,0,1)\}$$

**a.** Show that $\mathfrak{B}$ is a basis for $\mathbb{R}^4$.
**b.** Show that $\mathfrak{B}'$ is a basis for $\mathbb{R}^3$.
**c.** Find a matrix $M$ so that, for any vector $X$ in $\mathbb{R}^4$,

$$M\underline{X}_{\mathfrak{B}} = \underline{T(X)}_{\mathfrak{B}'}$$

← The coordinate vectors $\underline{X}_{\mathfrak{B}}$ and $\underline{T(X)}_{\mathfrak{B}'}$ are written as columns here.

**11.** Consider the following two bases for $\mathbb{R}^3$:

$$\mathfrak{B} = \{(1,1,0), (0,1,1), (0,0,1)\} \quad \text{and}$$
$$\mathfrak{B}' = \{(-1,2,3), (2,-3,-6), (1,-3,-2)\}$$

**a.** Find a matrix $M$ so that, for any vector $X$ in $\mathbb{R}^3$,

$$M\underline{X}_{\mathfrak{B}} = \underline{X}_{\mathfrak{B}'}$$

**b.** Find a matrix $N$ so that, for any vector $X$ in $\mathbb{R}^3$,

$$N\underline{X}_{\mathfrak{B}'} = \underline{X}_{\mathfrak{B}}$$

**c.** Calculate $MN$.

**12.** Show that

$$\mathfrak{M} = \left\{1, x, \frac{x(x-1)}{2}, \frac{x(x-1)(x-2)}{6}\right\}$$

is a basis for $\mathbb{R}_3[x]$.

← $\mathfrak{M}$ is called the "Mahler basis" for $\mathbb{R}_3[x]$, named for the German mathematician Kurt Mahler. What would be a good candidate for the Mahler basis of $\mathbb{R}_4[x]$?

**13.** Consider the following two bases for $\mathbb{R}_3[x]$:

$$\mathfrak{B} = \{1, x, x^2, x^3\} \quad \text{and}$$

$$\mathfrak{M} = \left\{1, x, \frac{x(x-1)}{2}, \frac{x(x-1)(x-2)}{6}\right\}$$

← $\mathfrak{B}$ is the standard basis and $\mathfrak{M}$ is the Mahler basis from Exercise 12.

**a.** Find a matrix $M$ so that, for any vector $F$ in $\mathbb{R}_3[x]$,

$$M\underline{F}_{\mathfrak{B}} = \underline{F}_{\mathfrak{M}}$$

**b.** Find a matrix $N$ so that, for any vector $F$ in $\mathbb{R}^3$,

$$N\underline{F}_{\mathfrak{M}} = \underline{F}_{\mathfrak{B}}$$

**c.** Calculate $MN$.

## 8.2 Building Bases

In Lesson 7.5, you learned that any two bases for a finite-dimensional vector space have the same size. This fact will allow you to create bases from a given set of vectors that only meets some of the necessary criteria.

←
In this chapter, all vector spaces will be finite-dimensional. The number of vectors in a basis for a vector space is its *dimension*.

**In this lesson, you will learn how to**

- build a basis for a vector space by *blowing up* a set of linearly independent vectors that is not a generating system into a basis, or by *sifting out* a basis from a generating system whose vectors are not linearly independent.

**Minds in Action** Episode 31

*In Episode 27, Derman and Sasha were discussing generating systems, but they hadn't quite finished that discussion.*

DERMAN: So what? It's still a generating system, right?

SASHA: I guess it is. But isn't it a nuisance having all that extra stuff? I'd think you'd want it to be as small as possible.

DERMAN: Small like your generating system, you mean. Do you know yours is the smallest? Could it be done with just two vectors instead of three?

SASHA: I have no idea.

DERMAN: But I think I *do* have an idea! It's pretty obvious that the vectors $x^2$, $x$, and 1 would generate all of $\mathbb{R}_2[x]$, but they wanted us to find *other* generating systems. The reason I listed so many vectors is that I was thinking about what parts I *could* use to "build" each polynomial in $\mathbb{R}_2[x]$. I know I don't need them all, but I do need to be able to make any $x^2$ part I want, so I need *something* with an $x^2$ in it. Same for $x$ and the constant: I need *something* with an $x$ in it and something with a number.

SASHA: Well, but *my* solution . . .

DERMAN: See? Because if I start with, say, the $x + 1$ vector I had before, that lets me build some things, but I'll never get any $x^2$ parts. So that's not enough! I can fix that by using my $3x^2$ vector.

SASHA: Yes, but . . .

DERMAN: But those two aren't enough because I still can't control the $x$ part and constant part *independently* of each other. Using my $3x^2$ vector, I can get any $x^2$ term I like, and using my $x + 1$ vector, I can get any $x$ term or any constant term I like, but no matter what I do with $x + 1$, the constant term and the coefficient of the $x$ term will be the same. I need another vector! Something that separates the $x$ term and the constant term. Like my $2x$ vector.

*Sasha wrinkles her nose in thought as Tony comes in.*

TONY: I like to translate the whole thing into coordinate vectors. Look, we know that $\mathfrak{B} = \{x^2, x, 1\}$ is a basis for $\mathbb{R}_2[x]$. If I express Sasha's vectors in terms of $\mathfrak{B}$, I get

$$\begin{aligned} \underline{x^2+1}_{\mathfrak{B}} &= (1,0,1) \\ \underline{x-1}_{\mathfrak{B}} &= (0,1,-1) \\ \underline{x^2+x+1}_{\mathfrak{B}} &= (1,1,1) \end{aligned}$$

By Exercise 9 from Lesson 8.1, Sasha's vectors form a basis of $\mathbb{R}_2[x]$ if and only if their coordinate vectors with respect to $\mathfrak{B}$ form a basis for $\mathbb{R}^3$.

SASHA: And we have a way to think about that: take the matrix whose columns are the coordinate vectors and reduce it to echelon form. If the echelon form is the identity, then the columns form a basis.

DERMAN: I just did it, and the echelon form is $I$. Bingo.

TONY: Very smooth.

### For You to Do

1. Find, if possible, a basis of the linear span of $\{99x + 18, 231x + 42\}$ in $\mathbb{R}_1[x]$.
2. Find, if possible, a basis for $\mathbb{R}_1[x]$ that contains $99x + 18$.
3. Find, if possible, a basis for $\mathbb{R}_1[x]$ that contains $99x + 18$ and $231x + 42$.

Derman, Sasha, and Tony are working on the problem of finding a basis for a vector space, starting from a predetermined set of vectors. Given such a set $\mathfrak{S}$, it can either *be* a basis already or fail for one or both of two reasons:

1. $\mathfrak{S}$ doesn't generate.
2. $\mathfrak{S}$ is linearly dependent.

For the rest of this lesson, you'll look at two techniques for building bases from deficient sets:

1. Starting with a linearly independent set of vectors, you'll learn how to "blow it up" to a basis.
2. Starting with a generating system, you'll learn how to "sift out" a basis.

← If your set of vectors is deficient for both reasons, as in the $\{99x + 18, 231x+42\}$ example above, you'll have to first remove the dependencies and then extend it to a basis.

### Blowing up

Suppose you have a linearly independent set of vectors that's not a basis. You can extend it to a basis via the method described in the next example.

### Example 1

**Problem.** Is the set $\left\{\begin{pmatrix}1&1\\0&0\end{pmatrix}, \begin{pmatrix}0&1\\1&0\end{pmatrix}\right\}$ a basis for $\text{Mat}_{2\times 2}$? If not, can you find a basis that contains it?

**Solution.** It's not a basis because $\dim(\mathrm{Mat}_{2\times 2}) = 4$, and this set only contains two vectors. But the set *is* linearly independent. So, it must not span. The idea is to find a matrix that is not in the linear span of $\left\{\begin{pmatrix}1 & 1\\ 0 & 0\end{pmatrix}, \begin{pmatrix}0 & 1\\ 1 & 0\end{pmatrix}\right\}$. Well,

$$a\begin{pmatrix}1 & 1\\ 0 & 0\end{pmatrix} + b\begin{pmatrix}0 & 1\\ 1 & 0\end{pmatrix} = \begin{pmatrix}a & a+b\\ b & 0\end{pmatrix}$$

So, there are plenty of matrices that are not in the linear span, $\begin{pmatrix}0 & 0\\ 0 & 1\end{pmatrix}$, for example. So, throw this in and consider the set

$$\left\{\begin{pmatrix}1 & 1\\ 0 & 0\end{pmatrix}, \begin{pmatrix}0 & 1\\ 1 & 0\end{pmatrix}, \begin{pmatrix}0 & 0\\ 0 & 1\end{pmatrix}\right\}$$

Is this a basis? Well, no, because it only has three things in it and a basis requires four. Is it independent? Yes, as you can check directly with a calculation. Or, you can also reason like this:

Suppose that

$$a\begin{pmatrix}1 & 1\\ 0 & 0\end{pmatrix} + b\begin{pmatrix}0 & 1\\ 1 & 0\end{pmatrix} + c\begin{pmatrix}0 & 0\\ 0 & 1\end{pmatrix} = \begin{pmatrix}0 & 0\\ 0 & 0\end{pmatrix}$$

If $c \neq 0$, then you could solve for $\begin{pmatrix}0 & 0\\ 0 & 1\end{pmatrix}$ and conclude that it is a combination of the original two matrices. But you chose $\begin{pmatrix}0 & 0\\ 0 & 1\end{pmatrix}$ to ensure that it's *not* such a combination, so $c = 0$. But then $a = b = 0$ because the original two matrices are linearly independent.

So, now you have three linearly independent vectors. It's not a basis, so it must not span. Look at a generic linear combination:

$$a\begin{pmatrix}1 & 1\\ 0 & 0\end{pmatrix} + b\begin{pmatrix}0 & 1\\ 1 & 0\end{pmatrix} + c\begin{pmatrix}0 & 0\\ 0 & 1\end{pmatrix} = \begin{pmatrix}a & a+b\\ b & c\end{pmatrix}$$

$\longleftarrow$ In every matrix of this form the entry in the first row, second column must be the sum of the entries in the first column.

Pick a vector that is not of this form, say $\begin{pmatrix}0 & 0\\ 1 & 1\end{pmatrix}$. Use the above line of reasoning to show that

$$\left\{\begin{pmatrix}1 & 1\\ 0 & 0\end{pmatrix}, \begin{pmatrix}0 & 1\\ 1 & 0\end{pmatrix}, \begin{pmatrix}0 & 0\\ 0 & 1\end{pmatrix}, \begin{pmatrix}0 & 0\\ 1 & 1\end{pmatrix}\right\}$$

is linearly independent. Finally, you have four independent vectors, so this is a basis.

---

This method is perfectly general. Starting from an independent set, you can keep adjoining vectors, each of which is not in the linear span of the previous ones, until you have enough to make a basis. And, each time you adjoin a vector that's not in the span of the previous ones, the resulting set is linearly independent, thanks to the following lemma.

### Lemma 8.1

*Suppose that the vectors $\{v_1, v_2, \ldots, v_n\}$ are linearly independent and that $v$ is not in the linear span of $\{v_1, v_2, \ldots, v_n\}$. Then $\{v_1, v_2, \ldots, v_n, v\}$ is also linearly independent.*

### For You to Do

**4.** Use the reasoning in Example 1 to prove Lemma 8.1.

Lemma 8.1 shows how to adjoin vectors, one at a time, to a linearly independent set so as to keep the larger set linearly independent. If the dimension of the space is finite, which is the assumption in this chapter, eventually, you get enough independent vectors to make a basis.

**Theorem 8.2 (The Blow-Up Theorem)**
*Any linearly independent set of vectors can be extended to form a basis.*

### Developing Habits of Mind

**Use different strategies to establish a proof.** The proof of Theorem 8.2 is implicit in the method used in Example 1, and Lemma 8.1 guarantees that this method will work. Such a proof is called **constructive**—it establishes that something happens by showing you how to make it happen.

### Sifting out

Suppose that you start with a generating system that's not a basis. You can "pare it down" to make a basis. The following, similar to the blowing-up algorithm, does that.

Suppose you have a set of vectors $\{v_1, v_2, \ldots, v_n\}$ that is a generating system for a vector space $V$, but it's not a basis. Then it must be linearly dependent. (Why?) So, there are scalars $c_i$, not all zero, so that

← In other words, $L\{v_1, v_2, \ldots, v_n\} = V$.

$$c_1 v_1 + \cdots + c_n v_n = \mathbf{0} \qquad (*)$$

Suppose, for example, that $c_n \neq 0$. Then, you can solve equation $(*)$ for $v_n$:

$$v_n = -\frac{1}{c_n}\left(c_1 v_1 + \cdots + c_{n-1} v_{n-1}\right)$$

By Exercise 8 from Lesson 8.1, $\{v_1, v_2, \ldots, v_{n-1}\}$ is still a generating system for $V$. If it is independent, you have a basis. If not, you can eliminate another vector in this same way and still keep the fact that the linear span is all of $V$. One at a time, you can eliminate extraneous vectors until you get an independent set that still generates. That will be your basis.

← In other words, $L\{v_1, v_2, \ldots, v_{n-1}\} = V$.

← See Exercise 7 from Lesson 8.1.

There may be several ways to sift a basis out of a generating system, eliminating at each step a vector that's a combination of the rest. In the end, though, any of the bases you construct in this way will contain the same number of vectors. (Why?) Another way to think about it is that this sifting process produces a **maximal linearly independent set**. See Exercise 8 for more on this.

# Exercises

1. For each of the given sets, blow it up to a basis.
   a. $\{(1,2,3)\}$ for $\mathbb{R}^3$
   b. $\{(1,2,3),(1,2,0)\}$ for $\mathbb{R}^3$
   c. $\{(1,1,0,0),(0,1,1,0)\}$ for $\mathbb{R}^4$
   d. $\left\{\begin{pmatrix}1 & 0\\ 0 & 1\end{pmatrix}, \begin{pmatrix}0 & 1\\ 1 & 0\end{pmatrix}\right\}$ for $\text{Mat}_{2\times 2}(\mathbb{R})$
   e. $\{(1,0,1)\}$ for $L\{(1,2,1),(1,0,1),(2,2,2),(0,2,0)\}$
   f. $\{x^2+1\}$ for $L\{(x+1)^2, x^2+1, 2(x^2+x+1), 2x\}$
   g. $\{(3,4,4,3),(2,3,4,1)\}$ for the row space of $\begin{pmatrix}1 & 1 & 0 & 2\\ 2 & 3 & 4 & 1\\ 1 & 0 & 1 & 0\\ 4 & 4 & 5 & 3\end{pmatrix}$
   h. $\left\{\begin{pmatrix}2\\5\\1\\8\end{pmatrix}, \begin{pmatrix}1\\7\\1\\9\end{pmatrix}\right\}$ for the column space of the matrix in part **g**.
   i. $\{(1,-1,1,2,0)\}$ for $\ker M = \ker \begin{pmatrix}1 & 3 & 0 & 1 & 5\\ 2 & 5 & 1 & 1 & 9\\ 1 & 2 & 1 & 0 & 4\end{pmatrix}$

$\longleftarrow$ $(1,-1,1,2,0)$ is in the kernel of the matrix, right?

2. For each of the given vector spaces, sift out a basis from the given generating system.
   a. $V = L\{(1,2,1),(1,0,1),(2,2,2),(0,3,0)\}$
   b. $V$ is the row space of $N = \begin{pmatrix}1 & 1 & 0 & 2\\ 2 & 3 & 4 & 1\\ 1 & 0 & 1 & 0\\ 4 & 4 & 5 & 3\end{pmatrix}$, starting with the rows.
   c. $V$ is the column space of $N$ from part **b**, starting with the columns.
   d. $V = L\{1, x, 3x+2, x^2+1, x^2-1\}$
   e. $V = L\{1, x, \frac{x(x-1)}{2}, \frac{x(x-1)(x-2)}{6}\}$

$\longleftarrow$ If the vector space is given as the linear span of a set of vectors, use those vectors as the generating system.

3. Suppose $M$ is an $m \times n$ matrix. Show that the kernel of $M$ is a subspace of $\mathbb{R}^n$.

$\longleftarrow$ See Exercise 16 from Lesson 7.3.

4. Suppose that $T : \mathbb{R}^5 \to \mathbb{R}^3$ is represented by matrix $M$ from Exercise 1**i**.
   a. Find a basis for $\ker M$.
   b. Blow this up to a basis of $\mathbb{R}^5$.
   c. Apply $T$ to each of your basis vectors by multiplying each of them by $M$. This gives you five vectors in $\mathbb{R}^3$. But some of them will be $O$. (Why?)
   d. Show that the nonzero vectors in your set from the previous part of this exercise form a basis for the column space of $M$.

$\longleftarrow$ See Lesson 5.5 for the definition. The column space of $M$ is what was called the *image* of $M$ (or of $T$) in Chapter 5.

**5.** Suppose that $F : \mathbb{R}^4 \to \mathbb{R}^4$ is represented by matrix $N$ from Exercise 2**b**.

**a.** Find a basis for $\ker N$.
**b.** Blow this up to a basis of $\mathbb{R}^4$.
**c.** Apply $F$ to each of your basis vectors by multiplying each of them by $N$. This gives you four vectors in $\mathbb{R}^4$. But some of them will be $O$. (Why?)
**d.** Show that the nonzero vectors in your set from the previous part of this exercise form a basis for the column space of $N$.

**6.** Suppose that $V$ is a vector space of dimension $n$. Show that any generating system for $V$ that contains $n$ vectors is a basis for $V$.

**7.** Suppose $V$ is a vector space and $S$ is a subspace. Show that

$$\dim(S) \leq \dim(V)$$

and that equality occurs if and only if $S = V$.

**8.** Suppose you have a generating system for a vector space $V$, say $\mathfrak{G} = \{v_1, \ldots, v_n\}$. Let $\mathfrak{B}$ be a subset of $\mathfrak{G}$ that is linearly independent and that has the property that any subset of $\mathfrak{G}$ that has more vectors in it than $\mathfrak{B}$ is linearly dependent. Show that $\mathfrak{B}$ is a basis for $V$.

$\longleftarrow$
$\mathfrak{B}$ is a maximal linearly independent subset of $\mathfrak{G}$.

## 8.3 Rank

In Exercise 61 in Lesson 7.5, you explored relationships between the dimensions of the row space, the column space, and the kernel of a matrix. This lesson will formalize the relationships among these dimensions.

**In this lesson, you will learn how to**

- find the *rank* of a matrix
- determine the relationship among the dimensions of the row space, column space, and kernel of a matrix.

### In-Class Experiment

Choose a matrix larger than $3 \times 3$ and not necessarily square. Pick your matrix so that either the rows or the columns are linearly dependent. For your matrix, find

1. the dimension of the row space
2. the dimension of the column space
3. the dimension of the kernel

### Minds in Action Episode 32

*Tony, Sasha, and Derman are passing the time, playing with matrices.*

SASHA: Hey, look at this: I made a $3 \times 3$ matrix whose rows are linearly dependent:

$$A = \begin{pmatrix} 1 & 2 & 3 \\ 4 & 5 & 6 \\ 5 & 7 & 9 \end{pmatrix}$$

DERMAN: The sum of the first two rows is equal to the third.

SASHA: And, without trying, the columns are linearly dependent, too:

$$1 \begin{pmatrix} 1 \\ 4 \\ 5 \end{pmatrix} - 2 \begin{pmatrix} 2 \\ 5 \\ 7 \end{pmatrix} + 1 \begin{pmatrix} 3 \\ 6 \\ 9 \end{pmatrix} = \begin{pmatrix} 0 \\ 0 \\ 0 \end{pmatrix}$$

TONY: That can't always work. Let's try it with another one, say

$$M = \begin{pmatrix} 1 & 2 & 3 & 4 \\ 2 & 3 & 4 & 5 \\ 3 & 5 & 7 & 9 \\ 1 & 1 & 1 & 1 \end{pmatrix}$$

DERMAN: This time, the row space has dimension 2:

$$M_{1*} + M_{2*} = M_{3*} \quad \text{and} \quad M_{2*} - M_{1*} = M_{4*}$$

So, the first two rows generate the row space.

SASHA: Ah, and look at the columns:

$$2M_{*2} - M_{*1} = M_{*3} \quad \text{and} \quad 3M_{*2} - 2M_{*1} = M_{*4}$$

TONY: Maybe the row space and column space always have the same dimension? That's weird.

DERMAN: This could be a great party trick. Make up a square matrix whose rows are dependent. Automatically the columns are, too.

SASHA: There are theorems to prove to make sure it always works before you can bring it to one of your parties.

---

Derman's party trick exposes a deep result about dimension: given any matrix, the dimensions of the row and column space are the same. You will prove this fact by the end of the lesson.

### Lemma 8.3

*Row-equivalent matrices have the same row space.*

***Proof.*** Suppose $A$ is row equivalent to $B$. Then you can get from $A$ to $B$ by performing some sequence of elementary row operations:

1. Switch two rows.
2. Multiply a row by a nonzero constant.
3. Replace $A_{i*}$ by $cA_{j*} + A_{i*}$ for some scalar $c$.

The row space is the linear span of the rows. Clearly, the first two operations don't change the row space. The last operation doesn't either, by Exercise 6 from Lesson 8.1. ∎

Applying Lemma 8.3 to each step of the reduction to echelon form results in a corollary.

### Corollary 8.4

*If $M$ is a matrix, the row space of $M$ is the same as the row space of* $\text{rref}(M)$.

### Definition

The **row rank** of a matrix is the dimension of its row space. The **column rank** of a matrix is the dimension of its column space.

### For You to Do

1. Find the row rank and column rank for

$$\begin{pmatrix} 1 & 3 & 2 & -1 & 2 \\ 1 & 1 & -1 & -1 & 3 \\ 1 & 0 & -1 & -1 & 3 \end{pmatrix}$$

### Developing Habits of Mind

**Reason about calculations.** Think about a matrix in echelon form. The nonzero rows are linearly independent because of the positions of the pivots. And the zero rows contribute nothing to the row space. This leads to a theorem.

**Theorem 8.5**

*The row rank of a matrix is the number of nonzero rows in its echelon form.*

⟵ And the column rank is the number of nonzero rows in the echelon form of the transpose, right?

### Minds in Action — Episode 33

TONY: And what about the kernel?

DERMAN: What *about* the kernel?

TONY: I mean, look at how we find the kernel. Here's an echelon form:

*Tony writes this matrix on the board:* $\begin{pmatrix} 1 & 0 & 0 & 3 & 2 \\ 0 & 1 & 0 & -4 & 2 \\ 0 & 0 & 1 & -5 & 8 \\ 0 & 0 & 0 & 0 & 0 \end{pmatrix}$

When I find the kernel, I can solve for the first three variables in terms of the last two. There will be two "free variables" that can be anything I want.

⟵ Tony is reasoning about the calculation without carrying it out completely.

DERMAN: So, the kernel will be all linear combinations of two things, right?

TONY: Right. It will be like

$$\begin{aligned}(-3w-2u, 4w-2u, 5w-8u, w, u) &= (-3w, 4w, 5w, w, 0) + (-2u, -2u, -8u, 0, u) \\ &= w(-3, 4, 5, 1, 0) + u(-2, -2, -8, 0, 1)\end{aligned}$$

so there will be a $w$ part and a $u$ part. So the dimension of the kernel is 2.

SASHA: Very smooth, Tony. This will always work. The dimension of the kernel is the number of free variables in the echelon form. And this is the number of nonpivot columns in the echelon form. And this is the number of columns minus the number of pivot columns in the echelon form. And this is the number of columns minus the number of nonzero rows in the echelon form. And this is the number of columns minus the row rank of the echelon form. And *this* is the number of columns minus the row rank of the original matrix.

DERMAN: Can you say that one more time?

---

Sasha's statement leads to the following theorem.

**Theorem 8.6 (Dimension of Kernel)**

*The dimension of the kernel of an $m \times n$ matrix whose row rank is $r$ is $n - r$.*

---

**For Discussion**

**2.** Help Derman unpack Sasha's sentences and prove Theorem 8.6.

---

**Minds in Action** Episode 34

TONY: I have an idea about how we might prove that Derman's party trick—that the row rank and column rank of any matrix is the same—always works. What if we could prove that, for an $m \times n$ matrix, the column rank—the dimension of the column space—is equal to $n$ minus the dimension of the kernel?

SASHA: That would do it. If the row rank is $r$, the dimension of the kernel is $n - r$, so, if what you say is true, the column rank would be $n - (n - r) \ldots$

DERMAN: And that's just $r$, so the column rank would be the same as the row rank. Things are heating up.

SASHA: But, Tony, how would you prove that for an $m \times n$ matrix, the column rank is equal to $n -$ (the dimension of the kernel)?

TONY: I have no idea.

DERMAN: Things are cooling down.

*They think about it and get nowhere, so they put it away for the day. The next morning, Sasha makes an announcement.*

SASHA: I've been thinking about this all night, and I have an idea. Remember Exercises 4 and 5 from Lesson 8.2? They hold the key. We just need to do what we did there, but make it general.

TONY: I think you're right. It will work—I know it.

DERMAN: I didn't do those problems.

TONY: Well, it goes like this, Derman. Suppose you have an $m \times n$ matrix $M$ and the dimension of the kernel is $k$. Take a basis for the kernel and blow it up to a basis for $\mathbb{R}^n$. You'll have $n$ vectors now, $k$ of them in the kernel. So, there are $n - k$ vectors in the basis for $\mathbb{R}^n$ that are *not* in the kernel. Could these be a basis for the column space?

DERMAN: No. They are in $\mathbb{R}^n$ and the columns are from $\mathbb{R}^m$.

SASHA: Right, but in Exercises 4 and 5, we then multiplied these vectors by the matrix—their images under $M$ are in $\mathbb{R}^m$.

DERMAN: I didn't do those problems.

---

Sasha, Tony, and Derman are onto something. The proof of the following theorem makes their method precise.

**Theorem 8.7 (Dimension of the Column Space)**
*Suppose $M$ is $m \times n$ and the dimension of $\ker(M)$ is $k$. Then the column rank of $M$ is $n - k$.*

***Proof.*** Think of $M$ as a linear mapping from $\mathbb{R}^n$ to $\mathbb{R}^m$.

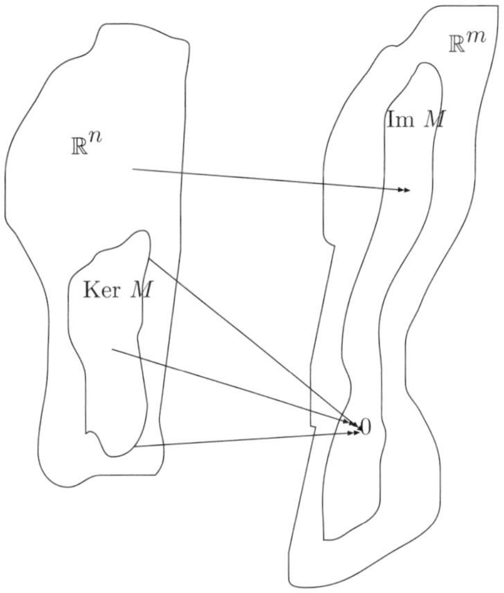


It might help to keep track of everything if you have a specific matrix in mind, say a $4 \times 5$ matrix whose echelon form is Tony's matrix from Episode 33:

← Follow along the proof using this specific $M$.

$$M = \begin{pmatrix} 1 & 0 & 0 & 3 & 2 \\ 0 & 1 & 0 & -4 & 2 \\ 0 & 0 & 1 & -5 & 8 \\ 0 & 0 & 0 & 0 & 0 \end{pmatrix}$$

In this example, $n = 5$, $m = 4$, and $k = 2$.

Take a basis for $\ker(M)$, say $\{v_1, \dots, v_k\}$, and blow it up to a basis for $\mathbb{R}^n$:

$$\{v_1, \dots, v_k, v_{k+1}, \dots, v_n\}$$

You want to prove that the dimension of the column space for $M$ is $n-k$, and you have $n - k$ vectors sitting there: $\{v_{k+1}, \dots, v_n\}$. But these are in $\mathbb{R}^n$, and the column space—the image of the mapping represented by $M$—is a subspace of $\mathbb{R}^m$.

Multiplying a vector in $\mathbb{R}^n$ by $M$ produces a vector in $\mathbb{R}^m$, so that's a way to get things to be the right size. Consider the set

$$\mathfrak{G} = \{Mv_1, \ldots, Mv_k, Mv_{k+1}, \ldots, Mv_n\}$$

Now $v_1$ through $v_k$ are in $\ker M$, so $M$ times each of them is $\mathbf{0}$. Hence, the nonzero vectors in $\mathfrak{G}$ form the set

$$\mathfrak{B} = \{Mv_{k+1}, \ldots, Mv_n\}$$

These vectors are in the right place ($\mathbb{R}^m$), and there are the right number of them ($n - k$), so the proof will be complete if you can show that $\mathfrak{B}$ is a basis for the column space of $M$. To be a basis, $\mathfrak{B}$ must have two properties:

1. **$\mathfrak{B}$ is linearly independent.** Suppose you have scalars $c_{k+1}, \ldots, c_n$ so that

   $$c_{k+1}Mv_{k+1} + c_{k+2}Mv_{k+2} + \cdots + c_nMv_n = \mathbf{0}$$

   Then
   $$M\left(c_{k+1}v_{k+1} + c_{k+2}v_{k+2} + \cdots + c_nv_n\right) = \mathbf{0}$$

   So the vector $v = c_{k+1}v_{k+1} + c_{k+2}v_{k+2} + \cdots + c_nv_n$ is in $\ker M$. But a basis for $\ker M$ is $\{v_1, \ldots, v_k\}$. This means that $v$ can be written as a linear combination of $v_1, \ldots, v_k$; so you have an equation like this:

   $$c_{k+1}v_{k+1} + c_{k+2}v_{k+2} + \cdots + c_nv_n = c_1v_1 + \cdots + c_kv_k$$

   for some scalars $c_1, \ldots, c_k$. Get everything on one side of this equation, and you have a linear combination of $\{v_1, \ldots, v_k, v_{k+1}, \ldots, v_n\}$ equal to $\mathbf{0}$. But $\{v_1, \ldots, v_k, v_{k+1}, \ldots, v_n\}$ is a basis, so it's linearly independent, so all the $c$'s (including the ones you care about) are 0.

   ⟵ This proof is typical of the constructive proofs in linear algebra—it's really a "generic example."

2. **$\mathfrak{B}$ is a generating system for the column space of $M$.** The column space of $M$ is the set of all linear combinations of the columns of $M$. By the Pick-Apart Theorem, every such combination is of the form $MX$ for a vector $X$ in $\mathbb{R}^n$. Write $X$ as a linear combination of the basis $\{v_1, \ldots, v_k, v_{k+1}, \ldots, v_n\}$:

   ⟵ The Pick-Apart Theorem is Theorem 4.8 from Lesson 4.5.

   $$X = c_1v_1 + \cdots + c_kv_k + c_{k+1}v_{k+1} + c_{k+2}v_{k+2} + \cdots + c_nv_n$$

   Multiply both sides by $M$:

   $$MX = c_1Mv_1 + \cdots + c_kMv_k + c_{k+1}Mv_{k+1} + c_{k+2}Mv_{k+2} + \cdots + c_nMv_n$$

   Since $v_1, \ldots, v_k$ are in the kernel of $M$, this becomes

   $$MX = c_{k+1}Mv_{k+1} + c_{k+2}Mv_{k+2} + \cdots + c_nMv_n$$

   and $MX$ is a linear combination of the vectors in $\mathfrak{B}$.

That does it: $\mathfrak{B}$ is a basis. ∎

As an immediate corollary, you can prove that the row rank is the same as the column rank.

**Theorem 8.8 (Row Rank = Column Rank)**
*The row rank and the column rank of a matrix are equal.*

***Proof.*** Suppose $M$ is $m \times n$ and its row rank is $r$. Then the dimension of the kernel is $n - r$ by Theorem 8.6. By Theorem 8.7, the column rank is

$$n - \dim(\ker M) = n - (n - r) = r \qquad \blacksquare$$

So, Derman's party trick always works. Pick a matrix, any matrix. The number of linearly independent rows is the same as the number of linearly independent columns.

**Facts and Notation**

From now on, there is no need to distinguish between row and column rank. You can simply refer to the **rank** of a matrix. If $M$ is a matrix, $r(M)$ will denote the rank of $M$.

The fact that the row and column ranks are equal is by no means obvious at the start, and many people find it surprising until they see a proof. The proof uses many of the important theorems and methods that have been developed over the course of this program.

The fact that the row rank and column rank are the same allows another refinement of the TFAE theorem that was first stated in Chapter 4 (Theorem 4.19 from Lesson 4.6) and refined once already in Chapter 7 (Theorem 7.14 from Lesson 7.5):

**Theorem 8.9 (The TFAE Theorem)**
*The following statements are all equivalent for an $n \times n$ matrix $A$:*

(1) *The columns of $A$ are linearly independent.*

(2) *The rows of $A$ are linearly independent.*

(3) *The rank of $A$ is $n$.*

(4) $\text{rref}(A) = I$.

(5) $A^{-1}$ *exists.*

(6) $AX = B$ *has a unique solution for any* $B$.

(7) $\ker(A) = O$.

(8) *The dimension of the column space for $A$ is $n$.*

(9) *The dimension of the row space for $A$ is $n$.*

## Exercises

**1.** Find the rank of each matrix.

**a.** $\begin{pmatrix} 1 & 2 & 3 \\ 4 & 5 & 6 \\ 7 & 8 & 9 \end{pmatrix}$ **b.** $\begin{pmatrix} 1 & 2 & 3 \\ 4 & 5 & 6 \\ 7 & 8 & 0 \end{pmatrix}$ **c.** $\begin{pmatrix} 1 & 2 & 3 & 1 \\ 4 & 1 & 0 & 2 \\ 5 & 3 & 3 & 3 \end{pmatrix}$

**d.** $\begin{pmatrix} 1 & 3 & 1 & 5 \\ 2 & 1 & 3 & 6 \\ 1 & 4 & 1 & 6 \\ 0 & 2 & 1 & 3 \end{pmatrix}$ **e.** $\begin{pmatrix} 3 & 1 & 2 \\ 4 & 0 & 1 \\ 6 & 1 & 3 \\ 2 & 1 & -1 \end{pmatrix}$ **f.** $\begin{pmatrix} 1 & 4 & 3 & 5 & 2 \\ 4 & 3 & -1 & 7 & 8 \\ 1 & 2 & 1 & 3 & 4 \\ 2 & 3 & 5 & 5 & -10 \end{pmatrix}$

**2.** For each condition, give an example, if possible, of a matrix that satisfies it. If there is no matrix, explain why not.

**a.** $A$ is $2 \times 3$ and $r(A) = 2$.
**b.** $A$ is $2 \times 4$ and $r(A) = 3$.
**c.** $A$ is $3 \times 3$ and $r(A) = 2$.
**d.** $A$ is $3 \times 3$ and $r(A) = 3$.
**e.** $A$ is $3 \times 4$ and $r(A) = 2$.
**f.** $A$ is $4 \times 3$ and $\ker(A)$ is infinite.
**g.** $A$ is $4 \times 4$ and $\ker(A)$ is infinite.
**h.** $A$ is $4 \times 3$ and $\ker(A) = \{\mathbf{0}\}$.
**i.** $A$ is $3 \times 4$ and $\ker(A) = \{\mathbf{0}\}$.
**j.** $A$ is $3 \times 4$ and the rows of $A$ are independent.
**k.** $A$ is $3 \times 4$ and the columns of $A$ are independent.

**3.** For each matrix, find its rank and the dimension of its kernel.

**a.** $\begin{pmatrix} 2 & 1 & 1 \\ 5 & 7 & 1 \\ 1 & 1 & 0 \\ 8 & 9 & 2 \end{pmatrix}$ **b.** $\begin{pmatrix} 1 & 2 & 1 & 3 \\ 4 & 6 & 2 & 8 \\ 2 & 2 & 0 & 2 \\ 5 & 6 & 1 & 7 \end{pmatrix}$ **c.** $\begin{pmatrix} 3 & 7 & -3 \\ 3 & 5 & 6 \\ 4 & 2 & -4 \end{pmatrix}$

**4.** For each set of conditions, give an example, if possible, of a matrix that satisfies them. If there is no matrix, explain why not.

**a.** $A$ is $3 \times 2$ and $r(A) = 2$.
**b.** $A$ is $3 \times 2$ and $r(A) = 3$.
**c.** $A$ is $3 \times 3$ and $\ker(A) = \{\mathbf{0}\}$.
**d.** $A$ is $3 \times 3$ and $\ker(A)$ is infinite.
**e.** $A$ is $4 \times 3$ and the rows of $A$ are independent.

**5.** For each set, show that it is a basis.

**a.** $\{1 + x, 1 - x, x^2\}$ for $\mathbb{R}_2[x]$

**b.** $\left\{ \begin{pmatrix} 1 & 1 \\ 0 & 0 \end{pmatrix}, \begin{pmatrix} 0 & 1 \\ 2 & 0 \end{pmatrix}, \begin{pmatrix} 0 & 0 \\ 3 & 4 \end{pmatrix}, \begin{pmatrix} 0 & 0 \\ 0 & 5 \end{pmatrix} \right\}$ for $\mathrm{Mat}_{2\times 2}(\mathbb{R})$

**6.** Let $G = \begin{pmatrix} 2 & 1 & 1 \\ 5 & 7 & 1 \\ 1 & 1 & 0 \\ 8 & 9 & 2 \end{pmatrix}$.

**a.** Find $r(G)$.
**b.** Find $\dim \ker(G)$.
**c.** Find a basis for $\ker(G)$.
**d.** Find a basis for the column space of $G$.

**7.** Let $H = \begin{pmatrix} 1 & 2 & 3 & 4 \\ 4 & 5 & 6 & 7 \\ 7 & 8 & 9 & 10 \end{pmatrix}$.

**a.** Find $r(H)$.
**b.** Find $\dim \ker(H)$.
**c.** Find a basis for $\ker(H)$.
**d.** Find a basis for the column space of $H$.

**8.** For each set of conditions, give an example, if possible, of an interesting matrix that satisfies them. If there is no matrix, explain why not.

**a.** $A$ is $4 \times 3$ and $r(A) = 4$.
**b.** $A$ is $4 \times 3$ and $r(A) = 3$.
**c.** $A$ is $4 \times 4$ and $A^{-1}$ exists.
**d.** $A$ is $4 \times 4$ and $\ker(A)$ is infinite.
**e.** $A$ is $3 \times 2$ and the rows of $A$ are independent.

**9.** For each set, explain why it cannot be a basis for the given vector space.

**a.** $\{(1,1,0,0), (0,1,1,0), (0,1,1,1)\}$ for $\mathbb{R}^4$

**b.** $\left\{ \begin{pmatrix} 1 & 0 \\ 0 & 0 \end{pmatrix}, \begin{pmatrix} 1 & 1 \\ 0 & 0 \end{pmatrix}, \begin{pmatrix} 0 & 0 \\ 1 & 0 \end{pmatrix}, \begin{pmatrix} 0 & 1 \\ 1 & 0 \end{pmatrix} \right\}$ for $\text{Mat}_{2\times 2}(\mathbb{R})$

**c.** $\{(1,3,2), (3,10,7), (1,4,3)\}$ for $\mathbb{R}^3$

**10.** Suppose that $A$ and $B$ are $n \times n$ matrices.

**a.** Show that the row space of $AB$ is contained in the row space of $B$.
**b.** Show that the column space of $AB$ is contained in the column space of $A$.
**c.** Show that $r(AB) \leq A$ and $r(AB) \leq r(B)$.

← See Exercise 7 from Lesson 8.2.

**11.** Let $M = \begin{pmatrix} 1 & 3 & 0 & 1 \\ 2 & 5 & 1 & 1 \\ 1 & 2 & 1 & 0 \end{pmatrix}$.

**a.** Find $r(M)$.
**b.** Find $\dim \ker(M)$.
**c.** Find dim(column space of $M$).
**d.** Find a basis for $\ker(M)$.
**e.** Find a basis for the row space of $M$.
**f.** Find a basis for the column space of $M$.

**12.** Let $N = \begin{pmatrix} 1 & 1 & 0 & 2 \\ 2 & 3 & 4 & 1 \\ 1 & 0 & 1 & 0 \\ 4 & 4 & 5 & 3 \end{pmatrix}$.

**a.** Find the row rank of $N$.
**b.** Find a basis for the column space of $N$.

**13.** Find a basis for $L\{(1,2,1),(1,0,1),(2,2,2),(0,2,0)\}$.

**14.** Find a basis for $\mathbb{R}^3$ that contains $(1,2,3)$ and $(1,2,0)$.

**15.** Find a basis for the row space of $\begin{pmatrix} 1 & 3 & 4 \\ 2 & 5 & 3 \\ 3 & 8 & 7 \\ 0 & -1 & -5 \end{pmatrix}$ that contains $(0,-1,-5)$.

## 8.4 Building and Representing Linear Maps

In Chapter 5, you saw that an $m \times n$ matrix could be used to define a mapping from $\mathbb{R}^n$ to $\mathbb{R}^m$ via matrix multiplication. Such mappings were called **linear maps**. In Chapter 7, you generalized the algebraic properties of $n$-tuples and defined a **vector space** to be any set of objects that satisfies the basic rules of $\mathbb{R}^n$.

⟵ By the end of Chapter 7, things came full circle: you saw how these arbitrary vector spaces could be made to "look like" $n$-tuples by picking a basis and working with coordinate vectors.

In this lesson, you'll generalize the notion of linear map to these arbitrary vector spaces. But how do you associate a matrix with a map from, say, $\mathbb{R}_3[x]$ to some other vector space, say, $2 \times 2$ symmetric matrices? The basic idea is to represent the vectors as $n$-tuples via the coordinate vectors and to work with those.The resulting matrix will therefore depend on the bases you pick for the domain and range.

⟵ But, as you'll see in this lesson and the next, all the different matrices associated with the same map will be related to each other in interesting ways.

**In this lesson, you will learn how to**

- find a formula for a linear map by its action on a basis
- find the matrix representation associated with a linear map with respect to a given pair of bases.

It would be good to define a general linear map in a way that doesn't depend on the way the map is represented but only on the way the map behaves. One way to do this is to use the properties of linear transformations listed in Theorem 5.1 from Lesson 5.2 as the *definition* of a linear map. This is the way most mathematicians think of it:

**Definition**

Suppose $V$ and $W$ are vector spaces. A mapping $T : V \to W$ is called a **linear map** if it satisfies two properties:

(1) $T(v_1 + v_2) = T(v_1) + T(v_2)$ for all vectors $v_1$, $v_2$ in $V$.

(2) $T(cv) = cT(v)$ for all vectors $v$ in $V$ and scalars $c$.

On the left-hand side of these equations, all of the operations take place in $V$; that is, $v_1 + v_2$ adds two vectors in $V$, and their sum is also in $V$, so $T$ is also acting on a vector in $V$. On the right-hand side, however, $T(v_1)$ and $T(v_2)$ are vectors in $W$, so the sum $T(v_1) + T(v_2)$ is a sum of vectors in $W$. Similarly $cT(v)$ is a scalar multiple of a vector in $W$.

**For You to Do**

1. Invent a linear map $T : \text{Mat}_{2\times 2} \to \mathbb{R}_3[x]$.
2. Invent a map $F : \text{Mat}_{2\times 2} \to \mathbb{R}_3[x]$ that is not linear.

### Developing Habits of Mind

**Ensure statements are consistent.** A linear map between vector spaces $V$ and $W$ is a function $T : V \to W$ that "respects the vector space structure."

This changes the definition of linear map from the one in Chapter 5. But Theorem 5.1 from Lesson 5.2 guarantees that a mapping that is linear in the Chapter 5 sense (it can be represented by a matrix) is linear in this new sense. And, as you'll see in this lesson, maps that are linear in this new sense can also be represented by matrices. But that will take a little extra work.

$\longleftarrow$ You have encountered many linear maps already: projections, functions defined by matrices, rotations of the plane, and the "coordinate map" that, given a basis $\mathfrak{B}$ of $V$, assigns a vector $\underline{v}_{\mathfrak{B}}$ in $\mathbb{R}^n$.

The following theorem states that a linear map "preserves linear combinations."

**Theorem 8.10 (Linearity)**

*Suppose that $T : V \to W$ is linear. If $v_1, \ldots, v_k$ are vectors in $V$ and $c_1, \ldots, c_k$ are scalars, then*

$$T(c_1 v_1 + \cdots + c_k v_k) = c_1 T(v_1) + \cdots + c_k T(v_k)$$

$\longleftarrow$ $V$ and $W$ could be polynomials, matrices, complex numbers, or subspaces of these.

### For You to Do

**3.** Prove Theorem 8.10.

### Example 1

**Problem.** Suppose $D : \mathbb{R}_3[x] \to \mathbb{R}_2[x]$ is linear and you know what it does to the vectors in the basis $\mathfrak{B} = \{1, 1+x, x+x^2, x^2+x^3\}$:

$$\begin{aligned} D(1) &= 0 \\ D(1+x) &= 1 \\ D(x+x^2) &= 1+2x \\ D(x^2+x^3) &= 2x+3x^2 \end{aligned}$$

$\longleftarrow$ Check that $\mathfrak{B}$ is a basis for $\mathbb{R}_3[x]$.

Find $D(4x^3 + 3x^2 - 2x + 3)$.

**Solution.** First write $4x^3 + 3x^2 - 2x + 3$ as a linear combination of the basis vectors:

$$4x^3 + 3x^2 - 2x + 3 = 4(1) + (-1)(1+x) + (-1)(x+x^2) + 4(x^2+x^3)$$

$\longleftarrow$ So, $\underline{4x^3 + 3x^2 - 2x + 3}_{\mathfrak{B}} = (4, -1, -1, 4)$—see Exercise 33 from Lesson 7.5.

So, applying Theorem 8.10,

$$\begin{aligned} D(4x^3 + 3x^2 - 2x + 3) &= D\left(4(1) + (-1)(1+x) + (-1)(x+x^2) + 4(x^2+x^3)\right) \\ &= 4\,D(1) + (-1)\,D(1+x) + (-1)\,D(x+x^2) + 4\,D(x^2+x^3) \\ &= 4 \cdot 0 + -1 \cdot 1 + (-1)(1+2x) + 4(2x+3x^2) \\ &= 12x^2 + 6x - 2 \end{aligned}$$

**For You to Do**

**4.** Find a formula for $D$ in Example 1 by getting an expression for

$$D(ax^3 + bx^2 + cx + d)$$

Theorem 8.10 and the method of Example 1 can be used to prove a very useful result.

**Theorem 8.11 (Extension by Linearity)**

*A linear map is determined by its action on a basis. More precisely, suppose $V$ and $W$ are vector spaces and $\mathfrak{B}$ is a basis for $V$. A linear map $T : V \to W$ is completely determined by what it does to the vectors in $\mathfrak{B}$ in the sense that if you know what it does to each basis vector, you know what it does to every vector in $V$.*

← You saw an example of extension by linearity in the *Developing Habits of Mind* in Lesson 5.3.

***Proof.*** Suppose $\mathfrak{B} = \{v_1, \ldots, v_k\}$. If $v$ is a vector in $V$, there is only one way to write $v$ as a linear combination of the vectors in $\mathfrak{B}$ (why?), say

$$v = c_1 v_1 + c_2 v_2 + \cdots + c_n v_k$$

Then, by Theorem 8.10, $T(v)$ is determined:

$$T(v) = c_1 T(v_1) + c_2 T(v_2) + \cdots + c_n T(v_k) \qquad \blacksquare$$

Theorem 8.11 is an example of a class of theorems that mathematicians prize. It says that a linear map on an infinite set, $V$, is determined by what it does to a *finite* set of inputs (the basis vectors).

This kind of "finiteness" theorem is pretty rare. In your other courses, you may have seen another one: a polynomial of degree $n$ is determined by $n + 1$ inputs.

**For You to Do**

**5.** Can you think of any other finiteness theorems, either from this course or others?

### Matrix representations

The goal is to associate a matrix with a linear map in a way that preserves the same properties as the matrix representations you met in Chapter 5. The most important properties of such representations are

← Such representations are said to be **faithful**.

**Bijective:** Every linear map is associated with a unique matrix and vice versa.

**Structure preserving:** Applying a linear map to a vector is accomplished by multiplying the matrix for that map by the vector.

But how do you multiply a matrix by a polynomial? You can't. However, you *can* multiply a matrix by the coordinate vector for the polynomial with respect to some basis. The idea is outlined in Exercises 10–13 from Lesson 8.1. The following outline shows how it's done in principle. It's worth reasoning through this outline first, so that the numbers don't get in the way. The outline will be followed by some numerical examples.

Suppose that

$$\text{a basis for } V \text{ is } \mathfrak{B} = \{v_1, v_2, v_3\} \quad \text{and}$$

$$\text{a basis for } W \text{ is } \mathfrak{B}' = \{w_1, w_2\}$$

And suppose that $T : V \to W$ is linear. You want to find a matrix $M$ that does the work of $T$; that is, find a matrix $M$ so that, for any vector $v$ in $V$,

$$M\underline{v}_{\mathfrak{B}} = \underline{T(v)}_{\mathfrak{B}'}$$

← $M$ will be $2 \times 3$. (Why?)

Pick some $v$ in $V$. The steps for seeing how $M$ is built are:

1. Write $v$ as a linear combination of vectors in $\mathfrak{B}$, say

$$v = av_1 + bv_2 + cv_3$$

   This gives you

$$\underline{v}_{\mathfrak{B}} = \begin{pmatrix} a \\ b \\ c \end{pmatrix}$$

2. And you want to express $T(v)$ in terms of $\mathfrak{B}'$, say

$$T(v) = dw_1 + ew_2$$

   so that

$$\underline{T(v)}_{\mathfrak{B}'} = \begin{pmatrix} d \\ e \end{pmatrix}$$

3. So, you are looking for a matrix $M$ so that

$$M \begin{pmatrix} a \\ b \\ c \end{pmatrix} = \begin{pmatrix} d \\ e \end{pmatrix}$$

4. By the Pick-Apart Theorem,

$$M \begin{pmatrix} a \\ b \\ c \end{pmatrix} = aM_{*1} + bM_{*2} + cM_{*3}$$

← The Pick-Apart Theorem is Theorem 4.8 from Lesson 4.5.

5. But by linearity (Theorem 8.10),

$$T(v) = aT(v_1) + bT(v_2) + cT(v_3)$$

   so that

$$\underline{T(v)}_{\mathfrak{B}'} = a\underline{T(v_1)}_{\mathfrak{B}'} + b\underline{T(v_2)}_{\mathfrak{B}'} + c\underline{T(v_3)}_{\mathfrak{B}'}$$

**Remember**

$v = av_1 + bv_2 + cv_3$ and $T$ is linear.

← . . . and $\underline{T(v)}_{\mathfrak{B}'} = \begin{pmatrix} d \\ e \end{pmatrix}$.

Now put it all together. You want $M\begin{pmatrix}a\\b\\c\end{pmatrix}$ to be $\begin{pmatrix}d\\e\end{pmatrix}$, which is $\underline{T(v)}_{\mathfrak{B}'}$.

And you know that

$$M\begin{pmatrix}a\\b\\c\end{pmatrix} = aM_{*1} + bM_{*2} + cM_{*3} \qquad \text{and}$$

$$\underline{T(v)}_{\mathfrak{B}'} = a\underline{T(v_1)}_{\mathfrak{B}'} + b\underline{T(v_2)}_{\mathfrak{B}'} + c\underline{T(v_3)}_{\mathfrak{B}'}$$

*And the punchline*: comparing those last two equations, you can take $M$ to be the matrix whose columns are the coordinate vectors: $\underline{T(v_1)}_{\mathfrak{B}'}$, $\underline{T(v_2)}_{\mathfrak{B}'}$, $\underline{T(v_3)}_{\mathfrak{B}'}$.

### Example 2

In Example 1, you considered the linear map $D : \mathbb{R}_3[x] \to \mathbb{R}_2[x]$ defined on the basis $\mathfrak{B} = \{1, 1+x, x+x^2, x^2+x^3\}$:

$$\begin{aligned} D(1) &= 0 \\ D(1+x) &= 1 \\ D(x+x^2) &= 1+2x \\ D(x^2+x^3) &= 2x+3x^2 \end{aligned}$$

**Problem.** Suppose the basis for $\mathbb{R}_2[x]$ is $\mathfrak{B}' = \{1, x, x^2\}$. Find a matrix $M$ that does the work of $D$; that is, for any $v$ in $\mathbb{R}_3[x]$,

$$M\underline{v}_{\mathfrak{B}} = \underline{D(v)}_{\mathfrak{B}'}$$

**Solution.** Take the columns of $M$ to be the coordinate vectors with respect to $\mathfrak{B}'$ of the image under $D$ of each vector in $\mathfrak{B}$:

$$\begin{aligned} \underline{D(1)}_{\mathfrak{B}'} &= \underline{0}_{\mathfrak{B}'} = (0,0,0) \\ \underline{D(1+x)}_{\mathfrak{B}'} &= \underline{1}_{\mathfrak{B}'} = (1,0,0) \\ \underline{D(x+x^2)}_{\mathfrak{B}'} &= \underline{1+2x}_{\mathfrak{B}'} = (1,2,0) \\ \underline{D(x^2+x^3)}_{\mathfrak{B}'} &= \underline{2x+3x^2}_{\mathfrak{B}'} = (0,2,3) \end{aligned}$$

So that $M = \begin{pmatrix} 0 & 1 & 1 & 0 \\ 0 & 0 & 2 & 2 \\ 0 & 0 & 0 & 3 \end{pmatrix}$.

**Check:** In Example 1, you saw that

$$D(4x^3 + 3x^2 - 2x + 3) = 12x^2 + 6x - 2$$

You should be able to get this by multiplying $M$ by the coordinate vector for this cubic with respect to $\mathfrak{B}$ and translating back from $\mathfrak{B}'$. Well, $\underline{4x^3 + 3x^2 - 2x + 3}_{\mathfrak{B}} = (4, -1, -1, 4)$ (see Exercise 33 from Lesson 7.5), and

$$\begin{pmatrix} 0 & 1 & 1 & 0 \\ 0 & 0 & 2 & 2 \\ 0 & 0 & 0 & 3 \end{pmatrix} \begin{pmatrix} 4 \\ -1 \\ -1 \\ 4 \end{pmatrix} = \begin{pmatrix} -2 \\ 6 \\ 12 \end{pmatrix}$$

and sure enough,

$$\underline{12x^2 + 6x - 2}_{\mathfrak{B}'} = (-2, 6, 12)$$

---

**Example 3**

Suppose $T : \mathbb{R}^3 \to \mathbb{R}^2$ is defined by $T(x, y, z) = (x + y, x - z)$. And suppose the bases of the domain and range are

$$\mathfrak{B} = \{(1, 1, 0), (0, 1, 1), (0, 0, 1)\}$$
$$\mathfrak{B}' = \{(1, 1), (0, 1)\}$$

**Problem.** Find a matrix $M$ so that, for any vector $v$ in $\mathbb{R}^3$,

$$M\underline{v}_{\mathfrak{B}} = \underline{T(v)}_{\mathfrak{B}'}$$

**Solution.** Apply $T$ to the basis vectors in $\mathfrak{B}$ and look at what you get in terms of $\mathfrak{B}'$:

$$\underline{T(1,1,0)}_{\mathfrak{B}'} = \underline{(2,1)}_{\mathfrak{B}'} = (2, -1)$$
$$\underline{T(0,1,1)}_{\mathfrak{B}'} = \underline{(1,-1)}_{\mathfrak{B}'} = (1, -2)$$
$$\underline{T(0,0,1)}_{\mathfrak{B}'} = \underline{(0,-1)}_{\mathfrak{B}'} = (0, -1)$$

So, $M$ is the matrix whose columns are these coordinate vectors:

$$M = \begin{pmatrix} 2 & 1 & 0 \\ -1 & -2 & -1 \end{pmatrix}$$

**Check:** Suppose $v = (x, y, z)$. Then

$$\underline{v}_{\mathfrak{B}} = (x, y - x, z - y + x)$$

and

$$\begin{pmatrix} 2 & 1 & 0 \\ -1 & -2 & -1 \end{pmatrix} \begin{pmatrix} x \\ y - x \\ z - y + x \end{pmatrix} = \begin{pmatrix} x + y \\ -y - z \end{pmatrix}$$

Sure enough,

$$(x + y)(1, 1) + (-y - z)(0, 1) = (x + y, x - z) = T(x, y, z)$$

---

**Minds in Action** Episode 35

*The three friends are thinking about Example* 3.

DERMAN: Wait a minute—if $T(x, y, z) = (x + y, x - z)$, I can just write down a matrix that does the work of $T$:

$$\begin{pmatrix} 1 & 1 & 0 \\ 1 & 0 & -1 \end{pmatrix} \begin{pmatrix} x \\ y \\ z \end{pmatrix} = \begin{pmatrix} x + y \\ x - z \end{pmatrix}$$

So the matrix is $N = \begin{pmatrix} 1 & 1 & 0 \\ 1 & 0 & -1 \end{pmatrix}$, not $M = \begin{pmatrix} 2 & 1 & 0 \\ -1 & -2 & -1 \end{pmatrix}$.

TONY: But the matrix $M$ wasn't designed to work from $\mathfrak{B}$ to $\mathfrak{B}'$. $N$ just works straight up, so to speak. It's kind of like $N$ works on vectors using the usual coordinates, but $M$ works on the same vectors using different coordinates. Or something like that.

SASHA: I think that's what's going on, Tony. The vector $(x, y, z)$ really is a coordinate vector—it's just that it's the coordinate vector with respect to the standard basis vectors

$$\mathfrak{E} = \{(1, 0, 0), (0, 1, 0), (0, 0, 1)\}$$

TONY: OK, let's try it. Suppose $\mathfrak{E} = \{(1, 0, 0), (0, 1, 0), (0, 0, 1)\}$ and $\mathfrak{E}' = \{(1, 0), (0, 1)\}$. Let's find the matrix that does the work of $T$ with respect to these bases.

DERMAN: OK. I'll set it up like before:

$$\begin{aligned} \underline{T(1,0,0)}_{\mathfrak{E}'} &= \underline{(1,1)}_{\mathfrak{E}'} = (1,1) \\ \underline{T(0,1,0)}_{\mathfrak{E}'} &= \underline{(1,0)}_{\mathfrak{E}'} = (1,0) \\ \underline{T(0,0,1)}_{\mathfrak{E}'} &= \underline{(0,-1)}_{\mathfrak{E}'} = (0,-1) \end{aligned}$$

Now I put the coordinate vectors into the columns and I get

$$\begin{pmatrix} 1 & 1 & 0 \\ 1 & 0 & -1 \end{pmatrix}$$

TONY: So, that's it. These new matrices from one basis to another are generalizations of the matrices we built in Chapter 5. Those Chapter 5 matrices are matrices from one standard basis to another standard basis, that's all. It's like we're using a different coordinate system here. The different bases give different sets of axes for the domain and range—the axes are determined by the vectors in the two bases . . .

SASHA: . . . at least for vector spaces like $\mathbb{R}^n$, where you can imagine "axes."

DERMAN: Didn't we make everything look like $\mathbb{R}^n$ in Chapter 7?

All of the examples and calculations in this lesson can now be summarized and brought to closure.

**Definition**

Suppose $V$ and $W$ are vector spaces with respective bases

$$\mathfrak{B} = \{v_1, \ldots, v_n\} \quad \text{and} \quad \mathfrak{B}' = \{w_1, \ldots, w_m\}$$

and suppose that $T : V \to W$ is a linear map. The **matrix for $T$ with respect to $\mathfrak{B}$ and $\mathfrak{B}'$**, written $M_{\mathfrak{B}}^{\mathfrak{B}'}(T)$, is the $m \times n$ matrix whose columns are

$$\underline{T(v_1)}_{\mathfrak{B}'}, \underline{T(v_2)}_{\mathfrak{B}'}, \ldots, \underline{T(v_n)}_{\mathfrak{B}'}$$

**Theorem 8.12 (Matrix Representation of a Linear Map)**
*With the notation of the above definition, $M_{\mathfrak{B}}^{\mathfrak{B}'}(T)$ does the work of $T$ in the sense that, if $v$ is any vector in $V$, then*

$$\underline{T(v)}_{\mathfrak{B}'} = M_{\mathfrak{B}}^{\mathfrak{B}'}(T)\underline{v}_{\mathfrak{B}}$$

***Proof.*** The proof is just a generic version of the calculations carried out. To keep the notation down, let $M = M_{\mathfrak{B}}^{\mathfrak{B}'}(T)$, so that, for any $i$ between 1 and $n$,

$$M_{*i} = \underline{T(v_i)}_{\mathfrak{B}'}$$

Write $v$ as a linear combination of the vectors in $\mathfrak{B}$:

$$v = c_1v_1 + c_2v_2 + \cdots + c_nv_n$$

so that

$$\underline{v}_{\mathfrak{B}} = \begin{pmatrix} c_1 \\ c_2 \\ \vdots \\ c_n \end{pmatrix}$$

Then

$$T(v) = c_1T(v_1) + c_2T(v_2) + \cdots + c_nT(v_n)$$

Then, taking coordinate vectors, you get

$$\underline{T(v)}_{\mathfrak{B}'} = c_1\underline{T(v_1)}_{\mathfrak{B}'} + c_2\underline{T(v_2)}_{\mathfrak{B}'} + \cdots + c_n\underline{T(v_n)}_{\mathfrak{B}'} \quad \text{(Theorem 7.10; see Lesson 7.5)}$$

$$= c_1M_{*1} + c_2M_{*2} + \cdots + c_nM_{*n}$$

$$= M\begin{pmatrix} c_1 \\ c_2 \\ \vdots \\ c_n \end{pmatrix} \quad \text{(the Pick-Apart Theorem)}$$

$$= M\underline{v}_{\mathfrak{B}}$$

■

One more generalization of previous results: Theorem 5.11 (see Lesson 5.5) generalizes to matrices with respect to arbitrary bases.

**Theorem 8.13 (Composite to Product)**
*Suppose that $V$, $W$, and $U$ are vector spaces and that $T$ and $S$ are linear maps:*

$$V \xrightarrow{T} W \xrightarrow{S} U$$

*Suppose further that $\mathfrak{B}$, $\mathfrak{B}'$, $\mathfrak{B}''$ are bases, respectively, for $V$, $W$, and $U$. Then*

$$M_{\mathfrak{B}}^{\mathfrak{B}''}(S \circ T) = M_{\mathfrak{B}'}^{\mathfrak{B}''}(S)\, M_{\mathfrak{B}}^{\mathfrak{B}'}(T)$$

***Proof.*** Suppose $v$ is any vector in $V$. Then

$$\begin{aligned} M_{\mathfrak{B}}^{\mathfrak{B}''}(S \circ T)\underline{v}_{\mathfrak{B}} &= \underline{S \circ T(v)}_{\mathfrak{B}''} \\ &= \underline{S\,(T(v))}_{\mathfrak{B}''} \\ &= M_{\mathfrak{B}'}^{\mathfrak{B}''}(S)\underline{T(v)}_{\mathfrak{B}'} \\ &= M_{\mathfrak{B}'}^{\mathfrak{B}''}(S)\left(M_{\mathfrak{B}}^{\mathfrak{B}'}(T)\underline{v}_{\mathfrak{B}}\right) \\ &= \left(M_{\mathfrak{B}'}^{\mathfrak{B}''}(S) M_{\mathfrak{B}}^{\mathfrak{B}'}(T)\right)\underline{v}_{\mathfrak{B}} \end{aligned}$$

← This is a challenging proof. Give reasons for each step.

■

Since this is true for every vector $\underline{v}_{\mathfrak{B}}$ in $\mathbb{R}^n$,

$$M_{\mathfrak{B}}^{\mathfrak{B}''}(S \circ T) = M_{\mathfrak{B}'}^{\mathfrak{B}''}(S)\, M_{\mathfrak{B}}^{\mathfrak{B}'}(T)$$

← See Exercise 12 from Lesson 5.1.

### Developing Habits of Mind

**Find associations between different concepts.** As you can see, composition of linear maps corresponds to matrix multiplication when the maps are represented by matrices.

## Exercises

1. For each mapping $F : \mathbb{R}^2 \to \mathbb{R}^2$, determine whether $F$ is linear.
   **a.** $F(x, y) = (2x, y)$
   **b.** $F(x, y) = (x^2, y)$
   **c.** $F(x, y) = (y, x)$
   **d.** $F(x, y) = (x, y + 1)$
   **e.** $F(x, y) = (0, y)$
   **f.** $F(x, y) = (\sqrt{3}x, \sqrt{3}y)$

2. For each mapping $F : \mathbb{R}^3 \to \mathbb{R}^2$, determine whether $F$ is linear.
   **a.** $F(x, y, z) = (x, x + y + z)$
   **b.** $F(x, y, z) = (0, 0)$
   **c.** $F(x, y, z) = (1, 1)$
   **d.** $F(x, y, z) = (2x + y, 3y - 4z)$
   **e.** $F(a, b, c) = a + c$
   **f.** $F(a, b, c) = a^2 + b^2$

3. For each mapping $F : \mathbb{R}_2(x) \to \mathbb{R}_2(x)$, determine whether $F$ is linear.
   **a.** $F(a_0 + a_1x + a_2x^2) = a_0 + (a_1 + a_2)x + (2a_2 - 3a_1)x^2$
   **b.** $F(a_0 + a_1x + a_2x^2) = 0$
   **c.** $F(a_0 + a_1x + a_2x^2) = (a_0 + 1) + a_1x + a_2x^2$

**4.** Let $T : \mathbb{R}^3 \to \mathbb{R}^2$ be a linear mapping such that

$$T\begin{pmatrix}1\\0\\0\end{pmatrix} = \begin{pmatrix}1\\1\end{pmatrix}, \quad T\begin{pmatrix}0\\1\\0\end{pmatrix} = \begin{pmatrix}3\\0\end{pmatrix}, \quad T\begin{pmatrix}0\\0\\1\end{pmatrix} = \begin{pmatrix}4\\-7\end{pmatrix}$$

Find $T\begin{pmatrix}1\\3\\8\end{pmatrix}$.

**5.** Show that complex conjugation

$$a + bi \mapsto a - bi$$

is a linear map from $\mathbb{C}$ to $\mathbb{C}$.

**6.** Let $T : \mathbb{R}^3 \to \mathbb{R}^2$ be defined by $(x, y, z) \mapsto (x + z, y - 2x)$, and let $S : \mathbb{R}^2 \to \mathbb{R}^3$ be defined by $(x, y) \mapsto (x + 2y, 3x, y - x)$. Let $M(T)$ be the matrix for $T$ with respect to the standard bases. Find

**a.** $T \circ S(1, 3)$
**b.** $S \circ T(2, 3, -1)$
**c.** $M(T)$
**d.** $M(S)$
**e.** $M(S \circ T)$
**f.** $M(T \circ S)$
**g.** $M(S)M(T)$
**h.** $M(T)M(S)$

**7.** Suppose $T : \mathbb{R}_3[x] \to L\{(1, 2, 0), (3, 1, 2)\}$ is defined by

$$\begin{aligned} T(1) &= (4, 3, 2) \\ T(1 + x) &= (2, -1, 2) \\ T(x + x^2) &= (5, 5, 2) \\ T(x^2 + x^3) &= (5, 0, 4) \end{aligned}$$

Let $M = M_{\mathfrak{B}}^{\mathfrak{B}'}(T)$, where

$$\mathfrak{B} = \{1, 1 + x, x + x^2, x^2 + x^3\} \quad \text{and} \quad \mathfrak{B}' = \{(1, 2, 0), (3, 1, 2)\}$$

**a.** Find $M$

Then use $M$ to find

**b.** $T(x)$
**c.** $T(x^3 - 3x^2 + 5x - 6)$
**d.** The set of all vectors $v$ so that $T(v) = (0, 0, 0)$
**e.** The set of all vectors $v$ so that $T(v) = (1, 2, 0)$

**8.** Let $T$, $M$, and $\mathfrak{B}$ be as in Exercise 7, and let

$$\mathfrak{B}'' = \{(8, 11, 2), (6, 7, 2)\}$$

If $N = M_{\mathfrak{B}}^{B''}(T)$, find

**a.** $r(M)$ and $r(N)$
**b.** $\ker(M)$ and $\ker(N)$

**9.** Suppose $T : V \to W$ is linear, and

$$\dim(V) = r \quad \text{and} \quad \dim(W) = s$$

What size is any matrix that represents $T$? Explain.

**10.** Suppose $T : \mathbb{R}^3 \to \mathbb{R}^3$ is linear and is given by

$$\begin{aligned} T(1,1,0) &= (-19,-6,-6) \\ T(0,1,1) &= (-60,-21,-15) \\ T(0,0,1) &= (-18,-7,-3) \end{aligned}$$

Find $M^{\mathfrak{B}}_{\mathfrak{B}}(T)$ if

**a.** $\mathfrak{B} = \{(1,1,0),(0,1,1),(0,0,1)\}$
**b.** $\mathfrak{B} = \{(1,0,0),(0,1,0),(0,0,1)\}$
**c.** $\mathfrak{B} = \{(12,5,3),(3,1,1),(2,1,0)\}$

← Here, the basis for the domain and the range is the same.

**11.** Suppose $V = \mathbb{R}_3[x]$ . Let $\mathfrak{E}$ be the standard basis for $\mathbb{R}^4$. Show that the map $C : V \to \mathbb{R}^4$ defined by $C(f) = \underline{f}_{\mathfrak{E}}$ is a linear map.

**12.** Suppose $V$ is a vector space of dimension $n$, and let $\mathfrak{B} = \{v_1, \dots, v_n\}$ be a basis for $\mathbb{R}^n$. Show that the map $C : V \to \mathbb{R}^n$ defined by $C(v) = \underline{v}_{\mathfrak{B}}$ is a linear map.

**13.** Suppose $T : V \to W$ is linear. Show that $T$ maps the zero vector in $V$ to the zero vector in $W$.

**14.** Suppose $T : \mathbb{R}^2 \to \mathbb{R}^2$ is linear. Show that $T$ maps a line to another line.

**15.** Suppose $T : \mathbb{R}^n \to \mathbb{R}^m$ is linear. Show that $T$ maps a line in $\mathbb{R}^n$ to a line in $\mathbb{R}^m$.

**16.** Suppose $T : \mathbb{R}^2 \to \mathbb{R}^2$ is a linear map defined by

$$T(x,y) = \left(5x + 6y, -3x - \tfrac{7}{2}y\right)$$

← See Exercise 14 from Lesson 5.6.

**a.** Find the matrix $M$ for $T$ with respect to the standard bases of $\mathbb{R}^2$.
**b.** Find two linearly independent vectors $v_1$ and $v_2$ that are scaled by $T$—that is, so that there are numbers $k_1$ and $k_2$ such that

$$\begin{aligned} T(v_1) &= k_1\, v_1 \quad \text{and} \\ T(v_2) &= k_2\, v_2 \end{aligned}$$

**c.** Find $M^{\mathfrak{B}}_{\mathfrak{B}}(T)$, where $\mathfrak{B} = \{v_1, v_2\}$.

The next several exercises use the following definition (which you will explore further in Lesson 8.5.

**Definition**

Let $V$ be a vector space. The **identity mapping** on $V$, denoted by $\text{id}(V)$, is the mapping $\text{id} : V \to V$ defined by $\text{id}(v) = v$ for every vector $v$ in $V$.

**17.** If $V$ is a vector space and $T : V \to V$, show that

$$T \circ \text{id} = \text{id} \circ T = T$$

**18.** Pick your favorite vector space $V$ and a basis $\mathfrak{B}$ for $V$. Calculate $M_{\mathfrak{B}}^{\mathfrak{B}}(\text{id})$.

**19.** Suppose $T : \mathbb{R}_5[x] \to \mathbb{R}_6[x]$, where $T(f)$ is obtained from $f$ by multiplying $f$ by $x$ and simplifying.

**a.** Show that $T$ is linear.
**b.** Suppose $\mathfrak{B} = \{1, x, x^2, x^3, x^4, x^5\}$ and $\mathfrak{B}' = \{1, x, x^2, x^3, x^4, x^5, x^6\}$. Find $M_{\mathfrak{B}}^{\mathfrak{B}'}(T)$.

**20.** Suppose $T : \mathbb{R}_5[x] \to \mathbb{R}_5[x]$, where $T(f)$ is obtained from $f$ by replacing $x$ by $x+1$ and expanding.

**a.** Show that $T$ is linear.
**b.** Suppose $\mathfrak{B} = \{1, x, x^2, x^3, x^4, x^5\}$. Find $M_{\mathfrak{B}}^{\mathfrak{B}}(T)$.

**21.** Suppose $T : \mathbb{R}_4[x] \to \mathbb{R}_3[x]$ is defined by $T(f) = g$, where

$$g(x) = f(x+1) - f(x)$$

**a.** Show that $T$ is linear.
**b.** Suppose $\mathfrak{B} = \{1, x, x^2, x^3, x^4\}$ and $\mathfrak{B}' = \{1, x, x^2, x^3\}$. Find $M_{\mathfrak{B}}^{\mathfrak{B}'}(T)$.
**c.** Suppose that

$$\mathfrak{B} = \left\{1, x, \frac{x(x-1)}{2}, \frac{x(x-1)(x-2)}{6}, \frac{x(x-1)(x-2)(x-3)}{24}\right\} \quad \text{and}$$

$$\mathfrak{B}' = \left\{1, x, \frac{x(x-1)}{2}, \frac{x(x-1)(x-2)}{6}\right\}$$

Find $M_{\mathfrak{B}}^{\mathfrak{B}'}(T)$.

**22.** If $V$ is a vector space and $\mathfrak{B}$ and $\mathfrak{B}'$ are two different bases for $V$, show that

$$M_{\mathfrak{B}}^{\mathfrak{B}'}(\text{id})\underline{v}_{\mathfrak{B}} = \underline{v}_{\mathfrak{B}'}$$

for any vector $v$ in $V$.

⟵ See Exercise 11 from Lesson 8.1.

**23.** Using the same notation as in Exercise 22, show that

$$\left(M_{\mathfrak{B}}^{\mathfrak{B}'}(\text{id})\right)^{-1} = M_{\mathfrak{B}'}^{\mathfrak{B}}(\text{id})$$

**24.** Suppose $V$ is a vector space, $T : V \to V$ is linear, $\mathfrak{B}$ and $\mathfrak{B}'$ are two bases for $V$, and $P = M_{\mathfrak{B}}^{\mathfrak{B}'}(\text{id})$. Show that

$$P^{-1} M_{\mathfrak{B}'}^{\mathfrak{B}'}(T) P = M_{\mathfrak{B}}^{\mathfrak{B}}(T)$$

**25.** Suppose $V = \mathbb{R}^3$, $v_1 = (2,1,2)$, $v_2 = (4,1,8)$, and $v_3 = v_1 \times v_2$.

**a.** Show that $\mathfrak{B} = \{v_1, v_2, v_3\}$ is a basis for $V$.
**b.** Let $E$ be the plane spanned by $v_1$ and $v_2$. Find a vector equation for $E$.

**c.** Suppose $T : V \to V$ is the map that projects a point onto $E$. Show that $T$ is linear and find

$$M_{\mathfrak{B}}^{\mathfrak{B}}(T)$$

$\longleftarrow$ $T(P)$ is the intersection with $E$ of the line through $P$ in the direction of $v_3$.

**26.** Combine your work in Exercises 16 and 24 to find, without a calculator, the value of

$$\begin{pmatrix} 5 & 6 \\ -3 & -\frac{7}{2} \end{pmatrix}^4$$

**27.** For each value of $n$, find the value of

$$\begin{pmatrix} 5 & 6 \\ -3 & -3.5 \end{pmatrix}^n$$

**a.** $n = 10$ **b.** $n = 20$ **c.** $n = 100$

## 8.5 Change of Basis

In the previous lesson, you learned how to represent a linear map by a matrix that depends on the bases of the domain and the range. When the domain and range are the same, as for a linear transformation, you will be able to use those results to switch coordinate systems. The goal of this lesson is to build such a matrix.

**In this lesson, you will learn how to**

- find the change of basis matrix for a vector space
- find the change of representation matrix for a given linear transformation and two given bases

In the previous lesson, you saw the following definition.

**Definition**

Let $V$ be a vector space. The **identity mapping** on $V$, denoted by $\text{id}(V)$, is the mapping $\text{id} : V \to V$ defined by $\text{id}(v) = v$ for every vector $v$ in $V$.

This definition leads to the following theorem.

**Theorem 8.14**

*If $V$ is a vector space and $\mathfrak{B}$ is a basis, then*

$$M_{\mathfrak{B}}^{\mathfrak{B}}(\text{id}) = I$$

*where $I$ is the identity matrix of the appropriate size.*

**For You to Do**

**1.** Prove Theorem 8.14.

What if you look at the matrix for the identity map with respect to two different bases?

Suppose $V$ is a vector space and $\mathfrak{B}$ and $\mathfrak{B}'$ are bases for $V$. By Theorem 8.12 (see Lesson 8.4), if $v$ is any vector in $V$,

$$M_{\mathfrak{B}}^{\mathfrak{B}'}(\text{id})\underline{v}_{\mathfrak{B}} = \underline{\text{id}(v)}_{\mathfrak{B}'} = \underline{v}_{\mathfrak{B}'}$$

⟵ See Exercise 22 from Lesson 8.4.

**Minds in Action** Episode 36

Tony: So, multiplying the coordinate vector for $v$ with respect to $\mathfrak{B}$ by this matrix $M_{\mathfrak{B}}^{\mathfrak{B}'}(\text{id})$ produces the coordinate vector for $v$ with respect to $\mathfrak{B}'$.

Sasha: So, the matrix changes the basis for you.

DERMAN: I need to try one.

SASHA: Good idea. Suppose $V = \mathbb{R}_2[x]$. Pick two bases.

DERMAN: OK. Let $\mathfrak{B} = \{1, x, x^2\}$ and let $\mathfrak{B}' = \{1+x, x+x^2, x^2\}$. Now I need a vector, I guess.

TONY: Let's do it in general. Suppose $v = ax^2 + bx + c$. Coordinates, anyone?

DERMAN: Sure: $\underline{v}_{\mathfrak{B}} = (c, b, a)$ and $\underline{v}_{\mathfrak{B}'} = (c, b-c, c-b+a)$.

SASHA: So, we want a matrix $M$ so that

← Check Derman's work here.

$$M \begin{pmatrix} c \\ b \\ a \end{pmatrix} = \begin{pmatrix} c \\ b-c \\ c-b+a \end{pmatrix}$$

TONY: I can find that. It's supposed to be $M_{\mathfrak{B}}^{\mathfrak{B}'}(\mathrm{id})$. So, we find the coordinate vectors with respect to $\mathfrak{B}'$ for each vector in $\mathfrak{B}$, and we make those the columns of a matrix:

$$\begin{aligned} \underline{\mathrm{id}(1)}_{\mathfrak{B}'} &= \underline{1}_{\mathfrak{B}'} = (1, -1, 1) \\ \underline{\mathrm{id}(x)}_{\mathfrak{B}'} &= \underline{x}_{\mathfrak{B}'} = (0, 1, -1) \\ \underline{\mathrm{id}(x^2)}_{\mathfrak{B}'} &= \underline{x^2}_{\mathfrak{B}'} = (0, 0, 1) \end{aligned}$$

Now make these the columns of a matrix:

$$M_{\mathfrak{B}}^{\mathfrak{B}'}(\mathrm{id}) = \begin{pmatrix} 1 & 0 & 0 \\ -1 & 1 & 0 \\ 1 & -1 & 1 \end{pmatrix}$$

Let's try it out.

DERMAN: The suspense is driving me crazy.

$$\begin{pmatrix} 1 & 0 & 0 \\ -1 & 1 & 0 \\ 1 & -1 & 1 \end{pmatrix} \begin{pmatrix} c \\ b \\ a \end{pmatrix} = \ldots$$

Yup. It works! And we could have used Pick-Apart to build this matrix to begin with:

$$\begin{pmatrix} c \\ b-c \\ c-b+a \end{pmatrix} = c \begin{pmatrix} 1 \\ -1 \\ 1 \end{pmatrix} + b \begin{pmatrix} 0 \\ 1 \\ -1 \end{pmatrix} + a \begin{pmatrix} 0 \\ 0 \\ 1 \end{pmatrix}$$

so $\begin{pmatrix} 1 & 0 & 0 \\ -1 & 1 & 0 \\ 1 & -1 & 1 \end{pmatrix} \begin{pmatrix} c \\ b \\ a \end{pmatrix} = \begin{pmatrix} c \\ b-c \\ c-b+a \end{pmatrix}$!

SASHA: Yes, it works. But I understand it better if I stand back and use the calculation in the book:

$$M_{\mathfrak{B}}^{\mathfrak{B}'}(\mathrm{id})\underline{v}_{\mathfrak{B}} = \underline{\mathrm{id}(v)}_{\mathfrak{B}'} = \underline{v}_{\mathfrak{B}'}$$

This shows that it will *always* work.

DERMAN: Right, but I have to see an example before I *really* believe it.

---

Tony, Sasha, and Derman have proved a theorem.

**Theorem 8.15 (Change of Basis Theorem)**

*Suppose that $V$ is a vector space and that $\mathfrak{B}$ and $\mathfrak{B}'$ are bases for $V$. Then for any vector $v$ in $V$,*

$$M_{\mathfrak{B}}^{\mathfrak{B}'}(\text{id})\underline{v}_{\mathfrak{B}} = \underline{v}_{\mathfrak{B}'}$$

**Definition**

Suppose that $V$ is a vector space and that $\mathfrak{B}$ and $\mathfrak{B}'$ are bases for $V$. The matrix $M_{\mathfrak{B}}^{\mathfrak{B}'}(\text{id})$ is called the **change of basis matrix** from $\mathfrak{B}$ to $\mathfrak{B}'$.

### Changing representations

The methods of the previous lesson are used when you want to represent a linear map $T$ from one vector space $V$ with basis $\mathfrak{B}$ to another vector space $W$ with basis $\mathfrak{B}'$. Of course, $V$ and $W$ might be the same. Then $T$ is a linear *transformation* of $V$, and any matrix that represents $T$ will be square (why?)

In Chapter 5, you represented a linear transformation $T$ of $\mathbb{R}^n$ by a square matrix. In the language of this chapter, matrix was $M_{\mathfrak{B}}^{\mathfrak{B}}(T)$, where $\mathfrak{B}$ is the standard basis of $\mathbb{R}^n$. Now, you can generalize to linear transformations defined on *any* vector space with respect to *any* basis. This means that for a particular transformation $T : V \to V$, there are many different matrix representations—there's an $M_{\mathfrak{B}}^{\mathfrak{B}}(T)$ for any choice of basis $\mathfrak{B}$. So, there are many different "faces" of the same linear map—different matrices that represent the same thing. In this section, you'll see how these different matrices are related. The key will be the Composite to Product Theorem (Theorem 8.13, see Lesson 8.4).

← Because the matrix is square, you can apply all of the results about square matrices, including the TFAE Theorem (Theorem 8.9; see Lesson 8.3.)

← The standard basis of $\mathbb{R}^n$ is $\{E_1, E_2, \ldots, E_n\}$, where $E_i$ is the $i^{\text{th}}$ row of the $n \times n$ identity matrix.

← Note that the basis for the domain ($V$) and the range ($V$) are taken to be the same in this section. This is how matrix representations of linear transformations are most often used.

**Minds in Action** Episode 37

*Derman, Sasha, and Tony are passing the time, talking about linear maps.*

TONY: Suppose you know the matrix for a linear transformation with respect to one basis. Can you use it to find the matrix of the same map with respect to a new basis?

DERMAN: Here's what I'd do: change from the new basis to the old one with a change of basis matrix. Then apply the map. Then convert from the old basis to the new one with another change of basis matrix.

SASHA: That sounds good. I think this can all be done with matrix multiplication. Let's get some notation down. Suppose we have a vector space $V$ with two bases $\mathfrak{B}$ and $\mathfrak{B}'$ and a linear map $T : V \to V$. And, suppose we know $M_{\mathfrak{B}}^{\mathfrak{B}}(T)$. We want $M_{\mathfrak{B}'}^{\mathfrak{B}'}(T)$. Here's my guess:

$$M_{\mathfrak{B}'}^{\mathfrak{B}'}(T) = M_{\mathfrak{B}}^{\mathfrak{B}'}(\text{id})M_{\mathfrak{B}}^{\mathfrak{B}}(T)M_{\mathfrak{B}'}^{\mathfrak{B}}(\text{id})$$

DERMAN: Isn't that what I said?

Tony, Sasha, and Derman have the right idea, as shown in the next theorem.

**Theorem 8.16 (Change of Representation)**
*Suppose $V$ is a vector space with two bases $\mathfrak{B}$ and $\mathfrak{B}'$, and suppose that $T : V \to V$ is a linear transformation. Then*

$$M_{\mathfrak{B}'}^{\mathfrak{B}'}(T) = M_{\mathfrak{B}}^{\mathfrak{B}'}(\mathrm{id})M_{\mathfrak{B}}^{\mathfrak{B}}(T)M_{\mathfrak{B}'}^{\mathfrak{B}}(\mathrm{id})$$

***Proof.*** The right side of the above equation is a product of three matrices. By the associative law for matrix multiplication, you can group things two at a time:

$$\begin{aligned}
M_{\mathfrak{B}}^{\mathfrak{B}'}(\mathrm{id})M_{\mathfrak{B}}^{\mathfrak{B}}(T)M_{\mathfrak{B}'}^{\mathfrak{B}}(\mathrm{id}) &= M_{\mathfrak{B}}^{\mathfrak{B}'}(\mathrm{id})\left(M_{\mathfrak{B}}^{\mathfrak{B}}(T)M_{\mathfrak{B}'}^{\mathfrak{B}}(\mathrm{id})\right) \\
&= M_{\mathfrak{B}}^{\mathfrak{B}'}(\mathrm{id})\, M_{\mathfrak{B}'}^{\mathfrak{B}}(T \circ \mathrm{id}) && \text{(Theorem 8.13; see Lesson 8.4)} \\
&= M_{\mathfrak{B}}^{\mathfrak{B}'}(\mathrm{id})\, M_{\mathfrak{B}'}^{\mathfrak{B}}(T) && \text{(Exercise 17 from Lesson 8.4)} \\
&= M_{\mathfrak{B}'}^{\mathfrak{B}'}(\mathrm{id} \circ T) \\
&= M_{\mathfrak{B}'}^{\mathfrak{B}'}(T)
\end{aligned}$$

∎

You can think of this equation

$$M_{\mathfrak{B}'}^{\mathfrak{B}'}(T) = M_{\mathfrak{B}}^{\mathfrak{B}'}(\mathrm{id})M_{\mathfrak{B}}^{\mathfrak{B}}(T)M_{\mathfrak{B}'}^{\mathfrak{B}}(\mathrm{id})$$

as a way to "cancel" $\mathfrak{B}$ and replace it with $\mathfrak{B}'$:

$$\begin{aligned}
M_{\mathfrak{B}}^{\mathfrak{B}'}(\mathrm{id})M_{\mathfrak{B}}^{\mathfrak{B}}(T)M_{\mathfrak{B}'}^{\mathfrak{B}}(\mathrm{id}) &= M_{\mathfrak{B}}^{\mathfrak{B}'}(\mathrm{id})M_{\cancel{\mathfrak{B}}}^{\mathfrak{B}}(T)M_{\mathfrak{B}'}^{\cancel{\mathfrak{B}}}(\mathrm{id}) \\
&= M_{\mathfrak{B}}^{\mathfrak{B}'}(\mathrm{id})M_{\mathfrak{B}'}^{\mathfrak{B}}(T \circ \mathrm{id}) \\
&= M_{\mathfrak{B}}^{\mathfrak{B}'}(\mathrm{id})M_{\mathfrak{B}'}^{\mathfrak{B}}(T) \\
&= M_{\cancel{\mathfrak{B}}}^{\mathfrak{B}'}(\mathrm{id})M_{\mathfrak{B}'}^{\cancel{\mathfrak{B}}}(T) \\
&= M_{\mathfrak{B}'}^{\mathfrak{B}'}(\mathrm{id} \circ T) \\
&= M_{\mathfrak{B}'}^{\mathfrak{B}'}(T)
\end{aligned}$$

**Minds in Action** Episode 38

DERMAN: I don't like the statement of the Change of Representation Theorem. It's really ugly looking. What does all that stuff mean, anyway?

TONY: I know what you mean, Derman. It does look pretty bad.

SASHA: What's wrong with it? It looks perfectly fine to me.

DERMAN: Aren't each of these things matrices? Why do we have to write things like $M_{\mathfrak{B}}^{\mathfrak{B}}(T)$ and $M_{\mathfrak{B}}^{\mathfrak{B}'}(\mathrm{id})$? It looks like the matrix, $M$, is being multiplied by something, like $T$.

TONY: I understand that part. It's like writing $f(x)$ to represent a function of $x$. That does not mean $f$ times $x$, Derman.

DERMAN: Yeah, I guess so. But why all the subscript and superscript $\mathfrak{B}$s? And why do we have to write *id* if it is a change of basis matrix? If we want to change from one basis to another, why not just write $M_{\mathfrak{B}}^{\mathfrak{B}'}$ or $M_{\mathfrak{B}'}^{\mathfrak{B}}$?

SASHA: Hmm, let me think about this.

*Sasha thinks for a moment, and then writes* $\begin{pmatrix} \cos\theta & -\sin\theta \\ \sin\theta & \cos\theta \end{pmatrix}$ *on a piece of paper.*

SASHA: What is this?

DERMAN: It's the rotation matrix for $\mathbb{R}^2$, but I don't understand what this has to do with my question.

SASHA: It's not *the* rotation matrix for $R^2$; it's *a* rotation matrix for $\mathbb{R}^2$.

DERMAN: What do you mean? It is the rotation matrix, isn't it?

TONY: Oh, I think I see! It is the most common rotation matrix, but that's only because we wrote, or constructed, that matrix relative to the standard basis for $\mathbb{R}^2$. But we could have picked a different basis and the matrix would have looked different!

SASHA: Right! It would have been a different matrix, but it would still represent the same rotation, just relative to a different basis.

DERMAN: I don't understand. How would we find the rotation matrix relative to a different basis?

TONY: You already told us how to, Derman! Change from the new basis to the old one, apply the standard rotation matrix, then convert back to the new basis!

SASHA: I wonder if we could build it the way we built the original rotation matrix, too.

TONY: What do you mean, Sasha?

SASHA: When we first built the rotation matrix, we made it by looking at what the transformation would do to $(1, 0)$ and $(0, 1)$. We picked those because that was the only basis we knew for $\mathbb{R}^2$ at the time.

TONY: Plus that basis is really nice and seems to make everything easier. But I think your way would work, too.

SASHA: Yeah, and the standard basis is how we usually represent vectors in $\mathbb{R}^2$. In this case, we would have to keep writing everything relative to our nonstandard basis.

DERMAN: What are you two talking about? Doesn't my way sound easier?

---

### For You to Do

**2.** Use Sasha's proposed method to find a matrix that represents a rotation in $\mathbb{R}^2$ counterclockwise through an angle $\theta$ relative to the basis vectors $(2, 1)$ and $(1, -1)$.

**3.** Use Derman's proposed method to find a matrix that represents a rotation in $\mathbb{R}^2$ counterclockwise through an angle $\theta$ relative to the basis vectors $(2, 1)$ and $(1, -1)$.

**4.** Let $\mathfrak{B} = \{(1,0),(0,1)\}$, the standard basis for $\mathbb{R}^2$, and let $\mathfrak{B}' = \{(2,1),(1,-1)\}$, the basis used in problems 2 and 3 above. Let the transformation $T$ be the counterclockwise rotation by $\theta$. Use the notation from the Change of Representation Theorem to notate the two rotation matrices for $\mathbb{R}^2$ used in problems 2 and 3 above.

## Exercises

**1.** Suppose $\mathfrak{B} = \{(1,0),(0,1)\}$ and $\mathfrak{B}' = \{(1,1),(-3,2)\}$.
- **a.** Find the change of basis matrix from $\mathfrak{B}$ to $\mathfrak{B}'$.
- **b.** Find the change of basis matrix from $\mathfrak{B}'$ to $\mathfrak{B}$.
- **c.** Find the product of these two matrices.

**2.** Suppose $\mathfrak{B} = \{(-3,5),(11,2)\}$ and $\mathfrak{B}' = \{(2,5),(7,1)\}$.
- **a.** Find the change of basis matrix from $\mathfrak{B}$ to $\mathfrak{B}'$.
- **b.** Find the change of basis matrix from $\mathfrak{B}'$ to $\mathfrak{B}$.
- **c.** Find the product of these two matrices.

**3.** Suppose

$$\mathfrak{B} = \{1, x, x^2\} \text{ and } \mathfrak{B}' = \{2 + x + 2x^2, 3 + x, 2 + x + x^2\}$$

- **a.** Find the change of basis matrix from $\mathfrak{B}$ to $\mathfrak{B}'$.
- **b.** Find the change of basis matrix from $\mathfrak{B}'$ to $\mathfrak{B}$.
- **c.** Find the product of these two matrices.

**4.** Suppose $\mathfrak{B} = \{(1,0,0,0),(0,1,0,0),(0,0,1,0),(0,0,0,1)\}$ and $\mathfrak{B}' = \{(-2,3,0,5),(3,-1,7,1),(0,4,-2,3),(1,-3,2,-4)\}$.
- **a.** Find the change of basis matrix from $\mathfrak{B}$ to $\mathfrak{B}'$.
- **b.** Find the change of basis matrix from $\mathfrak{B}'$ to $\mathfrak{B}$.
- **c.** Find the product of these two matrices.

**5.** A linear transformation $T$ is represented by the following matrix written relative to the standard basis for $\mathbb{R}^2$: $\begin{pmatrix} 5 & 0 \\ 0 & 2 \end{pmatrix}$
- **a.** Describe this transformation relative to the standard basis.
- **b.** Find a matrix that represents this transformation relative to the nonstandard basis $\mathfrak{B}' = \{(3,1),(-1,2)\}$.
- **c.** Find a matrix that represents this transformation relative to the nonstandard basis $\mathfrak{B}' = \{(-4,3),(7,1)\}$.
- **d.** When $T$ is considered relative to some basis for $\mathbb{R}^2$, the matrix that represents this transformation is $\begin{pmatrix} -1 & -18 \\ 1 & 8 \end{pmatrix}$. Find two bases for $\mathbb{R}^2$ that produce such a representation of $T$.

**6.** Suppose $T : \mathbb{R}^2 \to \mathbb{R}^2$ is linear and

$$\begin{aligned} T(1,0) &= (23, -12) \\ T(0,1) &= (40, -21) \end{aligned}$$

**a.** Find $M$, the matrix for $T$ with respect to the standard basis.
**b.** Find a basis $\mathfrak{B}' = \{v_1, v_2\}$ for $\mathbb{R}^2$ so that

$$T(v_1) = k_1 v_1 \quad \text{and} \quad T(v_2) = k_2 v_2$$

for numbers $k_1$ and $k_2$.
**c.** Find $N = M_{\mathfrak{B}'}^{\mathfrak{B}'}(T)$.
**d.** Find a matrix $P$ so that

$$N = P^{-1}MP$$

**7.** Suppose $T : \mathbb{R}_3[x] \to \mathbb{R}_3[x]$ is defined by

$$\begin{aligned} T(1) &= 15 - 2x + 3x^2 + 7x^3 \\ T(x) &= 20 - 3x + 7x^2 + 11x^3 \\ T(x^2) &= 22 - 4x + 10x^2 + 13x^3 \\ T(x^3) &= -34 + 6x - 10x^2 - 17x^3 \end{aligned}$$

Find $M_{\mathfrak{B}}^{\mathfrak{B}}(T)$, where

$$\mathfrak{B} = \{12 + 2x + 5x^2 + 9x^3, 2 - x + x^2 + x^3, 1 + x^2 + x^3, 1 + x + x^3\}$$

**8.** Suppose $T : \text{Mat}_{2\times 2} \to \text{Mat}_{2\times 2}$ is defined by

$$\begin{aligned} T\begin{pmatrix} 1 & 0 \\ 0 & 0 \end{pmatrix} &= \begin{pmatrix} 15 & -2 \\ 3 & 7 \end{pmatrix} \\ T\begin{pmatrix} 0 & 1 \\ 0 & 0 \end{pmatrix} &= \begin{pmatrix} 20 & -3 \\ 7 & 11 \end{pmatrix} \\ T\begin{pmatrix} 0 & 0 \\ 1 & 0 \end{pmatrix} &= \begin{pmatrix} 22 & -4 \\ 10 & 13 \end{pmatrix} \\ T\begin{pmatrix} 0 & 0 \\ 0 & 1 \end{pmatrix}) &= \begin{pmatrix} -34 & 6 \\ -10 & -17 \end{pmatrix} \end{aligned}$$

and suppose that

$$\mathfrak{B} = \left\{ \begin{pmatrix} 12 & 2 \\ 5 & 9 \end{pmatrix}, \begin{pmatrix} 2 & -1 \\ 1 & 1 \end{pmatrix}, \begin{pmatrix} 1 & 0 \\ 1 & 1 \end{pmatrix}, \begin{pmatrix} 1 & 1 \\ 0 & 1 \end{pmatrix} \right\}$$

Find $M_{\mathfrak{B}}^{\mathfrak{B}}(T)$.

**9.** Which linear mappings $T$ are one-to-one? For those that are, find a formula for $T^{-1}$.

**a.** $T(x, y, z) = (x - y, x + y)$
**b.** $T(x, y, z) = (x - y, x + y, 2z)$
**c.** $T(x, y) = (2x + y, x + y)$
**d.** $T(x, y, z) = (x + 2y + 3z, 4x + 5y + 6z, 7x + 8y)$

e. $T\begin{pmatrix} x & y \\ z & w \end{pmatrix} = (x+w, 2y, z+w, w)$

f. $T(ax^2+bx+c) = 2ax+b$

g. $T(x,y,z) = \begin{pmatrix} x & y & z \\ z & y & x \\ y & z & x \end{pmatrix}$

**10.** For those mappings in Exercises **9a**–**9e** that have inverses, find $M(T)$ and $M(T^{-1})$. Show that in each case,

$$M(T^{-1}) = (M(T))^{-1}$$

⟵ Here, all matrices are with respect to standard bases.

**11.** Suppose that $V$ is a vector space and that $T : V \to V$ is linear. Let $\mathfrak{B}$ be any basis for $V$, and let $M = M_{\mathfrak{B}}^{\mathfrak{B}}(T)$.

a. Show that if $T$ is one-to-one, then $M$ is invertible and

$$M_{\mathfrak{B}}^{\mathfrak{B}}(T^{-1}) = M^{-1}$$

⟵ Just as for matrices, the **kernel** of a linear map is the set of vectors that get mapped to **0**.

b. Show that $T$ is one-to-one if and only if $\ker(M)$ is just the zero vector.

# 8.6 Similar Matrices

In Lesson 8.5, you learned to find the change of representation matrix of a linear transformation for different bases. In this lesson, you will see that all those matrices, which are just different faces of the same map, are related in a particularly nice way.

**In this lesson, you will learn how to**

- determine when two matrices are similar
- find the eigenvalues and the eigenvectors of a matrix
- find a diagonal matrix, if it exists, similar to a given matrix

**Minds in Action** Episode 39

DERMAN: I still don't really understand what we are doing here. Why do we keep changing these transformations?

SASHA: We're not changing the transformations, Derman, just their representations. It's still the same transformation, it just looks different.

DERMAN: That doesn't make sense! The matrix *is* the transformation, isn't it?

TONY: Not exactly. Think about this . . .

*Tony sketches the following graph:*

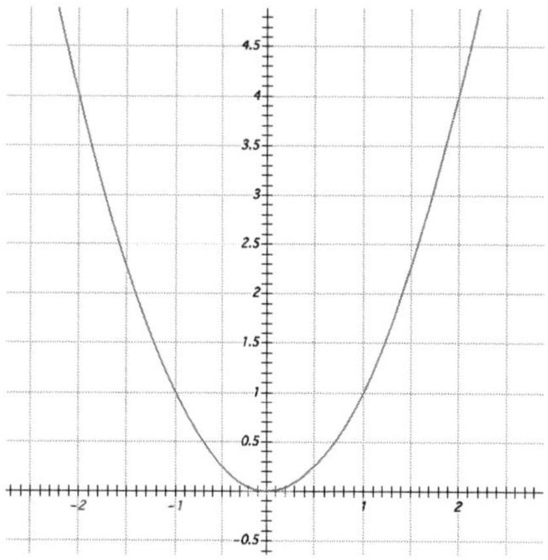


TONY: What is this?

DERMAN: It's the graph of the function defined by $y = x^2$.

TONY: But it's not actually the *function* defined by $y = x^2$; it's just the graph of it.

DERMAN: Right . . .

⟵ Derman is thinking of $y$ as a function of $x$ here.

TONY: So if we changed the axes on our coordinate plane, the graph would look different, but it still would be the graph of $y = x^2$. So, actually, this is only *a* graph of $y = x^2$. It's the most common, since this is what we generally set our axes to be, but if we halved the $x$-coordinates, the graph would get all skinny like this:

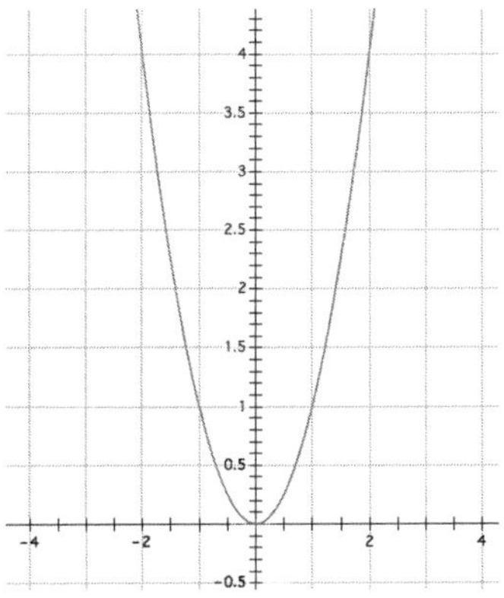


TONY: But this is still a graph of the same function. The function didn't change, just the way we are representing it on this graph.

SASHA: Right, but it would have been better if we had picked these coordinate axes:

*Sasha graphs the same function using coordinate axes where the $x$-axis is integer values but the $y$-axis is squares of integer values and the negative of squares of integer values.*

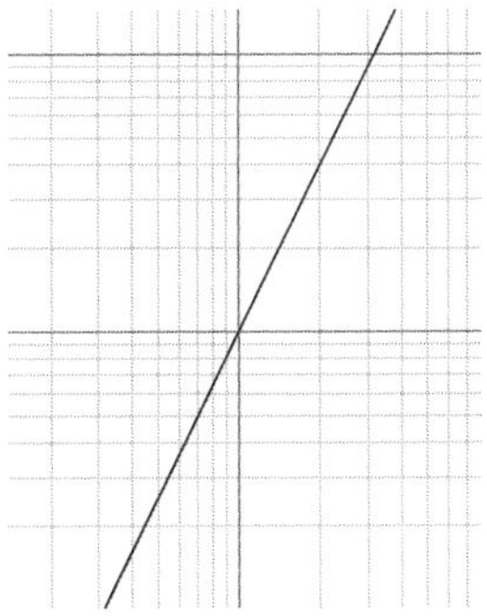

SASHA: Then the graph of our function would look really nice!

DERMAN: That's cheating!

SASHA: Not really. It's just changing the axes of the graph.

DERMAN: But now it looks like a different function!

TONY: But it's not a function at all—it's a graph. And the graph is just a different representation of the *same* function.

SASHA: Right. And if we can figure out what axes, or coordinate system, to use, we can make the graph of the function look really nice. In the same way, if we can figure out what basis to use, we can make the matrix that represents our transformation look really nice.

DERMAN: I guess so. But there's one more thing that still bothers me. Why do we need to put the *id* in for change of basis matrices in this formula: $M_{\mathfrak{B}'}^{\mathfrak{B}'}(T) = M_{\mathfrak{B}}^{\mathfrak{B}'}(\text{id})M_{\mathfrak{B}}^{\mathfrak{B}}(T)M_{\mathfrak{B}'}^{\mathfrak{B}}(\text{id})$? I asked you that yesterday, Sasha, and you didn't answer me.

SASHA: Oh, sorry about that, Derman. I must have gotten sidetracked. Well, it's related to what we were talking about then. We decided that to say a matrix represents a transformation is a little imprecise, right?

DERMAN: Yeah, we decided it had to represent a transformation relative to a particular basis.

SASHA: So that's why we write things like $M_{\mathfrak{B}}^{\mathfrak{B}}(T)$ or $M_{\mathfrak{B}}^{\mathfrak{B}'}(T)$; we're declaring what basis the transformation is occurring relative to.

DERMAN: But we don't want change-of-basis matrices to represent a transformation at all! They just change the basis.

SASHA: But why are those matrices so different? If every transformation matrix has to be written relative to a particular basis . . .

TONY: Oh! Then any change-of-basis matrix has to be written relative to a transformation!

DERMAN: But there isn't any transformation occurring!

TONY: Exactly! That's why we write *id*, the identity transformation. That way our vectors don't move under the transformation.

SASHA: Right.

DERMAN: I guess so. So *every* matrix represents a transformation relative to a particular basis?

SASHA: Yup!

DERMAN: Every matrix? What about $\begin{pmatrix} 1 & 0 \\ 0 & 1 \end{pmatrix}$? What transformation is that, and relative to what basis?

TONY: We already proved this back at Theorem 8.14, Derman! That's the identity transformation, *id*!

DERMAN: But what is the basis?

TONY: Any basis! It's the best transformation, because it has the same representation no matter what basis we are talking about. But now I'm interested in this equation from the theorem:

$$M_{\mathfrak{B}'}^{\mathfrak{B}'}(T) = M_{\mathfrak{B}}^{\mathfrak{B}'}(\text{id})M_{\mathfrak{B}}^{\mathfrak{B}}(T)M_{\mathfrak{B}'}^{\mathfrak{B}}(\text{id})$$

The "bookend" matrices on the right-hand side look a lot alike:

$$M_{\mathfrak{B}}^{\mathfrak{B}'}(\text{id}) \quad \text{and} \quad M_{\mathfrak{B}'}^{\mathfrak{B}}(\text{id})$$

One transitions from $\mathfrak{B}'$ to $\mathfrak{B}$ and the other transitions from $\mathfrak{B}$ to $\mathfrak{B}'$. I bet they're inverses of each other.

SASHA: Of course they are. Look:

$$\begin{aligned} M_{\mathfrak{B}}^{\mathfrak{B}'}(\mathrm{id})M_{\mathfrak{B}'}^{\mathfrak{B}}(\mathrm{id}) &= M_{\mathfrak{B}}^{\mathfrak{B}}(\mathrm{id}\circ\mathrm{id}) && \text{(Composite to Product)} \\ &= M_{\mathfrak{B}}^{\mathfrak{B}}(\mathrm{id}) && (\mathrm{id}\circ\mathrm{id}=\mathrm{id}) \\ &= I && \text{(Theorem 8.14; see Lesson 8.5)} \end{aligned}$$

Since the product is $I$, the matrices are inverses.

DERMAN: I need to try one.

### For You to Do

1. Help Derman work an example. Suppose $V = \mathbb{R}_2[x]$, $\mathfrak{B} = \{x^2, x, 1\}$, and $\mathfrak{B}' = \{2x^2+5x+3, 4x-1, x^2+x+2\}$. Calculate $M_{\mathfrak{B}}^{\mathfrak{B}'}(\mathrm{id})$ and $M_{\mathfrak{B}'}^{\mathfrak{B}}(\mathrm{id})$, and show that the two matrices are inverses.

⟵ If you arrange your calculations systematically, you'll find that expressing the elements of $\mathfrak{B}$ in terms of the elements of $\mathfrak{B}'$ will involve the same steps you used in Chapter 3 to find the inverse of a matrix.

In Episode 39, Sasha proves the next theorem.

#### Theorem 8.17

*Suppose $V$ is a vector space and $\mathfrak{B}$ and $\mathfrak{B}'$ are bases for $V$. Then*

$$\left(M_{\mathfrak{B}}^{\mathfrak{B}'}(\mathrm{id})\right)^{-1} = M_{\mathfrak{B}'}^{\mathfrak{B}}(\mathrm{id})$$

Theorem 8.17 allows the Change of Representation Theorem (Theorem 8.16; see Lesson 8.5) to be recast.

#### Corollary 8.18 (Change of Representation, Take 2)

*Suppose $V$ is a vector space with two bases $\mathfrak{B}$ and $\mathfrak{B}'$, and suppose that $T : V \to V$ is a linear transformation. Then there is a square matrix $P$ such that*

$$M_{\mathfrak{B}'}^{\mathfrak{B}'}(T) = P^{-1}M_{\mathfrak{B}}^{\mathfrak{B}}(T)P$$

*In fact, $P = M_{\mathfrak{B}'}^{\mathfrak{B}}(\mathrm{id})$, the change of basis matrix from $\mathfrak{B}'$ to $\mathfrak{B}$.*

#### Definition

Two $n \times n$ matrices $M$ and $N$ are **similar** if there is an invertible matrix $P$ such that

$$N = P^{-1}MP$$

⟵ **Notation:** If $N = P^{-1}MP$, one writes $N \sim M$.

So, using this definition, you can say that *matrices for the same transformation with respect to different bases are similar.*

You have met similar matrices (without calling them that) throughout this course. Similar matrices are related in several important ways (see Exercises 5 and 11, for example).

⟵ See Exercises 26–29 from Lesson 4.5 for more examples.

A major goal of much of linear algebra is, given a linear transformation $T$ on a vector space $V$, to find a basis $\mathfrak{B}$ of $V$ for which the matrix representation for $T$ is especially simple—a diagonal matrix, for example.

### Example 1

Suppose $T$ is a linear transformation on $\mathbb{R}^3$ whose matrix, with respect to the standard basis $\mathfrak{B}$, is

$$M = \begin{pmatrix} -14 & -3 & 33 \\ -76 & -21 & 174 \\ -16 & -4 & 37 \end{pmatrix}$$

**Problem.** Find $M_{\mathfrak{B}'}^{\mathfrak{B}'}(T)$, where $\mathfrak{B}' = \{(3,5,2), (-1,4,0), (2,1,1)\}$.

**Solution 1.** By Corollary 8.18, $M_{\mathfrak{B}'}^{\mathfrak{B}'}(T) = P^{-1}MP$, where $P = M_{\mathfrak{B}'}^{\mathfrak{B}}(\text{id})$. To build $P$, you need to express the elements of $\mathfrak{B}'$ in terms of the elements of $\mathfrak{B}$. That's pretty easy.

$\longleftarrow$ Since $\mathfrak{B}$ is standard, the columns of the change-of-basis matrix are just the the elements of $\mathfrak{B}'$.

$$\begin{aligned} (3,5,2) &= 3(1,0,0) + 5(0,1,0) + 2(0,0,1) \\ (-1,4,0) &= -1(1,0,0) + 4(0,1,0) + 0(0,0,1) \\ (2,1,1) &= 2(1,0,0) + 1(0,1,0) + 1(0,0,1) \end{aligned}$$

so

$$P = M_{\mathfrak{B}'}^{\mathfrak{B}}(\text{id}) = \begin{pmatrix} 3 & -1 & 2 \\ 5 & 4 & 1 \\ 2 & 0 & 1 \end{pmatrix}$$

It follows that $M_{\mathfrak{B}}^{\mathfrak{B}'}(\text{id})$ is

$\longleftarrow$ You can invert the matrix by hand or calculator.

$$P^{-1} = M_{\mathfrak{B}}^{\mathfrak{B}'}(\text{id}) = \begin{pmatrix} -4 & -1 & 9 \\ 3 & -1 & -7 \\ 8 & 2 & 17 \end{pmatrix}$$

So,

$$\begin{aligned} M_{\mathfrak{B}'}^{\mathfrak{B}'}(T) &= P^{-1}MP \\ &= \begin{pmatrix} -4 & -1 & 9 \\ 3 & -1 & -7 \\ 8 & 2 & 17 \end{pmatrix} \begin{pmatrix} -14 & -3 & 33 \\ -76 & -21 & 174 \\ -16 & -4 & 37 \end{pmatrix} \begin{pmatrix} 3 & -1 & 2 \\ 5 & 4 & 1 \\ 2 & 0 & 1 \end{pmatrix} \\ &= \begin{pmatrix} 3 & 0 & 0 \\ 0 & -2 & 0 \\ 0 & 0 & 1 \end{pmatrix} \end{aligned}$$

This is a much simpler matrix than $M$.

**Solution 2.** You could build $M_{\mathfrak{B}'}^{\mathfrak{B}'}(T)$ directly from the definition. This illustrates why the matrix came out so nicely.

$$\begin{pmatrix} -14 & -3 & 33 \\ -76 & -21 & 174 \\ -16 & -4 & 37 \end{pmatrix} \begin{pmatrix} 3 \\ 5 \\ 2 \end{pmatrix} = \begin{pmatrix} 9 \\ 15 \\ 6 \end{pmatrix} = 3 \begin{pmatrix} 3 \\ 5 \\ 2 \end{pmatrix}$$

$$\begin{pmatrix} -14 & -3 & 33 \\ -76 & -21 & 174 \\ -16 & -4 & 37 \end{pmatrix} \begin{pmatrix} -1 \\ 4 \\ 0 \end{pmatrix} = \begin{pmatrix} 2 \\ -8 \\ 0 \end{pmatrix} = -2 \begin{pmatrix} -1 \\ 4 \\ 0 \end{pmatrix}$$

$$\begin{pmatrix} -14 & -3 & 33 \\ -76 & -21 & 174 \\ -16 & -4 & 37 \end{pmatrix} \begin{pmatrix} 2 \\ 1 \\ 1 \end{pmatrix} = \begin{pmatrix} 2 \\ 1 \\ 1 \end{pmatrix} = 1 \begin{pmatrix} 2 \\ 1 \\ 1 \end{pmatrix}$$

So, $T$ *scales* each of the basis vectors in $\mathfrak{B}'$.

$$T \begin{pmatrix} 3 \\ 5 \\ 2 \end{pmatrix} = 3 \begin{pmatrix} 3 \\ 5 \\ 2 \end{pmatrix}, \quad T \begin{pmatrix} -1 \\ 4 \\ 0 \end{pmatrix} = -2 \begin{pmatrix} -1 \\ 4 \\ 0 \end{pmatrix}, \quad \text{and} \quad T \begin{pmatrix} 2 \\ 1 \\ 1 \end{pmatrix} = 1 \begin{pmatrix} 2 \\ 1 \\ 1 \end{pmatrix}$$

Or, to spell out all the details,

$$T \begin{pmatrix} 3 \\ 5 \\ 2 \end{pmatrix} = 3 \begin{pmatrix} 3 \\ 5 \\ 2 \end{pmatrix} + 0 \begin{pmatrix} -1 \\ 4 \\ 0 \end{pmatrix} + 0 \begin{pmatrix} 2 \\ 1 \\ 1 \end{pmatrix},$$

$$T \begin{pmatrix} -1 \\ 4 \\ 0 \end{pmatrix} = 0 \begin{pmatrix} 3 \\ 5 \\ 2 \end{pmatrix} - 2 \begin{pmatrix} -1 \\ 4 \\ 0 \end{pmatrix} + 0 \begin{pmatrix} 2 \\ 1 \\ 1 \end{pmatrix}, \quad \text{and}$$

$$T \begin{pmatrix} 2 \\ 1 \\ 1 \end{pmatrix} = 0 \begin{pmatrix} 3 \\ 5 \\ 2 \end{pmatrix} + 0 \begin{pmatrix} -1 \\ 4 \\ 0 \end{pmatrix} + 1 \begin{pmatrix} 2 \\ 1 \\ 1 \end{pmatrix}$$

Hence, by definition, $M_{\mathfrak{B}'}^{\mathfrak{B}'}(T) = \begin{pmatrix} 3 & 0 & 0 \\ 0 & -2 & 0 \\ 0 & 0 & 1 \end{pmatrix}$.

---

The basis $\mathfrak{B}'$ in the above example consists of vectors that are scaled by the transformation. Such vectors are called **eigenvectors** or **characteristic vectors** for the transformation, and the scale factors are called **eigenvalues**. If you can find a basis of eigenvectors for a transformation, the matrix with respect to that basis will be a diagonal matrix (with eigenvalues on the diagonal), and that makes it easy to apply the transformation.

← Note that the columns of the "diagonalizing matrix" $P$ in the above example are precisely the eigenvectors for $T$.

Geometrically, you can think of an eigenvector for $T$ as the generator of a line that is *invariant* under $T$. If you have a basis of eigenvectors, it's like having a set of axes that get scaled by the transformation.

So, if your basis consists of eigenvectors, the matrix representation is especially simple. Does such a basis exist? If so, how do you find it? These are the questions you'll study in the next chapter.

⟵ You've encountered eigenvectors before, without calling them that. See, for example, Exercises 29–31 from Lesson 4.6 or Exercises 13 and 14 from Lesson 5.6.

## Exercises

**1.** A linear transformation $T$ is represented by the following matrix written relative to the standard basis for $\mathbb{R}^2$:

$$\begin{pmatrix} 6 & 0 \\ 0 & 2 \end{pmatrix}$$

**a.** Describe the transformation relative to the standard basis.
**b.** Find a matrix that represents this transformation relative to the nonstandard basis $\mathfrak{B}' = \{(3,4),(1,1)\}$.
**c.** Find a matrix that represents this transformation relative to the nonstandard basis $\mathfrak{B}'' = \{(5,3),(4,3)\}$.
**d.** Find a matrix $P$ that shows that your answers to part **b** and part **c** above are, in fact, similar.
**e.** When this transformation is considered relative to some basis for $\mathbb{R}^2$, the matrix that represents it is $\begin{pmatrix} -2 & -4 \\ 8 & 10 \end{pmatrix}$. Find a basis for $\mathbb{R}^2$ that produces such a representation of $T$.

**2.** A linear transformation $T$ is represented by the following matrix written relative to the standard basis for $\mathbb{R}^2$:

$$\begin{pmatrix} -3 & 0 \\ 4 & 1 \end{pmatrix}$$

**a.** If $\mathfrak{B} = \{(1,0),(0,1)\}$, use the notation of this chapter to denote the given information of this problem.
**b.** Find a matrix that represents this transformation relative to the nonstandard basis $\mathfrak{B}' = \{(3,4),(1,1)\}$. Use the notation of this chapter to denote this answer.
**c.** Find a matrix that represents this transformation relative to the nonstandard basis $\mathfrak{B}''\{(5,3),(4,3)\}$. Use the notation of this chapter to denote your answer.
**d.** Find a matrix $P$ that shows your answers to part **b** and part **c** are, in fact, similar.
**e.** Find a diagonal matrix $D$ such that $D$ is similar to $M$.
**f.** Explain what your answer to part **e** means in terms of the transformation represented by $M$.

**3.** A linear transformation $T$ is represented by the following matrix written relative to the standard basis for $\mathbb{R}^2$:

$$\begin{pmatrix} \frac{1}{5} & \frac{12}{5} \\ \frac{12}{5} & -\frac{6}{5} \end{pmatrix}$$

**a.** If $\mathfrak{B} = \{(1,0),(0,1)\}$, use the notation of this chapter to denote the given information of this problem.

**b.** Find the image of the unit square under $T$. Try to describe the effect $T$ has on points in $\mathbb{R}^2$ geometrically.

**c.** Find the image of $(4, 3)$ under $T$. Describe the effect $T$ has on this point geometrically.

**d.** Find the image of $(-2, -1.5)$ under $T$. Describe the effect $T$ has on this point geometrically.

**e.** Find the image of $(4, 3)$ under $T^2$. Describe the geometric effect $T$ has on this point when applied twice.

**f.** Find the image of the line $X = k(4, 3)$ under $T$. Describe the effect $T$ has on this line geometrically. (This type of line contains many eigenvectors of $T$.)

**g.** Find another line $E = k(x, y)$ that is invariant under $T$. (Or, equivalently, find another eigenvector of $T$.)

**h.** Show that two eigenvectors for $T$ can form a basis for $\mathbb{R}^2$. Call this basis $\mathfrak{B}'$.

**i.** Find $M_{\mathfrak{B}'}^{\mathfrak{B}'}(T)$.

**4.** Show that the matrices $\begin{pmatrix} a & 0 \\ 0 & b \end{pmatrix}$ and $\begin{pmatrix} b & 0 \\ 0 & a \end{pmatrix}$ are similar.

**5.** **a.** Find two $2 \times 2$ matrices that are similar. Show that they have the same determinant.

**b.** Show that if any pair of $2 \times 2$ matrices are similar, they have the same determinant.

**6.** Are there any $2 \times 2$ matrices that have no eigenvectors in $\mathbb{R}^2$? If so, find one. If not, prove it.

⟵
In other words, are there any $2 \times 2$ matrices that don't fix any lines?

**7.** A linear transformation $T$ is represented by the following matrix $M$ written relative to the standard basis for $\mathbb{R}^3$:

$$\begin{pmatrix} 2 & 0 & 0 \\ 0 & -1 & 0 \\ 0 & 0 & 5 \end{pmatrix}$$

**a.** Find a matrix that represents this transformation relative to the nonstandard basis

$$\mathfrak{B}' = \{(1, -1, 1), (3, 2, -1), (2, 2, -1)\}$$

**b.** Find a matrix that represents this transformation relative to the nonstandard basis

$$\mathfrak{B}'' = \{(4, -1, 0), (-3, 1, 3), (1, -1, 2)\}$$

**c.** Find a matrix that shows your answers to part **b** and part **c** above are, in fact, similar.

**d.** When this transformation is considered relative to some basis for $\mathbb{R}^3$, the matrix that represents it is $\begin{pmatrix} 11 & -6 & 12 \\ 6 & 2 & 12 \\ -6 & 3 & -7 \end{pmatrix}$. Find a basis for $\mathbb{R}^3$ that produces such a representation of $T$.

**8.** Suppose that

$$M = \begin{pmatrix} 15 & 20 & 22 & -34 \\ -2 & -3 & -4 & 6 \\ 3 & 7 & 10 & -10 \\ 7 & 11 & 13 & -17 \end{pmatrix}$$

**a.** Find a $4 \times 4$ matrix $P$ so that $P^{-1}MP$ is a diagonal matrix.
**b.** How could you use part **a** to find $M^5$ without a calculator?

**9.** Suppose that $M$, $N$, and $Q$ are $n \times n$ matrices. Show that

**a.** $M \sim M$
**b.** if $M \sim N$, then $N \sim M$
**c.** if $M \sim N$ and $N \sim Q$, then $M \sim Q$
**d.** if $M \sim N$, then $M^k \sim N^k$ for any nonnegative integer $k$

**10.** Matrices for the same transformation with respect to different bases are similar. Is the converse true? That is, suppose $M$ and $N$ are, say, $3 \times 3$ matrices with the property that there is an invertible matrix $P$ so that

$$P^{-1}MP = N$$

Is there a linear transformation $T : \mathbb{R}^3 \to \mathbb{R}^3$ and a pair of bases $\mathfrak{B}$ and $\mathfrak{B}'$ so that

$$M_{\mathfrak{B}}^{\mathfrak{B}}(T) = M \quad \text{and} \quad M_{\mathfrak{B}'}^{\mathfrak{B}'} = N?$$

If so, what could you take as the bases, and what would the change of basis matrix be? If not, provide a counterexample.

**11.** Recall from Exercise 27 from Lesson 4.4 that the **trace** of a square matrix is the sum of its diagonal elements. Show that similar matrices have the same trace. Illustrate with an example.

# Chapter 8 Mathematical Reflections

These problems will help you summarize what you have learned in this chapter.

**1.** Is the set $\{(-2, 1, 3, 1), (1, 0, -4, 3)\}$ a basis for $\mathbb{R}^4$? If not, find a basis that contains it.

**2.** Let $M = \begin{pmatrix} 1 & 0 & -1 & -2 & 3 \\ 3 & 1 & -2 & 1 & 2 \\ 1 & 1 & 0 & 5 & -4 \end{pmatrix}$.

**a.** Find $r(M)$.
**b.** Find $\dim \ker(M)$.
**c.** Find dim(column space of $M$).
**d.** Find a basis for $\ker(M)$.
**e.** Find a basis for the column space of $M$.

**3.** Determine whether each mapping is linear.

**a.** $F : \mathbb{R}^2 \to \mathbb{R}^2$; $F(x, y) = (x - y, x + y)$
**b.** $F : \mathbb{R}^3 \to \mathbb{R}^2$; $F(x, y, z) = (x - y, z^2)$

**4.** Suppose $\mathfrak{B} = \{(1, 0, 0), (0, 1, 0), (0, 0, 1)\}$ and $\mathfrak{B}' = \{(1, 2, 4), (0, 1, 3), (1, 1, 0)\}$.

**a.** Find the change of basis matrix from $\mathfrak{B}$ to $\mathfrak{B}'$.
**b.** Find the change of basis matrix from $\mathfrak{B}'$ to $\mathfrak{B}$.
**c.** Find the product of these two matrices.

**5.** Let $M = \begin{pmatrix} -2 & 5 \\ -2 & 4 \end{pmatrix}$ and $N = \begin{pmatrix} 2 & -2 \\ 1 & 0 \end{pmatrix}$. Show that $M$ and $N$ are similar.

**6.** How is the dimension of the row space of a matrix related to the dimension of its kernel?

**7.** How are different coordinate systems in the domain and range of a linear transformation represented algebraically?

**8.** Let $M = \begin{pmatrix} 9 & 4 \\ -12 & -5 \end{pmatrix}$ and $D = \begin{pmatrix} 3 & 0 \\ 0 & 1 \end{pmatrix}$. Show $M$ is similar to $D$.

**Vocabulary**

In this chapter, you saw these terms and symbols for the first time. Make sure you understand what each one means, and how it is used.

- bijective linear map
- blow up to a basis
- change of basis matrix
- eigenvalue
- eigenvector, characteristic vector
- identity mapping
- invariant line
- linear map
- matrix for a transformation with respect to a basis
- maximal linearly independent set
- row and column rank
- similar matrices
- structure-preserving representation

# Chapter 8 Review

In Lesson 8.2, you learned to

- build a basis by expanding a set of linearly independent vectors so it generates
- build a basis by sifting out vectors from a set that generates but is not linearly independent

The following problems will help you check your understanding.

1. For each of the given sets, blow it up to a basis.
   **a.** $\{(1,1,-1),(-1,2,4)\}$ for $\mathbb{R}^3$
   **b.** $\{(0,1,1)\}$ for $L\{(3,2,-1),(0,1,1),(3,3,0),(-3,0,3)\}$
   **c.** $\{(2x^2-1)\}$ for $\mathbb{R}_2[x]$

2. For each of the given vector spaces, sift out a basis from the given generating system.
   **a.** $V = L\{2, 3x, x^2+x, x-4\}$
   **b.** $V = L\{(1,2,0),(3,1,4),(5,5,4),(2,-1,4),(4,1,-2)\}$
   **c.** $V$ is the row space of $N = \begin{pmatrix} 1 & -3 & -1 \\ 2 & 0 & 4 \\ 1 & 3 & 5 \end{pmatrix}$, starting with the rows.
   **d.** $V$ is the column space of $N = \begin{pmatrix} 1 & -3 & -1 \\ 2 & 0 & 4 \\ 1 & 3 & 5 \end{pmatrix}$, starting with the columns.

3. Is the set $\left\{\begin{pmatrix} 0 & 0 \\ 1 & 1 \end{pmatrix}, \begin{pmatrix} 1 & 0 \\ 1 & 0 \end{pmatrix}, \begin{pmatrix} 0 & 1 \\ 0 & 0 \end{pmatrix}, \begin{pmatrix} 1 & 1 \\ 1 & 0 \end{pmatrix}\right\}$ linearly independent and a basis for $\mathrm{Mat}_{2\times 2}(\mathbb{R})$? If not, remove any dependencies and then extend it to a basis.

In Lesson 8.3, you learned to

- find the dimension of the row and column spaces of a matrix
- find the dimension of the kernel and its relation with the dimensions of the row and column space

The following exercises will help you check your understanding.

4. For each matrix, find its rank and the dimension of its kernel.

   **a.** $\begin{pmatrix} 2 & -1 & 5 \\ 1 & 0 & -1 \\ 2 & 5 & -1 \end{pmatrix}$ **b.** $\begin{pmatrix} 2 & -1 & 5 \\ 1 & 0 & -1 \\ 2 & 5 & -1 \\ 4 & 4 & 4 \end{pmatrix}$

   **c.** $\begin{pmatrix} 2 & -1 & 5 \\ 2 & 5 & -1 \\ 4 & 4 & 4 \\ 8 & 2 & 14 \end{pmatrix}$ **d.** $\begin{pmatrix} 2 & -1 & 5 & -4 \\ 2 & 5 & -1 & 0 \\ 4 & 4 & 4 & -4 \end{pmatrix}$

**5.** For each set of conditions, give an example, if possible, of a matrix $A$ that satisfies them. If there is no matrix, explain why not.

**a.** $A$ is $3 \times 5$ and $r(A) = 3$.
**b.** $A$ is $5 \times 3$ and $r(A) = 2$.
**c.** $A$ is $5 \times 3$ and $r(A) = 4$.
**d.** $A$ is $3 \times 5$ and $\dim \ker(A) = 2$.

**6.** Let $N = \begin{pmatrix} 1 & 2 & 0 & 1 & -2 \\ 2 & 3 & 1 & 1 & 1 \\ 1 & 5 & 1 & 0 & 3 \\ 4 & -1 & 1 & 3 & -3 \end{pmatrix}$.

**a.** Find $r(N)$.
**b.** Find $\dim \ker(N)$.
**c.** Find a basis for $\ker(N)$.
**d.** Find a basis for the column space of $N$.

In Lesson 8.4, you learned to

- find a formula for a linear map knowing what it does to a basis
- build a matrix that does the work of a given linear map
- find the matrix representing a given linear map with respect to the bases of its domain and range

The following exercises will help you check your understanding.

**7.** Determine whether each mapping $F$ is linear.

**a.** $F : \mathbb{R}^3 \to R^3$; $F(x, y, z) = (-x, x + y, 3y)$
**b.** $F : \mathbb{R}^3 \to R^2$; $F(x, y, z) = (x + 2, 3y)$
**c.** $F : \mathbb{R}_2[x] \to \text{Mat}_{2\times 2}(\mathbb{R})$; $F(a + bx + cx^2) = \begin{pmatrix} 2a & b \\ -c & 1 \end{pmatrix}$

**8.** The linear map $D : R^2 \to R^3$ with respect to the standard bases is defined by $D(x, y) = (y, x, y + x)$.

**a.** Find $D(2, 3)$.
**b.** Find a matrix $M$ so that, for any vector $v$ in $R^2$, $M\underline{v}_{\mathfrak{B}} = \underline{D(v)}_{\mathfrak{B}'}$.
**c.** Use $M$ to find $D(2, 3)$.

**9.** Suppose the linear map $T : R^3 \to R_1[x]$ is defined by $T(a, b, c) = (b + c) + ax$. And suppose the bases of the domain and the range are

$$\mathfrak{B} = \{(1, 0, 1), (0, 1, 1), (1, 0, 0)\}$$
$$\mathfrak{B}' = \{1, 2 + x\}$$

**a.** Find $T(1, 2, 3)$.
**b.** Find a matrix $M = M_{\mathfrak{B}}^{\mathfrak{B}'}(T)$ so that, for any vector $v$ in $R^3$, $M\underline{v}_{\mathfrak{B}} = \underline{T(v)}_{\mathfrak{B}'}$.
**c.** Use $M$ to find $T(1, 2, 3)$.

**10.** Suppose the linear map $T : R^3 \to R^2$ is defined by $T(x, y, z) = (2(x + y), 2(x - z))$ and the linear map $S : R^2 \to R^2$ is defined by

$S(x, y) = (2x, -y)$. Let $M(T)$ and $M(S)$ be the matrices for $T$ and $S$ with respect to the standard bases. Find

**a.** $S \circ T(1, 2, 3)$
**b.** $M(S)$
**c.** $M(T)$
**d.** $M(S \circ T)$
**e.** $M(S)M(T)$
**f.** $M(S \circ T) \cdot \begin{pmatrix} 1 \\ 2 \\ 3 \end{pmatrix}$

In Lesson 8.5, you learned to

- find a matrix that will switch coordinate systems within the same vector space
- determine the change of representation for a linear transformation given two different bases on a vector space

The following problems will help you check your understanding.

**11.** Suppose $V = \mathbb{R}_2[x]$, let $\mathfrak{B} = \{1, x, x^2\}$, and let $\mathfrak{B}' = \{2, 1 + x, x - x^2\}$.

**a.** Find $M_{\mathfrak{B}}^{\mathfrak{B}'}(\text{id})$, the change-of-basis matrix from $\mathfrak{B}$ to $\mathfrak{B}'$.
**b.** If $v = 2 + 3x + x^2$, use this change-of-basis matrix to find $\underline{v}_{\mathfrak{B}'}$.

**12.** Suppose $\mathfrak{B} = \{(2, -1), (0, 3)\}$ and $\mathfrak{B}' = \{(1, 1), (2, 0)\}$.

**a.** Find the change-of-basis matrix from $\mathfrak{B}$ to $\mathfrak{B}'$.
**b.** Find the change-of-basis matrix from $\mathfrak{B}'$ to $\mathfrak{B}$.
**c.** Find the product of these two matrices.

**13.** A linear transformation $T$ is represented by the matrix $\begin{pmatrix} 1 & 0 & 0 \\ 0 & -1 & 0 \\ 0 & 0 & 2 \end{pmatrix}$ relative to the standard basis for $\mathbb{R}^3$.

**a.** Describe this transformation relative to the standard basis.
**b.** Find a matrix that represents this transformation relative to the nonstandard basis

$$\mathfrak{B}' = \{(1, 1, 0), (0, 1, 2), (1, 1, 1)\}$$

**c.** Let $v = (2, 3, 4)$. Show that $\underline{T(v)}_{\mathfrak{B}'} = T(\underline{v}_{B'})$.

In Lesson 8.6, you learned to

- determine if two matrices are similar
- find a diagonal matrix, if one exists, similar to a given matrix
- find the eigenvalues and eigenvectors of a given matrix

The following exercises will help you check your understanding.

**14.** Suppose $T : \mathbb{R}^3 \to \mathbb{R}^3$ is represented by the following matrix $M$ written relative to the standard basis $\mathfrak{B}$ for $\mathbb{R}^3$:

$$M = \begin{pmatrix} 8 & -9 & 6 \\ 6 & -7 & 4 \\ 0 & 0 & 1 \end{pmatrix}$$

and suppose that

$$\mathfrak{B}' = \{(3, 2, 0), (1, 1, 0), (-6, -4, 1)\}$$

**a.** Find $P = M_{\mathfrak{B}'}^{\mathfrak{B}}(\mathrm{id})$.
**b.** Find $P^{-1} = \left(M_{\mathfrak{B}'}^{\mathfrak{B}}(\mathrm{id})\right)^{-1}$.
**c.** Use your answers to parts **a** and **b** to find $M_{\mathfrak{B}'}^{\mathfrak{B}'}(T)$.

**15.** A linear transformation $T$ is represented by the following matrix written relative to the standard basis for $\mathbb{R}^2$:

$$\begin{pmatrix} 0 & -1 \\ 2 & 0 \end{pmatrix}$$

**a.** Describe this transformation relative to the standard basis.
**b.** Find a matrix that represents this transformation relative to the nonstandard basis $\mathfrak{B}' = \{(3, -1), (-4, 1)\}$.
**c.** Find a matrix that represents this transformation relative to the nonstandard basis $\mathfrak{B}'' = \{(-2, 0), (1, -1)\}$.
**d.** Find a matrix $P$ that shows your answers to part **b** and part **c** are similar.

**16.** A linear transformation $T$ is represented by the following matrix written relative to the standard basis for $\mathbb{R}^2$:

$$\begin{pmatrix} -26 & 10 \\ -75 & 29 \end{pmatrix}$$

When this transformation is considered relative to some nonstandard basis for $\mathbb{R}^2$, the matrix that represents it is $\begin{pmatrix} -1 & 0 \\ 0 & 4 \end{pmatrix}$. Find a basis for $\mathbb{R}^2$ that produces this representation of $T$.

# Chapter 8 Test

## Multiple Choice

**1.** Consider the set $\{(-1,1,2),(1,1,0)\}$. Which additional vector will blow it up to a basis for $\mathbb{R}^3$?

**A.** $(3,-3,-6)$
**B.** $(2,0,-2)$
**C.** $(1,-3,4)$
**D.** $(0,2,2)$

**2.** Let $N = \begin{pmatrix} -2 & 3 & 1 & -2 \\ 1 & 3 & -3 & 1 \\ -4 & 6 & 2 & -4 \end{pmatrix}$. What is the rank of $N$?

**A.** 1
**B.** 2
**C.** 3
**D.** 4

**3.** Let $T : \mathbb{R}^2 \to \mathbb{R}^3$ be a linear mapping such that

$$M_{\mathfrak{B}}^{\mathfrak{B}'}(T) = \begin{pmatrix} 1 & 1 \\ 0 & 2 \\ 1 & -2 \end{pmatrix} \quad \text{and } \mathfrak{B}' = \{(1,0,-1),(1,1,1),(0,0,1)\}$$

What is $T(4,-2)$?

**A.** $(-2,-4,2)$
**B.** $(-2,-4,-4)$
**C.** $(2,-4,8)$
**D.** $(2,4,2)$

**4.** Suppose $\mathfrak{B} = \{(1,0),(0,1)\}$ and $\mathfrak{B}' = \{(2,-1),(1,-3)\}$. What is the change-of-basis matrix from $\mathfrak{B}$ to $\mathfrak{B}'$?

**A.** $\begin{pmatrix} \frac{3}{5} & -\frac{1}{5} \\ \frac{1}{5} & -\frac{2}{5} \end{pmatrix}$
**B.** $\begin{pmatrix} \frac{3}{5} & \frac{1}{5} \\ -\frac{1}{5} & -\frac{2}{5} \end{pmatrix}$
**C.** $\begin{pmatrix} 1 & 2 \\ -3 & -1 \end{pmatrix}$
**D.** $\begin{pmatrix} 2 & 1 \\ -1 & 3 \end{pmatrix}$

**5.** Suppose $M_{\mathfrak{B}}^{\mathfrak{B}'}(id) = \begin{pmatrix} 1 & -3 \\ -2 & 4 \end{pmatrix}$. What is $M_{\mathfrak{B}'}^{\mathfrak{B}}(\mathrm{id})$?

**A.** $\begin{pmatrix} -2 & -\frac{3}{2} \\ -1 & -\frac{1}{2} \end{pmatrix}$
**B.** $\begin{pmatrix} -\frac{1}{10} & \frac{3}{10} \\ -\frac{1}{5} & -\frac{2}{5} \end{pmatrix}$

C. $\begin{pmatrix} \frac{2}{5} & \frac{3}{10} \\ \frac{1}{5} & \frac{1}{10} \end{pmatrix}$

D. $\begin{pmatrix} \frac{1}{2} & \frac{3}{2} \\ 1 & 2 \end{pmatrix}$

**6.** Suppose for some transformation $T$, $M_{\mathfrak{B}}^{\mathfrak{B}}(T) = \begin{pmatrix} 1 & 2 \\ 0 & -5 \end{pmatrix}$, where $\mathfrak{B}$ is the standard basis. Which matrix represents $M_{\mathfrak{B}'}^{\mathfrak{B}'}(T)$ when $\mathfrak{B}' = \{(-1,3),(2,4)\}$?

A. $\begin{pmatrix} -\frac{21}{5} & -\frac{52}{5} \\ -\frac{2}{5} & \frac{1}{5} \end{pmatrix}$

B. $\begin{pmatrix} -\frac{16}{5} & -\frac{7}{5} \\ -\frac{27}{5} & -\frac{4}{5} \end{pmatrix}$

C. $\begin{pmatrix} -5 & -8 \\ 0 & 1 \end{pmatrix}$

D. $\begin{pmatrix} -3 & -2 \\ -4 & -1 \end{pmatrix}$

## Open Response

**7.** Let $M = \begin{pmatrix} -1 & 1 & 0 & 1 \\ 2 & 1 & 3 & 4 \\ 1 & 2 & 3 & 5 \\ -3 & 0 & -3 & -3 \end{pmatrix}$. For each of the given vector spaces, sift out a basis from the given generating system.

**a.** $V$ is the row space of $M$, starting with the rows.
**b.** $V$ is the column space of $M$, starting with the columns.

**8.** Let $M = \begin{pmatrix} 1 & -2 & 5 \\ 3 & -1 & 0 \\ 2 & 1 & -5 \\ 5 & -5 & 10 \end{pmatrix}$.

**a.** Find $r(M)$.
**b.** Find $\dim \ker(M)$.
**c.** Find a basis for $\ker(M)$.
**d.** Find a basis for the row space of $M$.
**e.** Find a basis for the column space of $M$.

**9.** Suppose $D : \mathbb{R}^3 \to \mathbb{R}^2$ with respect to the standard bases is defined by $D(x,y,z) = (x-y, 2z)$.

**a.** Show that $D$ is linear.
**b.** Find a matrix $M$ so that, for any vector $v$ in $\mathbb{R}^3$, $M\underline{v}_{\mathfrak{B}} = \underline{D(v)}_{\mathfrak{B}'}$.
**c.** Let $\mathfrak{B} = \{(1,0,1),(0,-1,1),(1,-1,0)\}$ and $\mathfrak{B}' = \{(1,1),(0,2)\}$. Find $N = M_{\mathfrak{B}}^{\mathfrak{B}'}(D)$.

**10.** Suppose $\mathfrak{B} = \{(1,0),(0,1)\}$ and $\mathfrak{B}' = \{(4,-2),(1,-1)\}$. And, suppose $T : \mathbb{R}^2 \to \mathbb{R}^2$ is linear and

$$T(1,0) = (4,-5)$$

$$T(0,1) = (-2,3)$$

**a.** Find the change of basis matrix from $\mathfrak{B}$ to $\mathfrak{B}'$.
**b.** Find the change of basis matrix from $\mathfrak{B}'$ to $\mathfrak{B}$.
**c.** Find $M$, the matrix for $T$ with respect to the standard basis $\mathfrak{B}$, and $N$, the matrix that represents this transformation relative to the nonstandard basis $\mathfrak{B}'$.

**11.** Show that $M = \begin{pmatrix} -1 & 4 \\ 2 & 1 \end{pmatrix}$ and $N = \begin{pmatrix} 15 & 36 \\ -6 & -15 \end{pmatrix}$ are similar matrices.

CHAPTER

# 9 Determinants and Eigentheory

You have already met determinants of $2 \times 2$ and $3 \times 3$ matrices in several contexts.

- In Lesson 2.5, you learned that the cross product of two vectors $A$ and $B$ in $\mathbb{R}^3$ can be calculated with determinants of $2 \times 2$ matrices.
- In Lesson 5.4, you used what you knew about cross products to deduce that the area of the parallelogram spanned by two vectors in the plane must be equal to the absolute value of the determinant of the matrix formed by putting the two vectors in the columns of a matrix.
- In that same lesson, you learned that the determinant of a $2 \times 2$ matrix tells you what the associated linear map does to areas, and you proved many useful properties of $2 \times 2$ determinants.
- And if you did Exercise 14 from Lesson 5.4, you learned to calculate the volume of a parallelepiped determined by three vectors $A$, $B$, and $C$ by computing the absolute value of the determinant of a $3 \times 3$ matrix.

$\longleftarrow$
The cross product is orthogonal to both $A$ and $B$, and its length is equal to the area of the parallelogram spanned by the two vectors.

$\longleftarrow$
If you don't remember all of them, look back at Theorems 5.7, 5.8, and 5.10. For example, you learned that, for $2 \times 2$ matrices $A$ and $B$, $\det(AB) = \det A \det B$.

**Remember**

If the columns of a $3 \times 3$ matrix are the vectors $A$, $B$, and $C$ (in that order), then the determinant of the matrix is $C \cdot (A \times B)$.

Not surprisingly, it's time to use the extension program to define determinants for $n \times n$ matrices and to use that extension as the definition of volume in $\mathbb{R}^n$. In this chapter, you'll learn how to calculate determinants of any size and see which of the properties continue to hold in higher dimensions. And determinants will play a prominent role as you tie together some loose threads regarding fixed vectors and fixed lines for matrices.

**By the end of this chapter, you will be able to answer questions like these:**

1. How can you find a vector in $\mathbb{R}^n$ orthogonal to $n-1$ other vectors?
2. How can you extend the definition of volume to $\mathbb{R}^n$?
3. Diagonalize the matrix $A = \begin{pmatrix} 3 & -1 \\ 4 & -2 \end{pmatrix}$.

**You will build good habits and skills for ways to**

- use the familiar to extend
- use functional notation
- use general purpose tools
- pay close attention to definitions
- find ways to simplify
- use the equivalence of different properties
- plan a general proof
- look for connections and structural similarities

**Vocabulary and Notation**

- algebraic multiplicity
- characteristic polynomial
- determinant
- eigenvalue
- eigenvector
- generalized cross product
- geometric multiplicity
- height
- $M$-invariant subspace
- minors of a matrix
- parallelepiped
- probability vector
- special orthogonal vector

# 9.1 Getting Started

## Exercises

For Exercises 1–3, find the determinant of each matrix.

← As usual, there are hidden theorems in these calculations. What are they?

**1.** **a.** $\begin{vmatrix} a & c \\ b & d \end{vmatrix}$ **b.** $\begin{vmatrix} c & a \\ d & b \end{vmatrix}$ **c.** $\begin{vmatrix} a & b \\ c & d \end{vmatrix}$

**2.** **a.** $\begin{vmatrix} 3 & 0 & 1 \\ 1 & 4 & 0 \\ 2 & -3 & 1 \end{vmatrix}$ **b.** $\begin{vmatrix} 0 & 3 & 1 \\ 4 & 1 & 0 \\ -3 & 2 & 1 \end{vmatrix}$ **c.** $\begin{vmatrix} 1 & 0 & 3 \\ 0 & 4 & 1 \\ 1 & -3 & 2 \end{vmatrix}$

**3.** **a.** $\begin{vmatrix} a & d & 2a \\ b & e & 2b \\ c & f & 2c \end{vmatrix}$ **b.** $\begin{vmatrix} a & d & a+d \\ b & e & b+e \\ c & f & c+f \end{vmatrix}$

**4.** Find a nonzero vector that is orthogonal to the three given vectors $P_1 = (1, 1, 0, 1)$, $P_2 = (1, 0, 0, 1)$, and $P_3 = (0, 0, 1, 0)$.

**5.** Define the volume of the 4-dimensional "box" spanned by $P_1$, $P_2$, and $P_3$ to be volume of base (the volume of the parallelepiped spanned by $P_1$, $P_2$, and $P_3$) times height. Find a way to calculate the volume of the "box" spanned by the vectors from Exercise 4 and the vector $Q = (2, 7, -9, 3)$.

← How does this relate to Exercise 4 above? Look back at Exercise 14 from Lesson 5.4 for a hint.

**6.** Consider the vector of consecutive integers in $\mathbb{R}^n$:

$$A = (1, 2, 3, 4, \dots, n-1, n)$$

A **transposition** or **flip** is a switch of two adjacent entries. You can move 4 to the first position with three flips:

← Counting flips will be important in Lesson 9.3.

$$\begin{aligned}(1, 2, 3, 4, \dots, n-1, n) &\overset{(1)}{\to} (1, 2, \mathbf{4}, \mathbf{3}, \dots, n) \\ &\overset{(2)}{\to} (1, \mathbf{4}, \mathbf{2}, 3, \dots, n) \overset{(3)}{\to} (\mathbf{4}, \mathbf{1}, 2, 3, \dots, n)\end{aligned}$$

How many flips does it take to

**a.** move 9 to the first position?
**b.** interchange 1 and 9?
**c.** move $j$ to the first position?
**d.** interchange $i$ and $j$ (where $i < j$)?

**7.** Find $\lambda$ if the determinant of the given matrix is 0.

**a.** $\begin{vmatrix} 1-\lambda & 2 \\ -1 & 4-\lambda \end{vmatrix}$ **b.** $\begin{vmatrix} 1-\lambda & 0 & 0 \\ 0 & -\lambda & 1 \\ 0 & -4 & -4-\lambda \end{vmatrix}$

**8.** For the given matrices,

1. determine if the given matrix has any nonzero fixed vectors (that is, vectors where $AX = X$)

2. determine if any nonzero vectors satisfy $AX = 2X$
3. find all nonzero vectors that satisfy $AX = cX$ for any real number $c$

←
In part 3. such a vector would determine a fixed line for the matrix.

**a.** $A = \begin{pmatrix} 1 & 2 \\ -1 & 4 \end{pmatrix}$ **b.** $A = \begin{pmatrix} 1 & 0 & 0 \\ 0 & 0 & 1 \\ 0 & -4 & -4 \end{pmatrix}$

←
How do these matrices compare with the ones in Exercise 7? Hmm.

## 9.2 Determinants

You have learned how to find determinants of $2 \times 2$ and $3 \times 3$ matrices. It is now time to generalize the definition. The first step to generalizing the definition of determinants to $n \times n$ matrices is to understand better what's going on in the more familiar $2 \times 2$ and $3 \times 3$ cases.

**In this lesson, you will learn how to**

- extend the definition of determinant to $n \times n$ matrices
- use a recursive algorithm to find the determinant of a matrix
- evaluate a determinant along any row or column
- develop the basic rules for determinants

**Minds in Action** Episode 40

*Derman, Sasha, and Tony are talking about Exercise* 2 *from Lesson* 9.1.

TONY: Hey, did you guys notice that all three answers are the same, except for the sign?

DERMAN: Of course they are! The matrices are the same.

TONY: What do you mean? They're not the same. I mean, they're kind of the same, but . . .

SASHA: No, Derman's right. Remember when we first did determinants like this? We were finding the volume of the figure spanned by the three column vectors. The column vectors are the same, just switched around, so . . .

TONY: So the volumes would have to be the same because it's the same figure! Of course! That's why only the sign can change. The absolute value of the determinant can't change.

DERMAN: That's what I said.

TONY: But why does the sign change when two columns are switched?

SASHA: Well, look at Exercise 1. The same thing happened in the $2 \times 2$ case, and since we calculated with variables, that's actually a proof.

TONY: Yeah, but I still don't see why.

SASHA: Hmm. We learned that if the three columns of a matrix are the vectors $A$, $B$, and $C$, then we can calculate the determinant as $C \cdot (A \times B)$. What happens if we switch the first two columns to make a new matrix?

TONY: The new matrix would have columns $B$, $A$, and $C$, so its determinant is $C \cdot (B \times A)$. Of course! We already know that $A \times B = -(B \times A)$. But I don't see why switching the first and last column would change the sign. I guess we just have to calculate like we did for $2 \times 2$ matrices . . . work it out in general.

$\longleftarrow$ $A \times B = -(B \times A)$ because of the right-hand rule!

*Tony starts writing out some matrices.*

The three friends have figured out an important fact about calculating determinants. It's an idea that will form the basis for calculating more general determinants, so it deserves to be stated as a theorem.

**Theorem 9.1**

*Let $A = (a_1, a_2, a_3)$, $B = (b_1, b_2, b_3)$, and $C = (c_1, c_2, c_3)$ be three vectors. Form four matrices whose columns are these vectors:*

$$M_1 = (ABC) = \begin{pmatrix} a_1 & b_1 & c_1 \\ a_2 & b_2 & c_2 \\ a_3 & b_3 & c_3 \end{pmatrix}, \qquad M_2 = (BAC) = \begin{pmatrix} b_1 & a_1 & c_1 \\ b_2 & a_2 & c_2 \\ b_3 & a_3 & c_3 \end{pmatrix}$$

$$M_3 = (CBA) = \begin{pmatrix} c_1 & b_1 & a_1 \\ c_2 & b_2 & a_2 \\ c_3 & b_3 & a_3 \end{pmatrix}, \qquad M_4 = (ACB) = \begin{pmatrix} a_1 & c_1 & b_1 \\ a_2 & c_2 & b_2 \\ a_3 & c_3 & b_3 \end{pmatrix}$$

*Then $\det(M_1) = -\det(M_2)$, $\det(M_1) = -\det(M_3)$, and $\det(M_1) = -\det(M_4)$. In other words, interchanging two columns of a $3 \times 3$ matrix has the effect of multiplying the determinant by $-1$.*

***Proof.*** Sasha, Tony, and Derman have already provided a proof of the fact that $\det(M_1) = -\det(M_2)$. Here's the second case:

$$\begin{aligned} \det(M_1) &= C \cdot (A \times B) \\ &= c_1 \begin{vmatrix} a_2 & b_2 \\ a_3 & b_3 \end{vmatrix} - c_2 \begin{vmatrix} a_1 & b_1 \\ a_3 & b_3 \end{vmatrix} + c_3 \begin{vmatrix} a_1 & b_1 \\ a_2 & b_2 \end{vmatrix} \\ &= c_1(a_2b_3 - a_3b_2) - c_2(a_1b_3 - a_3b_1) + c_3(a_1b_2 - a_2b_1) \end{aligned}$$

On the other hand,

$$\begin{aligned} \det(M_3) &= A \cdot (C \times B) \\ &= a_1 \begin{vmatrix} c_2 & b_2 \\ c_3 & b_3 \end{vmatrix} - a_2 \begin{vmatrix} c_1 & b_1 \\ c_3 & b_3 \end{vmatrix} + a_3 \begin{vmatrix} c_1 & b_1 \\ c_2 & b_2 \end{vmatrix} \\ &= a_1(c_2b_3 - c_3b_2) - a_2(c_1b_3 - c_3b_1) + a_3(c_1b_2 - c_2b_1) \end{aligned}$$

Now, compare term by term: $\det(M_1)$ contains the term $a_1b_2c_3$, and $\det(M_3)$ contains the term $-a_1b_2c_3$; $\det(M_1)$ contains the term $-a_1b_3c_2$, and $\det(M_3)$ contains the term $a_1b_3c_2$; and so on. ∎

**For Discussion**

1. Finish the proof of Theorem 9.1. First, check that the terms appearing in the expansion of $\det(M_3)$ are exactly the negatives of the terms in the expansion of $\det(M_1)$. Then find a proof for the last claim—that $\det(M_1) = -\det(M_4)$.

The real utility of Theorem 9.1 is that it allows you to calculate the determinant of a $3 \times 3$ matrix using any column. If $A$, $B$, and $C$ are three vectors in $\mathbb{R}^3$, then there are three related matrices:

$$(ABC) \overset{\text{switch } B \text{ and } C}{\longrightarrow} (ACB) \overset{\text{switch } A \text{ and } B}{\longrightarrow} (BCA)$$

Theorem 9.1 says the determinants of these three matrices satisfy

$$\det(ABC) = -\det(ACB) = \det(BCA)$$

since two columns were interchanged to go from $(ABC)$ to $(ACB)$, and then two more columns were interchanged to go from $(ACB)$ to $(CBA)$.

Writing these three determinants out more explicitly gives

$$\begin{vmatrix} a_1 & b_1 & c_1 \\ a_2 & b_2 & c_2 \\ a_3 & b_3 & c_3 \end{vmatrix} = c_1 \begin{vmatrix} a_2 & b_2 \\ a_3 & b_3 \end{vmatrix} - c_2 \begin{vmatrix} a_1 & b_1 \\ a_3 & b_3 \end{vmatrix} + c_3 \begin{vmatrix} a_1 & b_1 \\ a_2 & b_2 \end{vmatrix}$$

$$= -b_1 \begin{vmatrix} a_2 & c_2 \\ a_3 & c_3 \end{vmatrix} + b_2 \begin{vmatrix} a_1 & c_1 \\ a_3 & c_3 \end{vmatrix} - b_3 \begin{vmatrix} a_1 & c_1 \\ a_2 & c_2 \end{vmatrix}$$

$$= a_1 \begin{vmatrix} b_2 & c_2 \\ b_3 & c_3 \end{vmatrix} - a_2 \begin{vmatrix} b_1 & c_1 \\ b_3 & c_3 \end{vmatrix} + a_3 \begin{vmatrix} b_1 & c_1 \\ b_2 & c_2 \end{vmatrix}$$

← The first row is $C \cdot (A \times B)$. The second row is $-B \cdot (A \times C)$. The third row is $A \cdot (B \times C)$.

So when you calculate a determinant of a $3 \times 3$ matrix, you can evaluate it along any of the three columns, as long as you are careful to keep the signs straight. This can often make the work of calculating $3 \times 3$ determinants much easier.

## Example 1

**Problem.** Find each determinant.

**a.** $\begin{vmatrix} 1 & 2 & 1 \\ 3 & 0 & 2 \\ 4 & 2 & 3 \end{vmatrix}$ **b.** $\begin{vmatrix} 3 & 0 & 1 \\ 4 & 1 & 0 \\ -3 & 2 & 1 \end{vmatrix}$

← In other courses, you may have learned other methods for evaluating some determinants. See, for example, Exercise 6.

**Solution.**

**a.** Notice that the matrix has a 0 in the second column. That means that if you expand along the second column, one of the terms will disappear, making the calculation shorter.

$$\begin{vmatrix} 1 & 2 & 1 \\ 3 & 0 & 2 \\ 4 & 2 & 3 \end{vmatrix} = -2 \begin{vmatrix} 3 & 2 \\ 4 & 3 \end{vmatrix} + 0 \begin{vmatrix} 1 & 1 \\ 4 & 3 \end{vmatrix} - 2 \begin{vmatrix} 1 & 1 \\ 3 & 2 \end{vmatrix} = -2(9-8) - 2(2-3) = 0$$

**b.** This matrix has zeros in both the second and third columns, making either one a good choice. Here's the calculation along the third column:

$$\begin{vmatrix} 3 & 0 & 1 \\ 4 & 1 & 0 \\ -3 & 2 & 1 \end{vmatrix} = 1 \begin{vmatrix} 4 & 1 \\ -3 & 2 \end{vmatrix} - 0 \begin{vmatrix} 3 & 0 \\ -3 & 2 \end{vmatrix} + 1 \begin{vmatrix} 3 & 0 \\ 4 & 1 \end{vmatrix} = (8+3) + (3-0) = 14$$

← Try calculating this determinant along the second column. Do you get the same answer?

Notice how the signs in the expansion are different in the two examples above. This "sign matrix" might help you keep track of which sign to use when you expand a determinant. Along the first and third columns, the expansion is $+, -, +$. Along the second column, it is $-, +, -$.

$$\begin{vmatrix} + & - & + \\ - & + & - \\ + & - & + \end{vmatrix}$$

← Why is this correct? Look back at the text before Example 1 if you're not sure.

### For You to Do

**2.** Calculate each determinant. Pick your columns wisely.

**a.** $\begin{vmatrix} 0 & 1 & 2 \\ 3 & 1 & -1 \\ 0 & 1 & 0 \end{vmatrix}$ **b.** $\begin{vmatrix} 1 & 2 & 0 \\ -1 & 3 & 0 \\ 1 & 4 & 0 \end{vmatrix}$

This method of "evaluating along a column" will become the definition of determinants for $n \times n$ matrices.

### Developing Habits of Mind

**The extension program.** Thinking back, the determinant of a $3\times3$ matrix was defined as the dot product of the third column with a "distinguished" vector orthogonal to the first two columns: if $A_{*1}$, $A_{*2}$, and $A_{*3}$ are columns of a $3 \times 3$ matrix $A$, then

$$\begin{aligned}\det A &= A_{*3} \cdot (A_{*1} \times A_{*2}) \\ &= A_{*3} \cdot \left( \begin{vmatrix} \bullet & \bullet \\ \bullet & \bullet \end{vmatrix}, - \begin{vmatrix} \bullet & \bullet \\ \bullet & \bullet \end{vmatrix}, \begin{vmatrix} \bullet & \bullet \\ \bullet & \bullet \end{vmatrix} \right)\end{aligned}$$

where each of the $2 \times 2$ matrices of bullets is obtained from $A$ in the usual way that you form the cross product—crossing out one row and one column.

One of the goals of this chapter is to see how you can use this same idea to define the determinant of, say, a $4 \times 4$ matrix whose columns are $A_{*1}$, $A_{*2}$, $A_{*3}$, and $A_{*4}$ as a dot product

$$-A_{*4} \cdot \left( \begin{vmatrix} \bullet & \bullet & \bullet \\ \bullet & \bullet & \bullet \\ \bullet & \bullet & \bullet \end{vmatrix}, - \begin{vmatrix} \bullet & \bullet & \bullet \\ \bullet & \bullet & \bullet \\ \bullet & \bullet & \bullet \end{vmatrix}, \begin{vmatrix} \bullet & \bullet & \bullet \\ \bullet & \bullet & \bullet \\ \bullet & \bullet & \bullet \end{vmatrix}, - \begin{vmatrix} \bullet & \bullet & \bullet \\ \bullet & \bullet & \bullet \\ \bullet & \bullet & \bullet \end{vmatrix} \right)$$

where the vector of $3 \times 3$ determinants is a vector orthogonal to $A_{*1}$, $A_{*2}$, and $A_{*3}$ and whose entries are obtained with the same method you used to construct cross products. And you'll generalize this to $n \times n$ matrices. This will be a long and somewhat technical process, so pull out your pencil.

Why the minus sign in front of $A_{*4}$? Stay tuned.

Actually, you'll see that there's nothing special about the fourth column here. You'll develop a method that will work for any column, always giving the same answer. See Theorem 9.1 for a preview.

The following definition for smaller matrices will prove useful moving forward.

#### Definition

Let

$$A = \begin{pmatrix} A_{11} & A_{12} & \cdots & A_{1n} \\ A_{21} & A_{22} & \cdots & A_{2n} \\ \vdots & \vdots & \ddots & A_{1n} \\ A_{n1} & A_{n2} & \cdots & A_{nn} \end{pmatrix}$$

be an $n \times n$ matrix. The **minors** of $A$ are the $(n-1) \times (n-1)$ square matrices formed by deleting one row and one column from $A$. The matrix $M_{ij}$ is the minor formed by deleting row $i$ and column $j$.

**Example 2**

**Problem.** Let $A = \begin{pmatrix} 1 & 2 & 0 \\ -1 & 3 & 0 \\ 1 & 4 & 0 \end{pmatrix}$. Find $M_{11}$ and $M_{23}$.

**Solution.** To find $M_{11}$, delete the first row and first column from $A$.

$$\begin{pmatrix} \square & \square & \square \\ \square & 3 & 0 \\ \square & 4 & 0 \end{pmatrix}, \text{ so } M_{11} = \begin{pmatrix} 3 & 0 \\ 4 & 0 \end{pmatrix}.$$

To find $M_{23}$, delete the second row and third column from $A$.

$$\begin{pmatrix} 1 & 2 & \square \\ \square & \square & \square \\ 1 & 4 & \square \end{pmatrix}, \text{ so } M_{23} = \begin{pmatrix} 1 & 2 \\ 1 & 4 \end{pmatrix}.$$

**For Discussion**

**3.** Find the minors $M_{13}$ and $M_{21}$ for the matrix $A$ given in Example 2 above.

**Facts and Notation**

The notion of matrix minors allows a more succinct description for calculating determinants of $3 \times 3$ matrices. Let $A = \begin{pmatrix} A_{11} & A_{12} & A_{13} \\ A_{21} & A_{22} & A_{23} \\ A_{31} & A_{32} & A_{33} \end{pmatrix}$. Calculating $\det(A)$ along the first column gives

$$\det(A) = A_{11}\,|M_{11}| - A_{21}\,|M_{21}| + A_{31}\,|M_{31}|$$

Calculating $\det(A)$ along the second column gives

$$\det(A) = -A_{12}\,|M_{12}| + A_{22}\,|M_{22}| - A_{32}\,|M_{32}|$$

Finally, calculating $\det(A)$ along the third column gives

$$\det(A) = A_{13}\,|M_{13}| - A_{23}\,|M_{23}| + A_{33}\,|M_{33}|$$

Notice that this agrees with the rule for calculating $2 \times 2$ determinants as well. Let $A = \begin{pmatrix} A_{11} & A_{12} \\ A_{21} & A_{22} \end{pmatrix}$. Calculating $\det(A)$ along the first column gives

$$\det(A) = A_{11}\,|M_{11}| - A_{21}\,|M_{21}| = A_{11}A_{22} - A_{21}A_{12}$$

Calculating $\det(A)$ along the second column gives

$$\det(A) = -A_{12}\,|M_{12}| + A_{22}\,|M_{22}| = -A_{12}A_{21} + A_{22}A_{11}$$

$\longleftarrow$ Why is $|M_{11}| = A_{22}$? Check the other minors as well.

### Developing Habits of Mind

**Notice similarities.** Note that each of the three expansions of the $3 \times 3$ determinant is a dot product.

Along the first column: $\det(A) = (A_{11}, A_{21}, A_{31}) \cdot (|M_{11}|, -|M_{21}|, |M_{11}|)$

Along the second column: $\det(A) = -(A_{12}, A_{22}, A_{32}) \cdot (|M_{12}|, -|M_{22}|, |M_{12}|)$

Along the third column: $\det(A) = (A_{13}, A_{23}, A_{33}) \cdot (|M_{13}|, -|M_{23}|, |M_{13}|)$

Each of these vectors of minors is a cross product. What do you cross to get each one?

This leads naturally to a method for calculating determinants for larger matrices: pick a column of the matrix (ideally one with nice numbers or zeros in it). Multiply each entry in that column by the determinant of the corresponding matrix minor. Add the terms together, with signs alternating as given in the "sign matrix" below:

⟵ Can you figure out a formula for $S_{ij}$ in terms of $i$ and $j$?

$$S = \begin{pmatrix} + & - & + & - & \\ - & + & - & + & \cdots \\ + & - & + & - & \\ - & + & - & + & \\ & & \vdots & & \ddots \end{pmatrix}$$

### Example 3

**Problem.** Find the value of

$$\begin{vmatrix} -1 & 0 & 2 & 1 \\ -1 & 1 & -4 & 1 \\ 2 & 1 & 0 & -2 \\ 0 & 3 & 5 & 3 \end{vmatrix}$$

**Solution.** Evaluating along the first column gives

$$-1\,|M_{11}| - (-1)\,|M_{21}| + 2\,|M_{31}| - 0\,|M_{41}|$$

The last term will vanish, so this simplifies to

$$-1\begin{vmatrix} 1 & -4 & 1 \\ 1 & 0 & -2 \\ 3 & 5 & 3 \end{vmatrix} + 1\begin{vmatrix} 0 & 2 & 1 \\ 1 & 0 & -2 \\ 3 & 5 & 3 \end{vmatrix} + 2\begin{vmatrix} 0 & 2 & 1 \\ 1 & -4 & 1 \\ 3 & 5 & 3 \end{vmatrix}$$

There are three determinants of a $3 \times 3$ matrix to evaluate, just like you've been doing.

$$\begin{vmatrix} 1 & -4 & 1 \\ 1 & 0 & -2 \\ 3 & 5 & 3 \end{vmatrix} = 4\begin{vmatrix} 1 & -2 \\ 3 & 3 \end{vmatrix} + 0\begin{vmatrix} 1 & 1 \\ 3 & 3 \end{vmatrix} - 5\begin{vmatrix} 1 & 1 \\ 1 & -2 \end{vmatrix}$$
$$= 4(3+6) - 5(-2-1) = 51$$

⟵ Why is the second column a good choice for this matrix?

$$\begin{vmatrix} 0 & 2 & 1 \\ 1 & 0 & -2 \\ 3 & 5 & 3 \end{vmatrix} = 0\begin{vmatrix} 0 & -2 \\ 5 & 3 \end{vmatrix} - 1\begin{vmatrix} 2 & 1 \\ 5 & 3 \end{vmatrix} + 3\begin{vmatrix} 2 & 1 \\ 0 & -2 \end{vmatrix}$$
$$= -1(6-5) + 3(-4-0) = -13$$

$$\begin{vmatrix} 0 & 2 & 1 \\ 1 & -4 & 1 \\ 3 & 5 & 3 \end{vmatrix} = 0\begin{vmatrix} -4 & 1 \\ 5 & 3 \end{vmatrix} - 1\begin{vmatrix} 2 & 1 \\ 5 & 3 \end{vmatrix} + 3\begin{vmatrix} 2 & 1 \\ -4 & 1 \end{vmatrix}$$
$$= -1(6-5) + 3(2+4) = 17$$

⟵ Work these out for yourself. You'll be glad that you did.

So the $4 \times 4$ determinant is $-1(51) + 1(-13) + 2(17) = -30$.

### Definition

Let $A$ be an $n$-by-$n$ square matrix.

$$A = \begin{pmatrix} A_{11} & A_{12} & \cdots & A_{1n} \\ A_{21} & A_{22} & \cdots & A_{2n} \\ \vdots & \vdots & \ddots & \vdots \\ A_{n1} & A_{n2} & \cdots & A_{nn} \end{pmatrix}.$$

Pick an arbitrary column of $A$, say $A_{*j}$. Then

$$\det(A) = (-1)^{j+1}\left(A_{1j}\,|M_{1j}| - A_{2j}\,|M_{2j}| + \cdots + (-1)^{n+1}A_{nj}\,|M_{nj}|\right).$$

### For Discussion

**4.** What is the purpose of the $(-1)^{j+1}$ in front of the sum above? What is the purpose of the $(-1)^{n+1}$ on the last term of the sum?

### Developing Habits of Mind

**Use functional notation.** It's useful to think of the determinant as a function defined on the columns of $A$, because they are vectors in $\mathbb{R}^n$ and there's a built-in arithmetic of vectors that you've developed over the last seven chapters. One goal of this chapter is to investigate the interaction of the determinant with that arithmetic.

### For You to Do

**5.** Calculate the determinants. Think about which columns are the most convenient to use.

**a.** $\begin{vmatrix} 2 & 0 & 0 & 0 \\ 5 & 3 & 0 & 1 \\ 4 & 4 & 1 & 0 \\ 7 & -3 & 2 & 1 \end{vmatrix}$

**b.** $\begin{vmatrix} 5 & 3 & 2 & 8 & 7 \\ 0 & 1 & 6 & -1 & 4 \\ 0 & 0 & 8 & 0 & 7 \\ 0 & 0 & 0 & 9 & 3 \\ 2 & 0 & 0 & 0 & 0 \end{vmatrix}$

**Minds in Action** Episode 41

SASHA: Something doesn't make sense.

DERMAN: Lots of things don't make sense. Like what does "easy as pie" mean?

SASHA: No, I mean about determinants. The definition seems to depend on the column you pick.

TONY: Let's try to do it two different ways and see what happens.

---

**For You to Do**

**6.** Calcuate $\begin{vmatrix} 2 & 0 & 0 & 0 \\ 5 & 3 & 0 & 1 \\ 4 & 4 & 1 & 0 \\ 7 & -3 & 2 & 1 \end{vmatrix}$ using a different column than you used above. Do you get the same result?

---

**Developing Habits of Mind**

**Ensure consistency.** Sasha's concern that the definition of determinants depends on the column you pick is important. In the next lesson, you'll prove that it makes no difference which column you expand along to calculate the determinant: no matter what, you'll get the same result. Assume this fact for now.

$\longleftarrow$ In other words, the determinant is *well-defined*.

---

Theorems 5.7 and 5.8 describe many properties of $2 \times 2$ determinants. If the definition above for $n \times n$ determinants is a good one, these properties should still hold. The next theorem establishes four important properties for $n \times n$ determinants.

**Theorem 9.2 (Basic Rules of Determinants)**
*Let $A$ be an $n \times n$ matrix:*

$$A = \begin{pmatrix} A_{11} & A_{12} & \cdots & A_{1j} & \cdots & A_{1n} \\ A_{21} & A_{22} & \cdots & A_{2j} & \cdots & A_{2n} \\ \vdots & \vdots & & \vdots & \ddots & \vdots \\ A_{n1} & A_{n2} & \cdots & A_{nj} & \cdots & A_{nn} \end{pmatrix}.$$

*Determinants have the following properties:*

(1) *Let $I$ stand for the $n \times n$ identity matrix. Then* $\det(I) = 1$.

(2) *If you replace a column $A_{*j}$ with a scalar multiple of that column $cA_{*j}$, the determinant is multiplied by the same scalar $c$. In other*

**Remember**
The identity matrix has 1 along the diagonal and 0 everywhere else.

*words, if*

$$A' = \begin{pmatrix} A_{11} & A_{12} & \cdots & cA_{1j} & \cdots & A_{1n} \\ A_{21} & A_{22} & \cdots & cA_{2j} & \cdots & A_{2n} \\ \vdots & \vdots & & \vdots & \ddots & \vdots \\ A_{n1} & A_{n2} & \cdots & cA_{nj} & \cdots & A_{nn} \end{pmatrix},$$

*then* $\det(A') = c\det(A)$.

← "scalars come out."

(3) *If you rewrite one column of $A$ as a sum of two column vectors, then* $\det(A)$ *breaks up as a sum of two determinants. Specifically, say* $A_{*j} = A'_{*j} + A''_{*j}$ *and let*

$$A' = \begin{pmatrix} A_{11} & A_{12} & \cdots & A'_{1j}, & \cdots & A_{1n} \\ A_{21} & A_{22} & \cdots & A'_{2j} & \cdots & A_{2n} \\ \vdots & \vdots & & \vdots & \ddots & \vdots \\ A_{n1} & A_{n2} & \cdots & A'_{nj} & \cdots & A_{nn} \end{pmatrix}, \text{ and}$$

$$A'' = \begin{pmatrix} A_{11} & A_{12} & \cdots & A''_{1j}, & \cdots & A_{1n} \\ A_{21} & A_{22} & \cdots & A''_{2j} & \cdots & A_{2n} \\ \vdots & \vdots & & \vdots & \ddots & \vdots \\ A_{n1} & A_{n2} & \cdots & A''_{nj} & \cdots & A_{nn} \end{pmatrix}.$$

*Then* $\det(A) = \det(A') + \det(A'')$.

← "det is linear."

(4) *Interchanging two columns* $A_{*i}$ *and* $A_{*j}$ *has the effect of multiplying the determinant by* $-1$.

← "det is alternating."

***Proof.*** You will prove parts ((1)), ((2)), and ((3)) in Exercises 10, 12, and 13.

The proof of part ((4)) will follow from the first three parts along with some intermediate results that are derived for them. Follow along with the next few lemmas that lead to a proof of part ((4)). ■

**Lemma 9.3**

*If one column of a matrix is $O$, the determinant is* 0.

***Proof.*** You'll prove this in Exercise 11. ■

**Lemma 9.4**

*If two adjacent columns are switched, the determinant changes by a sign.*

The proof of Lemma 9.4 is easier to think about than to write down. But a concrete example helps you see what's going on.

Try it with

$$A = \begin{vmatrix} 1 & 2 & 3 & 4 \\ 5 & 6 & 7 & 8 \\ 9 & 10 & 11 & 12 \\ 13 & 14 & 15 & 16 \end{vmatrix} \quad \text{and} \quad B = \begin{vmatrix} 1 & 2 & 4 & 3 \\ 5 & 6 & 8 & 7 \\ 9 & 10 & 12 & 11 \\ 13 & 14 & 16 & 15 \end{vmatrix}$$

← Don't just crank it out; delay the evaluation to help see how the structure of your calculations could be used in a general proof.

Expand $A$ along column 4 and $B$ along column 3. The coefficients are the same and the minors are the same. The only difference is a change of signs (as defined by the sign matrix).

The proof in general follows exactly the same idea.

**Corollary 9.5**

*If two adjacent columns are equal, the determinant is* $0$.

***Proof.*** Suppose $A$ has two adjacent equal columns. Switch the identical columns and $\det A$ changes by a sign (by Lemma 9.4). But the matrix stays the same. So, $\det(A) = -\det(A)$. This implies that $\det(A) = 0$. ■

**Corollary 9.6**

*If* any *two columns are equal, the determinant is* $0$.

***Proof.*** Switch adjacent columns, one at a time, to bring the two equal columns right next to each other. The determinant changes by at most a sign. But the determinant of the resulting matrix is 0, so the determinant of the original is too. ■

**Theorem 9.7 (Part ((4)) of Theorem 9.2)**

*If you switch any two columns, the determinant changes by a sign.*

***Proof.*** Given a matrix $A$, suppose you want to switch columns $i$ and $j$. Form the matrix whose $i^{\text{th}}$ and $j^{\text{th}}$ columns are each $A_{*i} + A_{*j}$—the sum of the two columns you want to switch. Its determinant is 0 by Corollary 9.6. Then use linearity to break the determinant into a sum of four determinants, two of which are 0 because they have two identical columns, and the other two are $A$ and the matrix obtained from $A$ by switching two columns:

← If the subscripts get in the way, trace out the proof for a 6-by-6 matrix, switching columns, say, 2 and 5.

$$\begin{aligned}
0 &= \det(A_{*1}, A_{*2}, \dots, A_{*i} + A_{*j}, \dots, A_{*i} + A_{*j}, \dots, A_{*n}) \\
&= \det(A_{*1}, A_{*2}, \dots, A_{*i}, \dots, A_{*i} + A_{*j}, \dots, A_{*n}) \\
&\quad + \det(A_{*1}, A_{*2}, \dots, A_{*j}, \dots, A_{*i} + A_{*j}, \dots, A_{*n}) \\
&= \det(A_{*1}, A_{*2}, \dots, A_{*i}, \dots, A_{*i}, \dots, A_{*n}) \\
&\quad + \det(A_{*1}, A_{*2}, \dots, A_{*i}, \dots, A_{*j}, \dots, A_{*n}) \\
&\quad + \det(A_{*1}, A_{*2}, \dots, A_{*j}, \dots, A_{*i}, \dots, A_{*n}) \\
&\quad + \det(A_{*1}, A_{*2}, \dots, A_{*j}, \dots, A_{*j}, \dots, A_{*n})
\end{aligned}$$

In this last display, the first and third determinants are 0, so

$$\begin{aligned}
0 &= \det(A_{*1}, A_{*2}, \dots, A_{*i}, \dots, A_{*j}, \dots, A_{*n}) \\
&\quad + \det(A_{*1}, A_{*2}, \dots, A_{*j}, \dots, A_{*i}, \dots, A_{*n})
\end{aligned}$$

and the result follows. ■

The proof of Theorem 9.2 is now complete.

### Developing Habits of Mind

**Use general-purpose tools.** The proof of part ((4)) of Theorem 9.2 may seem like a slick trick, but this kind of reasoning—using the linearity of the determinant function—is a general-purpose method that can be used to establish many key properties of determinants. The next examples show how these ideas can be used.

The rules for determinants can often make the process of computing complicated determinants easier, if you take a moment to examine the matrix first.

### Example

**Problem.** Find each determinant.

**a.** $\begin{vmatrix} 1 & 5 & 2 \\ 1 & -1 & 2 \\ -3 & 1 & -6 \end{vmatrix}$.  **b.** $\begin{vmatrix} -1 & 0 & 2 & 1 \\ -1 & 1 & -4 & 1 \\ 2 & 1 & 0 & -2 \\ 0 & 3 & 5 & 3 \end{vmatrix}$

**Solution.**

**a.** Notice that the last column is exactly twice the first column, so the determinant must be 0.

**b.** Here, the first and last columns are not identical, but they are close enough to make the computation much easier. First, rewrite the last column as a sum, and then split up the determinant.

$$\begin{vmatrix} -1 & 0 & 2 & 1+0 \\ -1 & 1 & -4 & 1+0 \\ 2 & 1 & 0 & -2+0 \\ 0 & 3 & 5 & 0+3 \end{vmatrix} = \begin{vmatrix} -1 & 0 & 2 & 1 \\ -1 & 1 & -4 & 1 \\ 2 & 1 & 0 & -2 \\ 0 & 3 & 5 & 0 \end{vmatrix} + \begin{vmatrix} -1 & 0 & 2 & 0 \\ -1 & 1 & -4 & 0 \\ 2 & 1 & 0 & 0 \\ 0 & 3 & 5 & 3 \end{vmatrix}.$$

The first determinant is 0. (Why? Look at the first and last columns.) The second one has a column with only one nonzero term, so it makes sense to evaluate along that column.

$$\begin{aligned} \begin{vmatrix} -1 & 0 & 2 & 0 \\ -1 & 1 & -4 & 0 \\ 2 & 1 & 0 & 0 \\ 0 & 3 & 5 & 3 \end{vmatrix} &= 3\begin{vmatrix} -1 & 0 & 2 \\ -1 & 1 & -4 \\ 2 & 1 & 0 \end{vmatrix} \\ &= 3\left(1\begin{vmatrix} -1 & 2 \\ 2 & 0 \end{vmatrix} - 1\begin{vmatrix} -1 & 2 \\ -1 & -4 \end{vmatrix}\right) \\ &= 3(-4-6) = -30 \end{aligned}$$

### Example

Suppose

$$A = \begin{pmatrix} 1 & 5 & 9 & 6 \\ 2 & 6 & 10 & 8 \\ 3 & 7 & 11 & 10 \\ 4 & 8 & 12 & 12 \end{pmatrix}$$

Notice that the columns of $A$ are linearly dependent. In fact,

$$A_{*4} = 2A_{*1} - A_{*2} + A_{*3}$$

← You could have figured out this dependence in Chapter 3.

Then, using the results from this section, $\det(A)$ can be expressed as the sum of three determinants, and you can calculate like this:

$$\begin{aligned} \det A &= |A_{*1}, A_{*2}, A_{*3}, 2A_{*1} - A_{*2} + A_{*3}| \\ &= |A_{*1}, A_{*2}, A_{*3}, 2A_{*1}| \\ &\quad + |A_{*1}, A_{*2}, A_{*3}, -A_{*2}| \\ &\quad + |A_{*1}, A_{*2}, A_{*3}, A_{*3}| \quad \text{(part ((3)) of Theorem 9.2)} \\ &= 2\,|A_{*1}, A_{*2}, A_{*3}, A_{*1}| \\ &\quad - |A_{*1}, A_{*2}, A_{*3}, A_{*2}| \\ &\quad + |A_{*1}, A_{*2}, A_{*3}, A_{*3}| \quad \text{(part ((2)) of Theorem 9.2)} \\ &= 0 \quad \text{(each of these determinants has two equal columns)} \end{aligned}$$

---

More generally,

**Theorem 9.8**

*If the columns of a matrix are linearly dependent, its determinant is* 0.

← You'll prove this theorem in Exercise 16.

## Exercises

**1.** Evaluate each of the following determinants.

**a.** $\begin{vmatrix} 1 & 2 & 1 \\ 3 & 0 & 7 \\ 1 & 2 & 1 \end{vmatrix}$ **b.** $\begin{vmatrix} 1 & 4 & 2 \\ 3 & 2 & 1 \\ 1 & -1 & 1 \end{vmatrix}$ **c.** $\begin{vmatrix} 1 & 4 & 2 \\ 1 & -1 & 1 \\ 3 & 2 & 1 \end{vmatrix}$

**d.** $\begin{vmatrix} 2 & 8 & 4 \\ 3 & 2 & 1 \\ 1 & -1 & 1 \end{vmatrix}$ **e.** $\begin{vmatrix} 5 & 3 & 8 & 1 \\ 0 & 1 & 7 & 6 \\ 0 & 0 & -2 & 3 \\ 0 & 0 & 0 & 4 \end{vmatrix}$ **f.** $\begin{vmatrix} 1 & 0 & 0 & 0 \\ 0 & 1 & 0 & 0 \\ 0 & 0 & 1 & 0 \\ 0 & 0 & 0 & 1 \end{vmatrix}$

**g.** $\begin{vmatrix} 3 & 0 & 0 \\ 0 & 4 & 0 \\ 0 & 0 & 5 \end{vmatrix}$ **h.** $\begin{vmatrix} 1 & -1 & 1 \\ 0 & 1 & 6 \\ 3 & -2 & 1 \end{vmatrix}$ **i.** $\begin{vmatrix} 1 & 0 & 3 \\ -1 & 1 & -2 \\ 1 & 6 & 1 \end{vmatrix}$

**j.** $\begin{vmatrix} 0 & 0 & 0 & 3 \\ 0 & 0 & 2 & 0 \\ 0 & 0 & -1 & 0 \\ 5 & 0 & 0 & 0 \end{vmatrix}$

**2.** Evaluate the determinant

$$\begin{vmatrix} 1 & 4 & 7 & 3 & 2 \\ 5 & 1 & 3 & 8 & 7 \\ 0 & 0 & 1 & -1 & 3 \\ 0 & 0 & 2 & 0 & 1 \\ 0 & 0 & 1 & -1 & 3 \end{vmatrix}$$

along

**a.** the first column
**b.** the second column

**3.** What values of $a$ make the following determinant 0?

$$\begin{vmatrix} 1 & 0 & a \\ 0 & 1 & 0 \\ a & 0 & 1 \end{vmatrix}$$

**4.** Calculate each determinant. Look for shortcuts.

**a.** $\begin{vmatrix} 1 & 2 & 4 \\ 3 & 0 & 3 \\ 1 & -1 & 6 \end{vmatrix}$ **b.** $\begin{vmatrix} 1 & 4 & 2 \\ 3 & 3 & 0 \\ 1 & 6 & -1 \end{vmatrix}$ **c.** $\begin{vmatrix} 1 & 2 & 8 \\ 3 & 0 & 6 \\ 1 & -1 & 12 \end{vmatrix}$

**5.** Calculate each determinant. Look for shortcuts.

**a.** $\begin{vmatrix} 1 & 3 & -1 & 3 \\ 0 & 1 & 0 & 0 \\ 1 & 0 & 0 & 5 \\ 2 & 4 & 5 & -1 \end{vmatrix}$ **b.** $\begin{vmatrix} 1 & -1 & 3 & 3 \\ 0 & 0 & 1 & 0 \\ 1 & 0 & 0 & 5 \\ 2 & 5 & 4 & -1 \end{vmatrix}$

**c.** $\begin{vmatrix} 2 & -1 & 3 & 3 \\ 0 & 0 & 1 & 0 \\ 2 & 0 & 0 & 5 \\ 4 & 5 & 4 & -1 \end{vmatrix}$ **d.** $\begin{vmatrix} 1 & 6 & -2 & 3 \\ 0 & 2 & 0 & 0 \\ 1 & 0 & 0 & 5 \\ 2 & 8 & 10 & -1 \end{vmatrix}$

**e.** $\begin{vmatrix} 1 & 3 & 6 & 3 \\ 0 & 1 & 2 & 0 \\ 1 & 0 & 0 & 5 \\ 2 & 4 & 8 & -1 \end{vmatrix}$ **f.** $\begin{vmatrix} 1 & 6 & -2 & 4 \\ 0 & 2 & 0 & 0 \\ 1 & 0 & 0 & 6 \\ 2 & 8 & 10 & 1 \end{vmatrix}$

**6.** You may have learned another technique for calculating determinants of $3 \times 3$ matrices, often called the "basket weave" method.

**a.** How does the basket weave method work?

**b.** Does it work for any matrices besides $3 \times 3$?

← If you don't know what the method is, you can find it on the Web—search for "basket weave determinants."

**7.** Suppose that $A = \begin{pmatrix} a & b & c \\ d & e & f \\ g & h & i \end{pmatrix}$ and that $\det(A) = 5$. Find each determinant.

**a.** $\det(4A)t$ **b.** $\begin{vmatrix} 2a & 3b & 4c \\ 2d & 3e & 4f \\ 2g & 3h & 4i \end{vmatrix}$

**c.** $\begin{vmatrix} b & c & a \\ e & f & d \\ h & i & g \end{vmatrix}$ **d.** $\begin{vmatrix} a & b & 2a \\ d & e & 2d \\ g & h & 2g \end{vmatrix}$

**e.** $\begin{vmatrix} a+2b & b & c \\ d+2e & e & f \\ g+2h & h & i \end{vmatrix}$ **f.** $\begin{vmatrix} a-b & b-c & 2c \\ d-e & e-f & 2f \\ g-h & h-i & 2i \end{vmatrix}$

**8.** Calculate the determinant.

$$\begin{vmatrix} a & a^2 & b & a^2+b \\ b & b^2 & c & ab+c \\ c & c^2 & d & ac+d \\ d & d^2 & a & ad+a \end{vmatrix}$$

**9.** Prove part ((1)) of Theorem 9.2.

← **Hint**: You already know it is true for $2 \times 2$ and $3 \times 3$ identity matrices. Show that if it's true for an $(n-1) \times (n-1)$ identity matrix, it must be true for an $n \times n$ identity matrix.

**10.** What is the determinant of a diagonal matrix such as $\begin{pmatrix} 3 & 0 & 0 \\ 0 & 5 & 0 \\ 0 & 0 & 7 \end{pmatrix}$?

**11.** Prove that if a matrix $A$ has a column of all 0's, then $\det(A) = 0$.

← What's a smart choice for a column on which to expand?

**12.** Prove part ((2)) of Theorem 9.2.

**13.** Prove part ((3)) of Theorem 9.2.

← For Exercises 12 and 13, try evaluating $\det(A)$ along the column $A_{*j}$.

**14.** Show that if one column of a matrix $A$ is a scalar multiple of another ($A_{*i} = cA_{*j}$, where $i \neq j$), then $\det(A) = 0$.

**15.** Suppose $A$, $A'$, and $A''$ are three $n$-by-$n$ matrices that are identical except in the $i^{\text{th}}$ column and, there,

$$A_{*i} = aA'_{*i} + bA''_{*i}$$

for some numbers $a$ and $b$. Then

$$\det A = a \det A' + b \det A''$$

**16.** Prove Theorem 9.8.

## 9.3 More Properties of Determinants

In Lesson 9.2, you used an algorithm to define the determinant of an $n \times n$ matrix. But there was a choice involved. In this lesson, you will make sure that the column you use to evaluate the determinant does not affect the outcome. **In this lesson, you will learn how to**

$\longleftarrow$ "Pick an arbitrary column of $A \ldots$"

- prove that the column you use to evaluate the determinant of a matrix does not affect the outcome
- expand a determinant along a row instead of a column
- compare the determinant of a matrix to the determinant of its transpose

### Developing Habits of Mind

**Pay close attention to definitions.** Mathematics is one of the only fields where you can create an object simply by writing down its definition. Mathematicians frequently invent new objects, defining them by listing their properties.

$\longleftarrow$ You've already seen several examples; for instance, vector spaces and matrices were probably new objects to you. In this book, they were "defined into being." But definitions are not arbitrary. The idea of vector space, for example, captured many commonalities among things with which you were already familiar.

Of course, mathematicians must take great care to make sure that their definitions make sense—that they actually define an object that could exist, and that the object in question is **well-defined**.

You've already seen an example of this issue. In defining the cross product of two vectors $A$ and $B$ in $\mathbb{R}^3$, one might be tempted to say, "$A \times B$ is a vector orthogonal to both $A$ and $B$." But then any reasonable person, after drawing a quick sketch, is likely to ask, "*Which* vector orthogonal to both $A$ and $B$? There are a whole bunch of them!"

In defining $A \times B$, mathematicians picked a nice example of a vector that is orthogonal to both $A$ and $B$: it is easy to compute with $2 \times 2$ determinants; it has the nice property that its length tells you something about the parallelogram spanned by $A$ and $B$; and you can geometrically describe its position in space with the right-hand rule.

$\longleftarrow$ You meet another aspect of showing that something is well-defined whenever you invoke the extension program. What is that aspect?

Now, you've run into this problem in the definition of determinant. If the definition doesn't say which column to choose, how do you know there's only one determinant?

The first thing to do is try some examples.

### For You to Do

**1.** Evaluate this determinant along each of the three columns. Do you get the same answer each time?

$$\begin{vmatrix} 0 & 2 & 1 \\ 1 & 0 & -2 \\ 3 & 5 & 3 \end{vmatrix}$$

**Theorem 9.9 (Determinant Is Well-Defined)**
*Let $A$ be an $n \times n$ square matrix.*

$$A = \begin{pmatrix} A_{11} & A_{12} & \cdots & A_{1n} \\ A_{21} & A_{22} & \cdots & A_{2n} \\ \vdots & \vdots & \ddots & \vdots \\ A_{n1} & A_{n2} & \cdots & A_{nn} \end{pmatrix}$$

*Let $A_{*i}$ and $A_{*j}$ be any two columns of $A$. Then*

$$\begin{aligned} &(-1)^{i+1}\left(A_{1i}\left|M_{1i}\right| - A_{2i}\left|M_{2i}\right| + \cdots + (-1)^{n+1}A_{ni}\left|M_{ni}\right|\right) \\ &\quad = (-1)^{j+1}\left(A_{1j}\left|M_{1j}\right| - A_{2j}\left|M_{2j}\right| + \cdots + (-1)^{n+1}A_{nj}\left|M_{nj}\right|\right) \end{aligned}$$

← In other words, the answer for $\det(A)$ is the same, no matter which column you use.

***Proof.*** This is a sketch of the main ideas for the proof.

← A formal proof of this theorem requires ideas from a field of mathematics called combinatorics.

First, look at what happens for $2 \times 2$ and $3 \times 3$ matrices to see what's going on.

$$\begin{vmatrix} A_{11} & A_{12} \\ A_{21} & A_{22} \end{vmatrix} = A_{11}A_{22} - A_{12}A_{21}$$

Each term in the sum looks like $A_{1x_1}A_{2x_2}$, where $x_1 \neq x_2$. In other words, in each term, each row and each column show up exactly once.

$$\begin{aligned} \begin{vmatrix} A_{11} & A_{12} & A_{13} \\ A_{21} & A_{22} & A_{23} \\ A_{31} & A_{32} & A_{33} \end{vmatrix} =& A_{11}A_{22}A_{33} + A_{12}A_{23}A_{31} + A_{13}A_{21}A_{32} \\ & - A_{11}A_{23}A_{32} - A_{12}A_{21}A_{33} - A_{13}A_{22}A_{31} \end{aligned}$$

In this case, each term in the sum looks like $A_{1x_1}A_{2x_2}A_{3x_3}$, where $x_1$, $x_2$, and $x_3$ are all different. Again, in each term each row and each column appear exactly once.

Now let $A$ be a $4 \times 4$ matrix, and suppose you evaluate $\det(A)$ along column $A_{*j}$. Then $A_{1j}\left|M_{1j}\right|$ is the first term in the formula for the determinant. Since $\left|M_{1j}\right|$ is a determinant of a $3 \times 3$ matrix, you know that each term in the sum looks like $A_{2x_2}A_{3x_3}A_{4x_4}$, where $x_2$, $x_3$, and $x_4$ are all different, and none of them equal $j$. So when you multiply these by $A_{1j}$, you get terms that look like $A_{1j}A_{2x_2}A_{3x_3}A_{4x_4}$, with each row and each column appearing exactly once.

← Remember that $M_{1j}$ is formed by deleting the first row and column $j$.

Similarly, $A_{2j}\left|M_{2j}\right|$ is a sum of terms $A_{1x_1}A_{2j}A_{3x_3}A_{4x_4}$, where all of the $x$'s are different and none of them equal $j$, so again each row and each column appears once in each term. The same can be said for $A_{3j}\left|M_{3j}\right|$, and $A_{4j}\left|M_{4j}\right|$.

← Convince yourself of this.

In this same way, you can see that for any $n \times n$ matrix, the terms in the expansion of $\det(A)$ all look like $A_{1x_1}A_{2x_2}\cdots A_{nx_n}$, where the $x$'s are all different. This is true no matter which column you choose to expand along.

The only thing to worry about, then, is the signs of these terms. The "sign matrix," from Lesson 9.2, tells you the sign attached to each minor, but then each minor is expanded according to its own sign matrix, all the way down to individual entries in the matrix. And each entry ends up with a + or a −. Is there any regularity to how the signs are attached? It may not be obvious, but there is. It involves looking at the $x_i$'s carefully and asking how many "flips" of two numbers it takes to rearrange the sequence $x_1, x_2, \ldots, x_n$ into the sequence $1, 2, 3, \ldots, n$.

← The sign matrix is this:
$$\begin{pmatrix} + & - & + & \\ - & + & - & \cdots \\ + & - & + & \\ & \vdots & & \ddots \end{pmatrix}$$

For example, in the $2 \times 2$ case, the positive term was $A_{11}A_{22}$. The $x_i$ sequence is $1, 2$, which is already in order, so it takes 0 flips to put it in order. The negative term was $A_{12}A_{21}$. Here the $x_i$ sequence is $2, 1$. To put it in order, you need one flip.

$$2, 1 \overset{1\leftrightarrow 2}{\longrightarrow} 1, 2.$$

In the $3 \times 3$ case, one of the positive terms requires no flips. Each of the others requires an even number of flips. Consider the term $A_{12}A_{23}A_{31}$. The sequence of $x_i$'s is $2, 3, 1$. So,

← This term requires no flips: $A_{11}A_{22}A_{33}$.

← Check that no matter how hard you try, you can't change $2, 3, 1$ into $1, 2, 3$ with an odd number of flips.

$$2, 3, 1 \overset{1\leftrightarrow 2}{\longrightarrow} 1, 3, 2 \overset{2\leftrightarrow 3}{\longrightarrow} 1, 2, 3$$

All of the negative terms require an odd number of flips. For example, look at $A_{11}A_{23}A_{32}$:

$$1, 3, 2 \overset{2\leftrightarrow 3}{\longrightarrow} 1, 2, 3.$$

### For You to Do

**2.** Check each of the other terms in the $3 \times 3$ determinant and make sure that the sign of each is given by the "even-odd" flip test.

← The **"even-odd" flip test**: If there are $k$ flips, the sign is $(-1)^k$.

This is promising. It gives a way to assign a sign to each term $A_{1x_1}A_{2x_2}\cdots A_{nx_n}$ in $\det(A)$ that doesn't depend on a choice of columns: the sign is "+ "if it takes an even number of flips and "− " if it takes an odd number of flips. But is that the same as the sign you get by expanding along any column? That comes next.

← Implicit here is the fact that, for any term, any way you pick to reorder the $x_i$'s will always have the same "parity"—even or odd. That's true, and you might try proving it.

The plan is to establish that, no matter which column you pick, the sign attached to $A_{1x_1}A_{2x_2}\cdots A_{nx_n}$ that results from repeatedly applying the sign matrix to the original determinant and all of its minors is exactly the number of flips it takes to move the $x_i$ into increasing order because *that*—the number of flips it takes to order the $x_i$'s—doesn't depend at all on which column you choose.

This is a bit tricky, but if you've followed along so far, the rest is downhill.

Stick with a $4 \times 4$ matrix—the following argument will work in general. Suppose you evaluate along column $j$. The first is term $A_{1j}\,|M_{1j}|$. Since $M_{1j}$ is a smaller matrix, you know that $\det(M_{1j})$ is a sum of terms that look like $A_{2x_2}A_{3x_3}A_{4x_4}$, where

← $M_{1j}$ is a $3 \times 3$ matrix.

- all the $x$'s are different and none of them are $j$
- the term is positive if it requires an even number of flips to put the sequence $x_2, x_3, x_4$ in increasing order
- the term is negative if it requires an odd number of flips

Suppose $A_{2x_2}A_{3x_3}A_{4x_4}$ is a positive term in the expansion of $|M_{1j}|$, and so $A_{1j}A_{2x_2}A_{3x_3}A_{4x_4}$ is a term in $\det(A)$. Should the term be positive or negative? Well, you need to see how many flips it takes to make the sequence $j, x_2, x_3, x_4$ look like $1, 2, 3, 4$. Here's one way to do it: first, move all the $x$'s into increasing order. You know that takes an even number of flips. Now move the $j$ into position:

← You know it takes an even number of flips because $A_{2x_2}A_{3x_3}A_{4x_4}$ was a *positive* term in the expansion of $|M_{1j}|$.

- If $j = 1$, you're done; the sequence is already in order. So the final term should be positive.
- If $j = 2$, your sequence looks like $2, 1, 3, 4$, so it takes one more flip to put it in order. That means it takes an odd number of flips overall, so the term should be negative.
- If $j = 3$, your sequence looks like $3, 1, 2, 4$, so it takes two flips to put it in order. That makes the total number of flips still even, so the term should be positive.
- If $j = 4$, it takes three flips (why?). So the total number of flips is odd, and the term should be negative.

The same reasoning works for the terms in $|M_{1j}|$ that are negative. In both cases, you conclude that you should multiply by $-1$ if $j$ is even, and 1 if $j$ is odd. And that's what the sign matrix predicts.

That explains the first row of the sign matrix. What about the second row? The exact same reasoning works, except now you need to count the number of flips it takes to put the sequence $x_1, j, x_3, x_4$ in order. After you put $x_1, x_2, x_3$ in order, you want to switch $j$ and 2 by a sequence of flips.

← The sign matrix is this:

$$\begin{pmatrix} + & - & + & \\ - & + & - & \cdots \\ + & - & + & \\ & \vdots & & \ddots \end{pmatrix}$$

If $j = 1$, the sequence looks like $2, 1, 3, 4$, so it takes one flip to put it in order. If $j = 2$, the sequence is already in order, so no flips are needed. If $j = 3$, the sequence looks like $1, 3, 2, 4$, so it takes one flip to finish putting it in order. And if $j = 4$, it takes two flips. (Why?) This means the signs attached to $A_{2j}$ should be positive if $j = 2$ or 4, and negative if $j = 1$ or 3.

The same kind of reasoning can explain why each row of the sign matrix does the right thing. And the same idea will work for $n \times n$ matrices. ∎

### Developing Habits of Mind

**Find ways to simplify** The above argument would need more precision to make it air-tight, but the basic idea is to use the fact that the determinant of an $n \times n$ matrix is well-defined by assuming that it is well-defined for $(n-1) \times (n-1)$ matrices and by showing that the sign of any term is determined by the "flip test."

There are many different ways to develop the theory of determinants—just take a glance at some other linear algebra books. But all of them have some sticky arguments at various points, because the notion of determinant is closely tied up with this idea of even/odd flips. This idea rears its head twice:

(1) When you try to figure out what happens if you switch two columns

(2) When you try to show that a determinant is well-defined

The treatment used in this course has managed to eliminate the stickiness from ((1)) by the proof used for Theorem 9.7, so that all the technicalities are buried in the proof of Theorem 9.9.

### Facts and Notation

The reasoning above has an interesting consequence. You will get exactly the same terms in the sum, with exactly the same signs attached to them, if you evaluate along rows instead of columns.

The proof considers terms of the form $A_{ij}\,|M_{ij}|$ separately, and shows that each term looks like $A_{1x_1}A_{2x_2}\cdots A_{nx_n}$, where the $x$'s are all different and where $x_i = j$. When you sum along column $j$, you get every possible term $A_{1x_1}A_{2x_2}\cdots A_{nx_n}$, where the $x$'s are all different. If you sum along row $i$ instead—adding and subtracting terms that look like $A_{i1}\,|M_{i1}|$, $A_{i2}\,|M_{i2}|$, and so forth—you will get exactly the same terms.

Consider evaluating along the first row of a $3 \times 3$ determinant:

$$\begin{aligned}\begin{vmatrix} A_{11} & A_{12} & A_{13} \\ A_{21} & A_{22} & A_{23} \\ A_{31} & A_{32} & A_{33} \end{vmatrix} &= A_{11}\,|M_{11}| - A_{12}\,|M_{12}| + A_{13}\,|M_{13}| \\ &= A_{11}A_{22}A_{33} + A_{12}A_{23}A_{31} + A_{13}A_{21}A_{32} \\ &\quad - A_{11}A_{23}A_{32} - A_{12}A_{21}A_{33} - A_{13}A_{22}A_{31}\end{aligned}$$

That means the definition of determinant could be stated in two equivalent ways. Let $A_{i*}$ be any row of an $n \times n$ matrix $A$, and let $A_{*j}$ be any column. Then

$$\begin{aligned}\det(A) &= (-1)^{i+1}\left(A_{i1}\,|M_{i1}| - A_{i2}\,|M_{i2}| + \cdots + (-1)^{n+1}A_{in}\,|M_{in}|\right) \\ &= (-1)^{j+1}\left(A_{1j}\,|M_{1j}| - A_{2j}\,|M_{2j}| + \cdots + (-1)^{n+1}A_{nj}\,|M_{nj}|\right)\end{aligned}$$

### Example

**Problem.** Evaluate this determinant along the first row, and along the third column. Check that the answers are the same.

$$\begin{vmatrix} 1 & 1 & 0 \\ 2 & 0 & 3 \\ 1 & -1 & 0 \end{vmatrix}.$$

**Solution.** Along the first row:

$$\begin{vmatrix} 1 & 1 & 0 \\ 2 & 0 & 3 \\ 1 & -1 & 0 \end{vmatrix} = 1\begin{vmatrix} 0 & 3 \\ -1 & 0 \end{vmatrix} - 1\begin{vmatrix} 2 & 3 \\ 1 & 0 \end{vmatrix} + 0\begin{vmatrix} 2 & 0 \\ 1 & -1 \end{vmatrix}$$
$$= 3 + 3 = 6$$

Along the third column:

$$\begin{vmatrix} 1 & 1 & 0 \\ 2 & 0 & 3 \\ 1 & -1 & 0 \end{vmatrix} = 0\begin{vmatrix} 2 & 0 \\ 1 & -1 \end{vmatrix} - 3\begin{vmatrix} 1 & 1 \\ 1 & -1 \end{vmatrix} + 0\begin{vmatrix} 1 & 1 \\ 2 & 0 \end{vmatrix}$$
$$= -3(-2) = 6$$

---

**Corollary 9.10 (Rows and Columns)**

*The determinant of a matrix can be evaluated along any row in the same way that it is evaluated along a column.*

It follows that the basic properties of determinants in Theorem 9.2 also hold for rows.

**Theorem 9.11 (Basic Rules of Determinants–Row Version)**

*Let $A$ be an $n \times n$ matrix. Then*

(1) *Interchanging two rows of $A$ has the effect of multiplying the determinant by $-1$.*

(2) *If you replace a row $A_{j*}$ with a scalar multiple of that row $cA_{j*}$, the determinant is multiplied by the same scalar $c$.*

(3) *If you rewrite one row of $A$ as a sum of two row vectors, then $\det(A)$ breaks up as a sum of two determinants. Specifically, say $A_{j*} = A'_{j*} + A''_{j*}$. Then*

$$\begin{vmatrix} A_{*1} \\ A_{*2} \\ \vdots \\ A'_{j*} + A''_{j*} \\ \vdots \\ A_{*n} \end{vmatrix} = \begin{vmatrix} A_{*1} \\ A_{*2} \\ \vdots \\ A_{j*} \\ \vdots \\ A_{*n} \end{vmatrix} + \begin{vmatrix} A_{*1} \\ A_{*2} \\ \vdots \\ A''_{j*} \\ \vdots \\ A_{*n} \end{vmatrix}$$

← "det is alternating."

← "scalars come out."

← "det is linear."

---

**For You to Do**

**3.** Evaluate the following determinant along each of the rows. Do you get the same answer each time? Do you get the same answer as when you evaluated along the columns in For You to Do problem 1?

$$\begin{vmatrix} 0 & 2 & 1 \\ 1 & 0 & -2 \\ 3 & 5 & 3 \end{vmatrix}$$

---

**Minds in Action** Episode 42

*Sasha is drawing pictures of parallelepipeds and looking worried while Tony and Derman eat their lunches.*

Sasha's Pictures

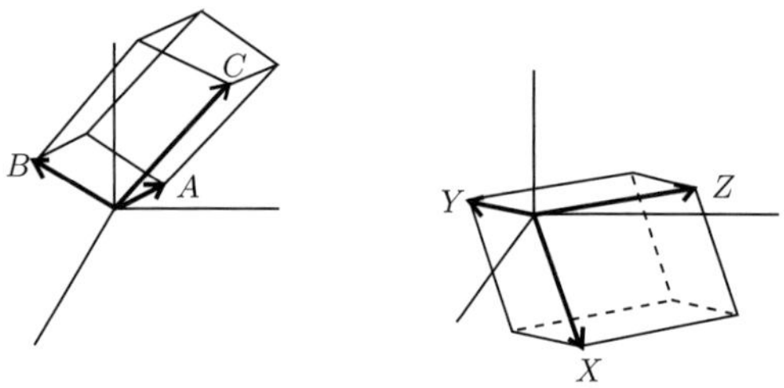


TONY: What are the pictures for?

SASHA: It seems weird to me that these boxes have the same volume.

TONY: What do you mean they have the same volume?

SASHA: Say I start with three vectors, like $A = (1, 1, 1)$, $B = (1, -1, 3)$, and $C = (-2, 1, 1)$. We learned a long time ago that if we make the matrix with those vectors as columns, like this . . .

*Sasha writes the matrix* $M = \begin{pmatrix} 1 & 1 & -2 \\ 1 & -1 & 1 \\ 1 & 3 & 1 \end{pmatrix}$

. . . then the volume of the box spanned by $A$, $B$, and $C$ is $|\det(M)| = |C \cdot (A \times B)|$.

*Derman loudly slurps his drink and looks thoughtful.*

TONY: Yeah, and if we switch the columns around, it's still the same box. But you have two different boxes drawn there.

SASHA: Well, we just learned that we can evaluate $\det(M)$ along the rows instead, right? So, lets call the rows $X = (1, 1, -2)$, $Y = (1, -1, 1)$, and $Z = (1, 3, 1)$. Then it seems like we just learned that $|\det(M)| = |Z \cdot (X \times Y)|$ if we . . .

TONY: . . . if we evaluate along the last row instead of the last column!

DERMAN: But that doesn't make sense. To get $|Z \cdot (X \times Y)|$, I should put those vectors as the columns of a matrix, like this . . .

*Derman puts down his drink, grabs Sasha's pencil, and writes a new matrix* $N = \begin{pmatrix} 1 & 1 & 1 \\ 1 & -1 & 3 \\ -2 & 1 & 1 \end{pmatrix}$.

TONY: But your matrix is just the transpose of Sasha's matrix.

SASHA: Wow! You're right. I feel a theorem coming on . . .

DERMAN: I'm not feeling it.

### For You to Do

4. Check that Sasha is right—that the determinant of her matrix $M$ and Derman's matrix $N$ are really the same. Remember

$$M = \begin{pmatrix} 1 & 1 & -2 \\ 1 & -1 & 1 \\ 1 & 3 & 1 \end{pmatrix} \text{ and } N = \begin{pmatrix} 1 & 1 & 1 \\ 1 & -1 & 3 \\ -2 & 1 & 1 \end{pmatrix}$$

Here's the theorem Derman, Sasha, and Tony discovered.

**Theorem 9.12**

*Let $A$ be an $n \times n$ matrix, and let $A^\top$ be its transpose. Then* $\det(A) = \det(A^\top)$.

***Proof.*** The proof will look similar to several you've seen already: show the statement is true for small matrices like $2 \times 2$ and $3 \times 3$. Then show that whenever it is true for all matrices of some particular size, say $(n-1) \times (n-1)$, then it must be true for all "next size bigger" matrices.

← This method of proof is so common in mathematics that it has a special name: *proof by mathematical induction*.

You've already seen that $\det A = \det A^\top$ for $2 \times 2$ and $3 \times 3$ matrices. Suppose you know it works for $(n-1) \times (n-1)$ matrices. Let $A$ be an $n \times n$ square matrix and $A^\top$ its transpose,

$$A = \begin{pmatrix} A_{11} & A_{12} & \cdots & A_{1n} \\ A_{21} & A_{22} & \cdots & A_{2n} \\ \vdots & \vdots & \ddots & \vdots \\ A_{n1} & A_{n2} & \cdots & A_{nn} \end{pmatrix} \text{ and } A^\top = \begin{pmatrix} A_{11} & A_{21} & \cdots & A_{n1} \\ A_{12} & A_{22} & \cdots & A_{n2} \\ \vdots & \vdots & \ddots & \vdots \\ A_{1n} & A_{2n} & \cdots & A_{nn} \end{pmatrix}.$$

Calculate $\det(A)$ by evaluating along the first column to get

$$\det(A) = A_{11}\,|M_{11}| - A_{21}\,|M_{21}| + \cdots + (-1)^{n+1} A_{n1}\,|M_{n1}|$$

Calculate $\det(A^\top)$ by evaluating along the first row to get

$$\begin{aligned} \det(A^\top) &= A_{11}\,|M_{11}^\top| - A_{21}\,|M_{21}^\top| + \cdots + (-1)^{n+1} A_{n1}\,|M_{n1}^\top| \\ &= A_{11}\,|M_{11}| - A_{21}\,|M_{21}| + \cdots + (-1)^{n+1} A_{n1}\,|M_{n1}| \\ &= \det(A) \end{aligned}$$

In the second line, use the fact that the minors have size $(n-1) \times (n-1)$. So for the minor $M_{ij}$, you already know that $\det(M_{ij}^\top) = \det(M_{ij})$. ∎

### For You to Do

5. Prove Theorem 9.11. Show Corollary 9.6 and Theorem 9.8 remain true if you replace "column" with "row."

← **Hint**: Use Theorem 9.12.

You've probably already noticed that calculating the determinant of a diagonal matrix is particularly simple. What about the determinants of matrices that are upper triangular or lower triangular? Try a few.

← Upper and lower triangular matrices were introduced in Lesson 4.3.

### For You to Do

**6.** Calculate these determinants:

**a.** $\begin{vmatrix} 1 & 2 & 3 \\ 0 & 4 & 5 \\ 0 & 0 & 6 \end{vmatrix}$ **b.** $\begin{vmatrix} 3 & 0 & 0 & 0 \\ 1 & 3 & 0 & 0 \\ -2 & 0 & -3 & 0 \\ 11 & -7 & 221 & 1 \end{vmatrix}$ **c.** $\begin{vmatrix} -2 & 2 & 2 & -1 & 1 \\ 0 & 4 & 2 & 1 & 8 \\ 0 & 0 & -8 & 0 & 1 \\ 0 & 0 & 0 & 2 & 1 \\ 0 & 0 & 0 & 0 & -1 \end{vmatrix}$

**Theorem 9.13**

*The determinant of an upper triangular matrix is the product of the diagonal entries, and the same is true for a lower triangular matrix.*

← The proof will look similar to several you've seen already: Show the statement is true for small matrices like $2 \times 2$ and $3 \times 3$. Then show that whenever it is true for all matrices of some particular size, say $(n-1) \times (n-1)$, then it must be true for all "next size bigger" matrices. This is *mathematical induction* again.

← You already proved one special case of this when you showed that $\det(I) = 1$ in Exercise 9 from Lesson 9.2.

***Proof.***

Here's the proof for upper triangular matrices: if

$$A = \begin{pmatrix} a & b \\ 0 & d \end{pmatrix}, \text{ then}$$

$$|A| = ad - b \cdot 0 = ad$$

which is exactly the product of the diagonal entries.

Now let

$$A = \begin{pmatrix} A_{11} & A_{12} & A_{13} \\ 0 & A_{22} & A_{23} \\ 0 & 0 & A_{33} \end{pmatrix}$$

be any upper triangular $3\times3$ matrix. Expand $\det(A)$ along the first column:

$$\det(A) = A_{11}\,|M_{11}| - 0\,|M_{21}| + 0\,|M_{31}| = A_{11}\,|M_{11}|$$

Since $M_{11}$ is an upper triangular $2\times2$ matrix, its determinant is the product of the diagonal entries $A_{22}$ and $A_{33}$. So $\det(A) = A_{11}A_{22}A_{33}$.

Now let $A$ be any $n \times n$ upper triangular matrix, and suppose you know that for every $(n-1) \times (n-1)$ upper triangular matrix, the determinant is the product of the diagonal entries. Expand $\det(A)$ along the first column to get:

$$\det(A) = A_{11}\,|M_{11}| - 0\,|M_{21}| + \cdots \pm 0\,|M_{n1}| = A_{11}\,|M_{11}|$$

Since $M_{11}$ is an upper triangular matrix of dimension $(n-1) \times (n-1)$, its determinant is the product of its diagonal entries, i.e., $|M_{11}| = A_{22}A_{33}\cdots A_{nn}$, and so

$$\det(A) = A_{11}A_{22}A_{33}\cdots A_{nn}$$

which is the product of the diagonal entries. ■

### For Discussion

**7.** Prove Theorem 9.13 for lower triangular matrices.

← What needs to change in the proof?

## Exercises

**1.** Evaluate $\begin{vmatrix} 1 & 3 & 1 \\ 2 & -1 & 4 \\ 0 & 1 & 3 \end{vmatrix}$

**a.** along the first column
**b.** along the last row.

**2.** Evaluate $\begin{vmatrix} -1 & 4 & 2 \\ 1 & 3 & 1 \\ 2 & 0 & 6 \end{vmatrix}$

**a.** along the last row
**b.** along the second column.

**3.** Evaluate $\begin{vmatrix} 1 & -1 & 3 & 2 \\ 1 & 4 & 1 & 0 \\ -1 & 3 & 1 & 2 \\ 0 & 1 & 6 & 1 \end{vmatrix}$

**a.** along row 2, and
**b.** along row 4.

**4.** Evaluate each determinant. Look for shortcuts.

**a.** $\begin{vmatrix} 1 & 2 & -3 \\ 5 & -1 & 1 \\ 2 & 4 & -6 \end{vmatrix}$ **b.** $\begin{vmatrix} 1 & -1 & 2 & 0 \\ 0 & 1 & 1 & 3 \\ 2 & -4 & 0 & 5 \\ 1 & 1 & -2 & 3 \end{vmatrix}$ **c.** $\begin{vmatrix} 1 & t & t^2 \\ t & t^2 & 1 \\ t^2 & t & 1 \end{vmatrix}$

**d.** $\begin{vmatrix} 1 & 1 & 1 & 1 \\ 1 & a & b & d \\ 1 & 1 & c & e \\ 1 & 1 & 1 & f \end{vmatrix}$ **e.** $\begin{vmatrix} 1 & a & b & c+d \\ 1 & c & d & a+b \\ 1 & c & b & a+d \\ 1 & a & d & c+b \end{vmatrix}$ **f.** $\begin{vmatrix} 8 & 2 & -1 \\ -3 & 4 & 6 \\ 1 & 7 & 2 \end{vmatrix}$

**g.** $\begin{vmatrix} k & -3 & 9 \\ 2 & 4 & 1 \\ 1 & k^2 & 3 \end{vmatrix}$ **h.** $\begin{vmatrix} 1 & 3 & 4 & 0 \\ 2 & 1 & 5 & 2 \\ 3 & 1 & 2 & 1 \\ 1 & 0 & -3 & -1 \end{vmatrix}$ **i.** $\begin{vmatrix} 2 & 3 & -1 \\ 1 & 5 & 6 \\ 3 & 4 & 7 \end{vmatrix}$

**j.** $\begin{vmatrix} 1 & 2 & 5 & 0 \\ 0 & 3 & -3 & 0 \\ 0 & 4 & 2 & 0 \\ 0 & -1 & 1 & 1 \end{vmatrix}$ **k.** $\begin{vmatrix} 5 & 3 & 2 & 8 & 7 \\ 0 & 1 & 6 & -1 & 4 \\ 0 & 0 & 8 & 0 & 7 \\ 0 & 0 & 0 & 9 & 3 \\ 0 & 0 & 0 & 0 & 2 \end{vmatrix}$ **l.** $\begin{vmatrix} 5 & 3 & 8 \\ 2 & 1 & 6 \\ 3 & 2 & 2 \end{vmatrix}$

**m.** $\begin{vmatrix} a & b & c \\ a+1 & b-3 & c+2 \\ a-2 & b+6 & c-4 \end{vmatrix}$ **n.** $\begin{vmatrix} 1 & 2 & 2 & 2 \\ 1 & 3 & 3 & 3 \\ 1 & 1 & 4 & 4 \\ 1 & 1 & 1 & 5 \end{vmatrix}$

**5.** Solve for $x$:

$$\begin{vmatrix} 3 & 4 & 2 \\ -1 & x & x \\ 2 & 5 & 7 \end{vmatrix} = 2$$

**6.** Prove that $\begin{vmatrix} 1 & 1 & 1 \\ a & b & c \\ a^2 & b^2 & c^2 \end{vmatrix} = (b-a)(c-a)(c-b)$.

←
This determinant is called the **Vandermonde determinant**, named for the violinist-mathematician Alexandre-Théophile Vandermonde.

**7.** Suppose $A = (A_{*1}, A_{*2}, A_{*3}, A_{*4})$ is a $4 \times 4$ matrix and $x_1, x_2, x_3, x_4$ are numbers. Express each determinant in terms of $\det(A)$ and the $x_i$.

←
In each problem, you are replacing a whole column of $A$ with a linear combination of the columns. You should probably try expanding the determinant along that column.

**a.** $\det\left((x_1A_{*1} + x_2A_{*2} + x_3A_{*3} + x_4A_{*4}), A_{*2}, A_{*3}, A_{*4}\right)$
**b.** $\det\left(A_{*1}, (x_1A_{*1} + x_2A_{*2} + x_3A_{*3} + x_4A_{*4}), A_{*3}, A_{*4}\right)$
**c.** $\det\left(A_{*1}, A_{*2}, (x_1A_{*1} + x_2A_{*2} + x_3A_{*3} + x_4A_{*4}), A_{*4}\right)$
**d.** $\det\left(A_{*1}, A_{*2}, A_{*3}, (x_1A_{*1} + x_2A_{*2} + x_3A_{*3} + x_4A_{*4})\right)$

# 9.4 Elementary Row Matrices and Determinants

Ever since Chapter 3, you have been using the three elementary row operations to put matrices in row reduced echelon form. Now elementary row operations will be a useful tool to prove some important properties of determinants.

**In this lesson, you will learn how to**

- use matrix multiplication as a way to reduce a matrix to echelon form
- relate the determinant of a matrix to the determinant of its echelon form
- add one more simple condition to the TFAE Theorem (Theorem 8.9 from Lesson 8.3)
- prove that the determinant of the product of two $n \times n$ matrices is the product of their determinants

**Remember**

You proved this for $2 \times 2$ matrices in Chapter 5 (Theorem 5.10 from Lesson 5.4).

Remember the three elementary row operations:

- interchange two rows
- replace one row with a nonzero multiple of that row
- replace one row with the sum of itself and some multiple of another row

**Minds in Action** Episode 43

*The three friends are talking about elementary row operations and determinants.*

DERMAN: I tried some examples, and it seems like that third elementary row operation—where you replace one row with the sum of itself and some multiple of another row—that doesn't do anything to the determinant.

TONY: How can that be? When you multiply a row by something, the determinant gets multiplied by that same constant.

DERMAN: I don't know why, but it's true. Check it out. Start with one of those easy triangular matrices like this:

$$\begin{vmatrix} 1 & 2 & 3 \\ 0 & 4 & 5 \\ 0 & 0 & 6 \end{vmatrix} = 24$$

right? Then take row 1, and subtract twice row 3 from it. You get this:

$$\begin{vmatrix} 1 & 2 & -9 \\ 0 & 4 & 5 \\ 0 & 0 & 6 \end{vmatrix} = 24$$

The determinant didn't change.

TONY: That's not fair, though. Because it's still triangular, and nothing on the diagonal changed.

SASHA: Well, that shouldn't matter. Because if the determinant stays the same when you do the row operation, it will always stay the same, whether you start with a triangular matrix or not. And if it changes somehow, it should still change in that case.

TONY: I guess so, but it still seems like too special a case. What if you took row 3, and subtracted twice row 1 from it?

DERMAN: OK, I'll do it in my head. You get $\begin{vmatrix} 1 & 2 & 3 \\ 0 & 4 & 5 \\ -2 & -4 & 0 \end{vmatrix} = 24$

TONY: Well, I guess I believe it then. But I still don't see why . . .

SASHA: Let's try that trick of "delayed evaluation," where you write something out but don't simplify everything too much. So we started with this determinant: $\begin{vmatrix} 1 & 2 & 3 \\ 0 & 4 & 5 \\ 0 & 0 & 6 \end{vmatrix}$.

And then we subtracted twice row 1 from row 3:

$$\begin{vmatrix} 1 & 2 & 3 \\ 0 & 4 & 5 \\ 0-2\cdot 1 & 0-2\cdot 2 & 6-2\cdot 3 \end{vmatrix}$$

TONY: Oh, this looks like part of that theorem on rules of determinants.

← Theorem 9.11, part ((3)).

SASHA: Yeah, we can split it up now as a sum of two determinants:

$$\begin{vmatrix} 1 & 2 & 3 \\ 0 & 4 & 5 \\ 0-2\cdot 1 & 0-2\cdot 2 & 6-2\cdot 3 \end{vmatrix} = \begin{vmatrix} 1 & 2 & 3 \\ 0 & 4 & 5 \\ 0 & 0 & 6 \end{vmatrix} + \begin{vmatrix} 1 & 2 & 3 \\ 0 & 4 & 5 \\ -2\cdot 1 & -2\cdot 2 & -2\cdot 3 \end{vmatrix}$$

TONY: That first one is just our original determinant. The second one . . .

SASHA: In the second one, the third row is a multiple of the first, and that's another one of the determinant rules. That determinant is 0.

← Sasha's comment follows from Theorem 9.11, part ((2)). (Why?)

TONY: So if you add a multiple of some row to a different row, it really doesn't change the determinant.

DERMAN: That's what I said!

---

**Theorem 9.14 (Effects of Elementary Row Operations)**

(1) *Interchanging two rows of a matrix multiplies the determinant of the matrix by* $-1$.

(2) *Replacing one row of a matrix* $A_{i*}$ *with a constant multiple of that row* $cA_{i*}$ *multiplies the determinant of the matrix by the same constant* $c$.

(3) *Replacing one row of a matrix with a sum of that row and some multiple of a different row does not change the determinant of the matrix.*

← In other words, replacing row $A_{i*}$ with $A_{i*} + cA_{j*}$ where $i \neq j$ doesn't change the determinant of the matrix $A$.

### For Discussion

1. Sasha, Derman, and Tony have outlined a proof of part ((3)). Turn their idea into a proof.

← You don't need to prove parts ((1)) and ((2)) because they are identical to statements in Theorem 9.11.

### Minds in Action Episode 44

*Sasha is still thinking about elementary row operations.*

SASHA: *Thinking aloud.* I wonder if there's some relationship between $\det(A)$ and $\det(\text{rref}(A))$.

TONY: What do you mean?

SASHA: Well, it's easy to calculate $\det(\text{rref}(A))$, right? If you start with a square matrix, then $\text{rref}(A)$ is a diagonal matrix. Either it's got all 1's on the diagonal, so $\det(\text{rref}(A)) = 1$, or . . .

← If $\text{rref}(A)$ has all 1's on the diagonal, then $\text{rref}(A) = I$, the identity matrix.

TONY: . . . or it's got a 0 somewhere on the diagonal, and then $\det(\text{rref}(A)) = 0$. Very smooth.

DERMAN: But does that really tell us anything about $\det(A)$?

SASHA: I don't know. But we know that we use elementary row operations to change $A$ into $\text{rref}(A)$. And we know what each of those operations does to $\det(A)$. It seems like somehow we should be able to keep track.

DERMAN: Where do you get these ideas?

As usual, Sasha is onto something. The first step in figuring out a relationship between $\det(A)$ and $\det(\text{rref}(A))$ is to understand the determinants of the **elementary row matrices**.

← See For You to Do problem 5 from Lesson 5.2 if you need a refresher about $2 \times 2$ elementary row matrices.

To form an elementary row matrix, perform an elementary row operation to the identity matrix. This table gives some examples for $3 \times 3$ matrices:

| Type | elementary row operation | elementary row matrix |
|---|---|---|
| Type 1 | switch rows 1 and 3 | $E = \begin{pmatrix} 0 & 0 & 1 \\ 0 & 1 & 0 \\ 1 & 0 & 0 \end{pmatrix}$ |
| Type 2 | multiply row 1 by $-5$ | $F = \begin{pmatrix} -5 & 0 & 0 \\ 0 & 1 & 0 \\ 0 & 0 & 1 \end{pmatrix}$ |
| Type 3 | replace row 2 with itself plus twice row 3 | $G = \begin{pmatrix} 1 & 0 & 0 \\ 0 & 1 & 2 \\ 0 & 0 & 1 \end{pmatrix}$ |

**For Discussion**

**2.** Use Theorem 9.14 and the fact that $\det(I) = 1$ to compute the determinants $|E|$, $|F|$, and $|G|$.

**For You to Do**

**3.** Start with $A = \begin{pmatrix} 0 & 1 & 2 \\ 3 & 4 & 5 \\ 6 & 7 & 8 \end{pmatrix}$, and find each matrix product.

**a.** $EA$ **b.** $FA$ **c.** $GA$

Describe in words what happens to the matrix $A$ in each case.

**Theorem 9.15 (Determinant and Effect of Elementary Row Matrix)**
*Let $A$ be any $n \times n$ matrix.*

(1) *Let $E$ be the $n \times n$ elementary row matrix obtained by interchanging rows $i$ and $j$ in the identity matrix $I$. Then*
 (a) $\det(E) = -1$, *and*
 (b) *$EA$ is exactly the same as matrix $A$ with rows $i$ and $j$ interchanged.*

(2) *Let $E$ be the $n \times n$ elementary row matrix obtained by multiplying row $i$ of the identity matrix by the nonzero constant $c$. Then*
 (a) $\det(E) = c$, *and*
 (b) *$EA$ is exactly the same as matrix $A$ with rows $i$ multiplied by $c$.*

(3) *Let $E$ be the $n \times n$ elementary row matrix obtained by replacing row $i$ of the identity matrix $I$ by itself plus $c$ times row $j$. Then*
 (a) $\det(E) = 1$, *and*
 (b) *$EA$ is exactly the same as matrix $A$ with row $i$ replaced by itself plus $c$ times row $j$.*

← If you do problems 2 and 3 above, the next proof will be much clearer.

***Proof.*** The statements about the determinants all come from how the elementary row matrices are formed. Here's the proof for the type 1 elementary row matrices: suppose $E$ is obtained from $I$ by switching two rows. Then, by part ((1)) of Theorem 9.14,

← You'll prove the other statements later.

$$\det(E) = -\det(I) = -1$$

To see that $EA$ is obtained from $A$ by switching rows $i$ and $j$, you need to use the Pick-Apart Theorem again.

← The Pick-Apart Theorem is Theorem 4.8 from Lesson 4.5, and you've used it many times since then.

The $k^{\text{th}}$ row of $EA$ is $E_{k*}A$, the $k^{\text{th}}$ row of $E$ times $A$; in symbols:

$$(EA)_{k*} = E_{k*}A$$

As long as $k$ is not equal to $i$ or $j$ (the two rows that got switched), the $k^{\text{th}}$ row of $E_{ij}$ looks just like the $k^{\text{th}}$ row of the identity matrix; that is,

$E_{k*} = I_{k*}$. It follows that

$$(EA)_{k*} = E_{k*}A = I_{k*}A = (IA)_{k*} = A_{k*}$$

So, the $k^{\text{th}}$ row of $EA$ is exactly the same as the $k^{\text{th}}$ row of $A$.

However, if you look at the $i^{\text{th}}$ row, then $E_{i*} = I_{j*}$ , so

$$(EA)_{i*} = E_{i*}A = I_{j*}A = (IA)_{j*} = A_{j*}$$

and the $i^{\text{th}}$ row of $EA$ is the $j^{\text{th}}$ row of $A$.

Similarly, if you look at the $j^{\text{th}}$ row, then $E_{j*} = I_{i*}$, so

$$(EA)_{j*} = E_{j*}A = I_{i*}A = (IA)_{i*} = A_{i*}$$

and the $j^{\text{th}}$ row of $EA$ is the $i^{\text{th}}$ row of $A$.

So every row of $E_{ij}A$ is identical to the same row of $A$, except rows $i$ and $j$ are switched. ■

**Remember**

$E$ is obtained from $I$ by switching rows $i$ and $j$.

⟵ Make sure you can supply a reason for each equality in the chain.

### For Discussion

**4.** Prove part ((3)) of Theorem 9.15. Show that if $E$ is a type 3 elementary row matrix, obtained from $I$ by replacing the $i^{\text{th}}$ row by itself plus $c$ times the $j^{\text{th}}$ row, the product $EA$ is obtained from $A$ by replacing $A_{i*}$ with $A_{i*} + cA_{j*}$.

⟵ See Exercise 18 from Lesson 4.6.

### For You to Do

**5.** Here are a few elementary row matrices of various sizes. For each one,

- describe the elementary row operation that is carried out when you multiply a matrix $A$ by the given matrix
- find the determinant of the given matrix
- find the inverse of the given matrix, or say why it doesn't exist

**a.** $\begin{pmatrix} 1 & 0 & 3 \\ 0 & 1 & 0 \\ 0 & 0 & 1 \end{pmatrix}$ **b.** $\begin{pmatrix} 1 & 0 & 0 & 0 & 0 & 0 \\ 0 & 0 & 0 & 0 & 0 & 1 \\ 0 & 0 & 1 & 0 & 0 & 0 \\ 0 & 0 & 0 & 1 & 0 & 0 \\ 0 & 0 & 0 & 0 & 1 & 0 \\ 0 & 1 & 0 & 0 & 0 & 0 \end{pmatrix}$ **c.** $\begin{pmatrix} 1 & 0 & 0 & 0 \\ 0 & 1 & 0 & 0 \\ 0 & 0 & -2 & 0 \\ 0 & 0 & 0 & 1 \end{pmatrix}$

Theorem 9.15 has an interesting consequence, which you might have already guessed.

### Theorem 9.16

*Every elementary row matrix $E$ is invertible; furthermore, $E^{-1}$ is also an elementary row matrix.*

***Proof.*** Here's a proof for one kind of elementary row matrix: suppose $E$ is obtained from the identity by replacing row $i$ with the sum of itself and

⟵ You'll take care of the other two cases yourself.

$c$ times row $j$ for some nonzero scalar $c$. So the rows of $E$ can be described this way:

$$(E)_{k*} = \begin{cases} I_{k*} & \text{whenever } k \neq i, \text{ and} \\ I_{i*} + cI_{j*} & \text{if } k = i \end{cases}$$

Consider the matrix $F$, which is obtained from the identity by replacing row $I_{i*}$ with $I_{i*} - cIj*$. From Theorem 9.15, multiplying any matrix $A$ by $F$ leaves every row alone except for row $i$, which it replaces with row $i$ minus $c$ times row $j$. Now think about the product $M = FE$; the rows of $M$ satisfy

- $M_{k*} = E_{k*}$ whenever $i \neq k$, since $F$ leaves row $k$ alone. So in particular $M_{k*} = I_{k*}$
- $M_{i*} = E_{i*} - cE_{j*}$. Using what you know about the rows of $E$ from above, this means

$$M_{i*} = (I_{i*} + cI_{j*}) - cI_{j*} = I_{i*}$$

So $M = I$, and $E^{-1} = F$. ∎

### Developing Habits of Mind

**Use clear language to unpack a proof.** In a way, the above proof masks the simplicity of the idea. Basically, every elementary row operation can be "undone" by another elementary row operation.

- Switching two rows is undone by doing it again.
- Multiplying a row by $c$ is undone by multiplying that row by $\frac{1}{c}$.
- Replacing the $i^{\text{th}}$ row by itself plus $c$ times the $j^{\text{th}}$ row is undone by replacing the $i^{\text{th}}$ row by itself plus $-c$ times the $j^{\text{th}}$ row.

If you have an elementary row matrix obtained from $I$ by an elementary row operation of a certain type, its inverse is obtained from $I$ by doing the inverse operation to $I$.

Elementary row matrices are especially useful, because they give a nice way to relate $A$ and $\text{rref}(A)$. Here it is.

**Theorem 9.17 (Echelon Form as Matrix Product)**
*There are elementary row matrices $E_1, E_2, \ldots, E_s$ such that*

$$E_1 E_2 \cdots E_s(\text{rref}(A)) = A. \tag{4}$$

*Furthermore, if $\text{rref}(A) = I$, then $A$ can be written as a product of elementary row matrices.*

← You've already seen one example of how this works, if you did Exercise 6 from Lesson 5.2.

***Proof.*** Notice that the last sentence follows immediately from the first one; just substitute $I$ for $\text{rref}(A)$ in equation (4). You know that you can turn $A$ into $\text{rref}(A)$ by doing elementary row operations, which is the same as multiplying by an elementary row matrix. So it looks like this:

$$\text{rref}(A) = F_s \cdots F_2 F_1 A$$

where the $F$'s are all elementary row matrices.

By Theorem 9.16 elementary row matrices are all invertible, so we can multiply by the inverses one at a time.

$$\begin{aligned} \text{rref}(A) &= F_s \cdots F_2 F_1 A \\ F_s^{-1}\,\text{rref}(A) &= F_{s-1} \cdots F_2 F_1 A \\ &\vdots \\ F_2^{-1} \cdots F_s^{-1}\,\text{rref}(A) &= F_1 A \\ F_1^{-1} F_2^{-1} \cdots F_s^{-1}\,\text{rref}(A) &= A \end{aligned}$$

From Theorem 9.16, you also know that the inverses are all elementary row matrices as well. So rename them this way: $F_1^{-1} = E_1$, $F_1^{-1} = E_1$, and so on. Then this becomes

$$E_1 E_2 \cdots E_s\,\text{rref}(A) = A$$

which is exactly what the theorem claimed. ∎

### The Product Rule

You have now shown that all but one rule you discovered for $2 \times 2$ determinants still hold for larger $n \times n$ determinants. The one rule that's still missing is the product rule: $\det(AB) = \det(A)\det(B)$. To begin, it makes sense to try an example to see if the product rule *might* continue to hold for larger matrices.

⟵ This was Theorem 5.10 from Lesson 5.4.

#### For You to Do

**6.** Verify that $\det(AB) = \det(A)\det(B)$ if

$$A = \begin{pmatrix} 1 & 3 & 0 \\ 2 & 1 & 4 \\ 3 & 4 & 5 \end{pmatrix} \text{ and } B = \begin{pmatrix} 0 & -1 & 1 \\ 1 & 0 & 2 \\ 5 & 1 & 2 \end{pmatrix}$$

First, you'll see that the product rule holds in a special case—when the first matrix is an elementary row matrix.

⟵ This will also help answer Sasha's question about the relationship between $\det(A)$ and $\det(\text{rref}(A))$.

#### Theorem 9.18

*If $A$ is any $n \times n$ matrix, and $E$ is an elementary row matrix of the same size, then* $\det(EA) = \det(E)\det(A)$.

***Proof.*** Suppose $E$ is an elementary row matrix formed by interchanging rows $i$ and $j$ of the identity matrix. You know two useful facts:

⟵ All the hard work for this proof has already been done. It's now just a matter of putting the pieces together.

- From Theorem 9.15, you know that multiplying $EA$ has the effect of switching rows $i$ and $j$ of $A$.
- From Theorem 9.14, you know that switching two rows of $A$ has the effect of multiplying $\det(A)$ by $-1$.

Together, these two facts say that $\det(EA) = -\det(A)$.

From Theorem 9.15, you also know that $\det(E) = -1$, so $\det(E)\det(A) = -\det(A)$ as well. Therefore, $\det(EA) = \det(E)\det(A)$, since they both equal $-\det(A)$. ■

### For Discussion

**7.** Prove Theorem 9.18 for type 2 elementary row matrices: show that if $E$ is an elementary row matrix formed by multiplying row $i$ of the identity matrix by some nonzero scalar $c$, then $\det(EA) = \det(E)\det(A)$.

⟵ You'll finish the proof of Theorem 9.18 by proving it for $E_{i+j(c)}$ in the Exercises.

### Minds in Action Episode 45

*The friends are still talking about* $\det(A)$ *and* $\det(\text{rref}(A))$.

SASHA: Oh, I get it now. We can write $A = E_1E_2\cdots E_s\,\text{rref}(A)$.

DERMAN: I still don't see how that helps.

TONY: Now we know that if you multiply on the left by an elementary row matrix, then the determinants multiply. So we can take determinants of both sides in Sasha's equation, and then expand it all out:

$$\begin{aligned}|A| &= |E_1E_2\cdots E_s\,\text{rref}(A)| \\ &= |E_1|\,|E_2|\cdots|E_s|\,|\text{rref}(A)|\end{aligned}$$

DERMAN: I still don't see how that helps.

SASHA: Well, we know that all of the determinants of the $E$'s are either 1 or $-1$ or some nonzero constant $c$. So if we multiply them all together, there's just some nonzero constant $k = |E_1|\,|E_2|\cdots|E_s|$ and $\det(A) = k\det(\text{rref}(A))$.

DERMAN: I *still* don't see how that helps.

TONY: Huh, yeah. Me, too. I mean, it seems like there's something there. But does that mean we should find $\text{rref}(A)$ and keep track of our steps to find $\det(A)$? I'm not sure that's any easier.

SASHA: No, but I think I see something even better. I think what this says is that if $\text{rref}(A) = I$, then $\det A \neq 0$. And if $\text{rref}(A) \neq I$, then ...

TONY: ... then it must have a 0 on the diagonal, so both $\det(\text{rref}(A))$ and $\det(A)$ will be 0.

SASHA: I think that means we have more stuff for our "TFAE Theorem."

DERMAN: I'm hungry. And I still don't see how this helps.

⟵ The most recent version of the "TFAE Theorem" is Theorem 8.9 from Lesson 8.3.

Embedded in the three friends' exchange is a theorem.

**Theorem 9.19**

*If $A$ is an $n \times n$ matrix, there is a nonzero number $k$ so that*

$$\det(A) = k \det(\text{rref}(A))$$

Theorem 9.19 allows you to add one more condition—a very important one—to the TFAE Theorem.

**Theorem 9.20 (The TFAE Theorem)**

*The following statements are all equivalent for an $n \times n$ matrix $A$.*

(1) *The columns of $A$ are linearly independent.*

(2) *The rows of $A$ are linearly independent.*

(3) *The rank of $A$ is $n$.*

(4) $\text{rref}(A) = I$.

(5) $A^{-1}$ *exists.*

(6) $AX = B$ *has a unique solution for any* $B$.

(7) $\ker(A) = O$.

(8) *The dimension of the column space for $A$ is $n$.*

(9) *The dimension of the row space for $A$ is $n$.*

(10) $\det(A) \neq 0$.

***Proof.*** Everything in this theorem has already been proved except the equivalence of the last statement. The idea now is to show that statement ((10)) is equivalent to statement ((4)).

← That's enough, since you already know statement ((4)) is equivalent to all the others.

Sasha has already explained what Theorem 9.19 states formally: $\det(A) = k \det(\text{rref}(A))$ for some nonzero constant $k$. If $\text{rref}(A) = I$, then $\det A = k \neq 0$. On the other hand, if $\text{rref}(A) \neq I$, then there is a zero somewhere on the diagonal of $\text{rref}(A)$. Since $\text{rref}(A)$ is a diagonal matrix with a zero on the diagonal, its determinant is 0. So $\det(A) = k \cdot 0 = 0$ as well. ∎

### Developing Habits of Mind

**Use the equivalence of different properties.** Theorem 9.20 says that the determinant of a matrix can provide a useful way to check all kinds of properties. For example:

- Is the rank of $A$ less than $n$ (where $A$ is an $n \times n$ matrix)? Well, $\text{rank}(A) = n$ is equivalent to $\det(A) \neq 0$. So calculate the determinant. If it's 0, then $\text{rank}(A) < n$, and otherwise $\text{rank}(A) = n$.
- Does $A$ have an inverse? Only if its determinant is not 0.
- Does $AX = 0$ have more than one solution? Only if $\det(A) = 0$.

**For Discussion**

**8.** List three other facts you can check about a matrix just by calculating whether the determinant is 0 or not.

**For You to Do**

**9.** Use determinants to find $t$ if $(3, 1, 2), (4, t, 2), (5, 1, 2t)$ are linearly dependent.

Now you're ready to put the last puzzle piece in place to see that even the product rule holds for determinants of any size square matrix.

**Theorem 9.21 (Product Rule for Determinants)**
*Let $A$ and $B$ be any two $n \times n$ matrices. Then* $\det(AB) = \det(A)\det(B)$.

***Proof.*** There are two possibilities: either $\det(A) = 0$, or $\det(A) \neq 0$. The very powerful TFAE Theorem will come in handy in dealing with each case.

First, suppose $\det(A) \neq 0$. Then $\text{rref}(A) = I$ from the TFAE Theorem. We know from Theorem 9.17 that in this case $A = E_1E_2 \cdots E_s$ is a product of elementary row matrices. So

$$AB = E_1E_2 \cdots E_sB$$

Taking determinants of both sides, we have

$$\begin{aligned}|AB| &= |E_1E_2 \cdots E_sB| \\ &= |E_1|\,|E_2 \cdots E_sB| \\ &\ \ \vdots \\ &= |E_1|\,|E_2| \cdots |E_s|\,|B| \\ &= |E_1E_2 \cdots E_s|\,|B| \\ &= |A|\,|B|\end{aligned}$$

Now suppose $\det(A) = 0$; then from the TFAE Theorem, you know that $\text{rank}(A) < n$. In Exercise 10 from Lesson 8.3, you showed that $\text{rank}(AB) \leq \text{rank}(A)$. So $\text{rank}(AB) < n$, which means (again, from the TFAE Theorem) that $\det(AB) = 0$. So

$$\det(A)\det(B) = 0 \cdot \det(B) = 0 = \det(AB).$$

And that takes care of both possibilities. ∎

**For Discussion**

**10.** Fill in reasons for each step in the derivation of $|AB|$ in the proof of Theorem 9.21.

## Exercises

**1.** Suppose that

$$A = \begin{pmatrix} a & b & c \\ d & e & f \\ g & h & i \end{pmatrix} \text{ and } B = \begin{pmatrix} j & k & l \\ m & n & p \\ q & r & s \end{pmatrix}.$$

Suppose also that $\det(A) = 5$ and $\det(B) = 3$. Find the following:

**a.** $\det(AB)$
**b.** $\det(2AB)$
**c.** $\det\left((AB)^{-1}\right)$
**d.** $\det\left(A^{-1}A^{-1}\right)$
**e.** $\det\left(A^{-1}B\right)$
**f.** $\det\left(A^{-1}BA\right)$
**g.** $\det\left(ABA^{-1}B^{-1}\right)$
**h.** $\det\left(A^{-1}B^4\right)$
**i.** $\det\left(\begin{pmatrix} a & b & c \\ g & h & i \\ d & e & f \end{pmatrix}\begin{pmatrix} j & k & l \\ 2m & 2n & 2p \\ q & r & s \end{pmatrix}\right)$

**2.** If $A$ is an $n \times n$ matrix and $A^2 = I$, show that $\det(A) = \pm 1$.

**3.** If $A$ is an invertible $n \times n$ matrix, show that $\det(A^{-1}) = \frac{1}{\det(A)}$.

$\longleftarrow$
Use the equation $A^{-1}A = I$.

**4.** If $M$ and $N$ are similar $n \times n$ matrices, use Exercise 3 to show that $\det(M) = \det(N)$.

**Remember**

$M$ and $N$ are *similar* if there's an invertible matrix $P$ so that $N = P^{-1}MP$. See Lesson 8.5.

**5.** If $A$ is an $n \times n$ matrix with real entries and $A^3 = I$, show that $\det(A) = 1$.

**6.** Show that there is no real $3{\times}3$ matrix $B$ so that $B^2 = \begin{pmatrix} -1 & 0 & 0 \\ 0 & 0 & 1 \\ 0 & -1 & 0 \end{pmatrix}$.

**7.** Suppose that

$$A = \begin{pmatrix} a & b & c \\ d & e & f \\ g & h & i \end{pmatrix} \text{ and } B = \begin{pmatrix} j & k & l \\ m & n & p \\ q & r & s \end{pmatrix}$$

Suppose also that $\det(A) = -7$ and $\det(B) = 5$. Find the following:

**a.** $\det(2A)$ **b.** $\det(B^{-1})$ **c.** $\det(AB)$ **d.** $\det(ABA^{-1})$

**e.** $\begin{vmatrix} a & b & c \\ g & h & i \\ d & e & f \end{vmatrix}$
**f.** $\begin{vmatrix} j & k & l \\ 3m & 3n & 3p \\ q & r & s \end{vmatrix}$

**g.** $\begin{vmatrix} a & b & c \\ d-a & e-b & f-c \\ g & h & i \end{vmatrix}$
**h.** $\begin{vmatrix} j & k & l \\ m & n & p \\ j+m & k+n & l+p \end{vmatrix}$

**i.** $\left|\begin{pmatrix} -2a & -2b & -2c \\ g & h & i \\ d & e & f \end{pmatrix}\begin{pmatrix} m & n & p \\ j & k & l \\ q+2m & r+2n & s+2p \end{pmatrix}\right|$

**8.** Find the value of $x$ if

$$\operatorname{rank}\begin{pmatrix} 1-x & -1 & -1 \\ 1 & 3-x & 1 \\ -3 & 1 & -1-x \end{pmatrix} < 3$$

**9.** Help finish the proof of Theorem 9.15.

**a.** If $E$ is obtained from $I$ by multiplying a row by $c$, show that $\det(E) = c$.

**b.** If $E$ is obtained from $I$ by adding a multiple of one row to another, show that $\det(E) = 1$.

**10.** Help finish the proof of Theorem 9.15 by showing that the product $E_{i(c)}A$ is exactly the same as matrix $A$ with row $i$ multiplied by $c$.

**11.** Finish the proof of Theorem 9.16 by finding the inverses of elementary row matrices of types 2 and 3, and showing that they really are the inverses.

**12.** Help finish the proof of Theorem 9.18: show that if $E$ is an elementary row matrix formed by replacing row $I_{i*}$ of the identity matrix with $I_{i*} + cI_{j*}$, then $\det(EA) = \det(E)\det(A)$.

## 9.5 Determinants as Area and Volume

Remember the philosophy of the extension program: take a familiar geometric idea in two and three dimensions, find a way to describe it with vectors, and then use the algebra as the definition of the idea in higher dimensions.

It's time to pull together some ideas relating determinants, area, and volume, and then extend those ideas to define volume for "boxes" in higher dimensions.

**In this lesson, you will learn how to**

- use Cramer's Rule to find a vector orthogonal to $n-1$ given vectors in $\mathbb{R}^n$
- extend the definition of cross product from the familiar two vectors in $\mathbb{R}^3$ to $n-1$ vectors in $\mathbb{R}^n$
- extend the definition of volume to a box spanned by $n$ vectors in $\mathbb{R}^n$
- use Cramer's Rule to find the solution to a system of linear equations

First, here's a summary of what you've seen so far, since it's the scaffolding for work in higher dimensions.

**Form the Geometric Object.** In $\mathbb{R}^2$, two vectors $P$ and $Q$ define a parallelogram whose other two vertices are $O$ and $P+Q$.

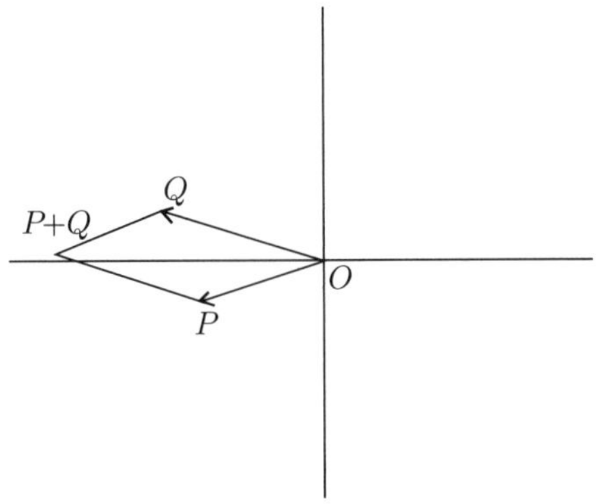


In $\mathbb{R}^3$, three linearly independent vectors $P$, $Q$, and $R$ determine a parallelepiped, whose other vertices are $O$, $(P+Q)$, $(P+R)$, $(Q+R)$, and $(P+Q+R)$. It looks like a (possibly slanted) box.

$\longleftarrow$ Note that $O$, $P$, and $Q$ determine a plane, and $P+Q$ lies in that same plane. The other four vertices are translations of these four points by the point $R$. What happens if the vectors are linearly dependent?

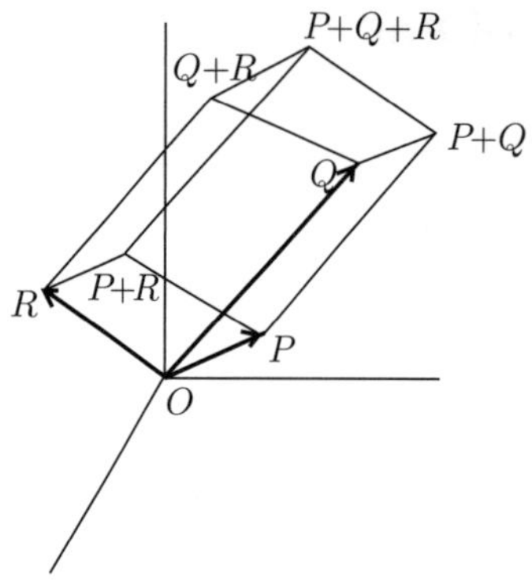


**Area/Volume Formula.** You know from geometry that

$$\text{area of a parallelogram} = (\text{length of base}) \cdot (\text{height})$$
$$\text{volume of a parallelepiped} = (\text{area of base}) \cdot (\text{height})$$

In $\mathbb{R}^2$, the base is one of the two vectors. In $\mathbb{R}^3$, the base is the parallelogram formed by two of the three vectors.

**Projection.** In both $\mathbb{R}^2$ and $\mathbb{R}^3$, you need to find the *height* of the figure, which is really just the projection of one vector onto a vector orthogonal to the others.

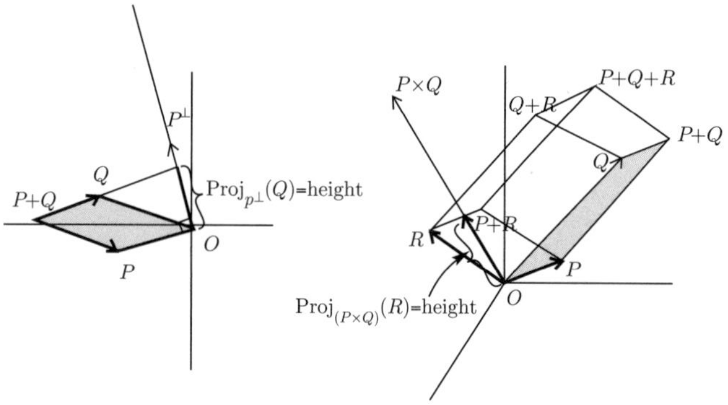


If $P = (a, b)$ in $\mathbb{R}^2$, then a vector orthogonal to $P$ is $P^\perp = (-b, a)$. It has the additional nice property that $\|P\| = \|P^\perp\|$.

⟵ Check that $P \cdot P^\perp = 0$.

In $\mathbb{R}^3$, you know that the cross product $P \times Q$ is orthogonal to both $P$ and $Q$, and has the additional nice property that $\|P \times Q\|$ is the area of the parallelogram spanned by $P$ and $Q$.

⟵ So in $\mathbb{R}^2$ and $\mathbb{R}^3$, you have a vector that is both orthogonal to the base of the figure and has a length numerically equal to the length or area of the base.

**Area/Volume as Dot Product.** Once you have a vector orthogonal to the base, you can find the height and then compute the area or volume.

$$\begin{aligned}
\text{height in } \mathbb{R}^2 &= \|\text{Proj}_{P^\perp} Q\| = \left\| \left( \frac{Q \cdot P^\perp}{P^\perp \cdot P^\perp} \right) P^\perp \right\| \\
&= \frac{|Q \cdot P^\perp|}{\|P^\perp\|^2} \|P^\perp\| = \frac{|Q \cdot P^\perp|}{\|P^\perp\|} \\
\text{height in } \mathbb{R}^3 &= \|\text{Proj}_{P\times Q} R\| = \left\| \left( \frac{R \cdot (P \times Q)}{(P \times Q) \cdot (P \times Q)} \right) (P \times Q) \right\| \\
&= \frac{|R \cdot (P \times Q)|}{\|(P \times Q)\|^2} \|(P \times Q)\| = \frac{|R \cdot (P \times Q)|}{\|(P \times Q)\|}
\end{aligned}$$

Now the area and volume calculations are easy.

$$\begin{aligned}
\text{area in } \mathbb{R}^2 &= (\text{length of base}) \cdot (\text{height}) \\
&= \|P\| \frac{|Q \cdot P^\perp|}{\|P^\perp\|} = |Q \cdot P^\perp| \\
\text{volume in } \mathbb{R}^3 &= (\text{area of base}) \cdot (\text{height}) \\
&= \|P \times Q\| \frac{|R \cdot (P \times Q)|}{\|(P \times Q)\|} = |R \cdot (P \times Q)|
\end{aligned}$$

**Remember**

One nice thing about $P^\perp$ is that $\|P^\perp\| = \|P\|$.

In both cases, the volume is the absolute value of a dot product between a special vector orthogonal to the base, and the one vector that's not part of the base.

**Area/Volume as Determinant.** Of course, you also know that if $P = (a, b)$ and $Q = (c, d)$, then the area of the parallelogram formed by $P$ and $Q$ is the absolute value of the determinant of a matrix, specifically the matrix where $P$ and $Q$ form the columns. You also started out calculating determinants of $3 \times 3$ matrices whose columns were vectors $P$, $Q$, and $R$ in $\mathbb{R}^3$ by calculating $R \cdot (P \times Q)$. So in both $\mathbb{R}^2$ and $\mathbb{R}^3$, the area and volume calculations boil down to finding the absolute value of the determinant of a matrix whose columns are $P$, $Q$, and $R$.

$\longleftarrow$ You should check that $Q \cdot P^\perp$ is exactly the same as $\begin{vmatrix} a & c \\ b & d \end{vmatrix}$.

It's hard to think about "parallelepipeds" and "volume" in dimensions greater than three. You can't draw a picture to help you figure out what's going on. So you need to use the extension program again: use the algebra to *define* the geometric objects. Then check that these objects have all the right properties.

$\longleftarrow$ The fancy name for parallelepipeds in dimension bigger than three is **parallelotope**.

### For Discussion

**1.** Here are four vectors in $\mathbb{R}^4$:

$$\begin{aligned}
P_1 &= (0, 1, 1, 1), \quad P_2 = (1, 0, 2, 1) \\
P_3 &= (2, 1, 1, 0), \quad P_4 = (1, 2, 1, 2)
\end{aligned}$$

**a.** What are the vertices of the "parallelepiped" they define?
**b.** What could be the "base" of the figure?

**c.** What would you call the "height" of this figure?

**d.** Without calculating it, describe what the "volume" of this figure might mean.

---

The first step on the extension program is to define the figures.

### Definition

Given $n$ vectors in $\mathbb{R}^n$, they span a **parallelepiped**. If the vectors are $P_1, P_2, \dots, P_n$, then there are $2^n$ vertices of the parallelepiped: $O$, the $n$ vectors themselves, and all possible sums of two vectors, three vectors, and so on, up to the sum of all $n$ vectors.

⟵ "The $n$ vectors themselves, and all possible sums of two vectors, three vectors, and so on, up to the sum of all $n$ vectors." Why does this give you $2^n$ vectors?

### Cramer's Rule

In $\mathbb{R}^2$ and $\mathbb{R}^3$, the area and volume calculations were made much easier by the fact that you could find a special vector that was orthogonal to the base and that had length numerically equal to the measure (length or area) of the base. Can you do the same trick in higher dimensions?

⟵ You already know what "orthogonal" *means* in any dimension. It means the dot product is 0. But can you find a nice way to do the calculation? And can you always get the special property of the length?

### Example 1

**Problem.** Given $P_1 = (1,1,0,3)$, $P_2 = (-1,1,0,0)$, and $P_3 = (0,1,1,1)$, find a vector $X$ that is orthogonal to all three.

**Solution.** Let $X = (x_1, x_2, x_3, x_4)$. You want to solve this system:

$$\begin{aligned} 1\cdot x_1 + 1\cdot x_2 + 0\cdot x_3 + 3\cdot x_4 &= 0\\ -1\cdot x_1 + 1\cdot x_2 + 0\cdot x_3 + 0\cdot x_4 &= 0\\ 0\cdot x_1 + 1\cdot x_2 + 1\cdot x_3 + 1\cdot x_4 &= 0 \end{aligned}$$

⟵ These equations come from expanding the equations

$$P_1 \cdot X = 0$$
$$P_2 \cdot X = 0$$
$$P_3 \cdot X = 0.$$

You can move the $x_4$'s to the other side :

$$\begin{aligned} 1\cdot x_1 + 1\cdot x_2 + 0\cdot x_3 &= -3\cdot x_4\\ -1\cdot x_1 + 1\cdot x_2 + 0\cdot x_3 &= 0\cdot x_4\\ 0\cdot x_1 + 1\cdot x_2 + 1\cdot x_3 &= -1\cdot x_4. \end{aligned}$$

⟵ This is like what Sasha and Tony did with a similar set of equations in $\mathbb{R}^3$. See Lesson 2.5.

This lets you write a matrix equation:

$$\begin{pmatrix} 1 & 1 & 0\\ -1 & 1 & 0\\ 0 & 1 & 1 \end{pmatrix}\begin{pmatrix} x_1\\ x_2\\ x_3 \end{pmatrix} = \begin{pmatrix} -3\\ 0\\ -1 \end{pmatrix} x_4$$

⟵ **Use New Tools.** In Lesson 2.5, the three friends didn't know how to write a system of equations as a matrix equation. They didn't even know about using the augmented matrix for a system to solve it. Look at how much they (and you) have learned since then.

Let $A = \begin{pmatrix} 1 & 1 & 0\\ -1 & 1 & 0\\ 0 & 1 & 1 \end{pmatrix}$. Expand along the last column to see that $\det(A) = 1\begin{vmatrix} 1 & 1\\ -1 & 1 \end{vmatrix} = 2 \neq 0$. So $A$ has an inverse. You can check that this is correct.

$$A^{-1} = \begin{pmatrix} \frac{1}{2} & -\frac{1}{2} & 0\\ \frac{1}{2} & \frac{1}{2} & 0\\ -\frac{1}{2} & -\frac{1}{2} & 1 \end{pmatrix}$$

That means

$$\begin{pmatrix} x_1 \\ x_2 \\ x_3 \end{pmatrix} = \begin{pmatrix} \frac{1}{2} & -\frac{1}{2} & 0 \\ \frac{1}{2} & \frac{1}{2} & 0 \\ -\frac{1}{2} & -\frac{1}{2} & 1 \end{pmatrix} \begin{pmatrix} -3 \\ 0 \\ -1 \end{pmatrix} x_4 = \begin{pmatrix} -\frac{3}{2} \\ -\frac{3}{2} \\ \frac{1}{2} \end{pmatrix} x_4$$

This equation gives the line orthogonal to the three vectors $P_1$, $P_2$, and $P_3$. It's clear from the form of the answer that a nice choice for a representative vector is $x_4 = 2$. Then you get $X = (-3, -3, 1, 2)$.

⟵ Notice that the convenient value for $x_4$ is exactly $\det(A)$. This is also what Sasha and Tony found in $\mathbb{R}^2$. See Lesson 2.5.

**For You to Do**

**2.** Use the same four points as in For Discussion problem 1.

$$P_1 = (0, 1, 1, 1), \quad P_2 = (1, 0, 2, 1)$$
$$P_3 = (2, 1, 1, 0), \quad P_4 = (1, 2, 1, 2)$$

**a.** Find a vector $X$ that is orthogonal to $P_2$, $P_3$, and $P_4$.
**b.** Find the projection of $P_1$ onto $X$.

For two vectors in $\mathbb{R}^3$, the equations to find a third vector orthogonal to both are simpler. It's a good idea to focus on that case first, to see if a general technique might emerge. If $A = (a_1, a_2, a_3)$ and $B = (b_1, b_2, b_3)$, the equations you want to solve are

$$a_1x_1 + a_2x_2 = -a_3x_3$$
$$b_1x_1 + b_2x_2 = -b_3x_3, \text{ or in matrix terms}$$
$$\begin{pmatrix} a_1 & a_2 \\ b_1 & b_2 \end{pmatrix} \begin{pmatrix} x_1 \\ x_2 \end{pmatrix} = \begin{pmatrix} -a_3 \\ -b_3 \end{pmatrix} x_3$$

You already know that one solution is

$$A \times B = \left( \det \begin{pmatrix} a_2 & a_3 \\ b_2 & b_3 \end{pmatrix}, -\det \begin{pmatrix} a_1 & a_3 \\ b_1 & b_3 \end{pmatrix}, \det \begin{pmatrix} a_1 & a_2 \\ b_1 & b_2 \end{pmatrix} \right)$$

⟵ See the first **Facts and Notation** in Lesson 2.5.

How does this solution relate to the above matrix equation? A little rewriting might make things more clear.

$$\begin{vmatrix} a_2 & a_3 \\ b_2 & b_3 \end{vmatrix} = -\begin{vmatrix} a_3 & a_2 \\ b_3 & b_2 \end{vmatrix} = \begin{vmatrix} -a_3 & a_2 \\ -b_3 & b_2 \end{vmatrix}$$
$$-\begin{vmatrix} a_1 & a_3 \\ b_1 & b_3 \end{vmatrix} = \begin{vmatrix} a_1 & -a_3 \\ b_1 & -b_3 \end{vmatrix}$$
$$\begin{vmatrix} a_1 & a_2 \\ b_1 & b_2 \end{vmatrix} = \begin{vmatrix} a_1 & a_2 \\ b_1 & b_2 \end{vmatrix}$$

Let

$$M = \begin{pmatrix} a_1 & a_2 \\ b_1 & b_2 \end{pmatrix} \quad \text{and} \quad B = \begin{pmatrix} -a_3 \\ -b_3 \end{pmatrix}.$$

Define new matrices $M_i$ where column $i$ is exactly the column vector $B$, and all other columns are identical to the columns of $M$. So

$$M_1 = \begin{pmatrix} -a_3 & a_2 \\ -b_3 & b_2 \end{pmatrix} \quad \text{and} \quad M_2 = \begin{pmatrix} a_1 & -a_3 \\ b_1 & -b_3 \end{pmatrix}$$

Then the cross product satisfies

$$A \times B = (\det(M_1), \det(M_2), \det(M))$$

### For Discussion

**3.** Provide reasons for each step of the calculation

$$\begin{vmatrix} a_2 & a_3 \\ b_2 & b_3 \end{vmatrix} = -\begin{vmatrix} a_3 & a_2 \\ b_3 & b_2 \end{vmatrix} = \begin{vmatrix} -a_3 & a_2 \\ -b_3 & b_2 \end{vmatrix}$$

$$-\begin{vmatrix} a_1 & a_3 \\ b_1 & b_3 \end{vmatrix} = \begin{vmatrix} a_1 & -a_3 \\ b_1 & -b_3 \end{vmatrix}$$

$$\begin{vmatrix} a_1 & a_2 \\ b_1 & b_2 \end{vmatrix} = \begin{vmatrix} a_1 & a_2 \\ b_1 & b_2 \end{vmatrix}$$

Does this generalize to higher dimensions? A natural thing to do is check it against an already worked example, such as the one in Example 1.

### Example 2

**Problem.** Given $P_1 = (1, 1, 0, 3)$, $P_2 = (-1, 1, 0, 0)$, and $P_3 = (0, 1, 1, 1)$, find a vector $X$ that is orthogonal to all three.

**Solution.** In this case, the matrix equation is

$$\begin{pmatrix} 1 & 1 & 0 \\ -1 & 1 & 0 \\ 0 & 1 & 1 \end{pmatrix} \begin{pmatrix} x_1 \\ x_2 \\ x_3 \end{pmatrix} = \begin{pmatrix} -3 \\ 0 \\ -1 \end{pmatrix} x_4, \text{ so } M = \begin{pmatrix} 1 & 1 & 0 \\ -1 & 1 & 0 \\ 0 & 1 & 1 \end{pmatrix} \text{ and } B = \begin{pmatrix} -3 \\ 0 \\ -1 \end{pmatrix}$$

Substitute $B$ for each column of $M$ to get three new matrices.

$$M_1 = \begin{pmatrix} -3 & 1 & 0 \\ 0 & 1 & 0 \\ -1 & 1 & 1 \end{pmatrix}, M_2 = \begin{pmatrix} 1 & -3 & 0 \\ -1 & 0 & 0 \\ 0 & -1 & 1 \end{pmatrix}, \text{ and } M_3 = \begin{pmatrix} 1 & 1 & -3 \\ -1 & 1 & 0 \\ 0 & 1 & -1 \end{pmatrix}$$

Form the vector

$$X = (\det(M_1), \det(M_2), \det(M_3), \det(M)) = (-3, -3, 1, 2)$$

It's a simple matter to check that $X$ is orthogonal to all three of the given vectors.

← In fact, this is the same vector $X$ that was found in Example 1 above, using the method of solving the matrix equation with inverses.

This suggests a general method.

### Developing Habits of Mind

**Plan a general proof.** Essentially, you want to mimic what you have done in the numerical examples so far.

Suppose you have $n-1$ linearly independent vectors in $\mathbb{R}^n$:

$$\begin{aligned} P_1 &= (p_{11}, p_{12}, \dots, p_{1n}) \\ P_2 &= (p_{21}, p_{22}, \dots, p_{2n}) \\ &\vdots \\ P_{n-1} &= (p_{(n-1)1}, p_{(n-1)2}, \dots, p_{(n-1)n}) \end{aligned}$$

Looking for a vector orthogonal to all of them amounts to finding a vector $X = (x_1, x_2, \dots, x_n)$ in $\mathbb{R}^n$ so that

$$\begin{aligned} P_1 \cdot X &= 0 \\ P_2 \cdot X &= 0 \\ &\vdots \\ P_{n-1} \cdot X &= 0 \end{aligned}$$

And this amounts to solving the system of equations

$$\begin{aligned} p_{11}x_1 + \cdots + p_{1(n-1)}x_{n-1} + p_{1n}x_n &= 0 \\ p_{21}x_1 + \cdots + p_{2(n-1)}x_{n-1} + p_{2n}x_n &= 0 \\ &\vdots \\ p_{(n-1)1}x_1 + \cdots + p_{(n-1)(n-1)}x_{n-1} + p_{(n-1)n}x_n &= 0 \end{aligned}$$

The idea is to treat one of the variables as a constant and to "move it over" to the other side. In the two previous examples, the last variable was chosen as the constant, and the square matrix that resulted from the remaining coefficients had nonzero determinant. That doesn't always happen. For example, the vectors $(2, 3, 5)$ and $(4, 6, 7)$ are linearly independent, but if we treat $x_3$ as a constant in the system,

$$\begin{aligned} 2x_1 + 3x_2 + 5x_3 &= 0 \\ 4x_1 + 6x_2 + 7x_3 &= 0 \end{aligned}$$

$$\begin{pmatrix} 2 & 3 \\ 4 & 6 \end{pmatrix} \begin{pmatrix} x_1 \\ x_2 \end{pmatrix} = -\begin{pmatrix} 5 \\ 7 \end{pmatrix} x_3$$

then the matrix $M = \begin{pmatrix} 2 & 3 \\ 4 & 6 \end{pmatrix}$ has determinant 0, and that's trouble.

⟵ Look back at the previous two examples to see where determinant 0 would have caused trouble.

But there's a save: if you start with $n-1$ independent vectors in $\mathbb{R}^n$, you know that the matrix

$$\begin{pmatrix} p_{11} & p_{12} & \cdots & p_{1(n-1)} & p_{1n} \\ p_{21} & p_{22} & \cdots & p_{2(n-1)} & p_{2n} \\ & \vdots & \ddots & & \\ p_{(n-1)1} & p_{(n-1)2} & \cdots & p_{(n-1)(n-1)} & p_{(n-1)n} \end{pmatrix}$$

has row-rank $n-1$. So, by TFAE (Theorem 9.20), it has column-rank $n-1$, too. That means that $n-1$ of the columns are linearly independent, and you can reorder the columns so that the linearly independent ones are the first $n-1$ columns. And the resulting coefficient matrix will have a nonzero determinant. So, the method of the examples will work after (possibly) a little column shuffling.

⟵ TFAE to the rescue again.

Hence, after a possible reordering of the columns, you can treat the last variable, $x_n$, as a constant, rearranging the system so that it is $(n-1) \times (n-1)$:

$$\begin{aligned} p_{11}x_1 + \cdots + p_{1(n-1)}x_{n-1} &= -p_{1n}x_n \\ p_{21}x_1 + \cdots + p_{2(n-1)}x_{n-1} &= -p_{2n}x_n \\ &\vdots \\ p_{(n-1)1}x_1 + \cdots + p_{(n-1)(n-1)}x_{n-1} &= -p_{(n-1)n}x_n \end{aligned}$$

This can be written as a matrix equation:

$$M \begin{pmatrix} x_1 \\ x_2 \\ \vdots \\ x_{n-1} \end{pmatrix} = \begin{pmatrix} -p_{1n} \\ -p_{2n} \\ \vdots \\ -p_{(n-1)n} \end{pmatrix} x_n$$

where

$$M = \begin{pmatrix} p_{11} & p_{12} & \cdots & p_{1(n-1)} \\ p_{21} & p_{22} & \cdots & p_{2(n-1)} \\ & \vdots & \ddots & \\ p_{(n-1)1} & p_{(n-1)2} & \cdots & p_{(n-1)(n-1)} \end{pmatrix}$$

Now, this isn't really a single system—you get a system for every choice of $x_n$. You could pick $x_n$ to be $1, 2, 345, \pi$, or anything. Because the coefficient matrix $M$ now has rank $n-1$, picking $x_n = 0$ would lead to $X = O$ (why?), certainly a vector orthogonal to each of the $P_i$, but not the one you want.

So, assume that $x_n = 1$. You can then write the system as

$$M \begin{pmatrix} x_1 \\ x_2 \\ \vdots \\ x_{n-1} \end{pmatrix} = \begin{pmatrix} -p_{1n} \\ -p_{2n} \\ \vdots \\ -p_{(n-1)n} \end{pmatrix}$$

Just as in the numericals, you can solve this for $X = (x_1, x_2, \ldots, x_{n-1})$. Then you can choose $x_n$ to make $X$ look pleasing, multiply the solution by that value, and get a nice vector orthogonal to each of the $P_i$. The results are in line with what you did in two and three dimensions, and they are detailed in the proof of the next theorem.

**Theorem 9.22 (Cramer's Rule)**
*Given $(n-1)$ linearly independent vectors in $\mathbb{R}^n$,*

$$\begin{aligned} P_1 &= (p_{11}, p_{12}, \ldots, p_{1n}) \\ P_2 &= (p_{21}, p_{22}, \ldots, p_{2n}) \\ &\vdots \\ P_{n-1} &= (p_{(n-1)1}, p_{(n-1)2}, \ldots, p_{(n-1)n}) \end{aligned}$$

*let*

$$M = \begin{pmatrix} p_{11} & p_{12} & \cdots & p_{1(n-1)} \\ p_{21} & p_{22} & \cdots & p_{2(n-1)} \\ & \vdots & \ddots & \\ p_{(n-1)1} & p_{(n-1)2} & \cdots & p_{(n-1)(n-1)} \end{pmatrix} \text{ and } B = \begin{pmatrix} -p_{1n} \\ -p_{2n} \\ \vdots \\ -p_{(n-1)n} \end{pmatrix}$$

*For each number $k = 1, \ldots, n-1$, define a matrix $M_k$ by the following rule: column $k$ of $M_k$ is equal to the column vector $B$, and all other columns are equal to the columns of $M$.*

← In other words,

$$(M_k)_{i*} = \begin{cases} M_{i*} & \text{if } i \neq k \\ B & \text{if } i = k \end{cases}$$

*Let $X$ be a vector in $\mathbb{R}^n$ defined by*

$$X = (\det(M_1), \det(M_2), \ldots, \det(M_{n-1}), \det(M)).$$

*Then $X$ is orthogonal to each of the vectors $P_1$, $P_2$, . . . , $P_{n-1}$.*

***Proof.***

After rearranging the independent vectors $P_i$, it's safe to assume that the rows of $M$ are linearly independent, and so $\det(M) \neq 0$. Hence, by TFAE (Theorem 9.20) there is a unique solution to the matrix equation $MY = B$. Let $Y = (y_1, y_2, \ldots, y_{n-1})$ be that solution.

Now,

$$\det(M_1) = \begin{vmatrix} -p_{1n} & p_{12} & \cdots & p_{1(n-1)} \\ -p_{2n} & p_{22} & \cdots & p_{2(n-1)} \\ & \vdots & \ddots & \\ -p_{(n-1)n} & p_{(n-1)2} & \cdots & p_{(n-1)(n-1)} \end{vmatrix}$$

Since $Y$ is a solution to the matrix equation, you know that

$$\begin{aligned} p_{11}y_1 + \cdots + p_{1(n-1)}y_{n-1} &= -p_{1n} \\ p_{21}y_1 + \cdots + p_{2(n-1)}y_{n-1} &= -p_{2n} \end{aligned}$$

. . . and so on. Substituting these equations in the first column of $M_1$ gives

$$\det(M_1) = \begin{vmatrix} p_{11}y_1 + \cdots + p_{1(n-1)}y_{n-1} & p_{12} & \cdots & p_{1(n-1)} \\ p_{21}y_1 + \cdots + p_{2(n-1)}y_{n-1} & p_{22} & \cdots & p_{2(n-1)} \\ & \vdots & \ddots & \\ p_{(n-1)1}y_1 + \cdots + p_{(n-1)(n-1)}y_{n-1} & p_{(n-1)2} & \cdots & p_{(n-1)(n-1)} \end{vmatrix}$$

Use the fact that determinant is linear to break this up as a sum of several determinants.

← This is the same calculation you did in Exercise 7**a** from Lesson 9.3.

$$\det(M_1) = \begin{vmatrix} p_{11}y_1 & p_{12} & \cdots & p_{1(n-1)} \\ p_{21}y_1 & p_{22} & \cdots & p_{2(n-1)} \\ & \vdots & \ddots & \\ p_{(n-1)1}y_1 & p_{(n-1)2} & \cdots & p_{(n-1)(n-1)} \end{vmatrix} + \begin{vmatrix} p_{12}y_2 & p_{12} & \cdots & p_{1(n-1)} \\ p_{22}y_2 & p_{22} & \cdots & p_{2(n-1)} \\ & \vdots & \ddots & \\ p_{(n-1)2}y_2 & p_{(n-1)2} & \cdots & p_{(n-1)(n-1)} \end{vmatrix} + \ddots + \begin{vmatrix} p_{1(n-1)}y_{n-1} & p_{12} & \cdots & p_{1(n-1)} \\ p_{22}y_{n-1} & p_{22} & \cdots & p_{2(n-1)} \\ & \vdots & \ddots & \\ p_{(n-1)(n-1)}y_{n-1} & p_{(n-1)2} & \cdots & p_{(n-1)(n-1)} \end{vmatrix}$$

$$= y_1 \begin{vmatrix} p_{11} & p_{12} & \cdots & p_{1(n-1)} \\ p_{21} & p_{22} & \cdots & p_{2(n-1)} \\ & \vdots & \ddots & \\ p_{(n-1)1} & p_{(n-1)2} & \cdots & p_{(n-1)(n-1)} \end{vmatrix} + y_2 \begin{vmatrix} p_{12} & p_{12} & \cdots & p_{1(n-1)} \\ p_{22} & p_{22} & \cdots & p_{2(n-1)} \\ & \vdots & \ddots & \\ p_{(n-1)2} & p_{(n-1)2} & \cdots & p_{(n-1)(n-1)} \end{vmatrix} + \ddots + y_{n-1} \begin{vmatrix} p_{1(n-1)} & p_{12} & \cdots & p_{1(n-1)} \\ p_{22} & p_{22} & \cdots & p_{2(n-1)} \\ & \vdots & \ddots & \\ p_{(n-1)(n-1)} & p_{(n-1)2} & \cdots & p_{(n-1)(n-1)} \end{vmatrix}$$

In this sum, each determinant except the first has two equal columns, so they all evaluate to 0. The first is exactly $\det(M)$. So $\det(M_1) = y_1 \det(M)$ and $y_1 = \det(M_1)/\det(M)$. Similar calculations show that $y_2 = \det(M_2)/\det(M)$, and so on for each component of the vector $Y$.

← See how important it is in this proof for $\det(M)$ to be nonzero?

So

$$M \begin{pmatrix} \det(M_1)/\det(M) \\ \det(M_2)/\det(M) \\ \vdots \\ \det(M_{n-1})/\det(M) \end{pmatrix} = B$$

This gives a formula for $Y$ so that $MY = B$. And, as you saw in the *Developing Habits of Mind* earlier in this lesson, you can find a vector $X = (x_1, \dots, x_{n-1}, x_n)$ orthogonal to each of the $P_i$ by scaling $Y$ by any number $x_n$ and setting $X = (y_1, \dots, y_{n-1}, x_n)$. But the above equation says that

$$M \begin{pmatrix} \det(M_1) \\ \det(M_2) \\ \vdots \\ \det(M_{n-1}) \end{pmatrix} = B \det(M)$$

So a clear choice is to let $x_n = \det(M)$. It follows that

$$X = (\det(M_1), \det(M_2), \dots, \det(M_{n-1}), \det(M))$$

is orthogonal to all of the vectors $P_1, \dots, P_{n-1}$. ■

⟵ This was the clear choice in $\mathbb{R}^2$, $\mathbb{R}^3$, and all the earlier examples as well.

### For You to Do

**4.** Again take the points

$$P_1 = (0,1,1,1), \quad P_2 = (1,0,2,1)$$
$$P_3 = (2,1,1,0), \quad P_4 = (1,2,1,2)$$

Use Cramer's Rule to find a vector orthogonal to $P_2$, $P_3$, and $P_4$.

⟵ Do you get the same answer as in For You to Do problem 2? If the vectors are not the same, how can they both be orthogonal to the three vectors?

Notice that Cramer's Rule actually gives you yet another method to solve certain matrix equations. Given a matrix equation $MX = B$ with $M$ a square matrix and $\det(M) \neq 0$, the proof of Cramer's Rule showed that the unique solution can be described by the following corollary.

**Corollary 9.23 (Cramer's Rule, System of Equations Version)**
*Suppose $M$ is a square matrix with nonzero determinant. Using the notation of Theorem 9.22, the unique solution to $MX = B$ is*

$$X = \left( \frac{\det(M_1)}{\det(M)}, \frac{\det(M_2)}{\det(M)}, \dots, \frac{\det(M_n)}{\det(M)} \right),$$

*where $M_i$ is the matrix obtained from $M$ by replacing $M_{*i}$ by $B$.*

### Example 3

**Problem.** Use Cramer's Rule to solve this matrix equation.

$$\begin{pmatrix} 0 & 1 & 1 & 0 \\ -1 & 2 & 1 & -1 \\ 0 & 3 & 1 & -1 \\ 0 & 0 & 4 & 1 \end{pmatrix} \begin{pmatrix} x \\ y \\ z \\ w \end{pmatrix} = \begin{pmatrix} 1 \\ 2 \\ 3 \\ 4 \end{pmatrix}$$

**Solution.** Evaluating along the first column, you can check that $\det(M) = 2$, so there is a unique solution. Define $M_i$ for $i = 1, 2, 3, 4$ by substituting the column vector $B$ for the appropriate column.

$$\det(M_1) = \begin{vmatrix} 1 & 1 & 1 & 0 \\ 2 & 2 & 1 & -1 \\ 3 & 3 & 1 & -1 \\ 4 & 0 & 4 & 1 \end{vmatrix} = 4$$

$$\det(M_2) = \begin{vmatrix} 0 & 1 & 1 & 0 \\ -1 & 2 & 1 & -1 \\ 0 & 3 & 1 & -1 \\ 0 & 4 & 4 & 1 \end{vmatrix} = -2$$

Similarly you can find that $\det(M_3) = 4$ and $\det(M_4) = -8$. So the solution is $X = (4/2, -2/2, 4/2, -8/2) = (2, -1, 2, -4)$.

⟵ Check that $X$ really is the solution to the original matrix equation.

### Developing Habits of Mind

**Look for structural similarities.** The two versions of Cramer's Rule really say the same thing if you look carefully at the systems of equations that arise. For example, the system in the above example can be written as

$$\begin{pmatrix} 0 & 1 & 1 & 0 & -1 \\ -1 & 2 & 1 & -1 & -2 \\ 0 & 3 & 1 & -1 & -3 \\ 0 & 0 & 4 & 1 & -4 \end{pmatrix} \begin{pmatrix} x \\ y \\ z \\ w \\ 1 \end{pmatrix} = \begin{pmatrix} 0 \\ 0 \\ 0 \\ 0 \\ 0 \end{pmatrix}$$

And this results in a vector orthogonal to the rows of the coefficient matrix.

### For You to Do

5. Use Cramer's Rule to solve this matrix equation.

$$\begin{pmatrix} 0 & 1 & 1 & 0 \\ -1 & 2 & 1 & -1 \\ 0 & 3 & 1 & -1 \\ 0 & 0 & 4 & 1 \end{pmatrix} \begin{pmatrix} x \\ y \\ z \\ w \end{pmatrix} = \begin{pmatrix} 0 \\ 2 \\ 2 \\ 0 \end{pmatrix}$$

## Generalized Cross Products

You already know a lot about cross products of two vectors in $\mathbb{R}^3$. Let $A$, $B$, and $C$ be vectors in $\mathbb{R}^3$ and let $c$ be a scalar: the following were all proved as part of the Basic Rules of Cross Product (Theorem 2.8):

(1) $(A \times B) \cdot A = 0$ and $(A \times B) \cdot B = 0$.

(2) $A \times B = -B \times A$.

(3) $A \times A = O$.

(4) $cA \times B = c(A \times B)$ and $A \times cB = c(A \times B)$.

(5) $A \times (B + C) = (A \times B) + (A \times C)$.

Any notion of cross product in higher dimensions should certainly preserve these properties. It turns out that the vector you get out of Cramer's Rule does just that.

### Definition

Given $(n-1)$ vectors in $\mathbb{R}^n$,

$$\begin{aligned} P_1 &= (p_{11}, p_{12}, \dots, p_{1n}) \\ P_2 &= (p_{21}, p_{22}, \dots, p_{2n}) \\ &\vdots \\ P_{n-1} &= (p_{(n-1)1}, p_{(n-1)2}, \dots, p_{(n-1)n}) \end{aligned}$$

let

$$M = \begin{pmatrix} p_{11} & p_{12} & \cdots & p_{1(n-1)} \\ p_{21} & p_{22} & \cdots & p_{2(n-1)} \\ & \vdots & \ddots & \\ p_{(n-1)1} & p_{(n-1)2} & \cdots & p_{(n-1)(n-1)} \end{pmatrix} \text{ and } B = \begin{pmatrix} -p_{1n} \\ -p_{2n} \\ \vdots \\ -p_{(n-1)n} \end{pmatrix}$$

For each number $k = 1, \dots, n-1$, let $M_k$ be the matrix

$$(M_k)_{i*} = \begin{cases} M_{i*} & \text{if } i \neq k \\ B & \text{if } i = k \end{cases}$$

The **generalized cross product** of the given vectors is

$$P_1 \times P_2 \times \cdots \times P_{n-1} = (\det(M_1), \det(M_2), \dots, \det(M_{n-1}), \det(M))$$

← Notice that the proof of Cramer's Rule required the vectors to be linearly independent, but this definition does not. The justification will come from the fact that when the vectors are linearly dependent, this cross product is the vector $O$.

### Facts and Notation

In dimensions higher than 3, the generalized cross product is sometimes denoted by $\wedge$, so that, in $\mathbb{R}^4$, for example, $A \times B \times C$ is denoted by $A \wedge B \wedge C$.

### For You to Do

**6.** Here are three vectors in $\mathbb{R}^4$. You'll use them to test some properties of cross product.

$$P_1 = (2, 3, 0, -2), P_2 = (0, -1, -5, -1), P_3 = (1, 0, 0, 2)$$

**a.** Find $P_1 \times P_2 \times P_3$.

**b.** Find $P_2 \times P_1 \times P_3$.

**c.** Find $P_1 \times P_2 \times P_1$.

**d.** Find $P_1 \times 3P_2 \times P_3$.

**Theorem 9.24 (Basic Rules of Generalized Cross Product)**
*Let $P_1, P_2, \ldots, P_{n-1}$ be vectors in $\mathbb{R}^n$ and let $c$ be a scalar. The cross product $P_1 \times P_2 \times \cdots \times P_{n-1}$ satisfies the following:*

(1) $(P_1 \times P_2 \times \cdots \times P_{n-1}) \cdot P_i = 0$ *for* $i = 1, 2, \ldots, n-1$.

(2) *Interchanging two vectors in the list $P_1, P_2, \ldots, P_{n-1}$ multiplies the cross product by $-1$.*

← So, for example, $P_2 \times P_1 \times P_3 \times \cdots \times P_{n-1} = -(P_1 \times P_2 \times P_3 \times \cdots \times P_{n-1})$.

(3) *If two of the vectors are equal, the cross product is $O$.*

(4) *Multiplying any of the vectors by the scalar $c$ multiplies the cross product by $c$.*

← So, for example, $P_1 \times cP_2 \times \cdots \times P_{n-1} = c(P_1 \times P_2 \times P_3 \times \cdots \times P_{n-1})$.

(5) *Cross product is linear in each of the vectors $P_i$. That is, if $P_i = P_i' + P_i''$, then*

$$\begin{aligned} &P_1 \times P_2 \times \cdots \times P_i \times \cdots \times P_{n-1} \\ &= (P_1 \times P_2 \times \cdots \times P_i' \times \cdots \times P_{n-1}) \\ &\quad + (P_1 \times P_2 \times \cdots \times P_i'' \times \cdots \times P_{n-1}) \end{aligned}$$

***Proof.*** For part ((1)), if the vectors are linearly independent, this follows immediately from Theorem 9.22 above. If they are linearly dependent, then it will follow from parts ((3)), ((4)), and ((5)).

Here's a proof for part ((2)). If you interchange vectors $P_i$ and $P_j$, then you interchange row $i$ and row $j$ in each of the matrices $M$, $M_1, \ldots,$ $M_{n-1}$. So each of the determinants is multiplied by $-1$, which means the cross product is multiplied by $-1$. ■

← You'll do the the rest in the Exercises.

**For Discussion**

**7.** Use parts ((2)), ((3)), ((4)), and ((5)) to prove that if any of the vectors is a linear combination of the others, then the cross product is $O$.

### Volume in Higher Dimensions

Recall that in $\mathbb{R}^2$, the area of a parallelogram spanned by vectors $P$ and $Q$ could be calculated as $|Q \cdot P^{\perp}|$, and in $\mathbb{R}^3$ the volume of the parallelepiped spanned by $P$, $Q$, and $R$ could be calculated as $|R \cdot (P \times Q)|$. The extension program suggests the following definition for higher dimensions.

**Definition**

Let $P_1, P_2, \cdots, P_n$ be vectors in $\mathbb{R}^n$. Define the **volume of the parallelepiped** spanned by these vectors to be $|P_n \cdot (P_1 \times P_2 \times \ldots \times P_{n-1})|$.

**Minds in Action** Episode 46

*The three friends are talking about this new definition of volume.*

TONY: I see how this definition makes sense. But it seems like when we actually calculated areas and volumes, we did it by calculating the determinant of a matrix.

DERMAN: Yeah, like if $P$ and $Q$ were two vectors in the plane, then we made the matrix with $P$ and $Q$ as columns. And if $P$, $Q$, and $R$ were three vectors in $\mathbb{R}^3$, then we made the matrix with those columns and calculated its determinant.

TONY: Yeah. With this new definition, we have to calculate a whole bunch of determinants to find the cross product, and *then* do the dot product with the last vector. It seems like more work.

SASHA: No, not really. Because if you calculate the determinant of the bigger matrix, you still have to calculate lots of determinants of the minors as you go along.

DERMAN AND TONY: Huh?

SASHA: Look, think about four dimensions so things don't get too complicated with the notation. Take four vectors $X, Y, Z, W$ and put them in a matrix as columns.

$$M = \begin{pmatrix} x_1 & y_1 & z_1 & w_1 \\ x_2 & y_2 & z_2 & w_2 \\ x_3 & y_3 & z_3 & w_3 \\ x_4 & y_4 & z_4 & w_4 \end{pmatrix}$$

Say you want to calculate the determinant along the last column. You have to do it like this:

$$\det(M) = -w_1 \begin{vmatrix} x_2 & y_2 & z_2 \\ x_3 & y_3 & z_3 \\ x_4 & y_4 & z_4 \end{vmatrix} + w_2 \begin{vmatrix} x_1 & y_1 & z_1 \\ x_3 & y_3 & z_3 \\ x_4 & y_4 & z_4 \end{vmatrix} - w_3 \begin{vmatrix} x_1 & y_1 & z_1 \\ x_2 & y_2 & z_2 \\ x_4 & y_4 & z_4 \end{vmatrix} + w_4 \begin{vmatrix} x_1 & y_1 & z_1 \\ x_2 & y_2 & z_2 \\ x_3 & y_3 & z_3 \end{vmatrix}$$

TONY: Oh, I see. So it looks like $W \cdot X$, where $X$ is a vector made up of a bunch of determinants of a $3 \times 3$ matrix.

DERMAN: But they're not the same determinants as $X \times Y \times Z$. Are they?

SASHA: They don't look like it. But . . . let's see . . . Let's just think about the first one. We want

$$-\begin{vmatrix} x_2 & y_2 & z_2 \\ x_3 & y_3 & z_3 \\ x_4 & y_4 & z_4 \end{vmatrix}$$

to be the same as

$$\begin{vmatrix} -x_4 & x_2 & x_3 \\ -y_4 & y_2 & y_3 \\ -z_4 & z_2 & z_3 \end{vmatrix}$$

since that's the first coordinate of $X \times Y \times Z$. If we switch columns two times, we get

$$\begin{vmatrix} -x_4 & x_2 & x_3 \\ -y_4 & y_2 & y_3 \\ -z_4 & z_2 & z_3 \end{vmatrix} = \begin{vmatrix} x_2 & x_3 & -x_4 \\ y_2 & y_3 & -y_4 \\ z_2 & z_3 & -z_4 \end{vmatrix}$$

TONY: Yeah, and then we can factor out the negative sign, so

$$\begin{vmatrix} -x_4 & x_2 & x_3 \\ -y_4 & y_2 & y_3 \\ -z_4 & z_2 & z_3 \end{vmatrix} = -\begin{vmatrix} x_2 & x_3 & x_4 \\ y_2 & y_3 & y_4 \\ z_2 & z_3 & z_4 \end{vmatrix}$$

SASHA: And then just take a transpose to get what we want. I bet the other ones work the same way. Very smooth.

---

Sasha, Tony, and Derman have figured out a very nice fact.

**Theorem 9.25 (Volume as Determinant)**
*Let $P_1, P_2, \ldots, P_n$ be vectors in $\mathbb{R}^n$, and let $M$ be the matrix defined by $M_{*j} = P_j$. Then*

$$P_n \cdot (P_1 \times P_2 \times \cdots \times P_{n-1}) = \det(M)$$

← In other words, the columns of $M$ are the vectors $P_1, P_2, \ldots, P_n$, in order.

***Proof.*** Here's a proof that the first term in each sum is always the same. Let

← The idea is essentially the same as what Sasha, Tony, and Derman describe. The only trick is keeping track of the negative signs.

$$\begin{aligned} P_1 &= (p_{11}, p_{12}, \ldots, p_{1n}) \\ P_2 &= (p_{21}, p_{22}, \ldots, p_{2n}) \\ &\vdots \\ P_{n-1} &= (p_{(n-1)1}, p_{(n-1)2}, \ldots, p_{(n-1)n}) \\ P_n &= (p_{n1}, p_{n2}, \ldots, p_{nn}) \end{aligned}$$

The first component of $P_1 \times P_2 \times \cdots \times P_{n-1}$ is

$$\begin{vmatrix} -p_{1n} & p_{12} & p_{13} & \cdots & p_{1(n-1)} \\ -p_{2n} & p_{22} & p_{23} & \cdots & p_{2(n-1)} \\ & \vdots & & \ddots & \\ -p_{(n-1)n} & p_{(n-1)2} & p_{(n-1)3} & \cdots & p_{(n-1)(n-1)} \end{vmatrix}$$

It takes $n-2$ switches to move the first column of this matrix to the last column, keeping all of the others in the same order, so this determinant is the same as

$$(-1)^{n-2} \begin{vmatrix} p_{12} & p_{13} & \cdots & p_{1(n-1)} & -p_{1n} \\ p_{22} & p_{23} & \cdots & p_{2(n-1)} & -p_{2n} \\ & \vdots & & \ddots & \\ p_{(n-1)2} & p_{(n-1)3} & \cdots & p_{(n-1)(n-1)} & -p_{(n-1)n} \end{vmatrix}$$

Then you can factor out $-1$ from the last column to get

$$(-1)^{n-1} \begin{vmatrix} p_{12} & p_{13} & \cdots & p_{1(n-1)} & p_{1n} \\ p_{22} & p_{23} & \cdots & p_{2(n-1)} & p_{2n} \\ & \vdots & & \ddots & \\ p_{(n-1)2} & p_{(n-1)3} & \cdots & p_{(n-1)(n-1)} & p_{(n-1)n} \end{vmatrix}$$

So the first term in the sum of $P_n \cdot (P_1 \times P_2 \times \cdots \times P_{n-1})$ is

$$(-1)^{n-1} p_{n1} \begin{vmatrix} p_{12} & p_{13} & \cdots & p_{1(n-1)} & p_{1n} \\ p_{22} & p_{23} & \cdots & p_{2(n-1)} & p_{2n} \\ & \vdots & & \ddots & \\ p_{(n-1)2} & p_{(n-1)3} & \cdots & p_{(n-1)(n-1)} & p_{(n-1)n} \end{vmatrix}$$

Taking the transpose of the matrix shows that this is the same as the first term in the expansion of $\det(M)$ along the last column. ■

### For You to Do

**8.** Start with the same four vectors as in For Discussion problem 1.

$$P_1 = (0, 1, 1, 1),\ P_2 = (1, 0, 2, 1),\ P_3 = (2, 1, 1, 0),\ P_4 = (1, 2, 1, 2)$$

Use determinants to find the four-dimensional volume of the box formed by these vectors.

Remember in Lesson 5.4 Sasha talked about how area should act, including:

1. The area of the unit square should be 1.

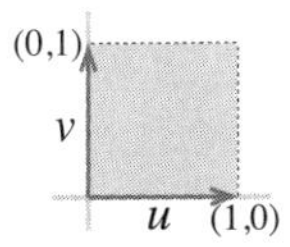


2. If you scale one side of a parallelogram by $c$, the area should scale by $|c|$.

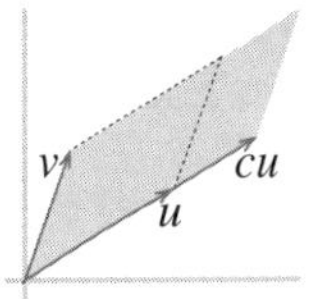


3. If you switch the order of the vectors that span the parallelogram, the area should stay the same.
4. If the vectors that span the parallelogram are scalar multiples of each other, the area should be 0.

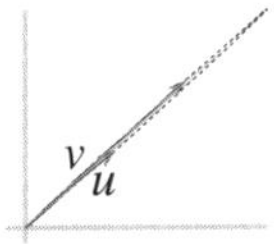


### For Discussion

**9.** Make a list like Sasha's for how volume should act. Use properties of determinants to prove that each one is true for volumes in higher dimensions as they've been defined in this lesson.

## Exercises

**1.** For each part, find a vector orthogonal to the given vectors.

**a.** $P_1 = (1, 0, 0, -2),\quad P_2 = (1, -1, 2, 0),\quad P_3 = (2, 0, 1, 1)$
**b.** $P_1 = (1, 2, 0, 1),\quad P_2 = (0, 1, 0, 2),\quad P_3 = (3, 2, 1, 1)$
**c.** $P_1 = (1, -2, 0, 1),\quad P_2 = (1, 2, 1, 0),\quad P_3 = (2, 0, -1, 1)$
**d.** $P_1 = (3, 1, 0, 1),\quad P_2 = (1, 1, 1, 1),\quad P_3 = (1, 0, 1, 3)$
**e.** $P_1 = (1, -1, 0, 1, 1),\quad P_2 = (1, 0, 2, 1, -1),$
$P_3 = (1, 1, 0, 0, 2),\quad P_4 = (1, -2, 1, 1, 0)$

**f.** $P_1 = (3,0,0,1,1)$, $P_2 = (2,1,1,1,0)$,
$P_3 = (-1,0,-1,1,1)$, $P_4 = (2,1,0,1,0)$

**2.** For each part, use Cramer's Rule to find a vector orthogonal to the given vectors.

**a.** $P_1 = (1,2,0,1)$, $P_2 = (0,1,0,2)$, $P_3 = (3,2,1,1)$
**b.** $P_1 = (1,-2,0,1)$, $P_2 = (1,2,1,0)$, $P_3 = (2,0,-1,1)$
**c.** $P_1 = (2,2,0,-1)$, $P_2 = (1,0,-1,0)$, $P_3 = (2,0,0,2)$
**d.** $P_1 = (1,-1,0,1)$, $P_2 = (2,0,1,1)$, $P_3 = (1,3,1,-1)$
**e.** $P_1 = (1,-1,0,1,1)$, $P_2 = (1,0,2,1,-1)$,
$P_3 = (1,1,0,0,2)$, $P_4 = (1,-2,1,1,0)$
**f.** $P_1 = (0,1,0,1,2)$, $P_2 = (2,0,1,-1,-1)$,
$P_3 = (-1,-1,0,2,0)$, $P_4 = (1,2,0,1,1)$

**3.** For each part below, use the cross product to find a vector orthogonal to the two vectors. Then use Cramer's Rule to find a vector orthogonal to them. Do both methods give you the same orthogonal vector?

**a.** $P_1 = (3,2,2)$, $P_2 = (2,1,0)$
**b.** $P_1 = (-1,0,1)$, $P_2 = (2,4,-1)$

**4.** For each given matrix equation, use Cramer's Rule to solve it.

**a.** $\begin{pmatrix} 1 & 2 \\ 3 & 2 \end{pmatrix} \begin{pmatrix} x \\ y \end{pmatrix} = \begin{pmatrix} 3 \\ 4 \end{pmatrix}$ **b.** $\begin{pmatrix} 1 & 0 \\ 2 & -1 \end{pmatrix} \begin{pmatrix} x \\ y \end{pmatrix} = \begin{pmatrix} 2 \\ 3 \end{pmatrix}$

**c.** $\begin{pmatrix} 1 & 2 & 1 \\ 1 & 0 & 2 \\ -1 & 0 & 1 \end{pmatrix} \begin{pmatrix} x \\ y \\ z \end{pmatrix} = \begin{pmatrix} 1 \\ 1 \\ 2 \end{pmatrix}$

**5.** For each part, find the volume of the parallelepiped spanned by the given vectors.

**a.** $P_1 = (1,2,0,1)$, $P_2 = (0,1,0,2)$,
$P_3 = (1,2,1,1)$, $P_4 = (1,0,0,1)$
**b.** $P_1 = (3,1,0,1)$, $P_2 = (1,1,1,1)$,
$P_3 = (1,0,1,3)$, $P_4 = (1,1,1,0)$
**c.** $P_1 = (3,2,0,0)$, $P_2 = (1,2,2,1)$,
$P_3 = (1,-2,1,1)$, $P_4 = (2,0,1,1)$
**d.** $P_1 = (2,-1,0,2)$, $P_2 = (1,0,0,0)$,
$P_3 = (1,0,-2,2)$, $P_4 = (-1,0,2,0)$
**e.** $P_1 = (0,1,0,1,2)$, $P_2 = (2,1,1,0,1)$,
$P_3 = (1,0,1,2,1)$, $P_4 = (1,1,1,2,0)$,
$P_5 = (1,3,1,1,1)$
**f.** $P_1 = (1,0,0,-2,1)$, $P_2 = (3,1,1,0,2)$,
$P_3 = (0,0,1,-1,0)$, $P_4 = (2,0,-1,0,1)$,
$P_5 = (1,2,0,0,2)$

**6.** Finish Sasha, Tony, and Derman's argument. Show that if you have four vectors $X, Y, Z, W$ in $\mathbb{R}^4$, then

$$W \cdot (X \times Y \times Z) = \det(M)$$

where $M$ is the matrix whose columns are $X, Y, Z, W$ in order.

**7.** Explain why there are $2^n$ vertices for an $n$-dimensional parallelepiped.

**8.** Suppose $T : \mathbb{R}^n \to \mathbb{R}^n$ is a linear map whose matrix with respect to the standard basis is $A$. If $P_1, P_2, \ldots, P_n$ are vectors in $\mathbb{R}^n$, show that the volume of the box spanned by $T(P_1), T(P_2), \ldots, T(P_n)$ is $\det(A)$ times the volume of the box spanned by $P_1, P_2, \ldots, P_n$.

**9.** Prove part ((3)) of Theorem 9.24.

**10.** Prove part ((4)) of Theorem 9.24.

**11.** Prove part ((5)) of Theorem 9.24.

**12.** Here's a picture to demonstrate Cramer's Rule in $\mathbb{R}^3$. Suppose $\mathbf{u}$, $\mathbf{v}$, and $\mathbf{w}$ are vectors in $\mathbb{R}^3$, so they form a parallelepiped. And suppose $\mathbf{q}$ is another vector that is a linear combination of those three. So $\mathbf{q} = x\mathbf{u} + y\mathbf{v} + z\mathbf{w}$.

Explain why the picture shows that

$$\det(\mathbf{q}, \mathbf{v}, \mathbf{w}) = x \det(\mathbf{u}, \mathbf{v}, \mathbf{w}), \text{ so}$$
$$x = \frac{\det(\mathbf{q}, \mathbf{v}, \mathbf{w})}{\det(\mathbf{u}, \mathbf{v}, \mathbf{w})}$$

⟵ Relate the determinants to volumes of objects in the picture. How are those objects related, and why?

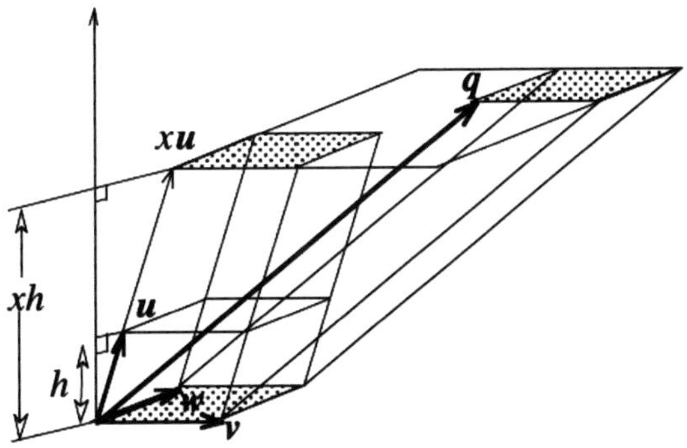

## 9.6 Eigenvalues and Eigenvectors

You have already seen several examples of matrices (and linear transformations) that have *fixed vectors*. In this lesson, you'll extend that notion to fixed lines and, more generally, fixed subspaces.

$\longleftarrow$ **Attend to precision.** A matrix or transformation fixes a line if the image of any point on the line is another point on the line. It doesn't have to map every point on the line to itself, just to some other point on the line.

**In this lesson, you will learn how to**

- find the *characteristic polynomial* of a matrix
- recognize the underlying geometry of the characteristic polynomial's real roots
- establish the relationship between the *eigenvalues*, *eigenvectors*, and characteristic polynomials of similar matrices
- find the *invariant subspaces* of a matrix or linear transformation

### Example 1

**Problem.** Decide if the following matrix has any fixed lines.

$$A = \begin{pmatrix} 1 & 2 \\ 3 & 2 \end{pmatrix}$$

$\longleftarrow$ You can think of $A$ as the matrix for a linear map $T : \mathbb{R}^2 \to \mathbb{R}^2$ with respect to, say, standard bases. In that case, $T(x, y) = (x + 2y, 3x + 2y)$.

**Solution.** If a line goes through the origin, then it has an equation like $X = tQ$ for some vector $Q$. That is, the line is all multiples of the vector $Q$. If the line is fixed by $A$, then it must be the case that $AQ$ is some other point on the line, so it must be a multiple of $Q$. Say $AQ = \lambda Q$. If that happens, you can take any other point on the line and calculate

$$A(tQ) = t(AQ) = t(\lambda Q) = (t\lambda)Q$$

$\longleftarrow$ The symbol $\lambda$ is the Greek letter "lambda," and is used here to stand for a real number.

So every multiple of $Q$ maps to a multiple of $Q$, and the whole line is fixed by $A$.

That means you need to look for vectors $Q$ where $A$ just multiplies $Q$ by some number. You could start by trying possible values for $\lambda$. Maybe start with $\lambda = 1$.

$$AQ = Q$$
$$\begin{pmatrix} 1 & 2 \\ 3 & 2 \end{pmatrix} \begin{pmatrix} x \\ y \end{pmatrix} = \begin{pmatrix} x \\ y \end{pmatrix}$$

This leads to two equations in two unknowns.

$$x + 2y = x$$
$$3x + 2y = y$$

The first equation says $2y = 0$, and the second says $3x + y = 0$. So the only solution is $Q = O$, which doesn't give a fixed line since all multiples of $O$ are just $O$.

Try $\lambda = -1$ instead.

$$AQ = -Q$$
$$\begin{pmatrix} 1 & 2 \\ 3 & 2 \end{pmatrix} \begin{pmatrix} x \\ y \end{pmatrix} = \begin{pmatrix} -x \\ -y \end{pmatrix}.$$

This leads to two equations in two unknowns.

$$\begin{aligned} x + 2y &= -x \\ 3x + 2y &= -y \end{aligned}$$

These equations simplify to

$$\begin{aligned} 2x + 2y &= 0 \\ 3x + 3y &= 0 \end{aligned}$$

Both of these yield $x = -y$ as a solution, so you can take $Q = (1, -1)$ as the vector, and see that the line with equation $X = tQ$ is fixed by $A$.

### For Discussion

**1.** Does the matrix $A = \begin{pmatrix} 1 & 2 \\ 3 & 2 \end{pmatrix}$ have any fixed lines other than the line with equation $X = t\begin{pmatrix} 1 \\ -1 \end{pmatrix}$? Split up the work and try a few other values of $\lambda$ to see if you can find any others.

← Focus on lines whose equations look like $X = tQ$ for some vector $Q$; that is, look for fixed lines that go through the origin.

### Minds in Action Episode 47

*Sasha, Tony, and Derman are in a group, working on the For Discussion problem, above.*

DERMAN: I don't get why we're just looking for lines through the origin. Why don't we look for fixed lines with equations like $X = P + tQ$? Those lines that are shifted off the origin?

SASHA: Hmm. Good question. What if there were a fixed line like that?

TONY: Well, $A$ would map any point on that line to another point on that line.

SASHA: Yeah, so let's start with the simplest version. $P$ is on the line, right? Because you just let $t = 0$.

TONY: But if $A$ fixes the line, we know that there is a number $c$ so that $AP = P + cQ$.

SASHA: OK. Let's try another: $P+Q$ is also on the line. That's kind of the next simplest point to think about. What if we multiply $A$ by that?

TONY: From what we did before, $A(P+Q) = AP + AQ = P + cQ + AQ$.

SASHA: Yeah, but if that's on the same line, then it has to look like $P + tQ$ for some number $t$. So, for some $t$, $P + cQ + AQ = P + tQ$. That means $cQ + AQ = tQ$ . . . it's just a multiple of $Q$.

TONY: Oh, and then $AQ = (t - c)Q$. So the line with equation $X = tQ$ through the origin would have to be fixed for this to even have the possibility of working. I guess it makes sense to focus on those lines, then.

**Definition**

Let $M$ be an $n \times n$ matrix. A nonzero vector $Q$ satisfying $MQ = \lambda Q$ for a real number $\lambda$ is called an **eigenvector** of $M$ with **eigenvalue** $\lambda$.

← Why does it make sense to require that an eigenvector is not $O$?

← The prefix "eigen" was adopted from German and means "self."

**For Discussion**

**2.** In the definition of eigenvector and eigenvalue, why is it necessary that $M$ is a square matrix?

This chapter deals with eigenvectors and eigenvalues for matrices. Because of the results in Chapter 8, all the definitions and theorems apply just as well if you replace "matrix" by "linear transformation." The "dictionary" that lets you translate from matrices to linear transformations is the assignment of coordinate vectors with respect to a basis. More about this as the lesson develops (and in the Exercises).

**Theorem 9.26 (Eigenvectors Scale)**

*If $Q$ is an eigenvector for a matrix $A$ with eigenvalue $\lambda$, then so is $tQ$ for every real number $t \neq 0$. Furthermore, the line with equation $X = tQ$ through the origin is fixed by the matrix $A$.*

***Proof.*** If $AQ = \lambda Q$, then $A(tQ) = t(AQ) = t(\lambda Q) = \lambda(tQ)$ (since real numbers commute with matrices and with each other). As long as $tQ \neq O$, it fits the definition of an eigenvector with eigenvalue $\lambda$.

Since every multiple of $Q$ is mapped by $A$ to another multiple of $Q$, and since $AO = O$, the line with equation $X = tQ$ is mapped to itself by $A$. ∎

**Minds in Action** Episode 48

*Sasha, Tony, and Derman are still working on the For Discussion problem 1, above.*

TONY: There's got to be a better way to do this. I'm tired of guessing values for $\lambda$ and checking if they work.

DERMAN: Yeah. And how will we even know when we're done? We could keep checking numbers forever . . . 1.2, 1.21, 1.212, . . .

SASHA: There's got to be some way to solve all the equations for all the $Q$'s at once. We want to know if we can find $Q$'s and $\lambda$'s so that

$$AQ = \lambda Q$$

DERMAN: It seems like we've got too many variables.

SASHA: I think this might be a time to write out the details. We want

$$\begin{pmatrix} 1 & 2 \\ 3 & 2 \end{pmatrix} \begin{pmatrix} x \\ y \end{pmatrix} = \lambda \begin{pmatrix} x \\ y \end{pmatrix} = \begin{pmatrix} \lambda x \\ \lambda y \end{pmatrix}$$

TONY: OK, that gives two equations

$$\begin{aligned} x + 2y &= \lambda x \\ 3x + 2y &= \lambda y \end{aligned}$$

which simplifies to

$$\begin{aligned} (1 - \lambda)x + 2y &= 0 \\ 3x + (2 - \lambda)y &= 0 \end{aligned}$$

Derman's right. We only have two equations, but we have three unknowns: $x$, $y$, and $\lambda$. And our system isn't linear. How can we solve that?

← What does Tony mean by "And our system isn't linear"?

SASHA: Hold on . . . what if we write this as a matrix equation again.

$$\begin{pmatrix} 1 - \lambda & 2 \\ 3 & 2 - \lambda \end{pmatrix} \begin{pmatrix} x \\ y \end{pmatrix} = \begin{pmatrix} 0 \\ 0 \end{pmatrix}$$

That'll always have solutions with $x = y = 0$. But we want a nonzero vector $Q$.

TONY: So we need the kernel of this new matrix to have more stuff in it than just $O$. But that can only happen if the matrix is not invertible. Hmm . . . I'm still stuck.

SASHA: It can only happen if the determinant is 0! We can set the determinant equal to 0 and solve the equation for $\lambda$.

*Derman looks into space and mumbles a bit.*

DERMAN: I get $\lambda = -1$, and $\lambda = 4$.

← Check that Derman is right. Find the determinant, set it equal to 0, and solve for $\lambda$.

← If you haven't done so already, find a vector $Q$ such that $AQ = 4Q$.

TONY: Cool. Now we can use those specific values in the original equation to figure out which vectors work for $Q$. And those are the only $\lambda$'s that can ever work. We'll know when we're done!

---

Sasha, Tony, and Derman have outlined a way to find the *eigenvalues* of a matrix. Once you find those, you can look for the corresponding *eigenvectors*. Here's another example of how it works.

### Example 2

**Problem.** Find all of the eigenvalues of this matrix. Then for each eigenvalue, find the corresponding eigenvectors.

$$M = \begin{pmatrix} 1 & 0 & 0 \\ 3 & -2 & 6 \\ 0 & 0 & 1 \end{pmatrix}$$

**Solution.** You want solutions to the matrix equation $MQ = \lambda Q$. The equation can be written like this:

$$\begin{pmatrix} 1 & 0 & 0 \\ 3 & -2 & 6 \\ 0 & 0 & 1 \end{pmatrix} \begin{pmatrix} x \\ y \\ z \end{pmatrix} = \begin{pmatrix} \lambda & 0 & 0 \\ 0 & \lambda & 0 \\ 0 & 0 & \lambda \end{pmatrix} \begin{pmatrix} x \\ y \\ z \end{pmatrix}$$

← Check that

$$\lambda \begin{pmatrix} x \\ y \\ z \end{pmatrix} = \begin{pmatrix} \lambda & 0 & 0 \\ 0 & \lambda & 0 \\ 0 & 0 & \lambda \end{pmatrix} \begin{pmatrix} x \\ y \\ z \end{pmatrix}$$

$$\begin{pmatrix} 1 & 0 & 0 \\ 3 & -2 & 6 \\ 0 & 0 & 1 \end{pmatrix} \begin{pmatrix} x \\ y \\ z \end{pmatrix} - \begin{pmatrix} \lambda & 0 & 0 \\ 0 & \lambda & 0 \\ 0 & 0 & \lambda \end{pmatrix} \begin{pmatrix} x \\ y \\ z \end{pmatrix} = \begin{pmatrix} 0 \\ 0 \\ 0 \end{pmatrix}$$

$$\begin{pmatrix} 1-\lambda & 0 & 0 \\ 3 & -2-\lambda & 6 \\ 0 & 0 & 1-\lambda \end{pmatrix} \begin{pmatrix} x \\ y \\ z \end{pmatrix} = \begin{pmatrix} 0 \\ 0 \\ 0 \end{pmatrix} \qquad (*)$$

Like Sasha said, you can only find nonzero vector solutions if the determinant of the matrix on the left is 0. Calculate the determinant.

**Remember**

Eigenvectors must be *nonzero*. It's part of the definition.

$$\begin{vmatrix} 1-\lambda & 0 & 0 \\ 3 & -2-\lambda & 6 \\ 0 & 0 & 1-\lambda \end{vmatrix} = (1-\lambda)^2(-2-\lambda)$$

So there are only two eigenvalues: 1 and $-2$. What are the eigenvectors? Plug in $\lambda = 1$, and equation $(*)$ becomes

$$\begin{pmatrix} 0 & 0 & 0 \\ 3 & -3 & 6 \\ 0 & 0 & 0 \end{pmatrix} \begin{pmatrix} x \\ y \\ z \end{pmatrix} = \begin{pmatrix} 0 \\ 0 \\ 0 \end{pmatrix}$$

The solution to this is a *plane*, not just a line! It has equation $x - y + 2z = 0$. What are the eigenvectors? Well, they're any (nonzero) vectors in this plane. Every vector in the plane can be described as a linear combination of two linearly independent vectors in the plane, say $Q_1 = (1, 1, 0)$ and $Q_2 = (0, 2, 1)$. So the eigenvectors with eigenvalue 1 look like $tQ_1 + sQ_2$, as long as $t$ and $s$ are not both 0.

← It's a good idea to check your answer. Multiply one of these vectors by the original matrix.

$$\begin{pmatrix} 1 & 0 & 0 \\ 3 & -2 & 6 \\ 0 & 0 & 1 \end{pmatrix} \begin{pmatrix} 1 \\ 1 \\ 0 \end{pmatrix} = \begin{pmatrix} 1 \\ 1 \\ 0 \end{pmatrix}$$

So, $X$ really is an eigenvector with eigenvalue 1. You check $Y$.

Now plug in $\lambda = -2$ to equation $(*)$:

$$\begin{pmatrix} -1 & 0 & 0 \\ 3 & 0 & 6 \\ 0 & 0 & -1 \end{pmatrix} \begin{pmatrix} x \\ y \\ z \end{pmatrix} = \begin{pmatrix} 0 \\ 0 \\ 0 \end{pmatrix}$$

The solutions to this equation satisfy $x = z = 0$ and $y$ is any real number. So one eigenvector with eigenvalue $-2$ is $Q_3 = (0, 1, 0)$. All eigenvectors with eigenvalue $-2$ look like nonzero multiples of this vector $Q_3$.

---

The general method looks like this:

**Step 1:** Rewrite $AQ = \lambda Q$ as $(A - \lambda I)Q = 0$.

**Step 2:** Find $\det(A - \lambda I)$, set it equal to 0, and solve for $\lambda$. These are your eigenvalues.

**Step 3:** Plug in your eigenvalues for $\lambda$ in the equation $(A-\lambda I)Q = 0$, and solve for $Q$ to find the eigenvectors.

### Definition

The determinant $\det(A - \lambda I)$ is a polynomial in $\lambda$. It is called the **characteristic polynomial** for the matrix $A$.

⟵ The **characteristic equation** for $A$ is the polynomial equation $\det(A - \lambda I) = 0$.

### For You to Do

**3.** For each matrix below, find the eigenvalues and the eigenvectors corresponding to each eigenvalue.

**a.** $\begin{pmatrix} 1 & 4 \\ 1 & 1 \end{pmatrix}$ **b.** $\begin{pmatrix} 1 & -1 & -1 \\ 0 & 0 & 1 \\ 0 & -2 & -3 \end{pmatrix}$

## Characteristic Polynomials

### Minds in Action Episode 49

*The three friends are talking about eigenvalues.*

DERMAN: You know what? I found a matrix with no eigenvalues!

TONY: That doesn't make any sense. Every example we've seen, the matrix always has eigenvalues . . . In fact, it seems like $2 \times 2$ matrices have 2 and $3 \times 3$ matrices have 3, and . . .

SASHA: Wait, that's not true. Remember Example 2 from Lesson 9.6? It was a $3 \times 3$ matrix, but it had only two eigenvalues: 1 and $-2$.

TONY: Oh, yeah. But 1 counted twice.

SASHA: What do you mean it "counted twice"?

TONY: The characteristic polynomial was $(1-\lambda)^2(-2-\lambda)$. So 1 counted twice, because the $(1 - \lambda)$ term is squared.

SASHA: Oh, I see what you mean. And also there was a whole plane of eigenvectors for $\lambda = 1$ instead of just a line like there was for $\lambda = -2$. So there really were three eigenvectors.

DERMAN: But that doesn't always happen. Look. Let $A = \begin{pmatrix} 0 & -1 \\ 1 & 0 \end{pmatrix}$. Then the characteristic polynomial is $\lambda^2 + 1$. And nothing makes that 0.

⟵ Is Derman right about the characteristic polynomial of $A$? Calculate it to check his work.

TONY: Whoa.

SASHA: Well, that's not totally true. If we let ourselves use complex numbers, then $i$ makes the characteristic polynomial 0.

DERMAN: Why would we let ourselves do that? You're changing the rules in the middle of the game.

As Sasha, Tony, and Derman are beginning to suspect, there's a lot going on with the characteristic polynomial. One way to think about Tony's claim that an $n \times n$ matrix has $n$ eigenvalues if you're allowed to count some of the eigenvalues twice is in terms of the degree of the characteristic polynomial.

**Theorem 9.27 (Coefficients of the Characteristic Polynomial)**
*Let $A$ be an $n \times n$ matrix. Then,*

(1) *The characteristic polynomial of $A$ has degree exactly $n$.*

(2) *The coefficient of $\lambda^n$ in the characteristic polynomial is $(-1)^n$.*

(3) *The coefficient of $\lambda^{n-1}$ in the characteristic polynomial is $(-1)^{n-1}$ times the trace of $A$.*

(4) *The constant term of the characteristic polynomial is exactly* $\det(A)$.

⟵ In other words, $\lambda^n$ is the highest power of $\lambda$ that appears in the polynomial.

**Remember**

The *trace* of a matrix is the sum of the diagonal entries.

***Proof.*** You'll prove part ((4)) in Exercise 13.

Let

$$A - \lambda I = \begin{pmatrix} A_{11}-\lambda & A_{12} & A_{13} & \cdots & A_{1n} \\ A_{21} & A_{22}-\lambda & A_{23} & \cdots & A_{2n} \\ & \vdots & & \ddots & \\ A_{n1} & A_{n2} & A_{n3} & \cdots & A_{nn}-\lambda \end{pmatrix}$$

Parts ((1))–((3)) all rely on the following. Since each term in any expansion of the determinant contains one entry from each row and one entry from each column, the determinant looks like:

$$\begin{aligned} \det(A - \lambda I) &= (A_{11}-\lambda)(A_{22}-\lambda)\cdots(A_{nn}-\lambda) \\ &\quad + \text{terms where at most } n-2 \text{ of the factors contain } \lambda \end{aligned}$$

⟵ Think about expanding along the first row. When you find the minor $M_{12}$, you delete row 1 and column 2. Since $A_{11} - \lambda$ and $A_{22} - \lambda$ both get deleted, at most $n - 2$ terms with $\lambda$ can show up in $A_{12}\det(M_{12})$. Can you see why the same thing is true of all the other terms except $(A_{11}-\lambda)\det(M_{11})$?

Parts ((1)) and ((2)) are now pretty clear. The highest degree term possible is when you expand out $(A_{11}-\lambda)(A_{22}-\lambda)\cdots(A_{nn}-\lambda)$. One term in the expansion comes from multiplying $n$ copies of $-\lambda$ together to get $(-1)^n\lambda$. Every other term will have smaller degree.

The $\lambda^{n-1}$ term also comes from expanding out

$$(A_{11}-\lambda)(A_{22}-\lambda)\cdots(A_{nn}-\lambda)$$

since all the other terms have degree $n-2$ or smaller. You get that term by taking $n-1$ copies of $-\lambda$ together with one of the $A_{ii}$ terms. You get one of those terms for each constant $A_{ii}$. So when you add them all together, you get $(-1)^{n-1}\lambda^{n-1}(A_{11}+A_{22}+\cdots+A_{nn})$. The sum in parentheses is exactly the trace of $A$. ■

⟵ If you're not sure why this works, try expanding out $(A_{11}-\lambda)(A_{22}-\lambda)$, $(A_{11}-\lambda)(A_{22}-\lambda)(A_{33}-\lambda)$, and $(A_{11}-\lambda)(A_{22}-\lambda)(A_{33}-\lambda)(A_{44}-\lambda)$.

## Developing Habits of Mind

**Look for connections.** Derman noticed that sometimes characteristic polynomials have no real roots. But a famous theorem in linear algebra says that every matrix has at least one eigenvalue. How can this be?

Well, you have to do what Sasha did and change the rules of the game a bit. Eigenvalues are just roots of a polynomial. A famous theorem from algebra says that every (nonconstant) polynomial has a root, as long as you allow the roots to be complex numbers. For an $n \times n$ matrix, the characteristic polynomial has degree $n$, meaning it's not constant. So it always has at least one (complex) root. In fact, the same theorem can be used to prove that a degree $n$ polynomial has exactly $n$ roots, as long as you count "with multiplicity" like Tony does. You'll learn more about this in the next lesson.

⟵ It's such a famous theorem that it's called "The Fundamental Theorem of Algebra."

Is there a geometric way to think about matrices that do have real eigenvalues versus those that do not? If a matrix has entries from $\mathbb{R}$ and has real eigenvalues, then it will have eigenvectors that have real entries as well. So the matrix will have a fixed line somewhere. Can you think of matrices that don't have any fixed lines at all?

Think about matrices as geometric transformations. If you picture rotating the plane around the origin, unless you rotate by some multiple of 180°, you won't have any fixed lines at all. That means most rotation matrices in $\mathbb{R}^2$ don't have any fixed lines, so they can't have real eigenvalues.

⟵ You learned about rotations and rotation matrices in Lesson 5.3.

In fact, Derman's matrix $A = \begin{pmatrix} 0 & -1 \\ 1 & 0 \end{pmatrix}$ is a rotation matrix. It rotates everything in $\mathbb{R}^2$ around $O$ by an angle of 90°. There aren't any fixed lines in this transformation, so the corresponding matrix can't have any real eigenvalues.

### For Discussion

**4.** Remember that a rotation in $\mathbb{R}^2$ about $O$ through an angle $\theta$ can be described by the matrix

$$R_\theta = \begin{pmatrix} \cos\theta & -\sin\theta \\ \sin\theta & \cos\theta \end{pmatrix}$$

Find the characteristic polynomial of $R_\theta$, and use algebra to find a condition for this polynomial to have real roots.

It should come as no surprise that matrices that are related to each other in some way have characteristic polynomials that are related to each other in some way. In the Exercises, you will explore the relationship between the characteristic polynomials of a matrix and its inverse, and the characteristic polynomials of a matrix and its transpose. Right now, you're going to think about similar matrices.

**Remember**

$M$ and $N$ are *similar* if there's an invertible matrix $P$ so that $N = P^{-1}MP$. See Lesson 8.5.

You already know that similar matrices have the same determinant (see Exercise 4 in Lesson 9.4). Does that mean they have the same eigenvalues? Do they have the same eigenvectors? As usual, the best thing to do is try some specific cases first.

## In-Class Experiment

Below there are several pairs of matrices $M$ and $P$. The matrix $P^{-1}$ is also given to you. For each pair of matrices,

(1) Find the eigenvalues and eigenvectors of $M$.

(2) Calculate the matrix $N = P^{-1}MP$.

(3) Decide if the eigenvalues of $M$ are also eigenvalues of $N$. If not, how do the eigenvalues compare?

(4) Decide if the eigenvectors of $M$ are also eigenvectors of $N$. If not, how do the eigenvectors compare?

← You should not have to do any calculations to find the eigenvalues of $M$. Ask yourself: What is special about these matrices?

**a.** $M = \begin{pmatrix} -1 & 5 \\ 0 & 1 \end{pmatrix}$, $P = \begin{pmatrix} 3 & -1 \\ -2 & 1 \end{pmatrix}$, $P^{-1} = \begin{pmatrix} 1 & 1 \\ 2 & 3 \end{pmatrix}$

**b.** $M = \begin{pmatrix} 2 & 0 \\ -11 & 2 \end{pmatrix}$, $P = \begin{pmatrix} -2 & 5 \\ 1 & -2 \end{pmatrix}$, $P^{-1} = \begin{pmatrix} 2 & 5 \\ 1 & 2 \end{pmatrix}$

**c.** $M = \begin{pmatrix} 1 & 2 \\ 0 & 3 \end{pmatrix}$, $P = \begin{pmatrix} 0 & 1 \\ -1 & 1 \end{pmatrix}$, $P^{-1} = \begin{pmatrix} 1 & -1 \\ 1 & 0 \end{pmatrix}$

**d.** $M = \begin{pmatrix} 0 & 2 \\ 0 & 1 \end{pmatrix}$, $P = \begin{pmatrix} 3 & 2 \\ 4 & 3 \end{pmatrix}$, $P^{-1} = \begin{pmatrix} 3 & -2 \\ -4 & 3 \end{pmatrix}$

From your work on the In-Class Experiment, you may have come up with the following theorem.

**Theorem 9.28 (Eigenvalues, Eigenvectors for Similar Matrices)**

(1) *Similar matrices have the same eigenvalues.*

(2) *If $M$ is a matrix with eigenvector $Q$ corresponding to eigenvalue $\lambda$, and if $P$ is invertible, then $P^{-1}Q$ is an eigenvector for the matrix $N = P^{-1}MP$ corresponding to the same eigenvalue $\lambda$.*

## For Discussion

**5.** Prove Theorem 9.28.

## Developing Habits of Mind

**Look for connections.** In the notation of Theorem 9.28, suppose that $N = P^{-1}MP$. In the language of Lesson 8.5, $N$ and $M$ can be considered different matrices for the same linear map, say $T$, with respect to different bases. So, Theorem 9.28 says that the eigenvalues of all the different matrix representations for $T$ are the same—*they are the eigenvalues for $T$*. And the eigenvectors for the different matrix representations for $T$ are just *the coordinate vectors for the eigenvectors for $T$* (with respect to the appropriate bases).

You've already learned that the eigenvalues of $M$ and any matrix similar to $M$ are the same. But even more is true: the characteristic polynomials are identical.

**Theorem 9.29 (Characteristic Polynomials for Similar Matrices)**
*If $M$ and $N$ are similar $n \times n$ matrices, then the characteristic polynomials of $M$ and $N$ are the same.*

***Proof.*** Let $M$ and $N$ be similar matrices, and let $P$ be a matrix such that $N = P^{-1}MP$. The key step in the proof is realizing that you can write $\lambda I$ as $P^{-1}(\lambda I)P$.

The characteristic polynomial of $N$ is

$$\begin{aligned}
\det(N - \lambda I) &= \det(P^{-1}MP - P^{-1}(\lambda I)P) \\
&= \det\left(P^{-1}(M - \lambda I)P\right) \\
&= \det(P^{-1})\det(M - \lambda I)\det(P) \\
&= \frac{1}{\det(P)}\det(M - \lambda I)\det(P) \\
&= \det(M - \lambda I)
\end{aligned}$$

The last line is the characteristic polynomial of $M$, so the two characteristic polynomials are, indeed, equal. ■

← If you're not sure about this step, look back at For You to Do problem 3 from Lesson 9.4.

**For Discussion**

**6.** Justify each step of the calculations in the last equations. In particular, prove if $P$ is an invertible matrix of the same size as an identity matrix $I$ and $\lambda$ is some real number, then $\lambda I = P^{-1}\lambda IP$.

### Invariant Subspaces

Recall that a *subspace* of a vector space $V$ is any subset of $V$ that itself forms a vector space. For example, a line through the origin in $\mathbb{R}^2$ is a vector space. So is a line (or a plane) through the origin in $\mathbb{R}^3$.

← You learned about subspaces in Lesson 7.3.

**Definition**

Given a square matrix $M$, a vector space $W$ is ***M*-invariant** if whenever a vector $Q$ is in $W$, so is the vector $MQ$. If $W$ is a subspace of some bigger vector space $V$, then $W$ is called an ***M*-invariant subspace of *V***.

← If $T : V \to V$ is linear, a subspace $W$ of $V$ is $T$-invariant if $T(w)$ is in $W$ whenever $w$ is in $W$.

**Example 3**

You've already seen several examples of invariant subspaces of $\mathbb{R}^2$ and $\mathbb{R}^3$. Here are some to remind you.

If $A = \begin{pmatrix} 1 & 2 \\ 3 & 2 \end{pmatrix}$, then

$$\begin{aligned} A\left(t\begin{pmatrix}1\\-1\end{pmatrix}\right) &= tA\begin{pmatrix}1\\-1\end{pmatrix} = t\begin{pmatrix}-1\\1\end{pmatrix} \\ &= -t\begin{pmatrix}1\\-1\end{pmatrix} \end{aligned}$$

So the line with equation $X = t\begin{pmatrix}1\\-1\end{pmatrix}$ is an $A$-invariant subspace of $\mathbb{R}^2$.

In Example 2 from this lesson, you saw that any vector $Q$ in the plane with equation $x - y + 2z = 0$ satisfies $MQ = Q$ for the matrix

$$M = \begin{pmatrix} 1 & 0 & 0 \\ 3 & -2 & 6 \\ 0 & 0 & 1 \end{pmatrix}$$

So that plane is an $M$-invariant subspace of $\mathbb{R}^3$.

⟵ Indeed, $M$ fixes the plane *point by point*.

### For Discussion

**7.** Let $M$ be a $4 \times 4$ matrix. Explain why $O$ and $\mathbb{R}^4$ are both $M$-invariant subspaces of $\mathbb{R}^4$.

### For You to Do

**8.** Let $M$ be an $n \times n$ matrix, and let $\lambda$ be an eigenvalue of $M$. Show that the set

$$\begin{aligned} E_\lambda &= \{Q \text{ a vector in } \mathbb{R}^n \text{ such that } MQ = \lambda Q\} \\ &= \{\text{all eigenvectors of } M \text{ with eigenvalue } \lambda \text{ together with } O\} \end{aligned}$$

is an $M$-invariant subspace of $\mathbb{R}^n$.

⟵ You have to show that $E_\lambda$ is a subspace *and* that it is $M$-invariant.

#### Definition

The subspace $E_\lambda$ in For You to Do problem 8 is called the **eigenspace** of $M$ associated to the eigenvalue $\lambda$.

### Example 4

**Problem.** Let

$$M = \begin{pmatrix} 0 & 1 & 0 \\ 1 & 0 & 0 \\ 0 & 0 & 1 \end{pmatrix}$$

Check that $\lambda = 1$ is an eigenvalue for $M$, and find the eigenspace $E_1$.

**Solution.** If $\lambda = 1$ is an eigenvalue, then it must be the case that $\det(M - I) = 0$.

$$\det(M - I) = \begin{vmatrix} -1 & 1 & 0 \\ 1 & -1 & 0 \\ 0 & 0 & 0 \end{vmatrix}$$

Since this matrix has a column of 0's, the determinant is 0.

To find the eigenspace for $\lambda = 1$, you need to solve $(M - I)Q = O$. The equation is

$$(M - I)Q = \begin{pmatrix} -1 & 1 & 0 \\ 1 & -1 & 0 \\ 0 & 0 & 0 \end{pmatrix} \begin{pmatrix} x \\ y \\ z \end{pmatrix} = \begin{pmatrix} 0 \\ 0 \\ 0 \end{pmatrix}$$

Any vector $(x, y, z)$ where $x = y$ will satisfy this equation. So the eigenspace is the plane with equation $x - y = 0$ in $\mathbb{R}^3$.

⟵ What are some eigenvectors corresponding to eigenvalue 1?

**For You to Do**

**9.** Check that $\lambda = -1$ is also an eigenvalue of the matrix $M$ in Example 4 above. Find the eigenspace $E_{-1}$.

The examples of $M$-invariant subspaces above are all eigenspaces. Is every $M$-invariant subspace an eigenspace? See Exercises 21–22 and 27–29.

## Exercises

**1.** Decide if the vector $\begin{pmatrix} -1 \\ 0 \\ 2 \end{pmatrix}$ is an eigenvector for each given matrix.

**a.** $\begin{pmatrix} -1 & 1 & 0 \\ 2 & 0 & 1 \\ 0 & 3 & -1 \end{pmatrix}$ **b.** $\begin{pmatrix} 0 & 0 & 1 \\ 2 & -1 & 1 \\ 3 & 2 & 1 \end{pmatrix}$

**c.** $\begin{pmatrix} 3 & 2 & 1 \\ 2 & 2 & -2 \\ -1 & 3 & 0 \end{pmatrix}$ **d.** $\begin{pmatrix} 1 & 4 & -1 \\ 0 & 0 & 0 \\ 2 & -1 & 4 \end{pmatrix}$

**2.** Decide if 3 is an eigenvalue for each matrix.

**a.** $\begin{pmatrix} -8 & -5 \\ 10 & 7 \end{pmatrix}$ **b.** $\begin{pmatrix} 4 & 1 \\ -1 & 2 \end{pmatrix}$ **c.** $\begin{pmatrix} 6 & 3 \\ -5 & -2 \end{pmatrix}$ **d.** $\begin{pmatrix} 10 & 5 \\ -14 & -7 \end{pmatrix}$

**3.** Find all eigenvalues of each given matrix. Then find the eigenvectors corresponding to those eigenvalues.

**a.** $\begin{pmatrix} -2 & -6 \\ 4 & 8 \end{pmatrix}$ **b.** $\begin{pmatrix} 1 & 0 \\ 3 & 1 \end{pmatrix}$ **c.** $\begin{pmatrix} 7 & -3 \\ 1 & 3 \end{pmatrix}$ **d.** $\begin{pmatrix} 2 & 0 \\ 0 & 2 \end{pmatrix}$

e. $\begin{pmatrix} 1 & 1 & -1 \\ -2 & 1 & 1 \\ -2 & 2 & 0 \end{pmatrix}$ f. $\begin{pmatrix} -1 & 0 & 0 \\ -6 & 2 & 0 \\ -6 & 0 & 2 \end{pmatrix}$ g. $\begin{pmatrix} 3 & 2 & 1 \\ -2 & -1 & -1 \\ 0 & 0 & 1 \end{pmatrix}$

h. $\begin{pmatrix} -1 & 1 & 4 \\ -2 & -4 & -8 \\ 2 & 2 & 3 \end{pmatrix}$ i. $\begin{pmatrix} -1 & -3 & -3 \\ 6 & 8 & 6 \\ -6 & -6 & -4 \end{pmatrix}$

**4.** For each matrix $A$,

(i) find all eigenvalues and eigenvectors for $A$
(ii) calculate $A^{-1}$
(iii) find all eigenvalues and eigenvectors for $A^{-1}$

**a.** $A = \begin{pmatrix} -2 & -6 \\ 4 & 8 \end{pmatrix}$ **b.** $A = \begin{pmatrix} 5 & 4 \\ -2 & -1 \end{pmatrix}$

**c.** $A = \begin{pmatrix} 1 & 1 & -1 \\ -2 & 1 & 1 \\ -2 & 2 & 0 \end{pmatrix}$ **d.** $A = \begin{pmatrix} 2 & 0 & 0 \\ 2 & 4 & -2 \\ 6 & 6 & -4 \end{pmatrix}$

Any conjectures?

**5.** Consider one of the eigenvectors for each eigenvalue you found for the matrix in Exercise **3e**. Check whether these vectors are linearly independent.

**6.** Consider one of the eigenvectors for each eigenvalue you found for the matrix in Exercise **3h**. Check whether these vectors are linearly independent.

**7.** Find the characteristic polynomial of each given matrix.

**a.** $\begin{pmatrix} 1 & 0 \\ 3 & 1 \end{pmatrix}$ **b.** $\begin{pmatrix} -1 & 1 \\ 1 & 2 \end{pmatrix}$ **c.** $\begin{pmatrix} 7 & -3 \\ 1 & 3 \end{pmatrix}$

**d.** $\begin{pmatrix} 1 & 1 & -1 \\ -2 & 1 & 1 \\ -2 & 2 & 0 \end{pmatrix}$ **e.** $\begin{pmatrix} -1 & 0 & 0 \\ -6 & 2 & 0 \\ -6 & 0 & 2 \end{pmatrix}$ **f.** $\begin{pmatrix} 1 & 0 & 1 \\ 3 & 1 & -1 \\ 1 & -1 & -1 \end{pmatrix}$

**8.** Here is an $n \times n$ diagonal matrix. What is its characteristic polynomial?

$$\begin{pmatrix} A_{11} & 0 & 0 & \cdots & 0 \\ 0 & A_{22} & 0 & \cdots & 0 \\ 0 & 0 & A_{33} & \cdots & 0 \\ \vdots & \vdots & \vdots & \ddots & 0 \\ 0 & 0 & 0 & \cdots & A_{nn} \end{pmatrix}$$

**9.** For each matrix $A$,

(i) find the characteristic polynomial of $A$
(ii) find $A^\top$
(iii) find the characteristic polynomial of $A^\top$

**a.** $A = \begin{pmatrix} 7 & -3 \\ 1 & 3 \end{pmatrix}$ **b.** $A = \begin{pmatrix} 5 & 4 \\ -2 & -1 \end{pmatrix}$

**c.** $A = \begin{pmatrix} -1 & -3 & -3 \\ 6 & 8 & 6 \\ -6 & -6 & -4 \end{pmatrix}$ **d.** $A = \begin{pmatrix} 1 & 0 & -2 \\ 0 & 1 & -1 \\ 1 & -1 & -2 \end{pmatrix}$

Any conjectures?

**10.** For each matrix $A$,

(i) find the characteristic polynomial of $A$
(ii) find $A^{-1}$
(iii) find the characteristic polynomial of $A^{-1}$

**a.** $\begin{pmatrix} -1 & 1 \\ 2 & 0 \end{pmatrix}$ **b.** $\begin{pmatrix} 9 & -4 \\ 12 & -5 \end{pmatrix}$ **c.** $\begin{pmatrix} -\frac{10}{3} & -8 \\ \frac{8}{3} & 6 \end{pmatrix}$

**d.** $\begin{pmatrix} 0 & 1 \\ -3 & \frac{7}{2} \end{pmatrix}$ **e.** $\begin{pmatrix} 1 & 1 & -1 \\ -2 & 1 & 1 \\ -2 & 2 & 0 \end{pmatrix}$ **f.** $\begin{pmatrix} -\frac{11}{2} & 2 & 2 \\ -6 & \frac{3}{2} & \frac{5}{2} \\ -6 & 3 & 2 \end{pmatrix}$

Any conjectures?

**11.** For each matrix, find all its eigenvalues and for each of these eigenvalues, describe the eigenspace.

**a.** $\begin{pmatrix} 2 & 0 \\ 0 & 2 \end{pmatrix}$ **b.** $\begin{pmatrix} -2 & 2 \\ -6 & 5 \end{pmatrix}$ **c.** $\begin{pmatrix} 1 & 1 & -1 \\ -2 & 1 & 1 \\ -2 & 2 & 0 \end{pmatrix}$

**d.** $\begin{pmatrix} -1 & 0 & 0 \\ -6 & 2 & 0 \\ -6 & 0 & 2 \end{pmatrix}$ **e.** $\begin{pmatrix} 3 & -2 & -4 \\ 2 & -1 & -5 \\ -1 & 1 & 3 \end{pmatrix}$ **f.** $\begin{pmatrix} 2 & 0 & -\frac{1}{2} \\ 1 & 1 & -\frac{1}{2} \\ 2 & 0 & 0 \end{pmatrix}$

**12.** For each matrix and eigenvalue pair, find the associated eigenspace.

**a.** $\begin{pmatrix} 3 & 2 & 1 \\ -2 & -1 & -1 \\ 0 & 0 & 1 \end{pmatrix}$, eigenvalue $= 1$

**b.** $\begin{pmatrix} -1 & 1 & 4 \\ -2 & -4 & -8 \\ 2 & 2 & 3 \end{pmatrix}$, eigenvalue $= -2$

**c.** $\begin{pmatrix} 7 & -4 & 4 \\ 6 & -4 & 6 \\ -2 & 1 & 1 \end{pmatrix}$, eigenvalue $= 3$

**d.** $\begin{pmatrix} -3 & -2 & 2 \\ 4 & 3 & -4 \\ 0 & 0 & -1 \end{pmatrix}$, eigenvalue $= -1$

**13.** Prove part ((4)) of Theorem 9.27.

←
**Hint:** What happens if $\lambda = 0$ in $\det(A - \lambda I)$? What happens if $\lambda = 0$ in the characteristic polynomial?

**14.** Suppose $A$ is an $n \times n$ matrix.

**a.** Show that $\det(A)$ is the product of the eigenvalues for $A$.
**b.** Show that the trace of $A$ is the sum of the eigenvalues for $A$.

**15.** Here is an $n \times n$ diagonal matrix. For this matrix, $A_{ii} = i$ and $A_{ij} = 0$ if $i \neq j$. What are the eigenvalues of this matrix? What are the eigenvectors?

$$\begin{pmatrix} 1 & 0 & 0 & \cdots & 0 \\ 0 & 2 & 0 & \cdots & 0 \\ & \vdots & & \ddots & 0 \\ 0 & 0 & 0 & \cdots & n \end{pmatrix}$$

**16.** Here is an $n \times n$ upper triangular matrix. What are its eigenvalues?

$$\begin{pmatrix} A_{11} & A_{12} & A_{13} & \cdots & A_{1n} \\ 0 & A_{22} & A_{23} & \cdots & A_{2n} \\ & \vdots & & \ddots & 0 \\ 0 & 0 & 0 & \cdots & A_{nn} \end{pmatrix}$$

**17.** Without calculating the determinant, explain why $x - y$ and $x - z$ are eigenvalues of this matrix.

⟵ Can you find the other eigenvalues?

$$M = \begin{pmatrix} x & y & z & w \\ y & x & z & w \\ y & z & x & w \\ y & z & w & x \end{pmatrix}$$

**18.** Let $M$ be a $3 \times 3$ matrix, and let $Q = \begin{pmatrix} 1 \\ 0 \\ 1 \end{pmatrix}$. Suppose $MQ = O$. Is $Q$ an eigenvector of $M$? If so, what is its associated eigenvalue? If not, why not?

**19.** Suppose $T : \mathbb{R}^2 \to \mathbb{R}^2$ is the map that reflects a point over the line with equation $x = y$. Find the eigenvalues and eigenvectors for $T$.

**20.** Suppose $T : \mathbb{R}^2 \to \mathbb{R}^2$ is the map that reflects a point over the line with equation $X = t(3, 4)$. Find the eigenvalues and eigenvectors for $T$.

**21.** Let $M = \begin{pmatrix} 1 & 1 & 0 \\ 0 & 1 & 1 \\ 0 & 0 & 0 \end{pmatrix}$. Decide if these statements are true.

**a.** The $x$-axis is an $M$-invariant subspace of $\mathbb{R}^3$.
**b.** The $y$-axis is an $M$-invariant subspace of $\mathbb{R}^3$.
**c.** The $z$-axis is an $M$-invariant subspace of $\mathbb{R}^3$.
**d.** The $x$–$y$ plane is an $M$-invariant subspace of $\mathbb{R}^3$.
**e.** The $x$–$z$ plane is an $M$-invariant subspace of $\mathbb{R}^3$.

**22.** Suppose $A$ is a $4\times4$ matrix, $Q_1$ is an eigenvector of $A$ with eigenvalue 3, and $Q_2$ is an eigenvector of $A$ with eigenvalue $-1$. Prove that the plane in $\mathbb{R}^4$ spanned by $Q_1$ and $Q_2$ is an $A$-invariant subspace of $\mathbb{R}^4$.

**23.** Prove that a matrix and its transpose have the same characteristic polynomial.

⟵ So $A$ and $A^{\top}$ have the same eigenvalues. Can you say anything about the eigenvectors?

**24.** Prove that a matrix has 0 as an eigenvalue if and only if the determinant of the matrix is 0. (So the matrix is not invertible.)

**25.** Suppose $Q$ is an eigenvector for an invertible matrix $A$ with eigenvalue $\lambda$. Prove that $Q$ is also an eigenvector for $A^{-1}$. What is its eigenvalue?

⟵ Use the equation $AQ = \lambda Q$.

**26.** For each given matrix $M$, compute its characteristic polynomial. Then plug $M$ into the polynomial as the value for $\lambda$ and simplify as much as possible.

⟵ For constant terms in the polynomial, multiply by the identity matrix. For example, if $M$ is a $2\times 2$ matrix, then $(2M-3)$ means $(2M-3I)$, where $I$ is the $2\times 2$ identity matrix.

**a.** $M = \begin{pmatrix} 1 & -1 \\ 1 & 3 \end{pmatrix}$ **b.** $M = \begin{pmatrix} 1 & -1 \\ -4 & 1 \end{pmatrix}$

**27. Take It Further.** Suppose $M$ is a matrix and $Q$ is a vector. The **$M$-cyclic subspace generated by $Q$** is the linear span of $\{Q, MQ, M^2Q, M^3Q, \dots\}$.

For each given matrix $M$ and vector $Q$, find the $M$-cyclic subspace generated by $Q$.

**a.** $M = \begin{pmatrix} 0 & 0 & -1 \\ 1 & 1 & 1 \\ 1 & -1 & 0 \end{pmatrix}, \quad Q = \begin{pmatrix} 1 \\ -1 \\ 0 \end{pmatrix}$

**b.** $M = \begin{pmatrix} 0 & 0 & -1 \\ 1 & 1 & 1 \\ 1 & -1 & 0 \end{pmatrix}, \quad Q = \begin{pmatrix} 1 \\ 1 \\ 1 \end{pmatrix}$

**c.** $M = \begin{pmatrix} 0 & 1 & -1 & 0 \\ 1 & 0 & 1 & 1 \\ 1 & 0 & 2 & 1 \\ 1 & 1 & 1 & 1 \end{pmatrix}, \quad Q = \begin{pmatrix} -1 \\ 0 \\ 1 \\ 0 \end{pmatrix}$

**28. Take It Further.** If $M$ is an $n \times n$ matrix, the dimension of the $M$-cyclic subspace generated by $Q$ is at most $n$. Explain how to find a basis for it.

**29. Take It Further.** If $M$ is an $n\times n$ matrix, show that the $M$-cyclic subspace generated by $Q$ is $M$-invariant.

## 9.7 Topics in Eigentheory

*I pray you, be gone: I will make an end of my dinner; there's pippins and cheese to come.*

—William Shakepeare, The Merry Wives of Windsor

---

Eigenvalues and eigenvectors have unexpected applications in a variety of fields besides mathematics: physics, engineering, statistics, probability, finance, optics, etc. The eigenvectors determine lines where a matrix function just acts by stretching or reducing a vector. The eigenvalues are the scale by which the stretch is applied. When there are enough linearly independent eigenvectors to form a basis, a change of basis leads to a diagonal matrix, with the eigenvalues as its diagonal entries, that is similar to the original matrix. This helps make some very messy calculations much simpler and direct.

**In this lesson, you will learn how to**

- find the algebraic and geometric multiplicity of the eigenvalues of a matrix
- determine which matrices can be diagonalized
- use a basis made of eigenvectors to create a change of basis matrix
- apply the diagonalization process to simplify calculations in probability theory and *dynamical systems*
- find the equation of a circle, ellipse, or hyperbola whose axes have been rotated by a certain angle

---

### Developing Habits of Mind

**Reflect and review.** This is the last lesson of the last chapter in the course. You've learned a great deal over the past chapters. It's a good idea now to go back to the beginning and to read the introduction. How would *you* describe linear algebra to someone just starting this program?

There's always so much more to learn in mathematics. Think of this final lesson as a preview of coming attractions. You'll get a taste of several directions for further study—things you can dig into as you learn more about linear algebra.

---

You've already seen situations where taking powers of a matrix is helpful to solve a problem. For example, in Minds in Action—Episode 12 from Lesson 4.4, Derman, Sasha, and Tony figured out that if a matrix $M$ represents the number of one-step paths between various destinations, then $M^2$ represents the number of two-step paths, $M^3$ represents the number of three-step paths, and so on.

⟵ Markov chains, yet another situation where finding powers of matrices comes in handy, are explored in Chapter 6.

You've also seen that finding powers of diagonal matrices is much easier than finding powers of matrices that are not diagonal. And in Example 1 from Lesson 8.6, you saw that—at least sometimes—you can find a change of basis so as to "diagonalize" a matrix.

⟵ For example, look back at Exercise 21 from Lesson 4.5.

In this lesson, you'll use the ideas of eigenvalues and eigenvectors to figure out when you can diagonalize a given matrix, and you'll see some of the useful consequences of diagonalization.

### Definition

An $n \times n$ matrix $M$ is **diagonalizable** if there is an invertible matrix $P$ so that $P^{-1}MP$ is a diagonal matrix.

⟵ In other words, $M$ is diagonalizable if $M$ is similar to a diagonal matrix.

### Example 1

**Problem.** Is the matrix $M = \begin{pmatrix} -14 & -3 & 33 \\ -76 & -21 & 174 \\ -16 & -4 & 37 \end{pmatrix}$ diagonalizable?

**Solution.** Well, suppose it is diagonalizable. What would that mean? You could find a matrix $P$ so that $P^{-1}MP$ is a diagonal matrix $D$, say

$$D = \begin{pmatrix} a_1 & 0 & 0 \\ 0 & a_2 & 0 \\ 0 & 0 & a_3 \end{pmatrix}$$

Write $P$ as a matrix of column vectors $P = (P_{*1}, P_{*2}, P_{*3})$, and rewrite the equation above as

$$MP = PD$$

$$M(P_{*1}, P_{*2}, P_{*3}) = (P_{*1}, P_{*2}, P_{*3}) \begin{pmatrix} a_1 & 0 & 0 \\ 0 & a_2 & 0 \\ 0 & 0 & a_3 \end{pmatrix}$$

$$(MP_{*1}, MP_{*2}, MP_{*3}) = (a_1 P_{*1}, a_2 P_{*2}, a_3 P_{*3})$$

⟵ This last step comes from the Pick-Apart Theorem, Theorem 4.8 from Lesson 4.5.

But look! You have three equations now.

$$MP_{*1} = a_1 P_{*1}$$
$$MP_{*2} = a_1 P_{*2}$$
$$MP_{*3} = a_1 P_{*3}$$

In other words, the columns of $P$ must be eigenvectors for the matrix $M$, and the diagonal entries of $D$ must be the eigenvalues of $M$.

In fact, you already saw in Example 1 from Lesson 8.6 that this matrix *is* diagonalizable. The "diagonalizing matrix" is

$$P = \begin{pmatrix} 3 & -1 & 2 \\ 5 & 4 & 1 \\ 2 & 0 & 1 \end{pmatrix} \text{ with } P^{-1} = \begin{pmatrix} -4 & -1 & 9 \\ 3 & -1 & -7 \\ 8 & 2 & 17 \end{pmatrix}$$

⟵ You can check that the columns of $P$ are all eigenvectors of $M$, and that $P^{1-}MP$ is a diagonal matrix.

So if you want to diagonalize a matrix, just make a matrix $P$ of eigenvectors, and then find $P^{-1}MP$. Well, almost. The matrix $P$ has to be invertible for this to work. So a $3 \times 3$ matrix needs to have three *linearly independent* eigenvectors in order to be diagonalizable. In fancier language: an $n \times n$ matrix $M$ is diagonalizable if and only if there is a *basis* of $\mathbb{R}^n$ made up of eigenvectors of $M$.

⟵ And a $2 \times 2$ matrix needs to have *two* linearly independent eigenvectors, and an $n \times n$ matrix needs to have . . .

### Example 2

**Problem.** Are these two matrices diagonalizable?

**a.** $A = \begin{pmatrix} 1 & 0 & 0 \\ 3 & -2 & 6 \\ 0 & 0 & 1 \end{pmatrix}$ **b.** $B = \begin{pmatrix} 1 & 2 & 3 \\ 0 & 2 & 1 \\ 0 & 1 & 2 \end{pmatrix}$

**Solution.** You saw matrix $A$ in Example 2 from Lesson 9.6. Its characteristic polynomial is $(1-\lambda)^2(-2-\lambda)$, so its eigenvalues are 1 and $-2$.

For $\lambda = 1$, you found two eigenvectors, $X = (1,1,0)$ and $Y = (0,2,1)$. For $\lambda = -2$, you found the eigenvector $Z = (0,1,0)$. If these three vectors are linearly independent, then this will be a diagonalizing matrix:

$$P = \begin{pmatrix} 1 & 0 & 0 \\ 1 & 2 & 1 \\ 0 & 1 & 0 \end{pmatrix}$$

You can calculate that $\det(P) = -1$, so the columns are, indeed, linearly independent. Then $A$ is diagonalizable and

$$P^{-1}AP = \begin{pmatrix} 1 & 0 & 0 \\ 0 & 1 & 0 \\ 0 & 0 & -2 \end{pmatrix}$$

⟵ The order of the eigenvalues corresponds to the order of the eigenvectors in the matrix $P$.

The characteristic polynomial for $B$ is $(1-\lambda)^2(3-\lambda)$, so the eigenvalues are 1 and 3. If you plug $\lambda = 1$ into $B - \lambda I$, you get the matrix

$$B - \lambda I = \begin{pmatrix} 0 & 2 & 3 \\ 0 & 1 & 1 \\ 0 & 1 & 1 \end{pmatrix}$$

You can check that the kernel of this matrix is the line with equation $X = t(1,0,0)$. For $\lambda = 3$, the matrix is

$$B - \lambda I = \begin{pmatrix} -2 & 2 & 3 \\ 0 & -1 & 1 \\ 0 & 1 & -1 \end{pmatrix}$$

which has kernel $X = t(5,2,2)$. You can find lots of eigenvectors: $(1,0,0)$, $(2,0,0)$, $(\pi,0,0)$, $(5,2,2)$, $(-15,-6,-6)$, and so on. But you can never find more than two *linearly independent* eigenvectors, because you only have two fixed lines to work with. So matrix $B$ is not diagonalizable.

### For You to Do

**1.** For each matrix $M$ below, do the following:

(i) Decide if $M$ is diagonalizable by checking if it has enough linearly independent eigenvectors.

(ii) If it is diagonalizable, find a diagonalizing matrix $P$.

(iii) Find the diagonal matrix $P^{-1}MP$.

←
Notice that you don't need to calculate $P^{-1}$ to do the last part. Why not?

**a.** $M = \begin{pmatrix} 1 & 2 \\ 2 & 1 \end{pmatrix}$ **b.** $M = \begin{pmatrix} 1 & 2 \\ 0 & 1 \end{pmatrix}$ **c.** $M = \begin{pmatrix} 1 & 2 & 1 \\ 0 & 2 & 1 \\ 0 & 0 & 3 \end{pmatrix}$

### Minds in Action Episode 50

*Derman is explaining his work on finding linearly independent eigenvectors.*

DERMAN: Whenever the eigenvalues are all different, you have enough linearly independent eigenvectors.

TONY: How do you figure that?

DERMAN: Well, each eigenvalue gets you at least one eigenvector. So then you'll have enough of them if all the eigenvalues are different.

SASHA: But *you* figured out that sometimes you don't have enough eigenvalues. Sometimes you don't have any.

DERMAN: No, I know. I mean . . . I guess I mean if you have $n$ different eigenvalues for an $n \times n$ matrix, then it works.

SASHA: OK, that makes more sense. But I still have a question. How do you know the eigenvectors you get are linearly independent?

DERMAN: Huh?

SASHA: You don't just need $n$ eigenvectors. You need them to be linearly independent.

DERMAN: I . . . uh . . . Oh, there's the bell for lunch.

This theorem answers Sasha's question.

**Theorem 9.30**

*Let $M$ be an $n \times n$ square matrix. Then eigenvectors corresponding to different eigenvalues are linearly independent.*

***Proof.*** Suppose you've got two eigenvectors $X_1$ and $X_2$, with different eigenvalues $\lambda_1$ and $\lambda_2$. And suppose you can find scalars $a_1$ and $a_2$ so that

$$a_1X_1 + a_2X_2 = O$$

Multiply both sides of this equation by the matrix $M$ and simplify.

$$M(a_1X_1 + a_2X_2) = MO$$
$$a_1(MX_1) + a_2(MX_2) = O$$
$$(a_1\lambda_1)X_1 + (a_2\lambda_2)X_2 = O$$

Combine this last equation with the first one above.

$$a_1X_1 + a_2X_2 = O$$
$$(a_1\lambda_1)X_1 + (a_2\lambda_2)X_2 = O$$

Multiply the top equation by $\lambda_2$ and subtract what you get from the bottom one to get

$$a_1(\lambda_1 - \lambda_2)X_1 = O$$

You know that $\lambda_1 - \lambda_2 \neq 0$, since the eigenvalues were different. You also know that $X_1 \neq O$, since it's an eigenvector. So then the only choice is that $a_1 = 0$, which means $a_2 = 0$ also. So the vectors are linearly independent.

Now you know that whenever you have two eigenvectors that correspond to two different eigenvalues, they are linearly independent. What if you have *three* eigenvectors that correspond to *three* different eigenvalues? Play the same game. Suppose you can write

$$a_1X_1 + a_2X_2 + a_3X_3 = O$$

Multiply both sides of this equation by the matrix $M$ and simplify.

$$M(a_1X_1 + a_2X_2 + a_3X_3) = MO$$
$$(a_1\lambda_1)X_1 + (a_2\lambda_2)X_2 + (a_3\lambda_3)X_3 = O$$

← Fill in the reasons for each step!

Now multiply

$$a_1X_1 + a_2X_2 + a_3X_3 = O$$

by $\lambda_3$ and subtract from

$$(a_1\lambda_1)X_1 + (a_2\lambda_2)X_2 + (a_3\lambda_3)X_3 = O$$

to get

$$a_1(\lambda_1 - \lambda_3)X_1 + a_2(\lambda_2 - \lambda_3)X_2 = O$$

This is a linear combination of $X_1$ and $X_2$ that yields $O$. But you know $X_1$ and $X_2$ are linearly independent, since they correspond to different eigenvalues, and the result for *two* vectors is already proved. That means the two coefficients in this last equation must both be 0. So $a_1 = a_2 = 0$, which means $a_3 = 0$ as well.

**Remember**

The eigenvalues are different, so $\lambda_1 - \lambda_3 \neq 0$ and $\lambda_2 - \lambda_3 \neq 0$.

The rest of the proof works just like this. Suppose you know that $k - 1$ eigenvectors corresponding to $k - 1$ different eigenvalues must be linearly independent. Do this same calculation in a little more generality to show that $k$ eigenvectors corresponding to $k$ different eigenvalues must be linearly independent. ■

← This is mathematical induction again.

An important consequence of Theorem 9.30 is the fact that Derman originally noticed, which is summarized in the following theorem.

**Theorem 9.31**

*If $M$ is an $n \times n$ matrix and $M$ has $n$ different eigenvalues, then $M$ is diagonalizable.*

**For Discussion**

**2.** Use Theorem 9.30 to prove Theorem 9.31.

**Minds in Action** Episode 51

*The three friends are still talking about Derman's idea.*

TONY: I wonder if Derman's theorem is one of those "if and only if" things.

DERMAN: What do you mean?

TONY: Well, we know that if we have $n$ different eigenvalues, then we can diagonalize the matrix. Maybe it's true that if we *don't* have $n$ different eigenvalues, then we *can't* diagonalize it.

SASHA: No way, that can't be right.

TONY: Why not?

SASHA: Well, you can make up really easy counterexamples. Just make a matrix that's already diagonal, but has repeated eigenvalues, like this:

**Habits of Mind**
Look for counterexamples.

$$\begin{pmatrix} 2 & 0 & 0 & 0 \\ 0 & 1 & 0 & 0 \\ 0 & 0 & 2 & 0 \\ 0 & 0 & 0 & 2 \end{pmatrix}$$

There are only two different eigenvalues, not four. But it's already diagonal, so it's certainly "diagonalizable."

Both Tony and Sasha have brought up the idea of "repeated" eigenvalues. It's time to figure out what happens when an $n \times n$ matrix has $n$ real eigenvalues, but some of them are "repeats."

**Definition**

Let $M$ be an $n \times n$ matrix, and suppose the characteristic polynomial of $M$ can be completely factored over the real numbers. So the characteristic polynomial looks like

$$(a_1 - \lambda)^{e_1}(a_2 - \lambda)^{e_2} \cdots (a_t - \lambda)^{e_t}$$

where the numbers $a_i$ are all different.

- The exponent $e_i$ is called the **algebraic multiplicity** of the eigenvalue $a_i$.

- The dimension of the eigenspace $E_{a_i}$ is called the **geometric multiplicity** of the eigenvalue $a_i$.

### Example 3

In Example 2, the characteristic polynomial of the matrix $A$ is

$$(1-\lambda)^2(-2-\lambda)$$

The eigenspace $E_1$ consists of the linear span of two linearly independent vectors: $Q_1 = (1, 1, 0)$ and $Q_2 = (0, 2, 1)$. The eigenspace of $E_{-2}$ is a line, $X = t(0, 1, 0)$. So 1 is an eigenvalue with algebraic multiplicity two and geometric multiplicity two, and $-2$ is an eigenvalue with algebraic multiplicity one and geometric multiplicity one.

For matrix $B$, the characteristic polynomial is $(1-\lambda)^2(3-\lambda)$. The eigenspace $E_1$ consists of the line with equation $X = t(1, 0, 0)$. The eigenspace $E_3$ is the line with equation $X = t(5, 2, 2)$. So 1 is an eigenvalue with algebraic multiplicity two and geometric multiplicity one, and 3 is an eigenvalue with algebraic multiplicity and geometric multiplicity both equal to one.

### For You to Do

**3.** For each matrix below,

(i) find the algebraic multiplicity and the geometric multiplicity of the eigenvalue $-2$

(ii) decide if the matrix is diagonalizable or not

a. $\begin{pmatrix} -2 & 1 \\ 0 & -2 \end{pmatrix}$ **b.** $\begin{pmatrix} 1 & 3 & 2 \\ 0 & -2 & 1 \\ 0 & 0 & -1 \end{pmatrix}$ c. $\begin{pmatrix} 1 & -3 & 3 \\ 3 & -5 & 3 \\ 6 & -6 & 4 \end{pmatrix}$

### Facts and Notation

From the examples you've seen so far, it seems reasonable to guess that for any eigenvalue $a$ of a matrix $M$,

$$1 \le \text{geometric multiplicity of } a \le \text{algebraic multiplicity of } a.$$

The first inequality must be true. Remember that the eigenvalue $a$ is defined so that $M - aI$ has something in the kernel besides $O$. So there must be at least one eigenvector $Q$. But then the whole line with equation $X = tQ$ is in the eigenspace $E_a$, and the geometric multiplicity is at least 1.

The formal proof of the second inequality—that the algebraic multiplicity is at least as large as the geometric multiplicity—is a little complicated. It's a good idea first to examine the case in $\mathbb{R}^3$, and then sketch the generalization. In $\mathbb{R}^3$, there are three possibilities for geometric multiplicity of an eigenvalue: 1, 2, or 3.

← The eigenspace $E_a$ is a subspace of $\mathbb{R}^3$, so it can have at most three dimensions.

If the geometric multiplicity is one, the algebraic multiplicity must be at least one. (Otherwise, $a$ is not an eigenvalue!) What if the geometric multiplicity is two? Let $Q_1$ and $Q_2$ be linearly independent eigenvectors with eigenvalue $a$. Expand those two vectors to a basis for $\mathbb{R}^3$ by adding a vector $Y$ such that $\{Q_1, Q_2, Y\}$ is a linearly independent set. Then, for some numbers $b_1$, $b_2$, and $b_3$,

$$\begin{aligned} MQ_1 &= aQ_1 \\ MQ_2 &= aQ_2 \\ MY &= b_1X_1 + b_2X_2 + b_3Y \end{aligned}$$

Using the Pick-Apart Theorem, you can rewrite the three equations above as a single matrix equation.

$$M\,(Q_1Q_2Y) = (Q_1Q_2Y)\begin{pmatrix} a & 0 & b_1 \\ 0 & a & b_2 \\ 0 & 0 & b_3 \end{pmatrix} \tag{5}$$

**Remember**

$(Q_1Q_2Y)$ means the matrix whose columns are the vectors $Q_1$, $Q_2$, and $Y$, in that order.

Let $P = (Q_1Q_2Y)$ and $N = \begin{pmatrix} a & 0 & b_1 \\ 0 & a & b_2 \\ 0 & 0 & b_3 \end{pmatrix}$. Since the columns of $P$ form a basis for $\mathbb{R}^3$, they are linearly independent. So $P$ is invertible, meaning equation (5) can be rewritten as $P^{-1}MP = N$. Theorem 9.29 says that similar matrices have the same characteristic polynomials. The characteristic polynomial of $N$ is easy to calculate if you expand the determinant on the first column; it's $(a-\lambda)^2(b_3-\lambda)$. So the algebraic multiplicity of $a$ is at least two.

⟵
"At least 2" because $b_3$ might also be equal to $a$.

If the geometric multiplicity of $a$ is three, the eigenvectors $Q_1, Q_2, Q_3$ already form a basis for $\mathbb{R}^3$. So equation (5) becomes

$$M\,(Q_1Q_2Q_3) = (Q_1Q_2Q_3)\begin{pmatrix} a & 0 & 0 \\ 0 & a & 0 \\ 0 & 0 & a \end{pmatrix}$$

You can run the same argument to see that the characteristic polynomial of $M$ is $(a-\lambda)^3$, which means $a$ has algebraic multiplicity three.

Suppose now that $M$ is an $n \times n$ matrix, and $a$ is an eigenvalue with geometric multiplicity $r$. Take $r$ linearly independent eigenvectors in $E_a$, say $Q_1, Q_2, \ldots, Q_r$, and then blow that set up to a basis for $\mathbb{R}^n$ by adding however many vectors you need. If the basis is $\{Q_1, Q_2, \ldots, Q_r, Y_{r+1}, \ldots, Y_n\}$, you can calculate that

$$\begin{aligned} &M\,(Q_1Q_2\cdots Q_rY_{r+1}\cdots Y_n) \\ &= (Q_1Q_2\cdots Q_rY_{r+1}\cdots Y_n)\begin{pmatrix} a & 0 & \cdots & 0 & b_{1(r+1)} & \cdots & b_{1n} \\ 0 & a & \cdots & 0 & b_{2(r+1)} & \cdots & b_{2n} \\ \vdots & & \ddots & \vdots & \vdots & \ddots & \vdots \\ 0 & 0 & \cdots & a & b_{r(r+1)} & \cdots & b_{rn} \\ 0 & 0 & \cdots & 0 & b_{(r+1)(r+1)} & \cdots & b_{(r+1)n} \\ \vdots & & \ddots & \vdots & \vdots & \ddots & \vdots \\ 0 & 0 & \cdots & 0 & b_{n(r+1)} & \cdots & b_{nn} \end{pmatrix} \end{aligned}$$

Expanding along the first column and arguing as in the $\mathbb{R}^3$ case, you can see that $(a - \lambda)^r$ will divide the characteristic polynomial of $M$. So the algebraic multiplicity of $a$ is at least $r$.

**Theorem 9.32 (Geometric-Algebraic Multiplicity)**
*If $M$ is a matrix with eigenvalue $a$, then*

$$1 \leq \text{geometric multiplicity of } a \leq \text{algebraic multiplicity of } a.$$

### Some Applications of Diagonalization

In Exercise 18 from Lesson 5.2, you were asked to figure out what the matrix

$$R = \frac{\sqrt{2}}{2}\begin{pmatrix} 1 & 1 \\ -1 & 1 \end{pmatrix}$$

does to the graph of the quadratic equation $x^2 - xy + y^2 = 1$. The algebra probably got a bit messy. Now that you know about diagonalization, you can tackle such problems in a more elegant way. First, notice that you can think of $x^2 - xy + y^2 = 1$ as a matrix equation.

$\longleftarrow$ At the time, you didn't know that $R$ is a rotation matrix. What is the angle $\theta$ for which $R = R_\theta$?

$$(x, y)A\begin{pmatrix} x \\ y \end{pmatrix} = 1, \quad \text{where } A = \begin{pmatrix} 1 & -1/2 \\ -1/2 & 1 \end{pmatrix}$$

#### For You to Do

**4.** Check that

$$(x, y)A\begin{pmatrix} x \\ y \end{pmatrix} = (x, y)\begin{pmatrix} 1 & -1/2 \\ -1/2 & 1 \end{pmatrix}\begin{pmatrix} x \\ y \end{pmatrix} = x^2 - xy + y^2$$

$\longleftarrow$ A polynomial like $x^2 - xy + y^2$ is called a **quadratic form**.

Also, check that the columns of the matrix

$$R = \frac{\sqrt{2}}{2}\begin{pmatrix} 1 & 1 \\ -1 & 1 \end{pmatrix}$$

are eigenvectors for the matrix $A$. What are the eigenvalues?

So you have a rotation matrix $R$ and a diagonal matrix $D$ so that $R^{-1}AR = D$, so $A = RDR^{-1}$. Of course, if $R$ is a rotation matrix, then so is $R^{-1}$; it just rotates in the opposite direction. If you have $A = R_\theta^{-1}DR_\theta$, then

$$\begin{aligned} (x, y)A\begin{pmatrix} x \\ y \end{pmatrix} &= 1 \\ (x, y)\left(R_\theta\right)^{-1} DR_\theta\begin{pmatrix} x \\ y \end{pmatrix} &= 1 \\ (x, y)\left(R_\theta\right)^{\top} DR_\theta\begin{pmatrix} x \\ y \end{pmatrix} &= 1 \\ \left(R_\theta\begin{pmatrix} x \\ y \end{pmatrix}\right)^{\top} D\left(R_\theta\begin{pmatrix} x \\ y \end{pmatrix}\right) &= 1 \end{aligned}$$

$\longleftarrow$ You proved in Theorem 5.5 that for a rotation matrix $R_\theta$, $(R_\theta)^{-1} = (R_\theta)^{\top}$.

So let

$$\begin{pmatrix} x' \\ y' \end{pmatrix} = R_\theta \begin{pmatrix} x \\ y \end{pmatrix}$$

Then the quadratic equation becomes

$$(x', y') D \begin{pmatrix} x' \\ y' \end{pmatrix} = 1 \tag{6}$$

In other words, it looks like $\lambda_1 x'^2 + \lambda_2 y'^2 = 1$. This looks like either a circle, an ellipse, or a hyperbola, depending on $\lambda_1$ and $\lambda_2$. Since $R_\theta$ is just a rotation, the original graph must have that same shape.

⟵ *Circle:* the eigenvalues are equal and positive. *Ellipse:* the eigenvalues are unequal and positive. *Hyperbola:* one eigenvalue is negative, and one is positive.

### For Discussion

**5.** You already found the eigenvalues of $\begin{pmatrix} 1 & -1/2 \\ -1/2 & 1 \end{pmatrix}$. So, is the graph of $x^2 - xy + y^2 = 1$ a circle, an ellipse, or a hyperbola?

The $(x', y')$ in equation (6) is the coordinate vector for $(x, y)$ with respect to the basis $\mathfrak{B}$ that consists of eigenvectors of $A$. $R_\theta$ is the transition matrix from the standard basis to $\mathfrak{B}$. And $D$ is none other than $M_{\mathfrak{B}}^{\mathfrak{B}}(T)$, where $T : \mathbb{R}^2 \to \mathbb{R}^2$ is the map whose matrix with respect to the standard basis is $A$.

⟵ The elements of the basis of eigenvectors are the columns of $R_\theta$.

So, in this example, the process of diagonalization allows you to change the axes of $\mathbb{R}^2$ in a way that transforms the equation $x^2 - xy + y^2 = 1$ so that it's easier to recognize its graph.

⟵ This "change of variable" doesn't change the graph—it just changes the point-tester used to define the graph.

### Minds in Action Episode 52

DERMAN: I was playing around with matrices, and I noticed something weird. Look at this matrix we were just using.

$$M = \begin{pmatrix} 1 & -3 & 3 \\ 3 & -5 & 3 \\ 6 & -6 & 4 \end{pmatrix}$$

It turned out the characteristic polynomial is $(-2 - \lambda)^2(4 - \lambda)$, right?

TONY: Sure, that's what I got.

DERMAN: Yeah, now watch. I'm going to plug in the matrix $M$ for $\lambda$. Then I get

$$(-2 - M)^2(4 - M) = \begin{pmatrix} -3 & 3 & -3 \\ -3 & 3 & -3 \\ -6 & 6 & -6 \end{pmatrix}^2 \begin{pmatrix} 3 & 3 & -3 \\ -3 & 9 & -3 \\ -6 & 6 & 0 \end{pmatrix}$$

⟵ By "$(-2 - M)$" Derman means the matrix $(-2I - M)$, where $I$ is the $3 \times 3$ identity matrix.

TONY: What am I supposed to be seeing here?

DERMAN: Don't you get it? If you multiply that all out, you get $O$!

TONY: You do? *(Tony checks it on his calculator.)* You do! But, what does that mean?

DERMAN: I don't know. But I tried a couple of examples. It even works for matrices that don't have real eigenvalues like $A = \begin{pmatrix} 0 & -1 \\ 1 & 0 \end{pmatrix}$.

SASHA: Where do you get these ideas?

DERMAN: Well, the characteristic polynomial is $\det(M - \lambda I)$. Replace $\lambda$ by $M$ and you get $\det(M - MI) = \det(O) = 0$.

TONY: Derman, $\lambda$ is a placeholder for a *number*, not a matrix. You can't do that.

DERMAN: But it works! It works, I tell you.

---

### For You to Do

**6.** Check that Derman is right about the matrix $A = \begin{pmatrix} 0 & -1 \\ 1 & 0 \end{pmatrix}$. Find the characteristic polynomial of $A$, and then plug $A$ in to that polynomial and do all the simplification. Do you get the $O$ matrix?

⟵ So, for example, for the polynomial $x^2 + 2x - 3$, you'd calculate $A^2 + 2A - 3I$.

---

### Facts and Notation

Derman's "proof" is flawed, but he has discovered the **Cayley-Hamilton Theorem**, which says that every $n \times n$ matrix is a root of its own characteristic polynomial. The proof in general can get kind of messy, but it's not hard to prove for matrices that are diagonalizable.

First, think about diagonal matrices. The $4 \times 4$ case is good enough to see the general principles at work. Let

$$D = \begin{pmatrix} a_1 & 0 & 0 & 0 \\ 0 & a_2 & 0 & 0 \\ 0 & 0 & a_3 & 0 \\ 0 & 0 & 0 & a_4 \end{pmatrix}$$

The characteristic polynomial of $D$ is $(a_1 - \lambda)(a_2 - \lambda)(a_3 - \lambda)(a_4 - \lambda)$. Plug in $D$ for $\lambda$, and you get $A_1 A_2 A_3 A_4$, where

$$A_1 = \begin{pmatrix} 0 & 0 & 0 & 0 \\ 0 & a_1 - a_2 & 0 & 0 \\ 0 & 0 & a_1 - a_3 & 0 \\ 0 & 0 & 0 & a_1 - a_4 \end{pmatrix}, \quad A_2 = \begin{pmatrix} a_2 - a_1 & 0 & 0 & 0 \\ 0 & 0 & 0 & 0 \\ 0 & 0 & a_2 - a_3 & 0 \\ 0 & 0 & 0 & a_2 - a_4 \end{pmatrix}$$

$$A_3 = \begin{pmatrix} a_3 - a_1 & 0 & 0 & 0 \\ 0 & a_3 - a_2 & 0 & 0 \\ 0 & 0 & 0 & 0 \\ 0 & 0 & 0 & a_3 - a_4 \end{pmatrix}, \quad A_4 = \begin{pmatrix} a_4 - a_1 & 0 & 0 & 0 \\ 0 & a_4 - a_2 & 0 & 0 \\ 0 & 0 & a_4 - a_3 & 0 \\ 0 & 0 & 0 & 0 \end{pmatrix}$$

Now analyze the calculation: the first row of $A_1$ has all 0's, so the same will be true of the first row of $A_1 A_2$, and in fact of $A_1 A_2 A_3 A_4$. Moreover, $(A_1)_{2*} \cdot (A_2)_{*j}$ has all 0's for every column $(A_2)_{*j}$. So the second row of $A_1 A_2$ is also all 0's, so the same will be true for the second row of $A_1 A_2 A_3 A_4$. In fact, all the rows will be 0 in the final product.

Now suppose $M$ is similar to a diagonal matrix. So for some matrix $P$, you have $P^{-1}MP = D$. By Theorem 9.29, $M$ and $D$ have the same characteristic polynomial, say $(a_1 - \lambda)(a_2 - \lambda)(a_3 - \lambda)(a_4 - \lambda)$. From the argument above, you know that

$$\begin{aligned}
O &= (a_1I - D)(a_2I - D)(a_3I - D)(a_4I - D) \\
&= (a_1I - P^{-1}MP)(a_2I - P^{-1}MP) \\
&\quad (a_3I - P^{-1}MP)(a_4I - P^{-1}MP) \\
&= (P^{-1}a_1IP - P^{-1}MP)(P^{-1}a_2IP^{-1} - P^{-1}MP) \\
&\quad (P^{-1}a_3IP - P^{-1}MP)(P^{-1}a_4IP - P^{-1}MP) \\
&= P^{-1}(a_1I - M)PP^{-1}(a_2I - M)P \\
&\quad P^{-1}(a_3I - M)PP^{-1}(a_4I - M)P \\
&= P^{-1}(a_1I - M)(a_2I - M)(a_3I - M)(a_4I - M)P
\end{aligned}$$

← In this equation, $O$ means the $4 \times 4$ matrix of all zeros.

Multiplying on each side by $P$ on the left and $P^{-1}$ on the right gives

$$O = (a_1I - M)(a_2I - M)(a_3I - M)(a_4I - M)$$

So $M$ is a root of its own characteristic polynomial.

**For Discussion**

**7.** Provide a reason for each step in the calculation in the above equations.

### Dynamics

A **dynamical system** is a system that changes over time, where what happens next depends on the current state of the system. One way to model many dynamical systems is through *iteration*. You apply a function to some starting value, and then take the output and put it back into the function as the next input.

← The random processes in Chapter 6 are examples of dynamical systems. Another example is the population of wolves in a national park. The future population depends on the current population—how many mature wolves can reproduce? Are there more wolves than the food supply and land can support? How many will die off? And so on.

**Example 4**

Here's an example of iteration with a matrix representing the function. Take the matrix

$$S = \begin{pmatrix} 1 & -1 \\ 0 & 1 \end{pmatrix} \quad \text{and the vector} \quad X = \begin{pmatrix} 0 \\ 1 \end{pmatrix}.$$

Find $SX$, $S(SX) = S^2X$, $S(S(SX)) = S^3X$, and so on. You get

$$SX = \begin{pmatrix} -1 \\ 1 \end{pmatrix}, S^2X = \begin{pmatrix} -2 \\ 1 \end{pmatrix}, S^3X = \begin{pmatrix} -3 \\ 1 \end{pmatrix}, \ldots$$

**For Discussion**

**8.** Using the matrix $S$ and the vector $X$ given above, find a general formula for $S^nX$, where $n$ is a positive integer. Explain why your formula is right. Draw a picture of what's happening to the vector $S^nX$ as $n$ gets really big.

So when a matrix represents the function, iteration amounts to finding powers of the matrix. This can be nice. If you want to know what happens in 100 steps, you don't have to figure out what happens at steps 1 through 99. Just raise the matrix to the hundreth power and apply it to your starting vector.

← Of course, that's assuming that you can find an efficient way to raise a matrix to some high power.

### For You to Do

**9.** Calculate $A^5$ and $B^5$.

**a.** $A = \begin{pmatrix} 1 & 1 & 0 \\ 1 & 0 & 1 \\ 0 & 1 & 1 \end{pmatrix}$ **b.** $B = \begin{pmatrix} 2 & 0 & 0 \\ 0 & -1 & 0 \\ 0 & 0 & 1 \end{pmatrix}$

Even though the powers of $A$ form a pretty nice pattern, you can probably imagine that calculating large powers of $B$ is much easier than calculating large powers of $A$. However, it turns out the characteristic polynomial of $A$ is $(2-\lambda)(-1-\lambda)(1-\lambda)$, and that if you let

$$P = \begin{pmatrix} 1 & 1 & -1 \\ 1 & -2 & 0 \\ 1 & 1 & 1 \end{pmatrix}, \quad \text{then} \quad P^{-1} = \begin{pmatrix} \frac{1}{3} & \frac{1}{3} & \frac{1}{3} \\ \frac{1}{6} & -\frac{1}{3} & \frac{1}{6} \\ -\frac{1}{2} & 0 & \frac{1}{2} \end{pmatrix}$$

and $P^{-1}AP = B$. Maybe that can help you calculate large powers of $A$.

**Theorem 9.33**

*If $P^{-1}AP = B$, then for any integer $n \geq 1$, $P^{-1}A^nP = B^n$.*

### Example 5

This theorem makes it much easier to calculate high powers of matrices, providing they are diagonalizable. Consider the matrices $A$ and $B$ above. You can see that

$$B^{10} = \begin{pmatrix} 1024 & 0 & 0 \\ 0 & 1 & 0 \\ 0 & 0 & 1 \end{pmatrix}, \text{ so } A^{10} = PB^{10}P^{-1} = \begin{pmatrix} 342 & 341 & 341 \\ 341 & 342 & 341 \\ 341 & 341 & 342 \end{pmatrix}.$$

### For Discussion

**10.** Prove Theorem 9.33.

Markov chains are a special kind of iteration that models lots of interesting processes, including how Google decides the order for displaying search results. Markov chains involve iterating special matrices applied to special vectors.

← Markov chains are explored in more detail in Chapter 6.

### Definition

A **probability vector** is a vector that has nonnegative entries that sum to 1. A **transition matrix** is a square matrix whose columns are probability vectors.

### For Discussion

**11.** Write down several examples of probability vectors and several examples of transition matrices.

**12.** **a.** Given the vector $X = \begin{pmatrix} 1 \\ 2 \end{pmatrix}$, can you find a multiple of $X$ that is a probability vector? If so, how? If not, why not?

**b.** Given the vector $X = \begin{pmatrix} 1 \\ -2 \end{pmatrix}$, can you find a multiple of $X$ that is a probability vector? If so, how? If not, why not?

### Example 6

Here's a $2 \times 2$ transition matrix:

$$T = \begin{pmatrix} 0.7 & 0.5 \\ 0.3 & 0.5 \end{pmatrix}$$

Here are some iterates $T^n X$ starting with the probability vector $X = \begin{pmatrix} 1 \\ 0 \end{pmatrix}$:

$$TX = \begin{pmatrix} 0.7 \\ 0.3 \end{pmatrix}, \quad T^2 X = \begin{pmatrix} 0.64 \\ 0.36 \end{pmatrix}$$
$$T^3 X = \begin{pmatrix} 0.628 \\ 0.372 \end{pmatrix}, \quad T^4 X = \begin{pmatrix} 0.6256 \\ 0.3744 \end{pmatrix}$$

The outputs of the dynamical system seem to be approaching some value. But what value?

### For You to Do

**13.** **a.** Check that .2 and 1 are both eigenvalues for the matrix

$$T = \begin{pmatrix} 0.7 & 0.5 \\ 0.3 & 0.5 \end{pmatrix}$$

**b.** You know that $T$ is similar to the diagonal matrix

$$D = \begin{pmatrix} 1 & 0 \\ 0 & 0.2 \end{pmatrix}$$

Find $D^2$, $D^3$, and $D^4$. Find a formula for $D^n$ and describe what happens for very large values of $n$.

**c.** Find the eigenvector associated with 1, and write it as a probability vector.
**d.** Find decimal approximations to your probability vector, and compare them to the outputs of the dynamical system in Example 6 above.

## Exercises

**1.** Determine whether each matrix is diagonalizable.

a. $\begin{pmatrix} 8 & 4 \\ -9 & -4 \end{pmatrix}$ b. $\begin{pmatrix} -7 & -3 \\ 18 & 8 \end{pmatrix}$

c. $\begin{pmatrix} 1 & 0 & -2 \\ 0 & 0 & 0 \\ -2 & 0 & 4 \end{pmatrix}$ d. $\begin{pmatrix} -1 & -4 & 0 \\ 0 & 3 & 0 \\ 1 & 5 & -1 \end{pmatrix}$

e. $\begin{pmatrix} -12 & -10 & -8 & 2 \\ 0 & 17 & 21 & 11 \\ 0 & 0 & 11 & -1 \\ 0 & 0 & 0 & 5 \end{pmatrix}$ f. $\begin{pmatrix} 1 & 0 & 0 & 0 & 0 & 0 & 0 \\ 0 & 2 & 0 & 0 & 0 & 0 & 0 \\ 0 & 0 & 1 & 0 & 0 & 0 & 0 \\ 0 & 0 & 0 & 1 & 0 & 0 & 0 \\ 0 & 0 & 0 & 0 & 2 & 0 & 0 \\ 0 & 0 & 0 & 0 & 0 & 2 & 0 \\ 0 & 0 & 0 & 0 & 0 & 0 & 1 \end{pmatrix}$

**2.** Diagonalize each matrix.

a. $\begin{pmatrix} -7 & -3 \\ 18 & 8 \end{pmatrix}$ b. $\begin{pmatrix} 13 & 16 \\ -8 & -11 \end{pmatrix}$ c. $\begin{pmatrix} -6 & -6 \\ 3 & 3 \end{pmatrix}$

d. $\begin{pmatrix} 1 & 0 & -2 \\ 0 & 0 & 0 \\ -2 & 0 & 4 \end{pmatrix}$ e. $\begin{pmatrix} 4 & 3 & 3 \\ -3 & -2 & -3 \\ -3 & -3 & -2 \end{pmatrix}$

f. $\begin{pmatrix} -12 & -10 & -8 & 2 \\ 0 & 17 & 21 & 11 \\ 0 & 0 & 11 & -1 \\ 0 & 0 & 0 & 5 \end{pmatrix}$

**3.** For each matrix, find the algebraic multiplicity and the geometric multiplicity of each real eigenvalue. Decide if the matrix is diagonalizable or not.

a. $\begin{pmatrix} -3 & 4 \\ -1 & -7 \end{pmatrix}$ b. $\begin{pmatrix} 0 & 0 \\ 0 & 0 \end{pmatrix}$ c. $\begin{pmatrix} 4 & 6 \\ -4 & -7 \end{pmatrix}$

d. $\begin{pmatrix} 4 & 4 & 2 \\ -5 & -5 & -2 \\ 0 & -1 & -1 \end{pmatrix}$ e. $\begin{pmatrix} 1 & 1 & 0 \\ 0 & 1 & 1 \\ 2 & 0 & 0 \end{pmatrix}$ f. $\begin{pmatrix} -1 & 1 & 0 \\ 0 & -1 & 0 \\ 2 & -1 & 0 \end{pmatrix}$

g. $\begin{pmatrix} -1 & 0 & 0 & 0 & 0 \\ 0 & 4 & 0 & 0 & 0 \\ 1 & 3 & 0 & 0 & 0 \\ 2 & 2 & -1 & -3 & 0 \\ 3 & 1 & -2 & -2 & 1 \end{pmatrix}$

**4.** For each of the following,

(i) determine the quadratic equation given by the matrix equation
(ii) find a rotation matrix that will diagonalize the given matrix
(iii) use the diagonalization to decide if the original equation described a circle, an ellipse, or a hyperbola

**a.** $(x, y)\begin{pmatrix} 5 & 2 \\ 2 & 2 \end{pmatrix}\begin{pmatrix} x \\ y \end{pmatrix} = 1$ **b.** $(x, y)\begin{pmatrix} 1 & -5 \\ -5 & 1 \end{pmatrix}\begin{pmatrix} x \\ y \end{pmatrix} = 1$

**Remember**

A rotation matrix looks like

$$R_\theta = \begin{pmatrix} \cos\theta & -\sin\theta \\ \sin\theta & \cos\theta \end{pmatrix}.$$

So its column vectors each have length 1, and the columns are orthogonal to each other. You'll need to make sure that's true of the matrix you use.

**5.** Suppose that $A = \begin{pmatrix} a & \frac{b}{2} \\ \frac{b}{2} & c \end{pmatrix}$. Show that

$$ax^2 + bxy + cy^2 = (x, y)A\begin{pmatrix} x \\ y \end{pmatrix}$$

**6.** For each matrix, find its characteristic polynomial, and then check that the matrix is a root of its own characteristic polynomial.

**a.** $A = \begin{pmatrix} 5 & 8 \\ -2 & -3 \end{pmatrix}$ **b.** $B = \begin{pmatrix} 6 & 10 \\ -4 & -6 \end{pmatrix}$

**c.** $C = \begin{pmatrix} 5 & 8 \\ -3 & -4 \end{pmatrix}$ **d.** $D = \begin{pmatrix} -3 & -1 & -2 \\ 2 & 0 & 2 \\ 3 & 2 & 1 \end{pmatrix}$

**7.** For each of the following, use the matrix $P$ to diagonalize the matrix $M$. Then use the diagonalization to compute $M^{11}$.

**a.** $M = \begin{pmatrix} -9 & 6 \\ -12 & 8 \end{pmatrix}$ and $P = \begin{pmatrix} 3 & 2 \\ 4 & 3 \end{pmatrix}$

**b.** $M = \begin{pmatrix} 1 & 2 & 2 \\ -3 & -4 & -2 \\ 3 & 3 & 1 \end{pmatrix}$ and $P = \begin{pmatrix} 0 & -1 & 1 \\ 1 & 1 & -1 \\ -1 & 0 & 1 \end{pmatrix}$

**8.** Suppose $M = P^{-1}DP$, $P = \begin{pmatrix} 1 & 2 \\ 1 & 1 \end{pmatrix}$, and $D^2 = \begin{pmatrix} -9 & 0 \\ 0 & -4 \end{pmatrix}$.

**a.** Calculate $M^6$.
**b.** Does $M$ have real eigenvalues? If so, what are they? If not, why not?

**9.** For each given matrix $M$, find a formula for $M^n$ for any integer $n \geq 1$.

**a.** $M = \begin{pmatrix} 3 & -4 \\ 2 & -3 \end{pmatrix}$ **b.** $M = \begin{pmatrix} 5 & -3 \\ 6 & -4 \end{pmatrix}$

**10.** For each given matrix $T$,

(i) choose a convenient probability vector $X$ and use a calculator to find the iterates $TX$, $T^2X$, $T^3X$, and $T^4X$
(ii) show that 1 is an eigenvalue of $T$, and find the other eigenvalue

**a.** $T = \begin{pmatrix} 0.6 & 0.2 \\ 0.4 & 0.8 \end{pmatrix}$ **b.** $T = \begin{pmatrix} 0.25 & 0.5 \\ 0.75 & 0.5 \end{pmatrix}$

**11.** If possible, give an example of two $3 \times 3$ matrices that have the same characteristic polynomial but are not similar to each other. If it's not possible, explain why not.

**12.** In Minds in Action—Episode 51, Sasha claims that any diagonal matrix is diagonalizable. Prove that what she said is true, based on the definition of "diagonalizable."

$\longleftarrow$ What is the "diagonalizing matrix" in this case?

**13.** Let $A = \begin{pmatrix} a & b \\ c & d \end{pmatrix}$. Can you find a condition on $a$, $b$, $c$, and $d$ that guarantees $A$ is diagonalizable?

**14.** Is it true that in a transition matrix the *rows* are probability vectors? If yes, explain why. If no, provide a counterexample.

**15.** Prove that a $2 \times 2$ transition matrix always has an eigenvalue 1.

**16.** Prove that a $3 \times 3$ transition matrix always has an eigenvalue 1.

# Chapter 9 Mathematical Reflections

These problems will help you summarize what you have learned in this chapter.

**1.** Let $A = \begin{pmatrix} 2 & -1 & 3 \\ 1 & 3 & 2 \\ -2 & -4 & 0 \end{pmatrix}$.

**a.** Determine the minors $M_{13}$, $M_{23}$, and $M_{33}$.

**b.** Use the results of part **a** to find the determinant of $A$.

**2.** Evaluate each determinant. Look for shortcuts.

**a.** $\begin{vmatrix} 3 & 1 & -4 \\ 2 & 0 & 0 \\ 5 & 1 & -2 \end{vmatrix}$

**b.** $\begin{vmatrix} 2 & -4 & 3 \\ 0 & -1 & 7 \\ 0 & 0 & 5 \end{vmatrix}$

**3.** Let $P_1 = (1, 0, 0, -2)$, $P_2 = (0, 1, 1, -1)$, $P_3 = (-2, -3, 0, 0)$, and $P_4 = (2, -1, 0, 1)$, and let $N$ be the matrix whose columns are $P_1$, $P_2$, $P_3$, $P_4$ in order.

**a.** Calculate the generalized cross product $P_1 \times P_2 \times P_3$.

**b.** Use the cross product to calculate the volume $V$ of the parallelepiped spanned by these vectors.

$$V = |P_4 \cdot (P_1 \times P_2 \times P_3)|$$

**c.** Calculate $V = |\det(N)|$. Are the results the same?

**4.** Suppose that $\lambda = 1$ is an eigenvalue of $M = \begin{pmatrix} 4 & 1 \\ 3 & 2 \end{pmatrix}$. What are the corresponding eigenvectors?

**5.** How can you find a vector in $\mathbb{R}^n$ orthogonal to $n - 1$ other vectors?

**6.** How can you extend the definition of volume to $\mathbb{R}^n$?

**7.** Under what conditions is an $n \times n$ matrix diagonalizable?

**8.** Diagonalize the matrix $A = \begin{pmatrix} 3 & -1 \\ 4 & -2 \end{pmatrix}$.

**Vocabulary**

In this chapter, you saw these terms and symbols for the first time. Make sure you understand what each one means, and how it is used.

- algebraic multiplicity
- characteristic polynomial
- eigenvalue
- eigenvector
- generalized cross product
- geometric multiplicity
- $M$-invariant subspace
- minors of a matrix
- parallelepiped
- probability vector

# Chapter 9 Review

In Lesson 9.2, you learned to

- extend the definition of determinant to $n \times n$ matrices
- use a recursive algorithm to find the determinant of a matrix
- evaluate a determinant along any row or column
- develop the basic rules for determinants

The following problems will help you check your understanding.

**1.** Evaluate each determinant. Look for shortcuts.

**a.** $\begin{vmatrix} 4 & -1 & 1 \\ 3 & 0 & 3 \\ 2 & 0 & -2 \end{vmatrix}$  **b.** $\begin{vmatrix} 1 & 3 & 5 \\ -1 & -3 & 2 \\ 2 & 6 & 7 \end{vmatrix}$

**c.** $\begin{vmatrix} 2 & -3 & 1 \\ -3 & 4 & 0 \\ 4 & -2 & -1 \end{vmatrix}$  **d.** $\begin{vmatrix} 2 & 1 & -3 \\ -3 & 0 & 4 \\ 4 & -1 & -2 \end{vmatrix}$

**2.** Evaluate the determinant

$$\begin{vmatrix} 0 & 2 & -1 & 4 \\ 1 & 3 & 0 & 1 \\ 0 & -2 & 1 & 4 \\ 0 & -3 & 1 & 0 \end{vmatrix}$$

along

**a.** the first column
**b.** the fourth column

**3.** Suppose that $A = \begin{pmatrix} a & b \\ c & d \end{pmatrix}$ and that $\det A = 6$. Find each determinant.

**a.** $\begin{vmatrix} 3a & b \\ 3c & d \end{vmatrix}$  **b.** $\begin{vmatrix} 2a & 2b \\ 2c & 2d \end{vmatrix}$

**c.** $\begin{vmatrix} b & a \\ d & c \end{vmatrix}$  **d.** $\begin{vmatrix} a+3b & b \\ c+3d & d \end{vmatrix}$

In Lesson 9.3, you learned to

- prove that the column you use to evaluate the determinant of a matrix does not affect the outcome
- expand a determinant along a row instead of a column
- compare the determinant of a matrix to the determinant of its transpose

The following problems will help you check your understanding.

**4.** Evaluate $\begin{vmatrix} -2 & 3 & 0 & -1 \\ 1 & 0 & 2 & -2 \\ 4 & -2 & 1 & 2 \\ 0 & 2 & 0 & -4 \end{vmatrix}$

**a.** along row 1
**b.** along row 4

**5.** Suppose that $A = \begin{pmatrix} a & b & c \\ d & e & f \\ g & h & i \end{pmatrix}$ and that $\det(A) = 5$. Find each determinant.

**a.** $\det A^{\top}$

**b.** $\det 3A$

**c.** $\begin{vmatrix} a & b & c \\ g & h & i \\ d & e & f \end{vmatrix}$

**d.** $\begin{vmatrix} a & b & c \\ 3d & 3e & 3f \\ g & h & i \end{vmatrix}$

**6.** Solve each equation for $x$. Look for shortcuts.

**a.** $\begin{vmatrix} 1 & -1 & 4 \\ 0 & 0 & x \\ 3 & x & 6 \end{vmatrix} = -4$

**b.** $\begin{vmatrix} 1 & -2 & -4 \\ 0 & 2 & x \\ 0 & 0 & x \end{vmatrix} = 6$

In Lesson 9.4, you learned to

- use matrix multiplication as a way to reduce a matrix to echelon form
- find in what ways the determinant of a matrix is changed by reducing it to its echelon form
- calculate the determinant of the product of two matrices
- understand how a nonzero determinant is equivalent to all other statements in the TFAE Theorem

The following problems will help you check your understanding.

**7.** Suppose $A$ is a $3 \times 3$ matrix and $\det(A) = 5$. For each elementary row matrix $E$,

- describe the elementary row operation that is carried out when you multiply matrix $A$ by $E$
- find the determinant of $E$
- find the determinant of the product $EA$

**a.** $E = \begin{pmatrix} 1 & 0 & 0 \\ 0 & 2 & 0 \\ 0 & 0 & 1 \end{pmatrix}$

**b.** $E = \begin{pmatrix} 0 & 2 & 0 \\ 1 & 0 & 0 \\ 0 & 0 & 1 \end{pmatrix}$

**c.** $E = \begin{pmatrix} 1 & 0 & 0 \\ 0 & 1 & 0 \\ 0 & 2 & 1 \end{pmatrix}$

**8.** Suppose that $A = \begin{pmatrix} a & b & c \\ d & e & f \\ g & h & i \end{pmatrix}$ and $B = \begin{pmatrix} j & k & l \\ m & n & p \\ q & r & s \end{pmatrix}$. Suppose also that $\det(A) = 4$ and that $\det(B) = -2$. Find the following:

**a.** $\det(AB)$ **b.** $\det(3AB)$
**c.** $\det(A^{-1})$ **d.** $\det(B^{-1})$
**e.** $\det((AB)^{-1})$ **f.** $\det(AB^{-1})$
**g.** $\begin{vmatrix} a & b & c \\ d & e & f \\ 5g & 5h & 5i \end{vmatrix}$ **h.** $\begin{vmatrix} a & b & c \\ d & e & f \\ 5a+g & 5b+h & 5c+i \end{vmatrix}$

**9.** Suppose $A = \begin{pmatrix} 1 & 0 & -2 \\ 3 & t & 4 \\ t & 1 & -1 \end{pmatrix}$. Use determinants to find $t$ if $\mathrm{rref}(A) \neq I$.

In Lesson 9.5, you learned to

- use Cramer's Rule to find a vector orthogonal to $n-1$ given vectors
- extend the definition of cross product to $n-1$ vectors in $\mathbb{R}^n$
- extend the definition of volume to a box spanned by $n$ vectors in $\mathbb{R}^n$
- use Cramer's Rule to find the solution to a system of linear equations

The following problems will help you check your understanding.

**10.** Let $P_1 = (1,1,1,2)$, $P_2 = (1,-1,3,0)$, and $P_3 = (0,1,0,1)$, and suppose $X$ is a vector that is orthogonal to all three given vectors.

**a.** Set up and solve a system of equations to find $X$.
**b.** Use Cramer's Rule to find $X$.

**11.** For each given matrix equation, use Cramer's Rule to solve it.

**a.** $\begin{pmatrix} 2 & -1 \\ 3 & 2 \end{pmatrix}\begin{pmatrix} x \\ y \end{pmatrix} = \begin{pmatrix} 10 \\ 8 \end{pmatrix}$ **b.** $\begin{pmatrix} 2 & 1 & -1 \\ -1 & 2 & 4 \\ 1 & 0 & -2 \end{pmatrix}\begin{pmatrix} x \\ y \\ z \end{pmatrix} = \begin{pmatrix} -3 \\ 4 \\ -2 \end{pmatrix}$

**12.** For each given set of vectors, find the volume of the parallelepiped spanned by them.

**a.** $P_1 = (0,1,2,0)$, $P_2 = (-1,2,3,0)$, $P_3 = (2,-1,0,1)$, $P_4 = (3,0,0,-1)$
**b.** $P_1 = (1,1,2,2)$, $P_2 = (3,-2,0,1)$, $P_3 = (1,2,2,3)$, $P_4 = (0,1,0,-1)$

In Lesson 9.6, you learned to

- find the *characteristic polynomial* of a matrix
- recognize the underlying geometry of the characteristic polynomial's real roots

- establish the relationship between the *eigenvalues*, *eigenvectors*, and characteristic polynomials of similar matrices
- find the *invariant subspaces* of a matrix or linear transformation

The following problems will help you check your understanding.

**13.** For each given matrix $A$,

(i) find the characteristic polynomial of $A$
(ii) find all eigenvalues and eigenvectors for $A$

**a.** $\begin{pmatrix} 1 & 2 \\ 6 & 5 \end{pmatrix}$ **b.** $\begin{pmatrix} 1 & -1 \\ 0 & 1 \end{pmatrix}$

**c.** $\begin{pmatrix} 1 & 2 & -2 \\ 0 & -1 & 1 \\ 0 & 0 & 2 \end{pmatrix}$ **d.** $\begin{pmatrix} 2 & 0 & 0 \\ 1 & 1 & -1 \\ 1 & 0 & 2 \end{pmatrix}$

**14.** Let $A = \begin{pmatrix} 1 & 1 & 1 \\ 0 & 2 & 0 \\ 0 & 2 & 1 \end{pmatrix}$.

**a.** Find all eigenvalues and eigenvectors for $A$.
**b.** Calculate $A^{-1}$.
**c.** Find all eigenvalues and eigenvectors for $A^{-1}$.

**15.** For each given matrix and eigenvalue pair, find the associated eigenspace.

**a.** $\begin{pmatrix} 1 & 2 \\ -1 & 4 \end{pmatrix}$, eigenvalue $= 3$

**b.** $\begin{pmatrix} 1 & 0 & -1 \\ 2 & 2 & -3 \\ 1 & 0 & 2 \end{pmatrix}$, eigenvalue $= 2$

**c.** $\begin{pmatrix} -1 & 1 & 1 \\ 0 & 1 & 2 \\ 0 & 0 & -1 \end{pmatrix}$, eigenvalue $= -1$

In Lesson 9.7, you learned to

- find the algebraic and geometric multiplicity of the eigenvalues of a matrix
- determine which matrices can be diagonalized
- use a basis of eigenvectors to create a change of basis matrix
- apply the diagonalization process to simplify calculations in probability theory and *dynamical systems*
- find the equation of a conic section whose axes have been rotated by a certain angle

The following problems will help you check your understanding.

**16.** For each matrix, find the algebraic multiplicity and the geometric multiplicity of each real eigenvalue. Is the matrix diagonalizable? Explain.

**a.** $\begin{pmatrix} 2 & -1 \\ 2 & 5 \end{pmatrix}$ **b.** $\begin{pmatrix} 1 & -1 \\ 1 & 3 \end{pmatrix}$ **c.** $\begin{pmatrix} -1 & 1 & 0 \\ 4 & 2 & 0 \\ 2 & 2 & -2 \end{pmatrix}$

**17.** For each of the following,

(i) determine the quadratic equation given by the matrix equation
(ii) find a rotation matrix that will diagonalize the given matrix
(iii) use the diagonalization to decide if the original equation described a circle, an ellipse, or a hyperbola

**a.** $(x, y)\begin{pmatrix} 4 & 1 \\ 1 & 4 \end{pmatrix}\begin{pmatrix} x \\ y \end{pmatrix} = 1$ **b.** $(x, y)\begin{pmatrix} 17 & 9 \\ 9 & -7 \end{pmatrix}\begin{pmatrix} x \\ y \end{pmatrix} = 1$

**18.** For each of the following, use the matrix $P$ to diagonalize the matrix $M$. Then use the diagonalization to compute $M^5$.

**a.** $M = \begin{pmatrix} -8 & 18 \\ -3 & 7 \end{pmatrix}$ and $P = \begin{pmatrix} 3 & 2 \\ 1 & 1 \end{pmatrix}$

**b.** $M = \begin{pmatrix} -1 & 4 & 11 \\ 0 & -10 & -24 \\ 0 & 4 & 10 \end{pmatrix}$ and $P = \begin{pmatrix} 1 & 1 & -1 \\ -2 & 0 & 3 \\ 1 & 0 & -1 \end{pmatrix}$

# Chapter 9 Test

## Multiple Choice

**1.** Suppose that $A = \begin{pmatrix} a & b & c \\ d & e & f \\ g & h & i \end{pmatrix}$ and that $\det(A) = -2$. What is the determinant of $B = \begin{pmatrix} 4a & 4b & 4c \\ 4g & 4h & 4i \\ 4d & 4e & 4f \end{pmatrix}$?

**A.** $-128$ **B.** $-8$ **C.** $8$ **D.** $128$

**2.** Suppose that $A = \begin{pmatrix} a & b \\ c & d \end{pmatrix}$ and $\det(A) = 7$. What is the determinant of $\begin{pmatrix} a-2c & b-2d \\ c & d \end{pmatrix}$?

**A.** $-14$ **B.** $-7$ **C.** $7$ **D.** $14$

**3.** Suppose that $A = \begin{pmatrix} a & b & c & d \\ e & f & g & h \\ i & j & k & l \\ m & n & p & q \end{pmatrix}$ has rank 3. Which statement must be true?

**A.** $\ker(A) = O$
**B.** $\det(A) = 0$
**C.** The dimension of the column space for $A$ is 4.
**D.** The columns of $A$ are linearly independent.

**4.** Let $P_1 = (1, 0, -1, 2)$, $P_2 = (-2, 1, 1, -3)$, $P_3 = (2, 2, 3, 1)$, and $P_4 = (0, 0, 1, -1)$. What is the volume of the parallelepiped spanned by these vectors?

**A.** 2 **B.** 3 **C.** 12 **D.** 15

**5.** Which is an eigenvector for $A = \begin{pmatrix} 1 & -1 & 2 \\ 0 & 1 & 3 \\ 0 & 0 & -2 \end{pmatrix}$?

**A.** $(-2, -2, 0)$
**B.** $(-1, 1, 0)$
**C.** $(1, 1, -2)$
**D.** $(2, 2, -2)$

**6.** Suppose $M = \begin{pmatrix} -2 & -1 \\ 1 & 0 \end{pmatrix}$. Let $a$ be the algebraic multiplicity and $g$ be the geometric multiplicity of the eigenvalue $-1$ of $M$. What are the values of $a$ and $g$?

**A.** $a = 1$ and $g = 1$
**B.** $a = 1$ and $g = 2$
**C.** $a = 2$ and $g = 1$
**D.** $a = 2$ and $g = 2$

## Open Response

**7.** Evaluate $\begin{vmatrix} 2 & -1 & 3 & 1 \\ 0 & 4 & 0 & 0 \\ 0 & -1 & 0 & -1 \\ 3 & 0 & -2 & -1 \end{vmatrix}$

**a.** along the third column
**b.** along the second row

**8.** Solve for $x$:

$$\begin{vmatrix} x & 2 & 1 \\ 4 & 1 & -2 \\ 2 & x & 1 \end{vmatrix} = -15$$

**9.** Let $P_1 = (1, -2, 0, 1)$, $P_2 = (1, 3, 1, 0)$, and $P_3 = (2, 0, -1, 1)$.

**a.** Find $X = P_1 \times P_2 \times P_3$.
**b.** Show that $X$ is orthogonal to each of the three given vectors.

**10.** Let $A = \begin{pmatrix} 1 & 0 & 0 \\ 2 & -2 & 1 \\ -1 & 0 & 1 \end{pmatrix}$.

**a.** Find the characteristic polynomial for matrix $A$.
**b.** Find all eigenvalues for $A$.
**c.** Find the eigenvectors corresponding to each eigenvalue.

**11.** Let $A = \begin{pmatrix} 7 & 4 \\ 2 & 5 \end{pmatrix}$.

**a.** Find a matrix $P$ that will diagonalize $A$.
**b.** Use $P$ to diagonalize $A$.

# Cumulative Review

**1.** Consider $S$, the set of $3 \times 2$ matrices of the form $\begin{pmatrix} a & b \\ c & d \\ e & f \end{pmatrix}$, where $e = -f$.

**a.** Show that $S$ is closed under addition.
**b.** Is $S$ a subspace of $V$, the set of all $3 \times 2$ matrices? Justify your answer.

**2.** Which of the following are subspaces of $\mathbb{R}_2[x]$, the set of polynomials with degree less than or equal to 2? Justify your answer.

**a.** All polynomials $a_2x^2 + a_1x + a_0$, where $a_1 = 0$
**b.** All polynomials $a_2x^2 + a_1x + a_0$, where $a_1 < 0$
**c.** All polynomials $a_2x^2 + a_1x + a_0$, where $a_0 + a_1 + a_2 = 0$

**3.** Determine whether $v$ is in $L\{(1, 2, -1), (3, 0, -4)\}$.

**a.** $v = (4, -4, -6)$
**b.** $v = (5, 3, -4)$
**c.** $v = (7, 2, -9)$

**4.** Find a generating system for each vector space.

**a.** Matrices in the form $\begin{pmatrix} a & b \\ 2a & a-b \end{pmatrix}$
**b.** The kernel of $\begin{pmatrix} 1 & -1 & 4 \\ 3 & 1 & 8 \\ 0 & 3 & -3 \end{pmatrix}$
**c.** The column space of $\begin{pmatrix} 1 & -1 & 4 \\ 3 & 1 & 8 \\ 0 & 3 & -3 \end{pmatrix}$

**5.** Determine whether the set is a basis for the given vector space. Justify your answer.

**a.** $\{(1, -2, 3), (2, 0, 5), (3, -4, 1)\}$ for $\mathbb{R}^3$
**b.** $\{3 - x, 1 + x^2\}$ for $\mathbb{R}_2[x]$
**c.** $\left\{ \begin{pmatrix} 1 \\ 3 \\ 0 \end{pmatrix}, \begin{pmatrix} -1 \\ 1 \\ 3 \end{pmatrix}, \begin{pmatrix} 4 \\ 8 \\ -3 \end{pmatrix} \right\}$ for the column space of $\begin{pmatrix} 1 & -1 & 4 \\ 3 & 1 & 8 \\ 0 & 3 & -3 \end{pmatrix}$

**6.** Let $V = L\{(1, -2, 3), (2, 0, 5), (0, -4, 1)\}$.

**a.** Find a basis for $V$.
**b.** What is the dimension of $V$?

**7.** Consider the basis $\mathfrak{B} = \{(1, 0, 1), (2, 1, 1), (0, 3, 2)\}$ for $\mathbb{R}^3$ and let $v = (2, -1, 3)$. Find $\underline{v}_{\mathfrak{B}}$.

**8.** Blow up $\{x^3 + x, x^2 - 3\}$ to a basis for $\mathbb{R}_3[x]$.

**9.** A generating system for vector space $V$ is the column space of $N = \begin{pmatrix} 1 & -1 & 2 \\ 2 & 1 & 1 \\ 0 & -1 & 1 \end{pmatrix}$. Starting with the columns, sift out a basis for $V$.

**10.** Let $M = \begin{pmatrix} 2 & -1 & 3 & 1 \\ 1 & -1 & 4 & 2 \\ 0 & 1 & -5 & -3 \\ 0 & 2 & -2 & 2 \end{pmatrix}$.

**a.** Determine the rank of $M$.
**b.** Determine the dimension of the kernel of $M$.
**c.** Find a basis for the kernel of $M$.

**11.** For each mapping $F : \mathbb{R}^2 \to \mathbb{R}_2[x]$, determine whether $F$ is linear.

**a.** $F(a, b) = ax + bx^2$
**b.** $F(a, b) = 1 + ax + bx^2$

**12.** Suppose $T : \mathbb{R}^2 \to \mathbb{R}^3$ is defined by

$$T(1, 0) = (1, 1, 2)$$
$$T(1, 1) = (2, 0, 3)$$

Let $M = M_B^{B'}(T)$, where

$$\mathfrak{B} = \{(1, 0), (1, 1)\} \text{ and } \mathfrak{B}' = \{(1, 1, 0), (0, 1, 0), (0, 1, 1)\}$$

**a.** Find $M$.
**b.** Use $M$ to find $T(3, 2)$.

**13.** Suppose

$$\mathfrak{B} = \{(1, 0, 0), (0, 1, 0), (0, 0, 1)\} \text{ and } \mathfrak{B}' = \{(1, 1, 1), (1, 1, 0), (-1, 0, 1)\}.$$

And, suppose that a linear transformation $T$ is represented by the matrix

$$\begin{pmatrix} 2 & 0 & 0 \\ 0 & -1 & 0 \\ 0 & 0 & 1 \end{pmatrix}$$

written relative to the standard basis $\mathfrak{B}$.

**a.** Find the change of basis matrix from $\mathfrak{B}$ to $\mathfrak{B}'$.
**b.** Find the change of basis matrix from $\mathfrak{B}'$ to $\mathfrak{B}$.
**c.** Describe the transformation $T$ relative to the standard basis $\mathfrak{B}$.
**d.** Find a matrix that represents this transformation relative to the nonstandard basis $\mathfrak{B}'$.

**14.** Let $M = \begin{pmatrix} 4 & -1 \\ 1 & 2 \end{pmatrix}$ and $N = \begin{pmatrix} 17 & -49 \\ 4 & -11 \end{pmatrix}$. Find a matrix $P$ that shows that $M$ and $N$ are similar.

**15.** Find the determinant of $A = \begin{pmatrix} 2 & -1 & 3 \\ 2 & 1 & 4 \\ 0 & 1 & -3 \end{pmatrix}$.

**16.** Let $M = \begin{pmatrix} 1 & 2 & -3 \\ x & 1 & -4 \\ 0 & 1 & x \end{pmatrix}$.
For what values of $x$ will the columns of $M$ be linearly dependent?

**17.** Let $P_1 = (1, -1, 0, -2)$, $P_2 = (0, -3, 1, 1)$, and $P_3 = (2, -1, -1, 0)$. Find a vector $X$ that is orthogonal to all three vectors.

**18.** Use Cramer's Rule to solve

$$\begin{pmatrix} 2 & -1 \\ 4 & 3 \end{pmatrix} \begin{pmatrix} x \\ y \end{pmatrix} = \begin{pmatrix} -4 \\ 2 \end{pmatrix}$$

**19.** What is the volume of the parallelepiped spanned by the vectors $P_1 = (1, 3, 0, 2)$, $P_2 = (0, -1, 1, 2)$, $P_3 = (2, 1, 0, 0)$, and $P_4 = (1, 0, -2, 3)$?

For problems 20–21, consider the matrix

$$M = \begin{pmatrix} -1 & 2 & 3 \\ 0 & -11 & -15 \\ 0 & 6 & 8 \end{pmatrix}$$

**20.** **a.** Find all eigenvalues and eigenvectors for $M$.
**b.** For each eigenvalue, describe the eigenspace and determine its algebraic multiplicity and geometric multiplicity.

**21.** **a.** Is $M$ diagonalizable? Explain.
**b.** If $M$ is diagonalizable, find the diagonalizing matrix $P$ and use $P$ to diagonalize $M$.

# Cumulative Test

## Multiple Choice

**1.** Let $V = \mathbb{R}^2$. Which of the following is a subspace of $V$?

**A.** The set of vectors of the form $(a, a+1)$
**B.** The set of vectors of the form $(a, 3a)$
**C.** The set of vectors $(a, b)$, where $a > b$
**D.** The set of vectors $(a, b)$, where $a + b = 1$

**2.** Which vector is in the linear span of $v_1 = (2, -3, 1)$ and $v_2 = (1, -1, 3)$?

**A.** $(3, -4, -2)$
**B.** $(3, -2, 2)$
**C.** $(4, -5, 7)$
**D.** $(4, -1, -5)$

**3.** Let $V$ be the set of all matrices of the form $\begin{pmatrix} a & b \\ 0 & c \\ a+c & 0 \\ 0 & a \end{pmatrix}$. What is the dimension of the vector space $V$?

**A.** 2 **B.** 3 **C.** 4 **D.** 6

**4.** What is the coordinate vector for $v = \begin{pmatrix} 2 & -1 \\ 1 & 3 \end{pmatrix}$ with respect to base $\mathfrak{B} = \left\{ \begin{pmatrix} 1 & 0 \\ 1 & 1 \end{pmatrix}, \begin{pmatrix} 1 & 0 \\ 1 & 0 \end{pmatrix}, \begin{pmatrix} 0 & 0 \\ 1 & 0 \end{pmatrix}, \begin{pmatrix} 0 & 1 \\ 1 & 0 \end{pmatrix} \right\}$?

**A.** $\begin{pmatrix} -2 & 2 \\ 1 & 3 \end{pmatrix}$ **B.** $\begin{pmatrix} 1 & 1 \\ -1 & 3 \end{pmatrix}$ **C.** $\begin{pmatrix} 2 & 1 \\ -1 & 1 \end{pmatrix}$ **D.** $\begin{pmatrix} 3 & -1 \\ 0 & -1 \end{pmatrix}$

**5.** Let $M = \{x^2 + 1, 2x\}$. What additional vector will blow up $M$ to a basis for $\mathbb{R}_2[x]$?

**A.** $x^2 + x$
**B.** $x^2 + x + 1$
**C.** $2x^2 + x + 2$
**D.** $2x^2 + 2$

**6.** Let $M = \begin{pmatrix} 1 & 2 & -1 & 5 \\ 0 & -1 & 3 & -2 \\ 2 & 1 & 1 & 4 \\ 0 & -3 & 3 & -6 \end{pmatrix}$. What is the rank of matrix $M$?

**A.** 1 **B.** 2 **C.** 3 **D.** 4

**7.** Which mapping $F : \mathbb{R}^3 \to \mathbb{R}^2$ is linear?

**A.** $F(x, y, z) = (x + y, z + 1)$
**B.** $F(x, y, z) = (x + y, z)$
**C.** $F(x, y, z) = (x^2, y + z)$
**D.** $F(x, y, z) = (x + y, \sqrt{z})$

**8.** Suppose $\mathfrak{B} = \{(1,0),(0,1)\}$ and $\mathfrak{B}' = \{(1,-3),(-1,4)\}$. Which is the change of basis matrix from $\mathfrak{B}$ to $\mathfrak{B}'$?

**A.** $\begin{pmatrix} \frac{4}{7} & \frac{1}{7} \\ -\frac{3}{7} & \frac{1}{7} \end{pmatrix}$ **B.** $\begin{pmatrix} \frac{4}{7} & -\frac{3}{7} \\ \frac{1}{7} & \frac{1}{7} \end{pmatrix}$ **C.** $\begin{pmatrix} 4 & 1 \\ 3 & 1 \end{pmatrix}$ **D.** $\begin{pmatrix} 4 & 3 \\ 1 & 1 \end{pmatrix}$

**9.** Suppose that $M = \begin{pmatrix} a & b & c \\ d & e & f \\ g & h & i \end{pmatrix}$ and that $\det(M) = 8$. What is the determinant of $N = \begin{pmatrix} b & 2a & 3c \\ e & 2d & 3f \\ h & 2g & 3i \end{pmatrix}$?

**A.** $-48$ **B.** $-40$ **C.** $40$ **D.** $48$

**10.** Suppose that $A$ is an $n \times n$ matrix and $\det(A) = 1$. Which of the following statements is true?

**A.** $\ker(A) \neq O$
**B.** $\operatorname{rref}(A) \neq I$
**C.** The rows of $A$ are linearly independent.
**D.** $A^{-1}$ does not exist.

**11.** In $\mathbb{R}^4$, let $P_1 = (0,0,2,0)$, $P_2 = (-1,1,3,-1)$, $P_3 = (2,-1,1,0)$, and $P_4 = (2,1,0,0)$. What is the volume of the parallelepiped spanned by these vectors?

**A.** 1 **B.** 8 **C.** 9 **D.** 24

**12.** Let $M = \begin{pmatrix} -1 & 2 \\ 4 & -3 \end{pmatrix}$. Which vector is an eigenvector for $M$ corresponding to the eigenvalue $\lambda = 1$?

**A.** $(1,-1)$ **B.** $(1,1)$ **C.** $(2,-4)$ **D.** $(2,4)$

## Open Response

**13.** Consider the set $V$ of $2 \times 2$ matrices, and define the operations of addition and scalar multiplication as usual.

**a.** Verify that $V$ is closed under addition and scalar multiplication.

**b.** If $V$ were the set of $2 \times 2$ matrices of the form $\begin{pmatrix} a & a \\ b & 0 \end{pmatrix}$, would $V$ be closed under addition and scalar multiplication? Explain.

**14.** Find a generating system for each vector space.

**a.** Ordered triples $(x,y,z)$ with $x + y = 0$

**b.** Matrices of the form $\begin{pmatrix} a & b \\ c & -c \\ a+c & b-c \end{pmatrix}$

**15.** Sift out a basis for the column space of $N = \begin{pmatrix} 1 & -1 & 2 \\ 3 & 2 & 4 \\ -1 & 5 & -7 \end{pmatrix}$, starting with the columns.

**16.** The linear map $D : \mathbb{R}^2 \to \mathbb{R}^3$ with respect to the standard bases is defined by $D(a,b) = (2a, a+b, 2a+b)$.

**a.** Find $D(3,5)$.

**b.** Find a matrix $M$, so that for any vector $v$ in $\mathbb{R}^2$, $M\underline{v}_{\mathfrak{B}} = \underline{D(v)}_{\mathfrak{B}'}$.

**c.** Use $M$ to find $D(3,5)$.

**17.** Show that $M = \begin{pmatrix} 3 & 2 \\ 1 & -1 \end{pmatrix}$ and $N = \begin{pmatrix} -5 & 3 \\ -10 & 7 \end{pmatrix}$ are similar matrices.

**18.** Let $A = \begin{pmatrix} -1 & 2 & 3 \\ 1 & -1 & 5 \\ 0 & 3 & 2 \end{pmatrix}$.

**a.** Determine the minors $M_{31}$, $M_{32}$, and $M_{33}$.

**b.** Use the results of part **a** to find the determinant of $A$.

**19.** Use Cramer's Rule to solve

$$\begin{pmatrix} 2 & -1 & 1 \\ 4 & 1 & -2 \\ 0 & 3 & -3 \end{pmatrix} \begin{pmatrix} x \\ y \\ z \end{pmatrix} = \begin{pmatrix} 0 \\ -10 \\ -6 \end{pmatrix}$$

**20.** Find all eigenvalues and eigenvectors for $A = \begin{pmatrix} 1 & -3 & 1 \\ 0 & -1 & 4 \\ 0 & 0 & 3 \end{pmatrix}$.

**21.** Suppose $M = \begin{pmatrix} 18 & -40 \\ 8 & -18 \end{pmatrix}$ and $P = \begin{pmatrix} -5 & 2 \\ -2 & 1 \end{pmatrix}$.

**a.** Use the matrix $P$ to diagonalize the matrix $M$.

**b.** Use the diagonalization to compute $M^7$.

# Index